DVD DEMYSTIFIED

DVD DEMYSTIFIED

JIM TAYLOR
MARK R. JOHNSON
CHARLES G. CRAWFORD

THIRD EDITION

McGraw-Hill
New York   Chicago   San Francisco   Lisbon   London   Madrid
Mexico City   Milan   New Delhi   San Juan   Seoul
Singapore   Sydney   Toronto

The McGraw·Hill Companies

Cataloging-in-Publication Data is on file with the Library of Congress

Copyright © 2006 by The McGraw-Hill Companies, Inc. All rights reserved. Printed in the United States of America. Except as permitted under the United States Copyright Act of 1976, no part of this publication may be reproduced or distributed in any form or by any means, or stored in a data base or retrieval system, without the prior written permission of the publisher.

1 2 3 4 5 6 7 8 9 0   DOC/DOC   0 1 0 9 8 7 6 5

P/N 0-07-142398-2
PART OF
ISBN 0-07-142396-6

The sponsoring editor for this book was Stephen S. Chapman, and the production supervisor was Richard C. Ruzycka. The art director for the cover was Anthony Landi; the cover designer was Margaret Webster-Shapiro.

Printed and bound by RR Donnelley

UMD, PlayStation, and PSP are trademarks of Sony Corporation.

McGraw-Hill books are available at special quantity discounts to use as premiums and sales promotions, or for use in corporate training programs. For more information, please write to the Director of Special Sales, McGraw-Hill Professional, Two Penn Plaza, New York, NY 10121-2298. Or contact your local bookstore.

Dedications

To my children, Anneke and Corwin, products of the digital age, who have grown up with DVDs and TiVo's and aren't quite sure what those clunky black rectangular things are for.

— Jim Taylor

To the only one who really matters, Eileen. Without you, there would simply be no point.

— Mark Johnson

To my wife and partner, Samantha Cheng, who brought the three of us together to slay the beast, and whose unswerving faith in us was the fuel that kept the flame burning, providing the power to see the project through.

— Chuck Crawford

Acknowledgments

The authors would like to thank everyone who helped in the creation of this third edition, which, like DVD, has grown in scope and complexity.

First, immense thanks and a group hug to Samantha Cheng, who herded the cats and banged the heads necessary to get the book and the disc finished. And who convinced me in the first place that I'd never get anything done without some help.

We are indebted to Michael Anderson for his invaluable help with figures, tables, and layout, and to Ralph Stice, our copyeditor, for making it all proper and readable.

Many kind people spent time reading and commenting on drafts: Cathy Guinan, Nick Otto, Mike Schmit, Kuni Takahashi, Tom Pratt, Rainer Brodersen, Greg Gewickey, Sean Hayes, Andrew Jewsbury, Allan Lamkin, Joe Rice, Graham Sharpless, Stacey Spears, Steve Vernon, Patrick Watson, Robert Zollo and Michael Zink. Thanks also to Bob Stuart, Jerry Pierce, Tom Holman, Geoff Tully, Charles Poynton, Roger Dressler, Dave Schnuelle, Andrew Rosen, Kilroy Hughes, the rest of the guys at Microsoft (including Kurt Hunter, who would never forgive me if I didn't mention him by name in this edition), and many others who have taken time over the years to explain many things.

Thanks to the gang at Deluxe-Song Yi Liu, Rohit Marble, and Randy Berg-for authoring and replicating the disc. Thanks to our bleary-eyed reviewers, Ari Zagnit, Tom Bennett, and Jay Simon, for checking the check disc. And a standing ovation for Raleigh Stewart, who generated the brilliant menus with groundbreaking animated highlights.

Thanks also and again to Steve Chapman at McGraw-Hill for his patience and good-natured advocacy. And once again Andy Parsons, like a consummate toastmaster, has started things off with a sagacious foreword. And thanks to Fleischman and Arthur, who have also been solidly supportive.

— Jim Taylor
Fox Island, Washington

A book like this cannot be written by one person alone (or even three). It exists because of the thoughtfulness and dilligence of dozens of friends and colleagues who have borne with us through the trials of not just this book, but the growth of an industry.

My greatest appreciation goes out to those who read, reviewed, and commented on the many drafts of this and prior editions, as well as those in the know who advised, including: Masaru Yamamoto, Ralf Ostermann, Nick Otto, Roger Dressler, Mike Ward, Joe Rice, and Phillip Maness. Thanks also to those who put long hours into the design and production of the sample disc that accompanies this book, in particular, Raleigh Stewart, Ari Zagnit, and Tom Bennett, as well as Randy Berg, Songyi Liu and the rest of the team at Deluxe Digital Studios who kindly authored this challenging disc at the peak of their season under an extreme schedule. Thank you all for your hard work and consultation.

Of course, this book would not exist if it were not for the perseverance, dedication, and impressive tolerance of Steve Chapman at McGraw-Hill, whose patience we most certainly tested, our producer E. Samantha Cheng, whose marriage we probably almost ended, and my co-writers Chuck Crawford and Jim Taylor, whose sense of humor made the most trying of times...well, still trying, but allowed us to get through.

— Mark Johnson
Pasadena, California

Without Jim Taylor this book could not have been written. Jim's insight and foresight are marvels and mysteries, and we are all wiser thanks to his imagination and curiosity. Mark Johnson brought a spirit to the enterprise that was youthful in its exuberance and solomon-like in its knowledge. I cannot say it often enough, we are all deeply indebted to Samantha Cheng, who remained focused and steadfast throughout, with her sheer force of will often being the only element that kept the project moving forward. To Ari and Raleigh, once again your contributions cannot be repaid! To Steve Chapman, you have the patience of Job, thank you. And, a very special thank you to Michael Anderson, a member of the family in more ways than kin, who took on challenges without question, contributing selflessly and unceasingly towards accomplishing the herculean task of finishing this book.

— Chuck Crawford
Washington, DC

Praise for DVD Demystified
Second Edition

"Jim Taylor's DVD Demystified is without a doubt the definitive reference book on DVD. DVD Demystified is aptly titled. Jim Taylor takes a subject that is still mired in confusion and lays out all the cards on the table. He then proceeds to describe each card in detailed but easy to understand style."

— One-to-One magazine (Bob Starett)

"A clear and intelligent guide to DVD, and a valuable reference for anyone who wants to take full advantage of the technology."

— Kamer Davis, Senior Vice President, Ogilvy Public Affairs Worldwide

"In DVD Demystified, Jim Taylor combines the technical expertise of an engineer with the imagination of a visionary. The book is a must-read for those who require factual information about the great potential of DVD technology for delivering digital video content."

— Dana Parker, Consultant, DVD Diva, and coauthor of the CD Recordable Handbook

"DVD Demystified should be read by everyone who is considering jumping into the DVD format. After reading Jim Taylor's definitive work on DVD, you will be prepared to enter the DVD market without losing your shirt in the process."

— Ralph LaBarge, President and Chief Technology Officer, NB Digital Solutions

"DVD Demystified is the clearest source of real information about DVD."

— Mike Schmit, Software Manager, CompCare/Zoran

"It was on April 17, 1992, when four bottles of Napa red wine were emptied between Warren Lieberfarb (President, Warner Home Video) and myself, that the very first shape of the DVD concept was originated. I was very excited but nevertheless seriously anxious about whether this new technology would ever see the light of the multimedia age.

In 1998, when I learned that DVD Demystified had been published, I began to feel there was proof that DVD was heading for success. This is the monumental book that gave me the confidence that we were doing the right thing. The new edition will surely add stability to the ever-growing DVD market.

DVD will be a pervasive technology for decades to come; DVD Demystified will continue for years as the benchmark reference of this technology."

— Koji Hase, Acting Chairperson of the DVD Forum, Vice President, Strategic Alliance Division, Digital Media Network Company, Toshiba Corporation

Foreword

There's an old saying that for a dog, a year of life is the same as seven human years. This rule of thumb is commonly referred to as a "dog year". I must tell you that having participated in the DVD business since its inception, I can sympathize with how dogs must feel: so much has happened with the format since late 1996 that it seems like seven normal years of growth and change are crammed into each "DVD year". Consider, for example, how different DVD is today compared with when it first launched: it took nearly two years after the first players appeared for all of the major Hollywood studios to commit to releasing DVD titles, and sales of early DVD players were anything but explosive. In fact, not long after the very first players started shipping in Japan in November 1996, I walked around Tokyo's Akihabara electronics retail district, watching in despair as huge crowds of people completely ignored street corner DVD player demonstrations. No one liked DVD! No one was even looking at it! The format was doomed!

Well, no. Within five years, DVD was being called the most successful consumer format ever launched, and with good reason: it turned out that everyone loves DVD. At the time of this writing, after only eight "DVD years", the Digital Entertainment Group informs us that over 140 million DVD players have been sold, with over 250 models currently available from more than 60 brands. More than 50,000 titles have been published on DVD, and approximately 200 more are released every week. In the U.S. alone, fully 80% of households are expected to have a DVD player by the end of 2005. In fact, many of those households have more than one player, which might take the shape of a laptop computer, game console or even a rear-seat entertainment system in the family car.

Since the last edition of *DVD Demystified* was published, DVD products have evolved at a breakneck pace. New DVD players can now be bought for as little as $35, which is less than the price of some DVD titles. DVD burners, a product that used to cost as much as a pretty decent car in 1998, can today be found as standard equipment in all but the very least expensive desktop computers. A raging battle between recordable formats has largely been settled — an unthinkable result only a few years ago — with multiple-format computer drives now the norm. Content has evolved as well: at the time of this writing, episodic television programming is the fastest growing segment of all title categories. TV shows you can barely remember (perhaps some that you don't want to remember) are coming to life once again, in pristine form. DVD's success has also been a windfall for film connoisseurs, with the release of some wonderful restorations of historic titles that have not been seen in their full glory for decades. The format has even provided a second life for films that weren't so historic at the box office, allowing them to find large, enthusiastic new audiences. On the production side, some directors, themselves DVD lovers, have said that they now shoot their films with DVD in mind, often simultaneously creating behind-the-scenes content for the in-depth bonus features that DVD buyers have come to expect.

DVD is also solving new problems. Not long ago, I heard a radio news anchor mention on the air that he had just finished transferring his family photos to recordable DVD in anticipation of a devastating earthquake someday striking Southern California. By making extra copies of his archive discs, he could also send them to other family members around the country for both enjoyment and further safekeeping. This scenario was almost unimaginable when the first recordable DVD products and authoring tools appeared on the market seven years ago.

Given all of DVD's progress and evolution, it's no wonder that an update was needed for the longtime bible of the industry, *DVD Demystified*. This third edition of the book has taken on this challenge in a number of successful ways. First, it was written by three authors: Jim Taylor, of course, the author of the first two editions, as well as Mark Johnson and Chuck Crawford. All three are battle-scarred DVD veterans, giving this enormous subject a broad, real-world perspective. Second, it's been updated and reorganized to reflect the many changes that have occurred in DVD during the past five years. Finally, the book's scope has also expanded beyond its thorough coverage of DVD as we've known it to also include an introduction to next generation, blue laser discs that will embrace the new high-definition video standards coming to life around the world. So much has changed, yet *DVD Demystified* has once again managed to keep up with it all.

By the way, I'm happy to report that something has not changed in *DVD Demystified*: the entertaining, always informative and sometimes irreverent style that characterized the previous editions. This is no dry, boring textbook — anyone who loves DVD should enjoy reading *DVD Demystified*. The only downside is that you may find that time has this peculiar way of speeding up as you get deeper into the world of DVD.

— Andy Parsons
Senior Vice President
Product Development
Pioneer Electronics (USA) Inc.

Table of Contents

Table of Contents

Table of Contents

Table of Contents

Table of Contents

Table of Contents

Table of Figures

Table of Tables

Table of Tables

Preface

They say the third time's the charm. Or three's company. As long as they don't say three strikes, you're out. This book is redolent with threes: three authors working on the third edition, covering the third version of shiny discs. But at least it's only two years late, not three. I would be remiss in the extreme if I didn't thank my two co-authors for saving me from being relegated forever to the "This book is not yet available. You may order it now and we will ship it to you when it arrives." aisle at Amazon.com. And the resultant timing is perfect. The supercession of DVD by blue laser formats is just beginning, although it will take many years for any new format to make a dent in the juggernaut that DVD has become, with almost a billion players and over 4 billion discs. Everyone wants to know what the new formats are like, what effect they will have, what will happen to them, and what will happen to DVD. Hopefully this book will succeed, as did its predecessors, in answering at least a few questions, opening at least a few eyes, and giving at least a few people a leg up in understanding both the old and new versions of a technology that has remade the face of entertainment, education, and computing.

— Jim Taylor
September 2005

DVD DEMYSTIFIED

Introduction

About Optical Discs

Who Needs to Know About Optical Discs?

"What are these guys trying to do?", you may be asking. Optical discs? Why not introduce the acronym, OD, and push to get the marketplace to accept that as shorthand for all technology that uses a laser to read/write data from a shiny disc? Alas, OD is already recognized as just such an acronym[1] and we'd rather not directly contribute to acronym overload, thank you very much.

We are now entering the third generation of modern disc technology (some might count this as the fourth, fifth or sixth generation if one were to include floppy disks and laserdiscs). CDs were the precursor to DVD, and the new-generation entities are disc flavors that use blue laser technology and fall into two incompatible groups, HD DVD (high-density DVD) and Blu-ray Disc (BD).

Yet, before we move into a blue laser world, red laser technology has morphed to include UMD™ (Universal Media Disc) for the Sony PlayStation® Portable aficionados and the burgeoning Asian markets are developing the idiosyncratic disc flavors, EVD and FVD.[2]

The expansion in disc development has resulted in an explosion of disc capacity. Advances in technology have moved us from the single-sided CD with 650 megabyte capacity to the imposing three-layer, single sided HD DVD with a 45 billion byte capacity, and onward to the astounding four-layer, single sided BD with a whopping 100 billion byte capacity, with a variety of disc, layer, and side combinations and capacities that span the range between those figures.

The number of people who are feeling the effect of DVD is truly astonishing. A large part of this multitude will require a working knowledge of DVD/UMD/HD DVD/BD, including the capabilities, strengths, and limitations of each media. This book provides a significant amount of this knowledge. Talk about acronym overload — we have yet to mention Microsoft's Windows Media Video 9 (WMV9) and WMA9 (audio) schemes.

Optical disc technology affects a remarkably diverse range of fields. Advances in disc applications, utilities, and programming, combined with breakthroughs in recording, display, and interactivity, have vaulted shiny discs into a leading role in a broad spectrum of uses. A few are illustrated in the following pages.

Movies

DVD has significantly raised the quality and enjoyability of home entertainment. And HD DVD and BD will provide astounding image and aural enhancements, comparable to that enjoyed in a relative handful of theaters and auditoriums, unlimited budgets notwithstand-

---

[1]Optical Disc is one of fifty-two recognized uses of the acronym "OD", according to www.acronymfinder.com. Other meanings range from Object Domain to Over Dose to Oxygen Depleted.

[2]China introduced EVD (Enhanced Versatile Disc) in 2003, and, not to be outdone, Taiwan debuted FVD (Forward Versatile Disc) in 2004. We'll have to see who does what with the letter "G".

ing. According to the Consumer Electronics Association (CEA), there were more than 33.3 million American homes with a home theater[3] system in the summer of 2005. Since their introduction, DVD players have become the central component of home theater systems, along with their acquisition as replacements for aging CD players.

Music and Audio

Because audio CD is well established and satisfies the needs of most music listeners, DVD has had less of an effect in this area. The introduction of DVD-Audio trailed DVD-Video by almost four years, and the DVD-Video format already incorporated higher-than-CD-quality audio and multichannel surround sound. DVD players also play audio CDs. All this left the DVD-Audio format with a tough sales job. However, the unprecedented success of DVD-Video ensured that DVD-Audio could come along for the ride, as players supported both formats and the distinctions between DVD and CD faded away.

DVD-Audio retains a special appeal for music labels because it includes content protection features that CD never had. Whether those protections prevail and whether they continue to be relevant in the face of Internet music distribution remains to be seen.

DVD is a boon for audio books and other spoken-word programs. Dozens of hours of stereo audio can be stored on one side of a single disc — a disc that is cheaper to produce and more convenient to use than cassette tapes or multi-CD sets.

Music Performance Video

Despite the success of MTV and the virtual prerequisite that a music group cut a video in order to be heard, music performance video on videotape or laserdisc did not do as well as expected. Perhaps it was because you could not pop a video in a player and continue to read or work around the house.

However, the added versatility of DVD has become the golden key that unlocks the gates for music performance video. With high-quality, long-playing video and multichannel surround sound, DVD music video appeals to a range of fans from opera to ballet to New Age to acid rock. Music albums on DVD have improved their fan appeal by adding such tidbits as live performances, interviews, backstage footage, outtakes, video liner notes, musician biographies, and documentaries. Sales of DVD-Video music titles have exceeded many industry expectations.

One of the targets for DVD player sales is the replacement CD player market. Shoppers looking for a new player are easily persuaded to get a device that also can play movies, with the costs of a standard DVD player oftentimes below that of a CD player. The natural convergence of the audio and video player will continue to bolster the success of combined music and video titles.

Training and Productivity

Until it is eventually replaced by streaming video on broadband Internet, DVD has become the medium of choice for video training. The low cost of hardware and discs, the widespread use of players, the availability of authoring systems, and a profusion of knowl-

---

[3]CEA defines a home theater system as a 25-inch or larger TV, hi-fi VCR or laserdisc player or DVD player, and a surround sound system with at least four speakers.

edgeable DVD developers and producers make DVD — in both DVD-Video and multimedia DVD-ROM form — ideal for industrial training, teacher training, sales presentations, home education, and any other application where full video and audio are needed for effective instruction. Especially popular are DVDs that connect to the Internet, which compensates for the fact that streaming video is too small, too slow, too fuzzy, and too unreliable to be of any use to the average learner.

Videos for teaching skills from accounting to TV repair to dental hygiene, from tai chi to guitar playing to flower arranging become vastly more effective when they take advantage of the on-screen menus, detailed images, multiple soundtracks, selectable subtitles, and other advanced interactive features of DVD. Consider an exercise video that randomly selects different routines each day or lets you choose the mood, the tempo, and the muscle groups on which to focus. Or, a first-aid training course that slowly increases the difficulty level of the lessons and the complexity of the practice sessions. Or, an auto-mechanic training video that allows you to view a procedure from different angles at the touch of a remote control (preferably one with a grease-proof cover). Or, a cookbook that helps you select recipes via menus and indexes and then demonstrates with a skilled chef leading you through every step of the preparation. All this on a small disc that never wears out and never has to be rewound or fast-forwarded.

DVD is cheaper and easier to produce, store, and distribute than videotape. Earlier products such as laserdisc, CD-i, and Video CD did well in training applications, but they required expensive or specialized players. As DVD became more established, almost every home and office now has either a standalone player or a DVD-capable computer.

Education

Filmstrips, 16-mm films, VHS tapes, laserdiscs, CD-i discs, Video CDs, CD-ROMs, and the Internet have all had roles in providing images and sound to supplement textbooks and teachers. But those technologies lacked the picture quality and clarity that is so important for classroom presentations. The exceptional image detail, high storage capacity, and low cost of DVD make it an excellent candidate for use in classrooms, especially since it integrates well with computers. Even though DVD-Video players still may not be widely adopted in education, DVD computers are becoming commonplace in the classroom. CD-ROM infiltrated all levels of schooling from home to kindergarten to college and is now passing the baton to DVD as new computers with built-in DVD-ROM drives are purchased. Educational publishers have discovered how to make the most of DVD, creating truly interactive applications with the sensory impact and realism needed to stimulate and inspire inquisitive minds.

Computer Software

CD-ROM has become the computer software distribution medium of choice. To reduce manufacturing costs, most software companies use CD-ROMs rather than expensive and unwieldy piles of floppy disks. Yet some applications are too large even for the multi-megabyte capacity of CD-ROM. These include large application suites, clip art collections, software libraries containing dozens of programs that can be unlocked by paying a fee and receiving a special code, specialized databases with hundreds of millions of entries, and massive software products such as network operating systems and document collections. Phone books that used to fill six or more CD-ROMs now fit onto a single DVD-ROM. Companies that distribute monthly updates of CD-ROM sets can ship free DVD-ROM drives to their customers and pay for them within a year with the savings on production costs alone.

Computer Multimedia

DVD-ROM products have been slow to appear, largely because the multimedia CD-ROM market is so tough. CD-ROM publishers struggling to produce a product within budget, convince distributors to carry it, and get it onto limited shelf space are reluctant to complicate their lives with a new format.

Still, many multimedia producers are stifled by the narrow confines of CD-ROM and yearn for the wide open spaces and liberating speed of DVD-ROM. Microsoft's Encarta encyclopedia has overflowed onto more than one CD-ROM but can expand for years to come without filling up its single DVD-ROM. The National Geographic collection that is spread across 31 CD-ROMs is much more usable on 4 DVD-ROMs.

In addition to space for more data, DVD brings along high-quality audio and video. Many new computers have hardware or software decoders that can be used to play DVD movies. These DVD-enabled computers have become even more effective for realistic simulations, games, education, and "edutainment." DVD has made blocky, quarter-screen computer video a distant, dismal memory.

Video Games

The capacity to add high-quality, real-life video and full surround sound to three-dimensional game graphics has proven to be highly attractive to video game manufacturers. The major game console manufacturers: Nintendo, Sega, and Sony — and Microsoft with its Xbox — are all using DVD in the newest versions of their systems. A combination video game/CD/DVD player is very appealing. Sony is the leading architect of Blu-ray technology, having focused its development with the PlayStation game consoles in mind.

Many past attempts to combine video footage with interactive games were met with yawns, but the technology has improved and now it finally clicks. Video games that make extensive use of full-screen video - even multimedia games traditionally available only for computers - are appearing in DVD-Video editions that play on any home DVD player and on DVD computers.

For its PlayStation Portable (PSP™), Sony has created a new disc format, Universal Media Disc (UMD). Games for the PSP will be exclusively developed on UMD, and studios have begun production and marketing of movies and music on UMD for the PSP.

Information Publishing

The Internet is a wonderfully effective and efficient medium for information publishing, but it lacks the bandwidth needed to do justice to large amounts of data rich with graphics, audio, and motion video. DVD, with storage capacity far surpassing CD-ROM and standardized formats for audio and video, is perfect for publishing and distributing information in our ever more knowledge-intensive and information-hungry world.

Organizations use DVD-Video and DVD-ROM to quickly and easily disseminate reports, training material, manuals, detailed reference handbooks, document archives, and much more. Portable document formats such as Adobe Acrobat and HTML are perfectly suited to publishing text and pictures on DVD-ROM. Recordable DVD is available and affordable for custom publishing discs created on the desktop.

Marketing and Communications

DVD-Video and DVD-ROM have proven well suited for carrying information from businesses to their customers and from businesses to businesses. A DVD can hold an exhaustive catalog able to elaborate on each product with full-color illustrations, video clips, demonstrations, customer testimonials, and more, at a fraction of the cost of printed catalogs. Bulky, inconvenient videotapes of product information have been replaced by thin discs containing on-screen menus that guide the viewer on how to use the product, with easy and instant access to any section of the disc.

The storage capacity of DVD-ROM can be exploited to put entire software product lines on a single disc that can be sent out to thousands or millions of prospective customers in inexpensive mailings. The disc can include demo versions of each product, with protected versions of the full product that can be unlocked by placing an order over the phone or the Internet.

And More . . .

Picture archives. Photo collections, clip art, and clip media have long since exceeded the capacity of CD-ROM. As recordable DVD has become affordable and easy to use, it has enabled personal media publishing. Anyone with a digital camera, some video editing software, and a recordable DVD drive can put pictures and home movies on a disc to send to friends and relatives who have a DVD player or a DVD computer.

Settop boxes, digital receivers, and personal video recorders. Savvy designers have combined DVD players with the boxes used for interactive TV, digital satellite, digital cable, hard-disk-based video recorders, and other digital video applications. All these systems are based on the same underlying digital compression technology and benefit from shared components.

WebDVD. Products that combine the multimedia capabilities of DVD-ROM and DVD-Video with the timeliness and interactivity of the Internet are readily available in the marketplace. Data-intensive media such as audio, video, and even large databases simply do not travel well over the Internet. Although broadband service is more widely available and affordable for Internet delivery of data, DVD remains the perfect candidate to deliver the lion's share of the content.

Home productivity and "edutainment." DVD-Video is used for reference products such as visual encyclopedias, fact books, and travelogues; training material such as music tutorials, arts and crafts lessons, and home improvement series; and education products such as documentaries, historical re-creations, nature films, and more, all with accompanying text, photos, sidebars, quizzes, and so on. DVD already covers the gamut from PlayStation to WebTV to home PC. And, the new disc technologies will continue to expand in these arenas.

About This Book

DVD Demystified is an introduction and reference for anyone who wants to understand DVD and the emerging variants of optical disc. It is not a production guide, nor is it a detailed technical handbook, but it provides an extensive technical grounding for anyone interested in the latest disc technology. This third edition is completely revised and expanded from the first two editions. This book is divided into sections, which are subdivided into chapters.

Section I - History and Background

Chapter 1, "The World Before DVD," provides historical context and background. Any top analyst or business leader will tell you that extrapolating from prior technologies is the best way to predict technology trends. This chapter takes a historical stroll through the developments leading up to the introduction of DVD.

Chapter 2, "DVD Arrives," describes the growing pains of DVD from its introduction to the establishment of DVD and its format cousins, DVD-Video, DVD-Audio, and recordable DVD. This chapter goes on to discuss the evolution of the newly introduced HD DVD and Blu-ray optical disc technologies.

Section II - Fundamentals and Features

Chapter 3, "Technology Primer," explains concepts such as aspect ratios, digital compression, progressive scan, and describes associated display and signal technologies. This chapter is a gentle technical introduction for nontechnical readers, and it will be useful for technical readers as well.

Chapter 4, "Features," covers the basic features of the various optical disc flavors, from DVD-Video to BD-ROM. In addition to feature details, topics such as interoperability, network capabilities, and programmability are discussed in this chapter.

Chapter 5, "Content Protection," details the schemata that have been developed to protect the disc assets of the content producers and intellectual property developers. Techniques range from content-scrambling to protection for recordable media and watermarking.

Section III - Formats

Chapter 6, "Overview of the Formats," contains descriptions of the optical disc families, from standard DVD to UMD to BD and beyond. This chapter also addresses similarities, differences, advantages, and disadvantages of the formats.

Chapter 7, "Red Laser Physical Disc Formats," breaks down the physical characteristics of the red laser optical disc technologies, from the mechanics to the data handling attributes.

Chapter 8, "Blue Laser Physical Disc Formats," does for blue laser technologies what chapter 7 did for red laser technologies.

Chapter 9, "Application Details," reveals the particulars of the video and audio specifications for optical discs, from standard DVD to BD to acronyms you'll be surprised exist. It lays out the data structures, stream composition, navigation information, and other elements of the application formats.

Section IV - Implementation

Chapter 10, "Players," provides information on disc players and their connections.

Chapter 11, "Myths," does a reality check on the myths, the myth-understandings, and the misunderstood characteristics of optical disc formats, their delivery systems, and some related gee-whiz gadgets and gee-gaws.

Chapter 12, "What's Wrong with DVD," explores the shortcomings of standard DVD and how these might be overcome as the technology develops.

Section V - Uses

Chapter 13, "New Interaction Paradigms," investigates the synergy front, with new user tools, Internet-connected players, and coalescing applications.

Chapter 14, "DVD in Home, Business, and Education," helps you decide what disc flavor is right for you. This chapter explains how the disc formats may be integrated in the home, in the workplace, and in training/teaching/learning situations.

Chapter 15, "DVD on Computers," explores the burgeoning universe of optical disc-enabled computers. Multimedia, -ROM, and WebDVD frontiers for all computer platforms are investigated in this chapter.

Chapter 16, "Production Essentials," dips into many of the mysteries of producing content for DVD-Video and DVD-ROM. This chapter provides a thorough grounding for anyone interested in creating DVDs or simply learning more about how optical discs are created.

Section VI - The Future

Chapter 17, "DVD and Beyond," is a peek into the crystal ball to see what possibilities lie ahead for optical discs, the world of digital video, and the far horizon. Woo-hoo!

Appendixes and Glossary

Appendixes provide reference details on the data and standards for the optical disc formats.

A glossary of esoteric, arcane and mundane words and terms is provided to aid those who know they saw a word somewhere but cannot remember what it means or where they saw it.

Units and Notation

DVD is a casualty of an unfortunate collision between the conventions of computer storage measurement and the norms of data communications measurement. The SI[4] abbreviations of k (thousands), M (millions), and G (billions) usually take on slightly different meanings when applied to bytes, in which case they are based on powers of 2 instead of powers of 10 (see Table I.1).

---

[4]Système International d'Unités - the international standard of measurement notations such as millimeters and kilograms.

Table I.1 Meanings of Prefixes

Symbol	Prefix SI	Prefix IEC	IEC Symbol	Common Use	Computer Use	Usage Difference
k or K	kilo	kibi	Ki	[k] 1,000 (10^3)	[K] 1024 (2^{10})	2.4%
M	mega	mebi	Mi	1,000,000 (10^6)	1,048,576 (2^{20})	4.9%
G	giga	gibi	Gi	1,000,000,000 (10^9)	1,073,741,824 (2^{30})	7.4%
T	tera	tebi	Ti	1,000,000,000,000 (10^{12})	1,099,511,627,776 (2^{40})	10%
P	peta	pebi	Pi	1,000,000,000,000,000 (10^{15})	1,125,899,906,842,624 (2^{50})	12.6%

The problem is that there are no universal standards for unambiguous use of these prefixes. One person's 4.7 GB is another person's 4.38 GB, and one person's 1.321 MB/s is another's 1.385 MB/s. Can you tell which is which?[5]

The laziness of many engineers who mix notations such as KB/s, kb/s, and kbps with no clear distinction and no definition compounds the problem. And since divisions of 1000 look bigger than divisions of 1024, marketing mavens are much happier telling you that a dual-layer DVD holds 8.5 gigabytes rather than a mere 7.9 gigabytes. It may seem trivial, but at larger denominations the difference between the two usages — and the resulting potential error — becomes significant. There is almost a five percent difference at the mega-level and more than a seven percent difference at the giga-level. If you are planning to produce a DVD and you take pains to make sure your data takes up just under 4.7 gigabytes (as reported by the operating system), you will be surprised and annoyed to discover that only 4.37 gigabytes fit on the disc. Things will get worse down the road with a ten percent difference at the tera-level.

Because computer memory and data storage, including optical disc ROM flavors and CD-ROM, usually are measured in megabytes and gigabytes (as opposed to millions of bytes and billions of bytes), this book uses 1024 as the basis for measurements of data size and data capacity, with abbreviations of KB, MB, and GB. However, since these abbreviations have become so ambiguous, the term is spelled out when practical. In cases where it is necessary to be consistent with published numbers based on the alternative usage, the words *thousand*, *million*, and *billion* are used, or the abbreviations k bytes, M bytes, and G bytes are used (note the small k and the spaces).

To distinguish kilobytes (1024 bytes) from other units such as kilometers (1000 meters), common practice is to use a large K for binary multiples. Unfortunately, other abbreviations such as M (mega) and m (micro) are already differentiated by case, so the convention cannot be applied uniformly to binary data storage. And in any case, too few people pay attention to these nuances.

---

[5]The first (4.7 GB) is the typical data capacity given for DVD: 4.7 billion bytes, measured in magnitudes of 1000. The second (4.38 GB) is the "true" data capacity of DVD: 4.38 gigabytes, measured using the conventional computing method in magnitudes of 1024. The third (1.321 MB/s) is the reference data transfer rate of the DVD-ROM measured in computer units of 1024 bytes per second. The fourth (1.385 MB/s) is DVD-ROM data transfer rate in thousands of bytes per second. If you do not know what any of this means, don't worry, by Chapter 4 it should all make sense.

In 1999, the International Electrotechnical Commission (IEC) produced new prefixes for binary multiples[6] (see Table I.1). Although the new prefixes may never catch on, or they may cause even more confusion, they are a valiant effort to solve the problem. The main strike against them is that they sound a bit silly. For example, the prefix for the new term *gigabinary* is *gibi-*, so a DVD can be said to hold 4.37 gibibytes, or GiB. The prefix for *kilobinary* is *kibi-*, and the prefix for *terabinary* is *tebi-*, yielding kibibytes and tebibytes. Jokes about "kibbles and bits" and "teletebis" are inevitable.

As if all this were not complicated enough, data transfer rates, when measured in bits per second, are almost always multiples of 1000, but when measured in bytes per second are sometimes multiples of 1000 and sometimes multiples of 1024. For example, a 1X DVD drive transfers data at 11.08 million bits per second (Mbps), which might be listed as 1.385 million bytes per second or might be listed as 1.321 megabytes per second. The 150 KB/s 1X data rate commonly listed for CD-ROM drives is "true" kilobytes per second, equivalent to 153.6 thousand bytes per second.

This book uses 1024 as the basis for measurements of byte rates (computer data being transferred from a storage device such as a hard drive or DVD-ROM drive into computer memory), with notations of KB/s and MB/s. For generic data transmission, generally measured in thousands and millions of bits per second, this book uses 1000 as the basis for bit rates, with notations of kbps and Mbps (note the small k). See Table I.2 for a listing of notations.

Keep in mind that when translating from bits to bytes, there is a factor of 8, and when converting from bit rates to data capacities in bytes, there is an additional factor of 1000/1024.

Table I.2 Notations Used in This Book

Notation	Meaning	Magnitude	Variations	Example
b	bit	(1)		
kbps	thousand per second	10^3	Kbps, kb/s, Kb/s	56 kbps modem
Mbps	million per second	10^6	mbps, mb/s, Mb/s	11.08 Mbps DVD data rate
B	byte	(8 bits)		
KB	kilobytes	2^{10}	Kbytes, KiB	2 KB per DVD sector
KB/s	kilobytes per second	2^{10}	KiB/s	150 KB/s CD-ROM data rate
MB	megabytes	2^{20}	Mbytes, MiB	650 MB in CD-ROM
M bytes	million bytes	10^6	Mbytes, MB	682 M bytes in CD-ROM
MB/s	megabytes per second	2^{20}	MiB/s	1.32 MB/s DVD data rate
GB	gigabytes	2^{30}	Gbytes, GiB	4.37 GB in a DVD
G bytes	billion bytes	10^9	Gbytes, GB	4.7 G bytes in a DVD
TB	terabytes	2^{40}	Tbytes, TiB	
T bytes	trillion bytes	10^{15}	Tbytes, TB	

[6]The new binary prefixes are detailed in *IEC 60027-2-am2 (1999-01): Letter symbols to be used in electrical technology. Part 2: Telecommunications and electronics, Amendment 2.*

Other Conventions

Spelling The word *disc*, in reference to optical discs, should be spelled with a c, not a k. The generally accepted rule is that optical discs are spelled with a c, whereas magnetic disks are spelled with a k. For magneto-optical discs, which are a combination of both formats, the word is spelled with c because the discs are read with a laser. The *New York Times*, after years of head-in-the-sand usage of k for all forms of data storage, revised its manual in 1999 to conform to industry practice. Standards bodies such as ECMA and the International Organization for Standardization (ISO) persist in spelling it wrong, but what can you expect from bureaucracies? Anyone writing about DVD who spells it as disk instead of disc immediately puts the reader on notice that the author is clueless.

PC The term *PC* means "personal computer." If a distinction is needed as to the specific type of hardware or operating system, brand names such as Windows, Mac OS, and Linux are added.

Title The word *title* fulfills two primary roles in optical disc parlance. In a general sense, title may be used to indicate the entire contents of a disc, as in, the disc accompanying this book is a DVD title. In a specific sense, title is the largest unit of a DVD-Video disc, and a single disc may contain up to 99 titles.

Aspect Ratios This book usually normalizes aspect ratios to a denominator of 1, with the 1 omitted most of the time. For example, the aspect ratio 16:9, which is equivalent to 1.78:1, is represented simply as 1.78. Normalized ratios such as 1.33, 1.78, and 1.85 are easier to compare than unreduced ratios such as 4:3 and 16:9. Note also that the ratio symbol (:) is used to indicate a relationship between width and height rather than the dimension symbol ($\times$), which implies a size.

Widescreen When the term *widescreen* is used in this book, it generally means an aspect ratio of 1.78 (16:9). The term *widescreen,* as traditionally applied to movies, has meant anything wider than the standard 1.33 (4:3) television aspect ratio, from 1.5 to 2.7. Since the 1.78 ratio has been chosen for DVD, digital television, and widescreen TVs, it has become the commonly implied ratio of the term *widescreen*.

Disc Format Names The term *DVD* is often applied both to the standard DVD family as a whole and specifically to the DVD-Video format. This book follows the same convention for simplicity and readability but only when unambiguous.

When a clear distinction is needed in discussing optical disc varieties, this book will strive to insure clarity in disc descriptions and their acronyms. For example, the book uses the terms *DVD-Video* (or DVD-V), *DVD-Audio* (or DVD-A), and *writable DVD* for the various record-once (DVD-R) and rewritable formats (DVD-RAM, DVD-RW, DVD+RW). The abbreviations *DVD-R(G)* and *DVD-R(A)* are used to distinguish "DVD-R for General" from "DVD-R for Authoring" respectively.

Television Systems There are basically two mutually incompatible television recording systems in common use around the world. Each is supported by corresponding digital encoding formats used by DVD. One system uses 525 lines scanned at 60 fields per second with NTSC color encoding and is used primarily in Japan and North America. The other system uses 625 lines scanned at 50 fields per second with PAL or SECAM color encoding and is used in most of the rest of the world. This book generally uses the technically correct terms of 525/60 (simplified from 525/59.94), and 625/50, but also uses the terms NTSC and PAL in the generic sense.

Colorspaces DVD uses the standard ITU-R BT.601 (formerly CCIR 601) component digital video colorspace of 4:2:0 $Y'C_bC_r$ nonlinear luma and chroma signals. Some DVD video playback systems include analog component video output in $Y'P_bP_r$ format, which is also correctly called $Y'/B'-Y'/R'-Y'$ but incorrectly called $Y'C_bC_r$ (it is analog, not digital). When a technical distinction is not critical or is clear from the context, this book uses the general term *YUV* to refer to the component video signals in nonlinear color-difference format. This book also uses the general term *RGB* to refer to nonlinear R'G'B' video.

Most of these terms and concepts are explained in further detail in Chapter 3.

Chapter 1
The World Before DVD

A Brief History of Audio Technology

In 1877, Thomas Edison recorded and played back the words "Mary had a little lamb" on a strip of tinfoil, presaging a profound change in the way we record events. Instead of relying on written histories and oral accounts, we began to capture audible information in a way that enables us to reproduce the events later. Since then, we have worked continuously to improve the verisimilitude of recording. See Figure 1.1 for a timeline of audio recording technology.

By the 1890s, 12-inch shellac gramophone disks that could play up to 4 1/2 minutes of sound at 78 rpm had become popular. The Victrola appeared in 1901. Radio followed soon after, with the first commercial broadcast in 1920. Acoustical recording (where sound vibrations were converted directly to wavy grooves on a wax disc) was replaced in the early 1920s by electrical recording (where sound vibrations are converted to electrical impulses that can be amplified and mixed before being used to drive an electromechanical cutting head to cut grooves). Performers no longer had to cluster around a large horn that gave too much emphasis to the most powerful instruments and the loudest voices. Electrical technology also was applied to sound reproduction, resulting in the birth of the loudspeaker.

Recording technology took a major leap in 1948 when the *long-playing record* (LP) was introduced by Columbia Records. New microgroove technology allowed 25 to 30 minutes of sound to fit on a 12-inch disk turning at the slower speed of 33 1/3 rpm. Columbia's LP was developed under the direction of Peter Goldman, who later developed the first commercial closed-circuit color television system. A year after the LP appeared, RCA Victor introduced a 7-inch disk that turned at 45 rpm and played for about eight minutes. These two new record types quickly replaced the unwieldy 5-minute 78-rpm records.

Magnetic recording appeared in the laboratory in the 1890s but was not integrated into an actual product until about 1940. The first systems recorded onto a thin wire, which later was replaced by polyester tape with a thin coating of magnetic particles. A major advantage of storing electrical sound impulses using the alignment of magnetic particles is that the recording process is as easy as the reproduction process, and the same head can be used for recording and playback.

Up to this point, sound recording had been monophonic, meaning it was recorded as a single-point source. The first patent on stereo sound was issued in 1931, but commercial stereophonic tape systems were not developed until 1956, with stereo phonographs following in 1958. These systems added a spatial component to the sound by incorporating a second channel, giving the reproduction a much greater sense of realism. Stereo technology debuted to an apathetic reception; many people felt that the slight improvement was not worth the added complexity, and some claimed that stereo recording would never be practical for mainstream applications. This was soon proven shortsighted as engineers became accustomed to the new technology and artists began to take advantage of it.

Figure 1.1 Timeline of Audio Technology

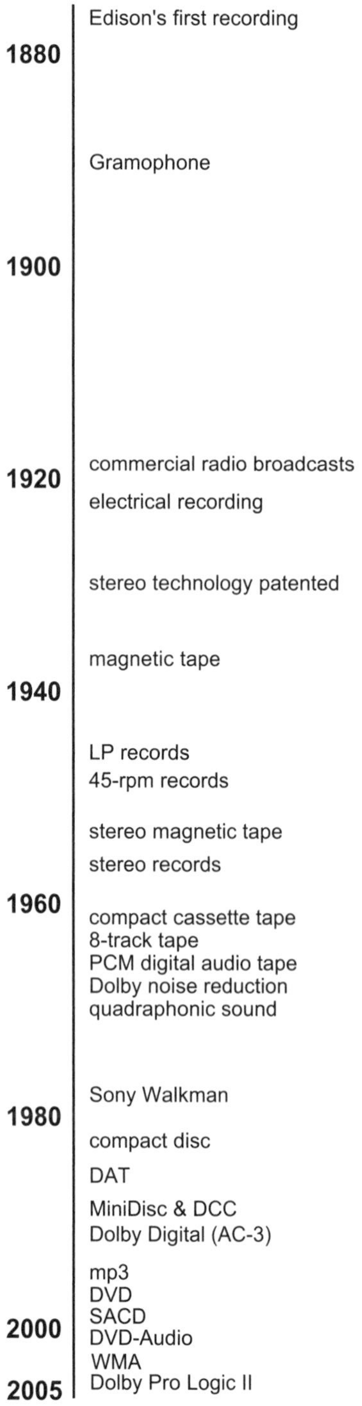

	Edison's first recording
1880	
	Gramophone
1900	
1920	commercial radio broadcasts
	electrical recording
	stereo technology patented
	magnetic tape
1940	
	LP records
	45-rpm records
	stereo magnetic tape
	stereo records
1960	compact cassette tape
	8-track tape
	PCM digital audio tape
	Dolby noise reduction
	quadraphonic sound
	Sony Walkman
1980	
	compact disc
	DAT
	MiniDisc & DCC
	Dolby Digital (AC-3)
	mp3
	DVD
	SACD
2000	DVD-Audio
	WMA
2005	Dolby Pro Logic II

The natural step beyond stereo was four-channel audio. Quadraphonic systems appeared in early 1970, but technical problems were compounded by three incompatible standards that confused and divided the marketplace and doomed the movement to quick extinction.

Audiocassettes using 1/8-inch tape were introduced by Philips in 1963, originally as a business dictation device that subsequently became popular for music. Prerecorded tapes were released in 1966 and steadily encroached on the sales of LPs. Eight-track (and less-successful four-track) audiotapes emerged in 1965 to enjoy a brief spurt of success mainly in automobile sound systems. Dolby noise reduction appeared in 1969 and did much to reduce the problem of tape hiss. The theories and techniques behind noise reduction have been refined steadily and today are the basis of the Dolby Digital surround-sound system used by DVD. In 1977, Sony attempted to launch Elcaset, a high-fidelity 1/4-inch audio cassette, but quality was not improved sufficiently to displace the established 1/8-inch cassette format, which became firmly entrenched after the introduction of the Sony Walkman in 1979.

By 1982, music sales on cassette surpassed vinyl. An attempt was again made to create a four-channel system, this time for cassette tapes. No one was able to agree on a standard, however, and the benefits did not seem to outweigh the technical complexity of extending the existing cassette format.

Philips again led the development of consumer electronics technology when it partnered with Sony to introduce *Compact Disc Digital Audio* (CD) in 1982. Digital audio recording is based on sampling. Sampling is the process of converting electrical signal levels that are measured over 44,000 times per second into discrete numbers represented by binary digits of zero and one, which are then stored as pulses. This *pulse-code modulation* (PCM) technology was invented in 1937 and was applied to communications theory by C. E. Shannon in 1949. The first digital PCM audiotape recorder was developed in 1967 at the NHK Technical Research Institute in Japan. PCM is used by most digital audio systems, including CD, *digital audiotape* (DAT), laserdisc, and DVD. Denon Records created the first digitally mastered audio recording for commercial release in 1974.

The advantages of recording and storing sound digitally include: exceptional frequency response (the rendition of frequencies covering the entire range of human hearing from 20 to 20,000 Hz), excellent dynamic range (the reproduction of very soft to very loud sounds with little or no extraneous noise), and no generational loss from copying. These improvements in quality over LP records (arguments of audio purists notwithstanding), plus other advantages, such as longer playing time, smaller size, near-instant track access, no wear from being played, and better resistance to dust and scratches, enabled CD sales to surpass LP sales around 1988.[1]

Of course, a major drawback to audio CD is that it does not record. All attempts to produce a recordable successor to CD have failed to become significant consumer successes. DAT, Philips's *digital compact cassette tape* (DCC), Sony's MiniDisc, and other technologies have languished or died for many reasons, including high cost, incompatibility with existing standards, limited manufacturer support, and politics — which come to the fore when the ability to make perfect digital copies increases the importance of copyright protection.

---

[1] It may seem reasonable that a similar comparison may be made to support the prediction that DVD will replace VHS videotape. However, the analogy must be tempered because VHS does not share the many deficiences of LP records, such as breakable disks, fragile tone arms, no protection from abuse, crackles and pops from dust and scratches, a tendency to skip or repeat, and the very high costs for high-fidelity sound.

Sony and Philips continued their work on the CD format, primarily for computer and multimedia applications. That part of the story is covered later in this chapter.

Many other companies are busy developing or promoting other audio storage techniques, each in the hope that its technology becomes the winning successor to tape or CD. These include magnetic cards, solid-state electronic cards, superdense miniature discs, rewritable CDs (CD-RW), DVD-audio, and the newest entry, the iPod. In addition, there is the MP3 audio format, which focuses less on quality and more on availability. Although DVD does little to advance the art of digital audio other than providing slightly higher quality and multichannel surround sound, many people have high hopes for the recordable version. If and when they become available, and if they garner sufficient support, DVD-Audio recorders may finally bring consumer audio recording completely into the digital realm. Of course, content protection and all its associated baggage will come along for the ride.

A Brief History of Video Technology

It is unknown when the principle of persistence of vision was first discovered. As it can be illustrated by waving a burning stick in the darkness, it undoubtedly has been known since prehistoric times. The phenomenon, caused by the brain holding an image for a fraction of a second after the optical stimulus is removed, led to some of the first optical illusions, such as tracing letters in the air with a bright light. Much later it was discovered that a quickly flickering light appears to be continuous. At low light levels the flicker threshold — also termed the *fusion frequency* — is around 50 times a second.[2]

Even later, in the 1830s, it was learned that a series of still images shown in rapid succession produces the illusion of motion (see Figure 1.2 for a timeline of the history of video technology).[3]

---

[2]Critical fusion frequency depends greatly on the ambient light level. In a dark movie theater, a flicker rate of 48 times per second is sufficient (each frame is flashed twice). European PAL television refreshes at 50 times a second (50 Hz), which is just at the limit for home viewing and is noticeable by some people, especially in well-lit environments or with bright video images. American NTSC television, at 60 Hz, is generally adequate for most environments. Many computer monitors flicker at 72 Hz or higher in order to abate eyestrain and mental fatigue in brightly lit offices.

[3]Most texts on video and motion pictures explain that persistence of vision is what makes it possible. Although there are conflicting opinions, this is apparently inaccurate — a myth that has passed through generations since Peter Mark Roget's definition in 1824. (Roget's term was misappropriated, since he was not describing perception of motion.) It is true that there is a physiologic mechanism by which the eye sees afterimages (some positive, some negative) after a light source ceases, but the scientific consensus is that it has little or nothing to do with flicker fusion or apparent motion. This has been understood since at least the beginning of the twentieth century. That humans perceive a series of changing still images as motion, without registering periods of blackness in between, is a psychologic characteristic of the brain (sometimes called the phi phenomenon), mostly unrelated to persistence of images on retinal photoreceptors or within the optic nerve. Research in high-speed filming and computer video games has shown that we perceive motion quality improvements at frame rates of 60 and 70 Hz and higher - far faster than the 24 Hz of film used for DVDs. If motion perception were based primarily on a buildup of afterimages, we would see blurry streaks or multiple images, especially when moving our eyes or head. For example, waving your hand in front of a strobe light demonstrates physiological persistence of vision: You see more than five fingers. However, a waving hand on a film or television screen produces no distortion and the hand looks normal, partly because the still images of the hand are blurred by motion captured during shutter exposure time, but mostly due to the psychological illusion of apparent motion.

Figure 1.2 Timeline of Visual Technology

1820	revolving picture toys
	photographs
1840	
	color photographs
1860	
	Muybridge's photographs
1880	Nipkow's scanning disc; Eastman's celluloid film
	Kinetograph
1900	mechanical TVs
1920	talking pictures; Farnsworth's electronic TV Baird's video disc, Technicolor commercial TV broadcasts in Germany and England
1940	cable TV, professional videotape color added to TV
1960	Ampex helical scan, color videotape Cinemascope PAL and SECAM adopted
	Sony Betamax, JVC VHS Dolby Stereo added to movies, video stores
1980	laserdisc HiFi VCRs, stereo TV broadcasts S-VHS, 8 mm HiVision, VideoCD
	digital satellite (DBS)
2000	digital videotape (DV), DVD HDTV (DTV), personal video recorders (PVRs)
	UMD™
2005	HD DVD, Blu-ray Disc (BD)

Revolving-picture toys with strange names such as *zoetrope*, *phenakistiscope*, *thaumatrope*, *praxinoscope*, and *fantascope* began to capitalize on this particular feature of human perception, which is what makes modern motion picture and video technology possible. Early experimenters discovered that a continuous stream of images merely causes blurring as the eye attempts to track the motion. They learned that each image has to be motionless long enough for the brain to acquire it. Most of the moving-image toys used slits through which the image was momentarily viewable (see Figure 1.3). Happily, the brain ignores the absence of the image. This technique of using a slit or shutter to briefly show a still picture as it moves is still used today in movie projectors.

Figure 1.3 Revolving Picture Toy

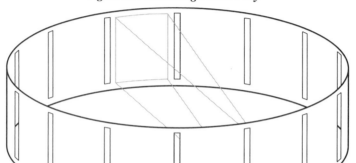

Many years later it was discovered that the eye employs high-frequency tremors to create snapshots as part of its image-acquisition mechanism. By rapidly jerking back and forth, the eye creates multiple still images that it then passes on to the brain for processing.[4]

Captured Light

Early motion picture toys used drawings, which were intriguing but fell short of the real thing. The first attempts at capturing a direct visual representation of reality began about the same time, in the 1820s, with Niépce and Daguerre's photographic plates. Black and white (or sepia and white) and hand coloring were the only options for about 40 years, until Scottish physicist James Clerk Maxwell studied vision and determined that all colors could be represented with combinations of red, green, and blue. In 1861, he produced the first color photograph from a three-color process. Photography steadily improved but remained motionless (pun intended) for almost 60 years after its invention. In 1877, the same year Edison invented the phonograph, photographer Eadweard Muybridge captured images of a

---

[4]Involuntary saccadic motion occurs about 10 to 25 times a second, typically over about 10 degrees of arc. Voluntary saccadic motion (such as when reading or during intense visual search) happens about 3 to 5 times a second.

moving horse as it tripped strings attached to 12 cameras in sequence.[5] He realized that the motion could be recreated by placing the photographs on a rotating wheel and projecting light through them. This led to another string of oddly named "magic lantern" gadgets such as the *zoopraxiscope*, *phantasmagoria*, *chronophotographe*, and *zoogyroscope*. The early recording process was very cumbersome, requiring that dozens of cameras be painstakingly set up.[6] In 1882, Étienne-Jules Marey was inspired by Muybridge's work to create a single camera, patterned after a rifle, that exposed 12 images in one second on a rotating glass plate. This was a great improvement, but it made for a very short viewing time and did not bode well for popcorn sales. It was not until seven years later that the celluloid roll film developed by George Eastman — the founder of Kodak — was used by Thomas Edison's engineers to create the Kinetograph camera and the Kinetoscope viewing box for movies lasting up to 15 seconds. Lessons from the past were apparently missed by these and other pioneers, who tried to use continuous film motion only to rediscover that repeated still images were required. Edison's lead engineer, William Dickson, shot the first film in the United States, *Fred Ott's Sneeze*.

The Lumière brothers were the first to project moving photographic pictures to a paying audience in 1895. Their first film, no less inspired than Dickson's, was the spine-tingling *Workers Leaving the Lumière Factory*. The motion picture industry was born and began to steadily improve the technology and the content. Projection speeds varied from 15 to 24 frames per second,[7] with 16 frames per second being the typical speed, requiring a three-bladed shutter to flash the picture 48 times per second to sufficiently reduce flicker.

Films at first were silent, although phonographs were used with film projectors even before the 1900s. Many methods of adding sound were tried, some less dismal than others, and by 1927, Warner Bros. (one of the modern-day contributors to the development of DVD) and Fox had developed a practical synchronized sound technology. Within two short years, most films had soundtracks. By this time the industry had settled on a film speed of 24 frames per second, with a two-bladed shutter to flash each frame twice. After the limited success of two-color film in the 1920s, three-color film was introduced by Technicolor in 1932, although it took another 25 years before most films were shot in color. At about the same time, in an effort to improve the movie theater experience and battle the threat of television, movies were made almost twice as wide. Cinemascope was introduced in 1953 with the biblical epic *The Robe*. Some widescreen systems used more than one projector, but the most successful early widescreen systems were based on anamorphic lenses that had been invented by Dr. Henri Chretien for tank periscopes in World War I. These lenses squeezed the image sideways to fit in a standard film frame and then unsqueezed it during projection. The same tech-

---

[5]The story goes that Muybridge was hired by a former governor of California in an attempt to win a bet that all four of a horse's hooves left the ground during a gallop. Muybridge's success left the governor $25,000 richer, although he may have paid Muybridge more than that to develop the experiments.

[6]The use of multiple still cameras arranged in sequence to capture complex motion formed the basis for the acclaimed "bullet time" visual effects used in the motion picture "The Matrix" (1999). The principal differences were simply that the cameras were digital and additional computer processing was used to augment the effect.

[7]Since there were no firm standards for film speed in cameras or projectors, films came with a cue sheet telling the projectionist the speed, in feet per second, at which to run the film. Unscrupulous theater owners would speed the films up to squeeze in more showings, even to the point of cutting the running time in half!

nique is still used occasionally for movies and is now used by DVD for widescreen video; the squeezing and unsqueezing are done by computers rather than glass lenses.

Less successful improvements to movies were attempted, such as Smell-O-Vision in the 1960s and many variations of three-dimensional images. Another disappointment, Sensurround, reappeared in a new form as a niche home theater product: motion transducers that attach to the frame of your couch to add extra kick to explosions and low-frequency sounds.

Other advances in motion picture film technology include stereo audio and 70-mm film — twice the width of standard 35-mm film. Prestige formats such as IMAX and OmniMax use 70-mm film with large square frames to achieve presentations of breathtaking impact.

In 1976, Dolby Laboratories created an affordable surround-sound system while using standard film soundtracks, and in 1992, it unleashed multichannel Dolby Digital (AC-3) for the movie *Batman Returns*. In 1993, *Jurassic Park* debuted the sonic realism of *Digital Theater Systems* (DTS) Digital Sound, which uses an optical synchronization track on the film coupled with multichannel digital audio from a CD. The primary multichannel audio system of DVD-Video is Dolby Digital. DTS is an optional audio system for DVD-Video, but has been made mandatory for the next-generation disc formats HD DVD and Blu-ray Disc, and is expected to be the primary audio system used for feature content on those discs. In 1999, Dolby and Lucasfilm THX added a new rear center channel option to Dolby Digital. Dubbed Surround EX, this new format first appeared on *Star Wars: Episode 1 — The Phantom Menace*. Not to be outdone, DTS came out with an *Extended Surround* (DTS-ES) rear center channel decoder that could decode Surround EX. DTS also released a new discrete 6.1-channel format, which provided a separate center surround channel instead of relying on matrix encoding.

Movie production has now entered the era where entire scenes or entire movies are generated inside a computer, never being touched by light to trace the images in silver halide crystals on a film negative. Pixar's *A Bug's Life* was the first motion picture feature to be transferred directly to DVD without being printed to film. The resulting picture quality was astounding, especially on a progressive-scan digital playback system such as a computer. *A Bug's Life* also pioneered the concept of reframing scenes and repositioning characters and objects to create a full-screen 4:3 version of the film. Digital movie projection, aka d-cinema, has been surprisingly successful in early tests, many of which used DVD-ROM to distribute the electronic movie files to theaters around the world.

The next-generation high-definition disc formats are now taking the technology to another level. Capable of holding a full length motion picture at very nearly the resolution of d-cinema and delivering up to 8 channels of audio, these next-generation formats redefine the term "home theater," with the media delivering a level of audio and visual quality that movie theaters could not offer only a few years ago. At the same time, film mastering is entering the digital age as material shot on film is being scanned into powerful digital systems that can manipulate the picture, apply color correction, film grain enhancement, and visual effects that become the basis for the print film, d-cinema, and home video releases. This *digital intermediary* process makes it possible to remain in the digital realm from the initial film scan all the way out to theater or home. With the advent of digital film cameras, such as the Grass Valley *Viper*, even the production out in the field can begin digitally. As the waves of digital convergence lap higher onto the shores of Hollywood, the digital nature of optical disc will become increasingly important for the best-quality distribution of movies.

Dancing Electrons

At about the same time as Muybridge was stretching strings across racetracks, scientists around the world were envisioning television. Once again, the new ideas were christened with fanciful names derived from familiar technology such as *radioscope, phonoscope, cinematophone, radiovision, telephone eye,* and the jaw-breaking *chronophotographoscope.* The word *television* is credited to science writer Hugo Gernsback[8], who published popular science journals beginning in 1908.

Many early television ideas developed almost simultaneously around the world and involved mechanical systems that transmitted each individual picture element on a separate connection. These were dropped in favor of rapidly scanning the image one line at a time. In 1884, Paul Nipkow applied for a patent in Germany on his image-scanning design that used a rotating disc with holes arranged in a spiral, essentially an updated version of the phenakistoscope. Television technology improved slowly, still based on clumsy mechanical designs, until the first experimental broadcasts were made in the 1920s. Meanwhile, a 14-year-old Utah farm boy named Philo Farnsworth was looking at plowed furrows and dreaming of deflecting beams of electrons in similar rows. Six years later, on January 7, 1927, he submitted the patent applications that established him as the inventor of the all-electronic television. The honor of creating the first working electronic television system also belongs to Kenjito Takayanagi of Tokyo, who transmitted an image of Japanese writing to a cathode-ray tube on Christmas Day, 1926.

The first nonexperimental television broadcasts began in 1935 in Germany and included coverage of the 1936 Olympics in Berlin. The German system had only 180 lines. In Britain, John Logie Baird steadily improved a mechanically scanned television system from 50 lines to 240, and the *British Broadcasting Company* (BBC) used it for experimental broadcasts. Improved "high definition" commercial broadcasts using a 405-line system were begun by the BBC in Britain in 1936, although they were shut down three years later by World War II (with a sign-off broadcast of a Mickey Mouse cartoon) and did not return until 1946 (with a repeat of the same cartoon). In the United States, the *National Broadcasting Company* (NBC) demonstrated television to entranced crowds at the 1939 World's Fair, and RCA's David Sarnoff declared that 10,000 sets would be sold before the end of the year. However, at a price equivalent to more than $2500 and with sparse programming from scattered transmitters, only a few thousand sets were sold. NBC began commercial broadcasts on July 1, 1941. *Columbia Broadcasting System* (CBS) also switched from experimental to commercial broadcasts at about the same time.

About 10,000 televisions were scattered around the United States in 1946, and when World War II production restrictions were lifted, another 10,000 were soon sold. Three years later, there were over 1 million sets, and after only two more years, in 1951, there were more than 10 million. In 1950, the *Federal Communications Commission* (FCC) selected a new "field sequential" color system from CBS that used a three-color wheel spinning inside each

---

[8]Hugo Gernsback may be familiar to science fiction fans as the eponym of the Hugo Award. His name also may be familiar to readers of *Popular Electronics* magazine, produced by Gernsback Publications.

TV. This was chosen over RCA's fully electronic version, which didn't have sufficient quality at the time. Luckily, circumstances and the Korean War delayed implementation long enough that CBS abandoned its mechanical system as RCA improved its electronic version. The official *National Television Standards Committee* (NTSC) color television system was implemented in 1954, and widespread purchase of color television sets began in 1964. The NTSC system piggybacked color information onto the black-and-white signal so that the millions of existing sets would still work with the new broadcasts. NTSC was adopted by Japan in 1960.

Meanwhile, television in Great Britain experienced the same rapid growth as in the United States after World War II. In 1967, when it came time to implement color television, Britain adopted the PAL system that had been developed in Germany. PAL was technologically superior to NTSC, but it was incompatible with existing British television sets and had to be phased in over many years. In 1967, France and the Soviet Union settled on the SECAM system. In 1968, the number of televisions surpassed 78 million in the United States and 200 million worldwide. By 1972, half of all American homes had a television set. At the end of the 1970s, almost every country had adopted one of the three standards, as television began to saturate the planet. NTSC is still the most widely used system, PAL is common in most of Europe, and SECAM holds a distant third.

As television was gaining a foothold, *Community Antenna TV* (CATV, the first version of cable television) was founded in rural Pennsylvania by a shopkeeper who built an antenna on a hill to improve reception in his store and then began connecting the receivers of his customers and neighbors. From its beginnings in 1949, cable TV's promise of more channels and better reception has attracted millions of customers, and in just under 40 years, over half the television homes in the United States were connected to cable.

The first television signal was beamed via satellite across the Atlantic Ocean in 1962, and in 1975, satellites were put in use to distribute programming to cable TV centers. Satellite dishes eventually were sold to consumers and ironically became a competitor to cable TV, especially with the advent of *direct broadcast satellite* (DBS) television, which uses the same digital video compression system as DVD.

Metal Tape and Plastic Discs

Visual information is hundreds of times more dense than audio information. Sound waves are simple vibrations that the human ear can detect from about 20 per second to over 20,000 per second (20 Hz to 20 kHz). Visible images, on the other hand, are formed from complex light waves that the eye can detect within the range of 430 to 750 trillion Hz. Clearly, recording a video signal is a much bigger challenge than recording audio.

For magnetic tape, raising the speed at which the tape moves past the head increases the amount of information that can be recorded in a given period of time, but the amount of tape required quickly becomes preposterous. Nevertheless, in 1951, Bing Crosby Enterprises demonstrated a commercial videotape recorder that used tape traveling at 100 inches per second (almost 6 mph) across 12 heads. A few similarly cumbersome and expensive systems were developed for television studios, but their price tags barred them from wider use. The "eureka factor" did not arrive for another 10 years, when engineers at Ampex hit on the idea of moving the head past the tape in addition to moving the tape past the head. A revolving head can record nearly vertical stripes on a slow-moving tape, greatly increasing the recording efficiency (see Figure 1.4). The first helical-scan videotape recorder appeared in 1961, and color versions followed within a few years.

Figure 1.4 Tape Head Comparison

linear audio tape; 1/8",1-7/8 ips

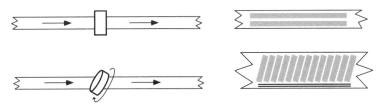

helical scan videotape; 1/2", 7/16 ips (EP), 1-3/8 ips (SP)

Videotape recording was first used only by professional television studios. A new era for home video was ushered in when Philips and Sony produced black-and-white, reel-to-reel videotape recorders in 1965, but at $3000 they did not grace many living rooms. Sony's professional 3/4-inch U-matic videocassette tape appeared in 1972, the same year as the video game Pong. More affordable color video recording reached home consumers in 1975 when Sony introduced the Betamax videotape recorder. The following year JVC introduced VHS, which was slightly inferior to Betamax but won the battle of the VCRs because of costs and extensive licensing agreements with equipment manufacturers and video distributors. The first Betamax VCR cost $2300 ($7600+ adjusted for inflation), and a 1-hour blank tape cost $16 (about $53 in today's dollars). The first VHS deck was cheaper at $885 ($2900 today), and Sony quickly introduced a new Betamax model for $1300, but both systems were out of the financial reach of most consumers for the first few years.

MCA/Universal and Disney studios sued Sony in 1976 in an attempt to prevent home copying with VCRs. Eight years later, the courts upheld consumer recording rights by declaring Sony the winner. Studios are fighting the same battle today with DVD but have switched to trying to change the technology and the law.

In 1978, Philips and Pioneer introduced videodiscs, which actually had been demonstrated in rudimentary form in 1928 by John Logie Baird. The technology had improved in 50 years, replacing wax discs with polymer discs and delivering an exceptional analog color picture by using a laser to read information from the disc. The technology got a big boost from General Motors, which bought 11,000 players in 1979 to use for demonstrating cars. Videodisc systems became available to the home market in 1979 when MCA joined the laserdisc camp with its DiscoVision brand. A second videodisc technology, called *capacitance electronic disc* (CED), introduced just over two years later by RCA, used a diamond stylus that came in direct contact with the disc — as with a vinyl record. CED went by the brand name SelectaVision and was more successful initially, but eventually failed because of its technical flaws. CED was abandoned in 1984 after fewer than 750,000 units were sold, leaving laserdisc to overcome the resulting stigma. JVC and Matsushita developed a similar technology called *video high density* or *video home disc* (VHD), which used a grooveless disc read by a floating stylus. VHD limped along for years in Japan and was marketed briefly by Thorn EMI in Great Britain, but it never achieved significant success.

For years, laserdisc was the Mark Twain of video technology, with many exaggerated reports of its demise as customers and the media confused it with the defunct CED. Adding to the confusion was the addition of digital audio to laserdisc, which in countries using the PAL television system required the relaunch of a new, incompatible version. Laserdisc persevered, but because of its lack of recording, the high price of discs and players, and the inability to show a movie without breaks (a laserdisc cannot hold more than one hour per side), it was never more than a niche success, catering to videophiles and penetrating less than 2 percent of the consumer market in most countries. Laserdisc systems did achieve modest success in education and training, especially after Pioneer's bar code system became a popular standard in 1987 and enabled printed material to be correlated to random-access visuals. In some Asian countries, laserdisc achieved as much as 50 percent penetration, almost entirely because of its karaoke features.

Despite the exceptional picture quality of both laserdisc and CED, they were quickly overwhelmed by the eruption of VHS VCRs in the late 1970s, which started out at twice the price of laserdisc players but quickly dropped below them. The first video rental store in the United States opened in 1977, and the number grew rapidly to 27,000 in the late 1990s. Direct sales to customers, known as *sellthrough*, began in 1980. Today, the home video market is a $26 billion business in the United States alone. More than 87 percent of households have at least one VCR, creating a market of over 100 million customers who, in 1996, rented about 3 billion tapes and bought more than 580 million tapes. By 2000 there were 25,000 VHS titles available in the United States, compared with 9,000 on laserdisc.[9]

In the late 1980s, a new video recording format based on 8-mm tape with metal particles was introduced. The reduced size and improved quality were not sufficient to displace the well-established VHS format, but 8-mm and Hi-8 tapes were quite successful in the camcorder market, where smaller size is more significant.

Around this time, minor improvements were made to television, with stereo sound added in the United States in 1985 and, later, closed captions.

In 1987, JVC introduced an improved Super VHS system called *S-VHS*. Despite being compatible with VHS and almost doubling the picture quality, S-VHS was never much of a success because there were too many barriers to customer acceptance. Players and tapes were much more expensive, and VHS tapes worked in S-VHS players but not vice versa. A special cable and an expensive TV with an s-video connector were required to take advantage of the improved picture, and S-VHS was not a step toward *high-definition television* (HDTV), which was receiving lavish attention in the late 1980s and was expected to appear shortly. It is interesting to note that DVD has many of the same strikes against it, even the specter of HDTV. The major contribution of S-VHS to the industry was the popularization of the s-video connector, which carried better-quality signals than the standard composite format. S-video connectors are still mistakenly referred to as S-VHS connectors.

---

[9]Despite the great success and ubiquity of VHS, it has been almost completely supplanted by DVD. As of this writing, it has been announced that several new major motion picture releases on home video will not be distributed on videotape. Due to the high rate of unsold returns, several major studios are now seriously considering abandonment of the format altogether.

The Digital Face-Lift

As television began to show its age, new treatments appeared in an attempt to remove the wrinkles. European broadcasters rolled out PALplus in 1994 as a stab at *enhanced-definition television* (EDTV) that maintains compatibility with existing receivers and transmitters. PALplus achieves a widescreen picture by using a letterboxed[10] image for display on conventional TVs and hiding helper lines in the black bars so that a widescreen PALplus TV can display the full picture with extended vertical resolution.

Other ways of giving television a face-lift include *improved-definition televisions* (IDTV) that double the picture display rate or use digital signal processing to remove noise and to improve picture clarity.

None of these measures are more than stopgaps while we wait for the old workhorse to be replaced by *high-definition television* (HDTV). HDTV was first demonstrated in the United States in 1981, and the process of revamping the "boob tube" reached critical mass in 1987 when 58 broadcasting organizations and companies filed a petition with the FCC to explore advanced television technologies. The *Advanced Television Systems Committee* (ATSC) was formed and began creeping toward consensus as 25 proposed systems were evaluated extensively and either combined or eliminated.

In Japan, a similar process was underway that was unfettered by red tape. An HDTV system called *HiVision*, based on MUSE compression, was developed quickly and put into use in 1991.[11] Ironically, the rapid deployment of the Japanese system was its downfall. The MUSE system was based on the affordable analog transmission technology of its day, but soon the cost of digital technology plummeted as its capability skyrocketed. Back on the other side of the Pacific, the lethargic US HDTV standards creation process was still in motion as video technology graduated from analog to digital. In 1993, the ATSC recommended that the new television system be digital. The Japanese government and the Japanese consumer electronics companies — which would be making most of the high-definition television sets for use in the United States — decided to wait for the American digital television standard and adopt it for use in Japan as well. A happy by-product of HiVision is that the technology-loving early Japanese adopters served as guinea pigs, supporting the development of high-resolution widescreen technology and paving the way for HDTV elsewhere. A European HDTV system, HD-MAC, was even more short-lived; it was demonstrated at the 1992 Albertville Olympics but was abandoned shortly thereafter.

In 1992, the ATSC outlined a set of proposed industry actions for documenting a standard. In 1993, the three groups that had developed the four final digital systems — AT&T and Zenith; General Instrument and MIT; and Philips, Thomson, and the David Sarnoff Research Center — formed a "grand alliance" and agreed to merge their best features into a final standard. The ATSC made its proposal for a *digital television* (DTV) standard to the FCC in November 1995. The motion picture industry and the computer industry had been aware of the proceedings but apparently had not bothered to become sufficiently involved.

---

[10]Letterboxing is the technique of preserving a wide picture on a less wide display by covering the gap at the top and bottom with black bars. Aspect ratios and letterboxing are covered in Chapter 3.

[11]The HiVision system uses 1125 video scan lines. In the charmingly quirky style of Japanese marketing, HiVision was introduced on November 25 so that the date (11/25) corresponded with the number of scan lines.

At the eleventh hour they both began to complain loudly that they were not happy with the choice of widescreen aspect ratio or the computer-unfriendly aspects of the proposal. In December of 1996, with this opposition in mind, the FCC approved all but the video format constraints (aspect ratios, resolutions, and frame rates) of the ATSC DTV proposal, setting these aside as a voluntary standard. This "grand compromise" freed broadcasters to begin implementation but wisely left the standard open for additional video formats. HDTV, originally promised for the early 1990s, finally concluded its long gestation period in December 1998, only to begin an even longer battle for ascendancy.

In the meantime, numerous new formats have been developed to store video digitally and convert it to a standard analog television signal for display. As with early videotape, the elephantine size of video is a problem. Uncompressed standard-definition digital television video requires at least 124 *million bits per second* (Mbps).[12] Compare this with hard drives, which run at around 25 to 100 Mbps; high-speed T-1 digital telephone lines, which run at 1.5 Mbps, and audio CD, which runs at a meager 1.4 Mbps. Obviously, some form of compression is needed. Many proprietary and incompatible systems have been developed, including Intel's DVI in 1988 (which had been developed earlier by the Sarnoff Institute). That same year, the *Moving Picture Experts Group* (MPEG) committee was created by Leonardo Chairiglione and Hiroshi Yasuda with the intent to standardize video and audio for CDs. In 1992, the *International Organization for Standardization* (ISO) and the *International Electrotechnical Commission* (IEC) adopted the standard known as MPEG-1. Audio and video encoded by this method could be squeezed to fit the limited data rate of the single-speed CD format. CD-i first used MPEG-1 video playback in 1992 to achieve near-VHS quality. The CD-i MPEG-1 Digital Video format was used as the basis for Karaoke CD, which became Video CD, a precursor to DVD. Video CD has done quite well in markets where VCRs were not already established. At the beginning of 2000, there were about 40 million Video CD players in Asia, but Video CD has not fared as well in Europe, and it may qualify as an endangered species in the United States. MPEG-1 is also commonly used for video and audio on personal computers (PCs) and over the Internet. The notorious MP3 format is a nickname for MPEG-1 Layer III audio.

The MPEG committee extended and improved its system to handle high-quality audio/video at higher data rates. MPEG-2 was adopted as an international standard in 1994 and is used by many digital video systems, including the ATSC's DTV. Direct broadcast satellite (DBS), with convenient 18-inch dishes and digital video, also appeared in 1994. With sales of more than three million units by the end of 1996, DBS was the most successful home entertainment product until DVD came along. DBS was introduced just as MPEG-2 was being finalized. Consequently, most early DBS systems used MPEG-1 but have since converted to MPEG-2. Digital videocassette tape (DV) appeared in 1996. With near-studio quality, it is aimed at the professional and "prosumer" market and is priced accordingly. New competitors to cable and satellite TV are also based on MPEG-2; these include video dialtone (video delivered over phone lines by the phone company), digital cable, and wireless

---

[12]124 Mbps is the data rate of active video at ITU-R BT.601 4:2:0 sampling with 8 bits, which provides an average of 12 bits per pixel (720 3 480 3 30 3 12 or 720 3 576 3 25 3 12). The commonly seen 270-Mbps figure comes from studio-format video that uses 4:2:2 sampling at 10 bits and includes blanking intervals. At higher 4:4:4 10-bit sampling, the data rate is 405 Mbps.

cable (terrestrial microwave video transmission). An updated version of Video CD, called Super Video CD, uses MPEG-2 for better quality. MPEG-2 is also the basis of DVD-Video, augmented with the Dolby Digital (AC-3) multichannel audio system, developed as part of the original work of the ATSC. DVD-Video is intended for the home video market, where it provides the highest resolution yet from a consumer format, but will soon be challenged by the new HD DVD and BD optical disc formats, which utilize the MPEG-4 Advanced Video Codec (MPEG-4 AVC) and SMPTE VC-1 to deliver high definition video at resolutions almost six times greater than DVD.

A Brief History of Data Storage Technology

Rewind about 200 years. In 1801, Joseph-Marie Jacquard devised an ingenious method for weaving complex patterns using a loom controlled by punched metal cards. The same idea was borrowed over 30 years later by Charles Babbage as the storage device for his mechanical computer, the Analytical Engine.[13] Ninety years after Jacquard, Herman Hollerith used a similar system of punched cards for tabulating the US Census after getting the idea from watching a train conductor punch tickets. Fifty years later, the first electronic computers of the 1940s also employed punched cards and punched tape — using the difference between a surface and a hole to store information. The same concept is used in modern optical storage technology. Viewed through an electron microscope, the pits and lands of a CD or DVD would be immediately recognizable to Jacquard or Hollerith (see Figure 1.5). However, the immensity and variety of information stored on these miniature pockmarked landscapes would truly amaze them. Hollerith's cards held only 80 characters of information and were read at a glacial few per second.[14]

Figure 1.5 Punched Card and Optical Disc

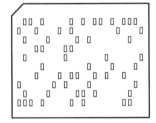 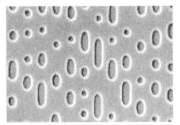

The other primary method of data storage — magnetic media — was developed for the UNIVAC I in 1951. Magnetic tape improved on the storage density of cards and paper tape by a factor of about 50 and could be read significantly faster. Magnetic disks and drums appeared a few years later and improved on magnetic tape, but cards and punched tape were still much cheaper and remained the primary form of data input until the late 1970s. At that

---

[13]Unlike Jacquard, whose system enjoyed widespread success, Babbage seemed incapable of finishing anything he started. He never completed any of his mechanical calculating devices, although his designs were later proven correct when they were turned into functioning models by other builders.

[14]Modern card readers of the 1970s still used only 72 to 80 characters per card but could read over 1000 cards per second.

time, flexible floppy disks appeared — first unwieldy 8-inch behemoths, then smaller 5.25-inch disks, and finally, in 1983, 3.5-inch diskettes small enough to fit in a shirt pocket. Each successive version held two or three times more data than its predecessor despite being smaller.

Optical media languished during the heyday of magnetic disks, never achieving commercial success other than a use for analog video storage (i.e., laser videodiscs). It was not until the development of compact disc digital audio in the 1980s that optical media again proved its worth in the world of bits and bytes, setting the stage for DVD.

Other innovations of the 1980s included removable hard disk cartridges, high-density floppy disks from IOmega based on the Bernoulli principle, and erasable optical media based on *magneto-optical* (MO) technology. Magneto-optical discs use a laser to heat a polyphase crystalline material that can then be aligned by a magnetic field. Features of MO technology were later adapted for DVD-RAM.

Innovations of CD

Sony and Philips reinvigorated optical storage technology when they introduced Compact Disc Digital Audio in 1982. This was known as the "Red Book standard" because of the red cover on the book of technical specifications. The first CD players cost around $1000 (over $1900 in 2004 dollars).

Three years later, a variation for storing digital computer data — CD-ROM — was introduced in a book with yellow covers. CD-ROM did not take hold immediately, especially since the first drives cost more than $1000, but many people understood the seeds of technological revolution that it carried. The first Microsoft CD-ROM conference in March 1986 had a sell-out crowd of a thousand people, and by 1992, the conference had metamorphosed into the Intermedia trade show, attended by hundreds of thousands. As CD-ROM entered the mainstream, original limitations such as slow data rates and glacial access times were overcome with higher spin rates, bigger buffers, and improved hardware. CD-ROM became the preeminent instrument of multimedia and the standard for delivery of software and data (see Figure 1.6).

A clever variation on audio CDs called *CD+G* was developed around 1986. A CD+G disc uses six previously unused bits in each audio sector (subcode channels R through W) to hold graphic bitmaps. The player collects snippets of a graphic from each sector as it plays the audio, and after a few thousand sectors, it has accumulated enough to display a picture. CD+G was never widely supported and is available mostly on karaoke CD players and CD-i players.

CD-Video (CDV) was developed in 1986 as a hybrid of laserdisc and CD. Part of the disc contained 5 to 6 minutes of standard audio tracks and could be played on a CD player, whereas the other part contained about 20 minutes of analog video and digital audio in laserdisc format. The standard was presented in a blue book but is not considered part of the canon of colors. The CDV format has mostly disappeared.

Figure 1.6 The CD Family Tree

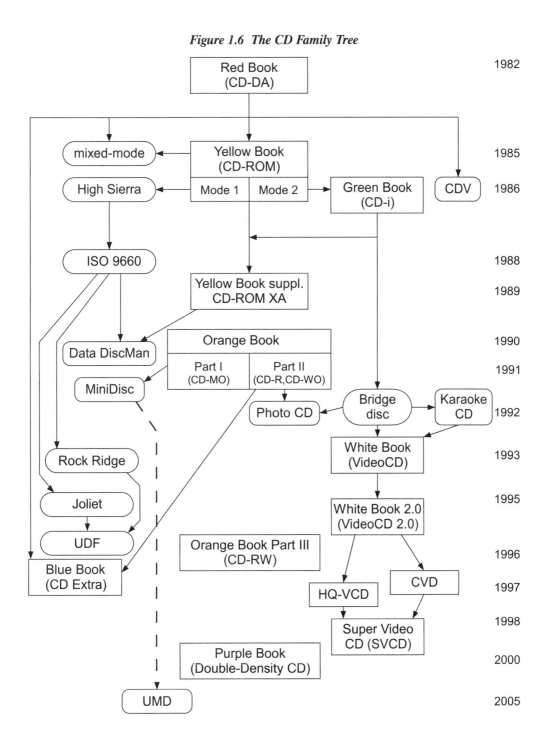

After introducing CD-ROM, Sony and Philips continued to refine and expand the CD family. In 1986, they produced *Compact Disc Interactive* (CD-i, the "Green Book standard"), intended to become the standard for interactive home entertainment. CD-i incorporated specialized file formats and custom hardware with the OS-9 operating system running on the Motorola 68000 microprocessor. Unfortunately, CD-i was obsolete before it was finished, and the few supporting companies dropped out early, leaving Philips to stubbornly champion it alone. After consuming more than a billion dollars in development and scattershot marketing, CD-i still limps along as a niche product. Philips, not one to give up easily, announced it would develop a DVD player that could accept a CD-i add-in card, but such a contraption has yet to appear.

Some early CD-ROM producers developed their own file systems to organize the data on the disc, requiring a specific device driver. This forced pioneering users to either reboot and load a different driver or use multiple CD-ROM drives, one for each title. In the face of proliferating proprietary CD-ROM file formats, a group of industry representatives met in 1986 at the High Sierra Hotel and Casino near Lake Tahoe, Nevada, and proposed a common structure that became known as the "High Sierra format." In 1988, ISO adopted this format, with a few modifications, as the ISO 9660 CD-ROM file interchange standard. Unfortunately, it was a lowest-common-denominator solution designed to support MS-DOS, so it did not properly support other systems such as the Apple Macintosh *Hierarchical File System* (HFS) and many UNIX systems. These operating systems could read ISO 9660 CDs but had to add their own incompatible extensions to make full use of the advantages of their native file systems. When Microsoft moved beyond DOS, it also rolled its own "Joliet" extensions into ISO 9660 for long filenames with multilingual characters.

In 1989, the CD-ROM "Yellow Book standard" was augmented with an updated format called CD-ROM XA (eXtended Architecture). XA was based on ideas developed for CD-i, including the interleaving of data, graphics, and ADPCM compressed audio. CD-ROM XA required newer hardware to read the mixed sector types on the disc and to take advantage of the interleaved streams. Most CD-ROM drives are now XA-compatible, but the interleaving features are rarely used. CD-ROM XA also introduced the so-called bridge disc, which is a disc with extra information that may be used in a CD-i player or a computer with a CD-ROM XA drive.

The "Orange Book standard" was developed in 1990 to support magneto-optical (MO) and *write-once* (WO) technology. MO has the advantage of being rewritable but is incompatible with standard CD drives and has remained expensive. CD-WO has been largely superseded by "Orange Book Part II," *recordable CD* (CD-R). CD-R can be written to once, which is sufficient for many uses. The Orange Book also added multisession capabilities to allow CDs to be written in chunks across multiple recording sessions. Recordable technology revolutionized CD-ROM production by enabling developers to create fully functional CDs for testing and submission to disc replicators. As prices dropped, CD-R also has become popular for business and even personal archiving. Although tape backup systems are cheaper, the familiarity of CD and the widespread availability of CD-ROM drives have made CD-R very appealing. Tape also lacks the quick random access of CD-R.

Around 1990, music CD singles became popular in Japan, and the format subsequently was adopted for use with Sony's Data Discman. The 8-centimeter discs had a capacity of 200 megabytes and commonly held modified CD-ROM XA applications. The Data Discman and its cousin the Bookman had piddling success outside Japan. Sony also developed the portable MMCD player based on an Intel 286 processor and an LCD display, which it concentrated on to the exclusion of CD-i. However, as with CD-i, heavy marketing failed to make the MMCD a success, and Sony finally gave up around 1994. The name was later reused for Sony's proto-DVD format.

Kodak and Philips developed the Photo CD format in 1992, based on the CD-ROM XA and Orange Book standards. Photo CDs are bridge discs that hold up to one hundred 35-mm photographs written in one or more sessions. Special Photo CD players and CD-i players can show the pictures on a television, but the concept never appealed much to home photographers. Nevertheless, because multisession CD-ROM XA drives can read Photo CDs, the format is popular for professional photo production work and for stock photography libraries.

In 1991, Sony introduced the 8-cm MiniDisc, a portable music format based on the rewritable Orange Book's MO standard and using Sony's ATRAC compression. The poor audio quality of MiniDisc earned it a bad reputation, but it has been improved considerably and is now in more than 10 percent of Japanese households, accounting for 30 percent of prerecorded music sales in Japan. MiniDisc had less success in other countries, and even the recent resurgence of support may be too late to save it if a recordable 8-cm DVD-Audio standard is developed.

MPEG-1 video from a CD was demonstrated by Philips and Sony on their CD-i system in 1991. About 50 movies were released in CD-i Digital Video format before a standardized version appeared as Video CD (VCD) in 1993, based on proposals by JVC, Sony, and Philips and documented as the "White Book specification." Not to be confused with CD-Video (CDV), Video CD uses MPEG-1 compression to store 74 minutes of near-VHS-quality audio and video on a CD-ROM XA bridge disc.[15] Philips developed a $200 Video CD add-on for its CD-i players. Hardware and software were made available for computers to play Video CDs. Video CD 2.0 was finalized in 1994. The format has done very well in countries where VHS was not well established. More than six million Video CD players were sold in China alone in 1996.

By 1994, drive speeds — and resulting data transfer rates — had risen to quadruple speed. Once again, the price for the new drives was around $1000. Two years later, 8X CD-ROM drives appeared, but the introductory price had dropped to $400. Later the same year, 12X drives appeared for about $250. Tests began to show that in many cases data transfer rates were not much better than those of 6X drives because the high speeds increased errors and required repeated reads.

---

[15]This is actually quite amazing when you consider the difference in data bandwidth between audio and video. VCD, using MPEG-1 compression, made it possible to store just as much video (and its accompanying audio) on a compact disc as CD audio can store for audio alone. The difference, of course, is that VCD uses data compression while CD audio does not.

After CD-ROM-based multimedia reached mass-market potential, music artists began to look for ways to add multimedia to their music. The idea was to create a CD that would play in a music player but also include software for use on a computer. The various techniques are grouped under the moniker of *Enhanced CD*. The first attempts, in the late 1980s, used mixed-mode CDs that put computer data on the first track and music on the remaining tracks. The downside was that some CD players attempted to play the computer data as if it were music, producing noise that could damage speakers. Stickers on CDs warning of potential equipment damage were not believed to be a good way to increase sales, so more ingenious methods were pursued. A trick of placing the computer data at index zero (in the pregap area before the beginning of the music) worked well with most players, which begin playing music at index one. It was still possible to use the rewind feature on some players to back up into the data, so clever implementations included an audio message warning the listener to stop the player before reaching the noise. Since CD-ROM drives begin at index zero, they are able to read the computer data. That is, they were until Microsoft released a new SCSI CD-ROM driver that began reading at index 1, thus skipping over the data intended for the computer.

By this time, Sony and Philips, with the help of Apple and Microsoft, were working on a more foolproof solution. The answer was stamped multisession, which uses a feature of the Orange Book format originally designed for recordable discs to put one session of Red Book audio data on the CD followed by a session of Yellow Book computer data. Combining red, yellow, and orange normally produces deep orange, but in this case blue covers were chosen for the book that documented the new CD Plus standard, released in 1995, which was almost immediately renamed CD Extra because of threatened lawsuits by the owner of the CD Plus trademark. The CD Extra format is completely compatible with audio CD players but does not work on all computers. Many CD-ROM drives are multisession-capable, but only about 50 percent of multimedia PCs have the requisite multisession software drivers. Fortunately, computers are easier to upgrade than CD players, and sales of Enhanced CDs are expected to do well. Because the DVD-Audio specification is built on top of the DVD-ROM standard, DVDs with both audio and computer data will be easy to produce and will be compatible with all DVD-Audio players and DVD computers.[16]

Erasable CD, part III of the "Orange Book standard," did not appear until 1997. The standard was finalized in late 1996, and the official name of *Compact Disc Rewritable* (CD-RW) was chosen in hopes of avoiding the disturbing connotations of "erasable." Unfortunately, CD-RW requires new drive hardware to read it. None of more than 700 million existing CD-ROM drives and CD players are able to read CD-RW discs, although it has succeeded well in the prolonged absence of recordable DVD.

An erasable CD format called *THOR* actually was announced by Tandy in 1988, eight years before CD-RW, but this ambitious technology — with its familiar-sounding promise of storing music, video, and computer data — was plagued by technical problems and never made it out of the laboratory. Tandy later joined with Microsoft to create the CD-based *Video Information System* (VIS), which also was mercifully short-lived.

---

[16]That's the theory, at least. In practice, sloppy engineering and cheap players result in problems with discs that contain anything other than video or audio.

Another notable failure was 1991's *Commodore Dynamic Total Vision* (CDTV), a consumer multimedia console along the lines of CD-i. CDTV employed a CD-ROM-equipped Amiga computer that was hooked up to a TV. Proprietary standards and the decline of Commodore's business doomed CDTV to museum shelves.

Around 1999, the prices of CD-R/RW drives and blank CD-R discs dipped enough to provoke their widespread use for recording custom music CDs. This, plus the proliferation of MP3 music files across the Internet, drove a worldwide demand for 1.3 billion CD-Rs in 1999, with an estimated 30 to 40 percent being used for music.

A later variations on the CD theme was from DataPlay, which in 2001 introduced miniature optical discs and players intended to compete with flash memory devices. DataPlay soon went out of business, was later resurrected by Ritek, but has yet to find a market.

Chapter 2
DVD Arrives

The Long Gestation of DVD

Shortly after the introduction of CDs and CD-ROMs, prototypes of their eventual replacement began to be developed. Systems using blue lasers achieved four times the storage capacity, but were based on large, expensive gas lasers because cheap semiconductor lasers at that wavelength were not available. In 1993, 10 years after the worldwide introduction of CDs, the then-next-generation prototypes neared realization. Nimbus first demonstrated a double-density CD proof of concept in January 1993. Optical Disc Corporation followed suit in October. Philips and Sony announced that they had a similar project underway, and later Toshiba claimed it had been working on a comparable design since 1993. Some of these first attempts simply used CD technology with smaller pits to create discs that could hold twice as much data. Although this far exceeded original CD tolerances, the optics of most drives were good enough to read the discs, and it would even be possible to connect the digital output of a CD player to an MPEG video decoder.[1] The reality that CD technology tended to falter when pushed too far, however, soon became apparent. Philips reportedly put its foot down and said it would not allow CD patent licensees to market the technology because it could not guarantee compatibility and a good user experience. When all things were considered, it seemed better to develop a new system (see Figure 2.1).

Hollywood Weighs In

The stage was set in September 1994 by seven international entertainment and content providers: Columbia Pictures (Sony), Disney, MCA/Universal (Matsushita), MGM/UA, Paramount, Viacom, and Warner Bros. (Time Warner), who called for a single worldwide standard for the new generation of digital video on optical media. These studios formed the Hollywood Digital Video Disc Advisory Group and requested the following:

- Room for a full-length feature film, about 135 minutes, on one side of a single disc
- Picture quality superior to high-end consumer video systems such as laserdisc
- Compatibility with matrixed surround and other high-quality audio systems
- Ability to accommodate three to five languages on one disc
- Content protection
- Multiple aspect ratios for wide-screen support
- Multiple versions of a program on one disc, with parental lockout

Preparations and proposals commenced, but two incompatible camps soon formed. Like antagonists in some strange mechanistic mating ritual, each side boasted of its prowess and attempted to enlist the most backers. At stake was the billion-dollar home video industry, as well as millions of dollars in patent licensing revenue.

---

[1]Pioneer commercialized its Alpha system for karaoke in Japan using higher-density CDs. Nimbus created the Presto prototype video player using MPEG-2 video on CDs.

Figure 2.1 Timeline of DVD Development

1995
Hollywood proposal

Sony/Philips MMCD proposal
SD Alliance proposal

computer industry objectives announced

UDF recommended
reconciliation announced

1996
combined DVD standard announced
product announced for fall 1996

Digital Recording Act attempted

copy protection agreement (Macrovision) announced

DVD-ROM and DVD-Video specification version 1.0 published

1997
copy protection agreement (CSS) announced
players appear in Japan
drives appear in Japan

players appear in U.S.
movie titles appear in U.S.

1998
Warner takes U.S. movie title distribution nationwide
DVD-R 1.0 drives appear
DVD consortium changes to DVD Forum, opens to all companies
first hack for copying DVD on PCs
Dolby Digital included on 625/50 (PAL) format; *DVD Demystified* published
first Macintosh DVD upgrade kit

soft launch of DVD in Europe
Divx trial program begins; DVD-Audio 0.9 spec published
DVD-RAM 1.0 drives appear

Divx goes nationwide; DMCA becomes law

1999
first DTS DVDs ship
DVI final draft released

SACD players released in Japan

2000
Divx shuts down; DVDA formed
DVD-R 1.9 drives appear
DVD-RAM 2.0 drives appear
progressive-scan players
DVD wins an Emmy; DeCSS released; official DVD launch in Europe
DVD becomes best-selling CE product; first DVD-18 ships
DVD-Audio players and DVD-RW recorders appear in Japan
DVD Demystified 2nd Edition published
Sony Playstation® 2 released in Japan

2005
DVD-Audio players appear in U.S.
MPAA wins DeCSS trial in New York
10 millionth DVD player sold in U.S.
Sony Playstation 2 released in U.S.
Sony Playstation® Portable released in U.S.

Dissension in the Ranks

On December 16, 1994, Sony and its partner Philips independently announced their own standard: a red laser based, single-sided, 3.7-billion-byte *Multimedia CD* (MMCD, renamed from HDCD, which, it turned out, was already taken). The remaining cast of characters jointly proposed a different red laser standard one month later on January 24, 1995. Their *Super Disc* (SD) standard was based on a double-sided design holding five billion bytes per side.

The SD Alliance was led by seven companies — Hitachi, Matsushita (Panasonic), Mitsubishi, Victor (JVC), Pioneer, Thomson (RCA/GE), and Toshiba (business partner of Time Warner) — and attracted about 10 other supporting companies, primarily home electronics manufacturers and movie studios.[2] Philips and Sony assembled a rival gang of about 14 companies, largely peripheral hardware manufacturers.[3] Neither group had support from major computer companies. At this stage, the emphasis was on video entertainment, with computer data storage as a subordinant goal.

Two days after the SD announcement from the rival group, Sony head Norio Ohga told the press that Sony "may make concessions but . . . will not join" and indicated that Sony would hold out for a third standard incorporating more of its own specifications, even to the point of asking the Ministry of International Trade and Industry to arbitrate the unification. Sony and Philips held the lion's share of CD technology patents and hoped to include as many of them as possible in the new format. The companies in the SD Alliance, including Sony's arch rival Matsushita, planned to use their own newly patented technology and slow the flow of patent revenue to the competition.

Sony and Philips played up the advantages of MMCD's single-layer technology, such as lower manufacturing costs and CD compatibility without the need for a dual-focus laser, but the SD Alliance was winning the crucial support of Hollywood with its dual-layer system's longer playing time. On February 23, 1995, Sony played catch-up by announcing a two-layer, single-side design licensed from 3M that would hold 7.4 billion bytes.

The Referee Shows Up

The scuffling continued, and the increasing emphasis on data storage by consumer electronics companies began to worry the computer industry. At the end of April 1995, five computer companies — Apple, Compaq, HP, IBM, and Microsoft — formed a technical working group that met with each faction and urged them to compromise. The computer companies flatly stated that they did "not plan to choose between these proposed new formats" and provided a list of nine objectives for a single standard:

- One format for both computers and video entertainment

- A common file system for computers and video entertainment

- Backward compatibility with existing CDs and CD-ROMs

---

[2]Other companies supporting SD included MCA (then owned by Matsushita), MGM/UA (owned by Turner), Nippon Columbia, Samsung, SKC, Turner Home Entertainment (independent at the time, now merged with Time Warner), WEA (Time Warner's giant CD-ROM manufacturing arm), and Zenith.

[3]Others backing the MMCD format included Acer, Aiwa, Alps, Bang & Olufsen, Grundig, Marantz, Mitsumi, Nokia, Ricoh, TEAC, and Wearnes.

- Forward compatibility with future writable and rewritable discs
- Costs similar to current CD media and CD-ROM drives
- No mandatory caddy or cartridge
- Data reliability equal to or better than CD-ROM
- High data capacity, extensible to future capacity enhancements
- High performance for video (sequential files) and computer data

Sony refused to budge, and a month later said there would be "no adjustment" in its MMCD format. Norio Ohga said, " . . . a split on the standard is unavoidable because we are in a world of democracy." He then rejected the possibility of a third standard and defended his decision on the grounds of "liberalism and democracy." Both sides invited the other to surrender, Toshiba inviting "Sony/Philips to engage in serious discussion to resolve this issue," and Sony pronouncing that "we would, of course, be happy to discuss our proposal with all interested parties." Meanwhile, the SD group announced on May 11 that Matsushita had developed a transparent bonding technology that allowed both substrates to be read from a single side. It then tried to stack the deck by announcing the development of record-able SD technology.

On August 14, 1995, the computer industry group, now up to seven members with the addition of Fujitsu and Sun, concluded that the most recent versions of the two formats essentially met all their requirements except the first — a single, unified standard. In order to best support the requirements for a cross-platform file system and read/write support, the group recommended adoption of the *Universal Disk Format* (UDF) developed by the *Optical Storage Technology Association* (OSTA). OSTA had already agreed to refine the UDF standard for interchange compatibility between read-only and nonsequential read/write applications for both television products and computers and had held the first of a set of technical meetings on July 25. IBM reportedly told Sony and Philips that it intended to settle on the SD format and gave them a few weeks to produce a compromise. Faced with the peril of no support from the computer industry or even worse prospect of a standards war reminiscent of Betamax versus VHS, the two camps announced at the Berlin IFA show that they would discuss the possibility of a combined standard. The companies officially entered into negotiations on August 24. The computer companies expressed their preference that the MMCD data storage method and dual-layer technology be combined with SD's bonded substrates and better error-correction method.

Reconciliation

At the end of August 1995, Philips and Sony proposed a new MMCD combined format to the SD Alliance. Conflicting reports appeared in the press, some saying a compromise was imminent and others claiming that officials from Sony and Toshiba had denied both the compromise news and the rumors that MMCD would be standardized for computer data applications, while SD would be used for video. On September 15, 1995, the SD Alliance announced that "considering the computer companies' requests to enhance reliability," it was willing to switch to the Philips/Sony method of bit storage despite a capacity reduction from 5 billion to 4.7 billion bytes. It complained that circuit designs would have to be changed and that "reverification of disc manufacturability" would be required, but conceded that it could be done. It did not concede the naming war, however, proposing that the SD name be retained. Sony and Philips made a similar conciliatory announcement, and thus, almost a year after they began, the hostilities officially ended.

Henk Bodt, executive vice president of Philips, said that the next step was to publish in October a comprehensive specification. Mr. Bodt made some notable predictions, stating that the new players would be "substantially more expensive" than VHS players, at around $800. He also thought that issues of data compression would make recordability realistic only in the professional field: "Certainly I don't think that these players will replace the videocassette recorder."

On September 25, 1995, OSTA announced establishment of the UDF file system interchange standard, a vast improvement over the old ISO 9660 format, finally implementing full support for modern operating systems along with recording and erasing. Work on support for application of the UDF format to DVD was still underway.

Seeing an opportunity to make its own recommendations, the *Interactive Multimedia Association* (IMA) and the *Laser Disc Association* (LDA, which later changed its name to the *Optical Video Disc Association* — OVDA) held a joint conference on October 19 to determine requirements for "innovative video programming" based on years of experience with laserdisc and CD-ROM.[4] The consensus was that better video and audio were insufficient and that DVD movie players required interactivity in order to appeal to the worldwide mass market. The group recommended that baseline interactivity be required in all DVD-Video players and that the design allow optional addition of proven features such as player control using printed bar codes, on-the-fly seamless branching under program control, and external control. Some of the group's recommendations, such as random access to individual frames, graphic overlay, and multiple audio tracks, were largely supported by the tentative DVD standards, but the remaining recommendations were essentially ignored.

On October 30, 1995, OSTA announced completion of an appendix to the UDF file system specification that described the restrictions and requirements for DVD media formatted with UDF. This simplified version, which became known commonly as *MicroUDF*, allowed DVD-Video players to implement low-cost circuitry for locating and reading movie files.

Undeterred by the prospect of retooling its standard, Toshiba announced on November 7 a prototype SD-ROM drive and claimed that its data transfer rate of nine times CD-ROM speed had not yet been achieved with CD-ROM technology.

The two DVD groups continued to hammer out a consensus that finally was announced on December 12, 1995. The new format covered the basic DVD-ROM format and video

---

[4]At this meeting, one of the authors, the humble yet foresighted one, predicted that DVD would not appear in the United States in meaningful numbers before 1997, despite manufacturer claims that it would be ready in early 1996. It turned out that DVD did not appear in the United States in any numbers at all before 1997.

standards, taking into account the recommendations made by movie studios and the computer industry.[5] A new alliance was formed — the DVD Consortium — consisting of Philips and Sony, the big seven from the SD camp, and Time Warner. When all was said and done, Matsushita held 25 percent of the approximately 4000 patents; Pioneer and Sony each had 20 percent; Philips, Hitachi, and Toshiba were left with 10 percent of the pie; Thomson had 5 percent; and the remaining members — Mitsubishi, JVC, and Time Warner — held negligible slivers (see Figure 2.2). On top of DVD-specific patents, the MPEG LA organization controversially claimed 44 essential patents from 12 companies: Columbia University, Fujitsu, General Instruments, Kokusai Denshin Denwa, Matsushita, Mitsubishi, Philips, Samsung, Scientific Atlanta, Sony, Toshiba, and Victor; Dolby, of course, had a finger in the pie with Dolby Digital (AC-3) patents. Fraunhofer, Thomson, and others held MPEG audio patents. Additional fundamental optical disc technology patents are held by Pioneer, Discovision, and others. All this led to a complex dance of cross-licensing that worked reasonably well for the major contributors, but left other companies with no recourse but to pay licensing royalties.

Figure 2.2 The DVD Patent Pie

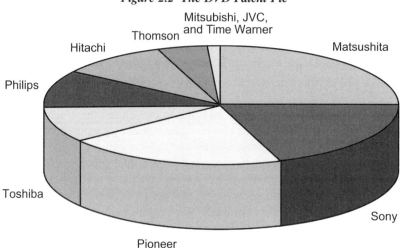

---

[5]The SD physical format of two 0.6-mm bonded substrates was selected, with both Matsushita's dual-layer system (one on each substrate) from the SD format and 3M's dual-layer technology (a second photopolymer layer on a single substrate) from the MMCD format. Toshiba's 8/15 modulation was replaced by MMCD's more reliable EFMPlus 8/16 modulation. SD's more robust Reed-Solomon Product Code error correction was chosen with 32-kilobyte blocks instead of 23-kilobyte blocks. The maximum pit length was reduced from 2.13 to 1.87 microns for higher data density, and the scanning velocity was raised from 3.27 to 3.49 meters per second. These last two changes resulted in the channel bit rate improving from SD's 25.54 Mbps to 26.16 Mbps and helped compensate for the increased modulation (see Table A.7 for more details).

Product Plans

In January 1996, companies began to announce their DVD plans for the coming year. Philips targeted late 1996 for its hardware release. Most other companies pegged a slightly earlier fall release. Thomson stunned everyone by announcing that its player would be available by summer for $499 — $100 less and months sooner than the rest.[6]

The road ahead looked smooth and clear until the engineers in their rose-colored glasses, riding their forecast-fueled marketing machines, crashed headlong into the protectionist paranoia of Hollywood.

As the prospect of DVD solidified, the movie studios began to obsess about what would happen when they released their family jewels in pristine digital format with the possibility that people could make high-quality videotape copies or even perfect digital dubs. Rumors began to surface that DVD would be delayed because of copyright worries, but Toshiba and others confidently stated that everything was on track for a fall release. On March 29, the *Consumer Electronics Manufacturers Association* (CEMA) and the *Motion Picture Association of America* (MPAA) announced that they had agreed to seek legislation that would protect intellectual property and consumers' rights concerning digital video recorders. They hoped their proposal would be included in the Digital Recording Act of 1996 that was about to be introduced in the US Congress.

Their recommendations were intended to:

- allow consumers to make home video recordings from broadcast or basic cable television.
- allow analog or digital copies of subscription programming, with the qualification that digital copies of the copy could be prevented.
- allow copyright owners to prohibit copying from pay-per-view, video-on-demand, and prerecorded material.

The two groups hailed their agreement as a landmark compromise between industries that often had been at odds over copyright issues. They added that they welcomed input from the computer industry, and input they got! A week later an agitated industry group fired off a "list of critiques" of the technical specifications that had been proposed by Hollywood and the consumer electronics companies. The *Information Technology Industry* (ITI) *Council*, a group of 30 computer and communications companies including Apple, IBM, Intel, Motorola, and Xerox was less than thrilled with MPAA and CEMA attempting to unilaterally dictate hardware and software systems that would keep movies from being copied onto personal computers. "No way will we simply accept it as is," said IBM's Dr. Alan Bell, chair of the Technical Working Group, in reference to the proposal. The computer industry said it preferred voluntary standards for content protection and objected to being told exactly how to implement things. "Any mandatory standard that was legislated and then administered by a government body is anathema to the computer industry," said an ITI spokesperson, who also pointed out that legal protection for copyrights "should cover all digital media including

---

[6]Almost exactly a year later, Thomson reannounced its RCA-brand players at $599 and $699 for limited spring 1997 release, with full availability delayed until fall. The Thomson-designed players were built by Matsushita.

records, movies, images, and texts" rather than focusing on motion pictures, as MPAA and CEMA had proposed. ITI announced that it would have a counterproposal ready for an April 29 meeting, recognizing that it was "an important issue to Hollywood, and we don't want to take money out of the studios' pockets." Hollywood countered that the computer industry had been invited to participate early on but did not, either due to laziness or arrogance.

Turbulence

The news media was filled with reports of DVD being "stalled," "embattled," and "derailed." A few weeks later, on May 20, Toshiba's executive vice president, Taizo Nishimuro, was reported to have said that a content protection deal between the computer industry's Technical Working Group and the DVD Consortium would be signed at a June 3 meeting. Apparently, however, the agreement had only been that all three parties would refrain from introducing content protection legislation before the June 3 meeting, and Toshiba officials later denied the reports of settlement. Despite all this, executives at Thomson said that they were confident the issues would be resolved and that they were ready for launch as early as summer.

As DVD progress was being attacked by the popular press, standardization efforts were encountering their own problems. Sony threw a monkey wrench into the process by proposing a completely new DVD-Audio format, *Direct Stream Digital* (DSD). Sony claimed that its single-bit format was not tied to specific sampling frequencies and sizes, thus eliminating downsampling and oversampling and resulting in less noise. The ARA did not agree and continued to push for PCM. Then, in the May 1996 issue of *CD-ROM Professional*, guest columnist Hugh Bennet publicized a serious flaw that had apparently gone unrecognized or was being ignored by DVD engineers. The dye used in recordable CD media (CD-R) did not reflect the smaller-wavelength laser light used in DVD, thus rendering the discs invisible. He pointed out that more than two million CD recorders were expected to be in use by the end of 1996, having written between 75 and 100 million CD-R discs. A new type II CD-R that would work with CD and DVD had been proposed but would take years to supersede the current format, leaving millions of CD-R users out in the cold when it came time to switch to DVD readers.

Glib Promises

At a June 1996 DVD conference, participants glossed over the technical difficulties and downplayed the content protection schism, expecting a solution to be announced before the end of the month. Toshiba executive Toshio Yajima said, "It was a misunderstanding between industries." Sixty executives from the three sides were committed to meeting once a week to resolve the copyright protection disputes. At the same conference, one of DVD's most ardent cheerleaders, Warner Home Video President Warren Lieberfarb, told attendees in his keynote speech that 250 movie titles would be ready for a fall launch. Lieberfarb also gave projections that player sales would be from 2.8 million to 3.7 million units in the first year. Other first-day speakers were equally optimistic, and everyone was assured that the fall launch was on schedule, that the 10 companies of the DVD Consortium had agreed to establish a one-stop agency for licensing, and that the preliminary DVD 0.9 specification book was immediately available for a paltry $5000. A few days into the conference, Thomson admitted that prospects for a fall debut of its player were only "50/50," and Toshiba revealed that it would delay its product launch until October.

Other attendees at the conference were even less sanguine. Many thought December was a more realistic target date, even though it would be too late for the Christmas buying spree. An industry executive was quoted as saying, "What the DVD industry needs is a healthy dose of realism. Somebody ought to just stand up and announce that it's a '97 product, and all this speculation would come to an end." Over the next few months, however, realism did not make much of an appearance, and speculation refused to leave the stage.

In anticipation of the imminent release of DVD, and with content protection details still unresolved, the MPAA and CEMA announced draft legislation called the *Video Home Recording Act* (VHRA), designed to legally uphold content protection. The two groups had been working on the legislation since 1994, without directly consulting the computer industry. Consequently, ITI and the Technical Working Group reportedly filed a letter of protest against the proposed legislation, requesting that an encryption and watermarking scheme be used. The original legislation covered the insertion of a 2-bit identifier in the video stream to mark content as copy-prohibited. The content protection technology eventually became the much more complex CSS, and the legislation mutated into the rather ambiguous *Digital Millennium Copyright Act* (DMCA).

Sliding Deadlines and Empty Announcements

On June 25, Toshiba announced that the 10 companies in the consortium had agreed to integrate standardized content protection circuits in their players, including a regional management system to control the distribution and release of movies in different parts of the world. Many people understood this to mean that the content protection issue had been laid to rest, not realizing that this content protection agreement dealt only with analog copying by using the Macrovision signal-modification technology to prevent recording on a VCR. The manufacturers, still clutching their dreams of DVD players under Christmas trees and hoping to keep enthusiasm high, conveniently failed to mention that Hollywood was holding out for digital content protection as well.

On July 11, Matsushita announced that it would begin shipping DVD players in Japan in September, with machines reaching the United States in October. Matsushita blamed copyright protection and licensing issues for the delay. Again, more pragmatic viewpoints appeared, such as that of Jerry Pierce, director of MCA's Digital Video Compression Center, who flatly stated that "the DVD launch will be in 1997." By this time Philips — the most conservative of the bunch — had moved its release date to the spring of 1997. Matsushita also said its DVD-ROM drives for PCs would not be introduced until early 1997, but that within two to three years every PC would have a DVD drive.

At the end of July, more than a month after a content protection settlement was to have been made, the DVD Consortium — the hardware side of the triangle — agreed to support a content protection method proposed by Matsushita. The content protection Technical

Working Group (CPTWG), representing all three sides of the triangle, agreed to examine the proposal, which used content encryption to prevent movie data from being copied directly using a DVD-ROM drive.

Not unexpectedly, licensing discussion had by now disintegrated. On August 2, Philips made a surprise announcement that it had been authorized by Sony to begin a licensing program for their joint DVD technology: "In an effort to avoid further undesirable delays, Philips and Sony have decided to move forward in the best interest of the DVD system and its future licensees." They called on other companies to join in pooling patents, but Thomson, for one, declined. Many less-than-responsible journalists had a field day, reporting that the DVD standard had been sabotaged or that Philips and Sony were trying to steal patent revenue from the other companies.

By the end of August, there was still no content protection agreement. At the gigantic CeBIT Home exhibition in Hanover, Germany, Jan Oosterveld of Philips explained that key issues such as content protection, regional coding, and software availability were unresolved. "The orchestra is assembled, the musicians have their positions, but they have not decided which tune they will play and how much time they need to rehearse and tune their instruments," he said. He elaborated that content protection was complicated by export and import restrictions and that key technology was still unavailable to non-Japanese companies. The regional control issue, on the table for almost a year, still "created confusion on how the world should be divided." The software supply for DVD-ROM drives also looked bleak, with only 12 companies working on DVD-ROM titles. Oosterveld reiterated that the music industry was expected to take the lead with DVD-Audio and that there was no decision on whether to use phase-change or magneto-optical technology for DVD-RAM. Despite this, Philips was optimistic that by the year 2000 around 10 percent of the 250 million existing optical drives, would be based on DVD. Compared with other predictions, this was rather pessimistic, but given Philips's slightly better track record at forecasting, it may have been the most realistic.

At the same time, Sakon Nagasaki of Matsushita told the press that completion of the DVD specification and resolution of the encryption problem were two separate issues, saying that encryption "is a problem in the United States, not in Japan." Matsushita announced that it could not wait any longer for a content protection agreement, that its two movie player models would be available in Japan on November 1 for 98,000 and 79,800 yen, and that if a content protection agreement were reached, the players would be introduced a few weeks later in the United States for $699 and $599.[7] Although Sony announced that it was delaying the release of its players until spring, Hitachi, Pioneer, and Toshiba followed Matsushita's lead and promised players before Christmas.

A few weeks later, on September 12, 1996, Toshiba announced its first home PCs and reaffirmed its commitment, despite ongoing copyright protection issues, to release DVD-Video and DVD-ROM players, stating that PC makers "should have the first units with our implementation of the copyright protection toward the end of September, and we think we can start shipping those units in volume by mid-November."

---

[7]When the Matsushita (Panasonic) players were finally released in March 1997, the price of the DVD-A100 was unchanged, but that of the DVD-A300 had risen by $50 to $749.95.

The following week was the IMA Expo, with its special focus on DVD. The glowing news from Toshiba was that every DVD-ROM PC would have hardware or software to play DVD movies, DVD-RAM would be ready within a year, and DVD-ROM would be such a success that Toshiba would no longer be making CD-ROM drives by the year 2000. Estimates from International Data Corporation (IDC) placed DVD-ROM drive sales at 10 million in 1997 and 70 million in 2000. During the Expo, Pioneer announced November 22 as the release date in Japan of a combination laserdisc and DVD player at 133,000 yen and a DVD-only player at 83,000 yen, plus two other DVD karaoke players. US prices were expected to be $1200 and $750. By this time, the DVD Consortium had managed to fit most of the world into six geographic regions for release-control purposes, but Mexico, Australia, and New Zealand were still bouncing from region to region. According to Warner Advanced Media, there were supposedly 30 DVD-ROM software titles in development for early 1997. Matsushita representatives at the Expo privately admitted that their players would not appear in the United States until February.

At about this time, the supposedly final 1.0 version of the $5000 DVD-ROM and DVD-Video technical specifications appeared. The books listed a publication date of August, but it took awhile for them to be shipped. A few features, such as NTSC Closed-Caption support, had been changed, and a complex country-specific ratings system had reverted to a simpler version from the earlier spec. Content protection details had been left for a later appendix, which, when it finally appeared, only covered the superficial basics because the decision had been made to keep content protection separate from the technical specifications.

Pioneer announced on September 27 that it would sell a mix-and-match system of DVD players and receivers beginning in the second half of October, bringing its total to five models for sale in 1996. European introduction of the mix-and-match models was planned for spring 1997, with the United States to follow in the summer.

On October 5, 1996, Samsung announced that its DVD player would debut on November 1 in Korea and cited predictions of a global market for 400,000 DVD players before the end of 1996. The affiliated Samsung Entertainment Group expected to release at least 10 DVD movie titles by year's end, with more than 100 in 1997 and over 500 in the year 2000. Samsung also expected to commercialize its DVD-ROM drive by the end of 1996.

Hopes for a DVD-Audio specification before the end of 1996 dwindled. Philips announced that it would team with Sony to develop bitstream-based Direct Stream Digital (DSD) technology for DVD.

The Birth of DVD

The big news finally arrived on October 29, 1996: the Copy Protection Technical Working Group announced that a tentative agreement had been reached. The modified copy protection technology developed by Matsushita and Toshiba, called *Content Scramble System* (CSS), would be licensed through a nonprofit entity. On the same day, Pioneer confessed that it would delay US shipment of DVD players until January, pushing back the December release date it had announced just the day before. At this point only Toshiba and Matsushita were still promising to make DVD hardware available in December, although most of the press and even more of the public were confused, thinking release dates in Japan applied to the United States.

Matsushita and Toshiba delivered DVD players in Japan as promised. News reports from November 1 described a dismal rainy day with lackluster player sales and a handful of discs, mostly music videos. A student was quoted as saying, "Prerecorded discs are not particularly appealing, but it would be great if you can record your favorite films on the disc as many times as you want," which undoubtedly did not help convince the studios to rush their movies onto DVD.

The PR machines did not rest, though. Fujitsu claimed on November 6 that it was the first company to market a DVD-ROM-equipped computer; Toshiba stated on November 7 that it expected rewritable DVD-RAM to be available within a year or soon thereafter; and, Matsushita announced on November 11 the development of the world's first DVD-Internet linkage system for playing video from an Internet Web page through the use of a DVD-ROM drive. Toshiba's news on November 18 was more down to earth: postponement of its DVD player release in the United States until late January or early February.

Most other announcements were put on hold for Comdex, the huge computer technology exhibition in Las Vegas to be held in January of the coming year. At Comdex, dozens of announcements regarding hardware and software were made, including the expected worldwide release of Toshiba's DVD-ROM drive in late January 1997. Intel CEO Andy Grove's keynote speech included a DVD demo of *Space Jam* using Compcore's software-only playback system.

November 1996 came and went with no sightings of DVD players for sale in North America. On December 2, Akai announced its own DVD player to be launched in Japan at the end of January and worldwide in April.

By December 13, 1996, Toshiba's Precia PC, which was to have been rolled out in Japan in November, had its debut pushed back to January. Toshiba blamed the delay on bureaucratic problems with content protection chips. "We're hoping it's resolved any day now, but we've been hoping for that for two or three weeks," went the now-too-familiar refrain. Other sources claimed the real reason for the delays was that the Japanese government had barred Toshiba from releasing hardware to the United States because of concerns over the export of encryption technology. The US delivery date for Toshiba's DVD-equipped PCs slipped to March.

Things began to look up on December 20 when Warner Home Video began sales in Japan of four major movie titles — *The Assassin*, *Blade Runner*, *Eraser*, and *The Fugitive* — with another four announced for January release.

By this time, the delay in the United States was being blamed on lack of titles rather than on content protection issues, but in the first week of January 1997, at the Consumer Electronics Show (CES) in Las Vegas, DVD finally began to resemble a real product. At least six studios announced the release of more than 60 DVD-Video titles for March and April: New Line Home Video, Warner Home Video, Sony's Columbia TriStar Home Video, Sony Music Entertainment/Sony Wonder, MGM Home Entertainment, and Philips' Polygram. Perennial favorites *Casablanca* and *Singin' in the Rain* were announced by two studios each.

Also at the Consumer Electronics Show, DVD players were demonstrated or announced by Akai, Denon, Faroudja, Fisher, JVC, Meridian, Mitsubishi, Panasonic, Philips/Magnavox,

Pioneer, RCA, Samsung, Sony, Toshiba, Yamaha, and Zenith, most with a US release date between March and summer. This brought the lineup of player manufacturers to at least 20, including others who had previously announced DVD movie plans: Goldstar, Hitachi, Hyundai, and Sharp. Compared with typical industry support of new launches of now-ubiquitous products such as telephones, televisions, VCRs, and CDs, this was an astonishing endorsement that contradicted the doom-and-gloom predictions for DVD as a consumer video product.

And on the computer side, where most people expected DVD to be a slam dunk, there were now more than 45 companies developing hardware or playback software. As it turned out, many of them abandoned their efforts within a year.

Players from Panasonic and Pioneer finally began to appear in the United States in February as more movies and music performance videos were announced from Lumivision, Warner Bros. Records, and others. Eager customers bought the players, only to discover that no titles other than those from Lumivision were scheduled to appear before the end of March. Warner, the primary supplier, limited its release to seven test cities. Pioneer announced that its $11,500 DVD-R recorder would be available, complete with 40 to 50 blank discs, around June 1997. Computer makers geared up for both hardware and software DVD-Video playback, but once again, the content protection issue settled on the scene like a wet blanket: Decryption/descrambling licenses were available for hardware but not yet allowed for software. The studios were unconvinced that DVD players implemented in software would completely protect their assets. They were right, of course, but that did not mollify the computer makers and software developers who were counting on cheap software playback.

DVD had finally embarked, late and lacking some of its early luster, on the rocky road to acceptance.

Disillusionment

The two final competing DVD-RAM proposals were merged and an official format was announced on April 14, 1997. The approval procedure was allegedly changed so that members of the original SD camp could steamroll their combined proposal over the objections and abstentions of Sony and Philips, who were promoting a format more closely related to their own CD-RW. The veneer of cooperation was cracking, revealing the possibility of a competing recordable DVD format. Toshiba, which apparently announced the DVD-RAM arrangement before it was approved, made almost comical claims that drives would be available before the end of 1997 for only $350. More sober predictions targeted 1998 and prices of $800 or more. At about the same time, Pioneer revealed that its DVD-R drive would cost $6000 more than originally expected, raising the price to about $17,000. The good news was that DVD-R was expected to be compatible with most players and drives; the bad news was that DVD-RAM would not be compatible with anything. The worse news, surfacing about a month later, was that Sony and Philips were apparently striking off on their own with a renegade recordable format, along with partner Hewlett Packard. A Philips spokeswoman said, "We feel the DVD-RAM spec 1.0 agreed upon by the DVD Consortium is not an ideal format. Our format will bring far better compatibility with the CD products that are already on the market." NEC also announced its own competing recordable format. The DVD family was losing unity. To top it off, various technology announcements were issued from DVD companies announcing research on high-density DVD. To DVD manufacturers, these were forward-looking announcements, necessary in the jostling for position that would precede work on the next generation of DVD. To new customers just learning about DVD, especially those used to the slow turnover of traditional technologies such as TVs and record players, news hinting at the imminent obsolescence of the new DVD format was disconcerting.

Rockley Miller, editor of the *Multimedia Monitor* newsletter, characterized the acceptance cycle of new technologies. He posited that every new technology hits an initial peak of euphoria, often before it is released to market, where wildly exaggerated forecasts are piled on claims of vast improvements. Of course, the new devices never live up to the impossible expectations, leading to Miller's second cycle: disillusionment. Sales do not meet expectations, equipment does not work perfectly, and development evolves more slowly than predicted. The press always jumps in at this point, sometimes even proclaiming the failure of the format. From the trough of disillusionment, however, comes the gradual rise to the third phase of the cycle: real product and real work. Bugs are eliminated, improved products are released, and people begin to figure out how to use the new technology for what it does best.

In the middle of 1997, DVD was in the trough of disillusionment. On the home video front, Time Warner and player manufacturers had not expanded beyond the six trial cities, although discs and players usually could be purchased via mail order. Many of the major studios held to their wait-and-see stance, with some sources claiming that they were pressuring the hardware makers for a kickback. Meanwhile, rumors spread of a pay-per-view version of DVD called *ZoomTV*, supposedly under development for use by the holdout studios such as Disney and DreamWorks. Home consumers who heard the rumors were less than thrilled at the prospect of having their movie-viewing habits held hostage to the vagaries of telephone systems and remote authentication servers. Amid the litany of boredom and frustration, one of the few bright spots was the initial success of disc sales. By the end of April, less than one month after the release of the first 40 or so DVD titles in the United States, over 50,000 copies had been sold, exceeding most expectations. The title wave of DVD had begun to swell. More than 30,000 players had been purchased in the first 15 weeks of US sales, and over 150,000 players had been sold in Japan since DVD's introduction six months earlier. Initial reviews of bad in-store demos were supplanted by glowing reports from early purchasers who made side-by-side comparisons of DVD and laserdisc and proclaimed DVD to be the clear winner.

On the computer side, however, little was happening. Creative Labs' DVD-ROM upgrade kit with hardware DVD-Video playback appeared in April but showed signs of a rushed release. Other DVD-ROM products failed to make their already delayed debuts, and both Apple and Microsoft pushed back release dates for DVD support in their operating systems. There were still no movie decryption licenses allowed for software playback, and industry observers began to doubt that DVD computers would be available in more than miniscule numbers before 1998. This was increasingly irksome to those who simply wanted a DVD-ROM data drive and had no interest in using their computers to play movies. Many in this group began hoping for a quick death for DVD-Video so that they could proceed with the real business of DVD-ROM.

The audio-only initiative met yet another setback in June when Philips and Sony announced that they were working on the Super Audio CD format based on their DSD audio coding process. Undaunted, the DVD Forum's *Working Group 4* (WG-4) was moving ahead on DVD-Audio, based on the recommendations of the *International Steering Committee* (ISC) formed by the *Recording Industry Association of America* (RIAA), Europe's *International Federation of Phonographic Industries* (IFPI), and the *Recording Industry Association of Japan* (RIAJ) (See Table 2.1.).

Table 2.1 Recommendations of the ISC for DVD-Audio

Very high quality multichannel audio with downmixing

Studio mastering and archiving with no loss of quality

Playing time of at least 74 minutes

Additional data such as video, lyrics, and still pictures

On-screen navigation menus

Simple navigation, similar to audio CD players

Copyright protection and anti-piracy measures

Conditional access

Compatibility with DVD-Video format wherever possible

Compatibility with existing audio CD players

No caddy, no proscribed package size

Durability equal to or better than CD

Single-sided, 12-cm discs preferred

In June, Hitachi demonstrated a home DVD video recorder containing a DVD-RAM drive, a hard disk drive acting as a buffer, two MPEG-1 encoders, and one MPEG-2 decoder. No production date was disclosed.

In July 1997, the MPEG LA (licensing authority) opened for business in Denver, Colorado, as a one-stop shop for MPEG-2 patents, given that they represented about 90 percent of the MPEG-2 patent holders. Because of the delay in establishing a patent royalty program, and due to the precedent set when MPEG-1 patents were not enforced, some people felt that the MPEG-2 patent holders had lost the right to make money from their patents. Most of the consumer electronic companies appeared on the list of MPEG LA's licensees, but computer hardware and software companies remained conspicuously absent.

Flickers of good news surfaced, as Universal Home Video announced in July that they would enter the DVD market, and finally, in August, Warner broadened distribution of discs from the original seven test cities to countrywide. Image Entertainment, a longtime supporter of laserdiscs, realized that the only way to recover from a precipitous drop in laserdisc sales was to embrace the DVD format with more alacrity. Panasonic announced the first DVD notebook PC.

Divx: A Tale of Two DVDs

In September 1997, ZoomTV surfaced, sporting a new name and heavy-duty financial backing. Circuit City and the Hollywood law firm of Ziffren, Brittenham, Branca & Fischer announced the formation of Digital Video Express, a company that planned to release Divx, a "DVD rental" format, in the summer of 1998. Furious controversy erupted, with backers hailing it as an innovative approach to video rental featuring cheap discs you could obtain almost anywhere and keep for later viewings while detractors spurned it as an insidious evil scheme by greedy studios to control what you see in your living room (Table 2.2), although most of the opponents already owned DVD players and were annoyed that they were not able to play Divx discs on them.

Table 2.2 Pros and Cons of Divx

Advantages of Divx

- Viewing could be delayed, unlike rentals.
- Discs need not be returned. No late fees.
- Movie could be watched again for a small fee. Initial cost of "owning" a disc was reduced.
- Discs could be unlocked for unlimited viewing (Divx Silver), an inexpensive way to preview before deciding to purchase.
- Unlike rental discs, the discs were new and undamaged.
- The "rental" market was opened to other retailers, including mail order.
- Studios gained more control over the use of their content.
- Special offers from studios delivered to your Divx mailbox.

Disadvantages of Divx

- Higher player cost.
- Although discs did not have to be returned, the viewer still had to go to the effort of purchasing the disc. Cable/satellite pay-per-view is more convenient.
- Higher cost than for regular DVD rental.
- Casual, quick viewing (looking for a name in the credits, playing a favorite scene) required paying a fee.
- Most Divx titles did not come in widescreen format and did not include extras such as foreign language tracks, supplemental video, and commentaries.
- The player had to be hooked to a phone line (but only once a month or after about 10 viewings).
- Divx players included a "mailbox" for companies to send you unsolicited offers.
- Those who did not password-protect their Divx player could receive unexpected bills when children or visitors played Divx discs.
- Divx discs would not play in regular DVD players or DVD computers.
- Unlocked (DivxSilver) discs only worked in players on the same account. Playback in a friend's Divx player would incur a charge.
- You could not give Divx discs as gifts unless you were sure that the recipient owned a Divx player.
- There was little market for used Divx discs.
- Divx players were never available outside the United States and Canada.

- The Divx central computer ostensibly collected information about viewing habits. Of course, cable/satellite pay-per-view services and rental chains do the same.

Contrary to many claims, Divx was not a competing format — it was a pay-per-viewing-period variation of DVD. The discs were designed to be purchased from any retail outlet at a price close to rental fees. The $4.50 purchase price covered the first viewing. Once inserted into a "Divx-enhanced" DVD player, the disc would play normally for the next 48 hours, allowing the viewer to pause, rewind, and even put in another disc before finishing the first disc. Once the 48 hours were up, the "owner" of the disc could pay $3.25 to unlock it for another 48 hours. Divx DVD players, which initially cost about $100 more than a regular player but later dropped to a premium of only $50, included a built-in modem so that they could call a toll-free number during the night to upload billing information. Since the player only called once a month or so, it did not have to be connected permanently to a phone line. There was a DivxSilver program that allowed most discs to be converted to unlimited play for an additional fee of about $20. The idea was to let buyers purchase a disc at low cost, try it out, and then pay full price if they liked it enough to watch repeatedly. Unlimited-playback DivxGold discs were announced but never shipped.

Divx players were able to play regular DVD discs, but Divx discs were designed not to play in standard DVD players. Each Divx disc was serialized with a bar code in the standard burst cutting area so that it could be uniquely identified and tracked by the player for billing and encryption purposes. In addition to normal CSS copy protection, Divx discs used modified channel modulation (making normal drives physically incapable of reading protected data from the disc), triple DES encryption (three 56-bit keys), and watermarking of the video. Ironically, the company had nothing to fear from pirates who made bit-for-bit copies, because copied discs would still require authorization and billing. A pirate outfit would simply be providing free Divx replication services. Actual cracking of the encryption for copying video to standard DVD discs was a potential threat, but Divx technicians claimed that no one ever got through even the first layer of protection.

Limited trials of Divx players began June 8, 1998, in San Francisco and Richmond. The only available player was from Zenith (which at the time was in Chapter 11 bankruptcy), and the promised 150 movies had dwindled to 14. The limited nationwide rollout (with one Zenith player model and 150 movies in 190 stores) began on September 25, 1998. By the end of 1998, about 87,000 Divx players (among four models) and 535,000 Divx discs were sold (among about 300 titles). These numbers were quite impressive, representing 10 percent of all player sales. By March 1999, 420 Divx titles were available, compared with more than 3500 open DVD titles. Despite the small numbers, it was again an impressive feat to produce over 400 titles in only six months. Support from studios and player manufacturers increased, motivated by incentives in the form of guaranteed licensing payments totaling more than $110 million. Initial backers Disney (Buena Vista), DreamWorks SKG, Paramount, and Universal were joined by MGM and Twentieth Century Fox, although Time Warner and Sony's Columbia/Tristar actively combated Divx by creating rental programs. Harman/Kardon, JVC, Kenwood, and Pioneer announced their intention to join Matsushita (Panasonic), Thomson (Proscan/RCA), and Zenith in making Divx players. Thomson even demonstrated a high-definition version of Divx, which appealed to studios that were even more fearful of releasing movies on high-definition DVD than on standard DVD. Retailer support was weak, partly because chains were reluctant to sell a format that would pad the pockets of competitor Circuit City.

Meanwhile, the vitriolic reaction to Divx showed no signs of abating. Dozens of anti-Divx Websites peppered the Internet. The major psychological hurdle was that many people were uncomfortable buying a disc that they did not control. Despite a promising beginning, the company could not shake its bad press. In March, shares of Circuit City rose more than 12 percent from speculation about the future of Divx and rumors of investment by Blockbuster, among others. The investments fell through, however, allegedly scuttled by studios that backed Divx but did not want to be beholden to Blockbuster. Richard Sharp, the CEO of Circuit City who originally brokered the Divx deal, finally yielded to pressure from stock-holders and others. On June 16, 1999, less than a year after initial product trials, Digital Video Express announced that it was shutting down operations. E-cheers rang out across the Internet, with the anti-Divx sites claiming victory in torpedoing a foe. A popular DVD news site, *The Digital Bits*, received 628 reader e-mails about the death of Divx by 11 A.M. on the day of the announcement. The swell of resentment on the Internet undoubtedly contributed to the demise of Divx. In an auto-obituarial press release, Richard Sharp grumbled that "unfortunately, we have been unable to obtain adequate support from studios and other retailers." Industry analyst Ted Pine was much closer to the mark when he said, "This is the first technology to run up against the *vox populi* of the Internet." Circuit City took a $114 million after-tax loss. *Variety* magazine estimated the total loss to be $337 million. Digital Video Express provided $100 rebates, making Divx players a great deal, especially since they generally had excellent features and quality. The Divx computer was shut down on June 30, 2001, after which Divx discs were no longer playable, though the players could continue to play standard DVDs.

When all was said and done, Divx did not confuse or delay development of the DVD market nearly as much as many people had predicted.[8] In fact, it probably helped by stimulating Internet rental companies to provide better services and prices, by encouraging manufacturers to offer more free discs with player purchases, and by motivating studios to develop rental programs. The world will never know if Divx could have been a success, but if Circuit City had managed to hold on for a few more years to the point where Divx discs were available as impulse purchases at supermarket checkout counters and corner convenience stores, it might have become the primary DVD format.

Ups and Downs

Meanwhile, the DVD roller coaster ride continued in the fall of 1997. Disney announced that it would finally enter the market with DVD discs, but the planned titles did not include any of the coveted animated features. Four days later, Disney announced that it would also release discs in the then-existing Divx format.

In October, the 10-member DVD Consortium changed its name to the DVD Forum and opened membership to all interested companies. By the time the first DVD Forum general meeting was held in December, the organization had grown to 120 members.

---

[8]Truthful disclosure time: one of the authors of this book, Jim Taylor, complained about Divx from time to time, usually from the perspective that it would confuse customers and hold back DVD. His point was that it should have been included in the DVD format from the beginning so that all DVD owners would have the option to purchase Divx discs. The Divx developers did propose Divx to the DVD Consortium while DVD was still under development, but they were turned down.

The Pioneer 3.95G DVD-R drives finally appeared. In spite of the $17,000 price tag, DVD developers desperate for an easy way to test their titles snapped them up. After all, compared to the $150,000 price of the first CD-R recorders, DVD-R recorders were a bargain.

In November 1997, the first public breach of DVD content protection occurred. A program called softDVDcrack was posted to the Internet, allowing digital copies of movies to be made on computers. The press reported that CSS encryption had been cracked, but in actuality the program hacked the Zoran software DVD player to get to the decrypted, decompressed video. Zoran plugged the leak, but pirated copies of the earlier Zoran player were now available on the Internet. Even so, the process was so complicated that only the most dedicated hackers were interested in spending half a day copying a two-hour movie.

A continuing difficulty in getting equipment to encode multichannel MPEG audio finally became such a problem that on December 5, 1997, the DVD Forum Steering Committee voted to amend the DVD specification to add Dolby Digital as one of the mandatory audio format options on 625/50 (PAL/SECAM) discs. Previously, PAL/SECAM discs had to have either PCM or MPEG audio tracks. The change allowed disc producers to exclusively use Dolby Digital soundtracks. Philips, the primary supporter of MPEG audio, and its partner Sony voted against the change, but they were overruled by the other eight companies. There was scattered cheering among European DVD enthusiasts who preferred the much wider selection of Dolby Digital-equipped audio gear. There was grumbling from Philips, which said, "The decision made is based on incorrect information. Philips will continue to support MPEG-2 multichannel audio for PAL/SECAM countries."

In December, Microsoft released DirectShow 5.2 (also called DirectShow 2.0), an extension to the Windows operating system that supported DVD playback. Up to this point, vendors of DVD hardware and software for Windows PCs had used the old MCI system, each implementing things slightly differently. DirectShow promised to provide a robust, standard interface for developers of DVD software, along with support for the full DVD-Video feature set. Unfortunately, most DVD software decoder vendors were more interested in signing OEM deals to get their software bundled into new PCs than in supporting Microsoft's standard. The shift from proprietary decoders to DirectShow-compatible decoders, which should have happened in less than a year, had barely reached the three-quarters point three years later. The resulting lack of a single DVD development target for Windows contributed to the retarded growth of multimedia DVD software.

At the end of December 1997, 340,000 DVD players had been sold in the US. This was far below "expectations" of 1.2 million. Dozens of reports appeared that disparaged DVD's performance. Few bothered to mention that most of the expectations had been unreasonably optimistic. Given that DVD had launched only nine months earlier, with the first five months limited to seven test cities, player sales were actually quite impressive. Rockley Miller's predicted cycles of euphoria leading to disillusionment were holding true.

The Second Year

1998 began with many predictions that it would be "the year of DVD." These predictions were about a year early, even though prospects began to brighten for the fledgling format. The DVD specification was treated to a minor freshening: version 1.01 of DVD-ROM and version 1.1 of DVD-Video. A draft version of DVD-Audio was announced. The first dual-layer discs, DVD-9s, were produced, even though yields, the number of usable discs in a replication batch, reached only 30 percent. Warner Home Video noted that it made more than $50 million in revenue from DVD sales. E4 finally shipped the first DVD upgrade kit for Macintosh computers. Panasonic announced the DVD-L10, the first portable player, which was a big hit with travelers and those doing DVD presentations. Pioneer announced that its LD-V7200 industrial-strength player would be available in the spring. With features such as bar code control, external control, mouse and keyboard input, video blackboard, genlock, and high reliability, the player met the needs of specialized applications such as museum installations, video walls, training centers, and kiosks.

Sonic Solutions announced DVDit[9], a simple, low-cost production tool to convert PowerPoint presentations, HyperCard stacks, and Premiere projects into DVD-Video titles. Initial excitement faded as the product underwent numerous metamorphoses before finally materializing more than two years later.

Announcements of DVD titles steadily rolled in, many from smaller studios. A few of the major Hollywood studios had yet to take off their hat, let alone throw it in the ring, although Fox tipped its bowler slightly and announced it would release Divx discs.

Because DVD still had not been officially launched outside of Japan and the US, a market for imported discs began to grow in other countries, fed by fans who bought region 1 players from the US. In February, the UK *Federation Against Copyright Theft* (FACT), raided a High Street record outlet and seized imported region 1 discs. A pair of British entrepreneurs tried to get around restrictions by setting up a region 1 DVD import business in France, but they were sentenced and fined. However, because many European countries had weak or no restrictions on imported movies, and since most countries, even the UK, allowed individuals to legally purchase titles from outside the country, international Internet DVD mail order business flourished. Plans for a large-scale launch of DVD in Western Europe fizzled, resulting in a "soft launch" in May 1998, which amounted to a few manufacturers and studios releasing a limited number of players and titles to join the existing trickle.

In March 1998, in spite of half-hearted opposition from the DVD-RAM camp, the DVD Forum officially adopted the DVD-RW format, developed primarily by Pioneer as a rewritable variation of DVD-R. The recordable side of the DVD family grew more complicated. In April, the DVD Forum publicly requested that HP, Philips, and Sony stop using the letters DVD in their product names. Philips responded that no one owned the letters. A few months later, on June 16th, HP, Mitsubishi, Philips, Sony, and Yamaha announced formation of the DVD+RW Compatibility Alliance to promote the DVD+RW format. Sony announced that its DVD+RW drive would be available at the end of 1998. Also in June, the DVD Forum released a preliminary version 0.9 of the DVD-Audio specification, with expectations that players would be available by the year's end. By this time, jaded DVD technology watchers knew to add six months to any announced timetable, but for DVD+RW and DVD-Audio, adding 24 months would still not have been a sufficient reality adjustment.

---

[9]Sonic puts an exclamation mark after the name of the product. Such abuses of punctuation by out-of-control marketing departments must be rigorously opposed whenever possible, which is why the mark does not appear here.

Philips and Sony continued to work half in and half out of the Forum, announcing that SACD would be licensed to existing CD licensees at no additional charge. They promoted features such as watermarking and a hybrid disc that would also play in CD players.

In April, one of the last holdout studios, Paramount, announced that it would begin releasing movies on DVD.

New 350-MHz and 400-MHz Intel Pentium II processors had reached the point where the CPU could handle most of the work for DVD decoding and processing. Low-cost, high quality, software-only DVD playback thus became feasible. In June, DVD-RAM drives appeared, only six months late. Prices were reasonable (around $800), with a surprise entry from Creative Labs at $500. 2X and 3X DVD-ROM drives also debuted, which at speeds equivalent to 18X and 24X CD-ROM drives at prices under $175, became slightly more competitive. Sony and others had already announced that 5X drives would be available soon. Still, the DVD-ROM market was not taking off as most had expected, causing many software developers to abandon DVD-ROM products and plans. Some ventures that had focused heavily on DVD-ROM, such as Divion, went out of business completely.

The Second Wind

In June 1998, as noted previously, Divx was launched. Time Warner and others had already introduced new rental programs in anticipation. Sears announced that it was canceling plans to carry Divx players, apparently due to negative publicity. Thomson (maker of RCA and ProScan brands) joined the Divx program, thus diluting image problems caused by Zenith's Chapter 11 bankruptcy. At the July Video Software Dealers Association conference, retailers were down on Divx but were moving quickly to build up DVD sales and rental programs based on reports of record player sales. At the show, Warner Home Video, DVD's biggest supporter, announced that it had generated more than $110 million in revenue from DVD sales in the first 6 months of 1998. NetFlix, the first Internet-based DVD rental store, had rented over 20,000 copies since opening its e-doors in April. JVC's announcement of its D-VHS digital videotape system, a potential competitor to DVD, did little to dampen DVD enthusiasm, especially because D-VHS did not include any provision for prerecorded tapes.

In August, Fox announced it would produce "open DVDs" along with Divx versions. This left DreamWorks SKG as the only major studio not committed to open DVD. A month later DreamWorks bowed to the inevitable and announced non-Divx titles. Still, the blockbuster movies from Steven Spielberg (the S in SKG) were nowhere to be seen. A DVD release of the Robert Zemeckis film, *Back to the Future,* was announced, only to be withdrawn a week later; it would not reappear for over two years. Digital Theater Systems (DTS) announced that titles supporting the optional DTS audio format would finally appear by year's end. As it turned out, the direct-to-video *Mulan* disc was the only one to appear on schedule.

At the September 1998 DVD Forum conference in San Francisco, Warren Lieberfarb, defender of the DVD faith, predicted that DVD would be "one of the hottest consumer electronics products to be launched in the last few decades." Dan Russel of Intel predicted that more than 40 million DVD-ROM drives would be available by the end of 1999. Lieberfarb's crystal ball was crystal clear, if not quite optimistic enough, while Russel's crystal ball turned out to be running a bit fast.

The new Toshiba SD-7108 progressive scan DVD players, able to display almost twice the resolution as standard players from the same DVD, didn't make it out of the warehouse. Concerns about lack of content protection on progressive scan output kept them there for another year. A few weeks later, Genesis Microchips announced a new deinterlacing chip that would find its way into most of the progressive scan DVD players released in 1999 and 2000.

In September, the Chinese government announced the new *Super Video CD* (SVCD) format, similar to DVD with MPEG-2 video, onscreen menus, and multilanguage subtitles, but using standard CDs, in part to avoid the high royalties demanded by DVD patent holders.

Hitachi announced that its DVD-RAM camcorder would debut by the end of 1999. The reality distortion field of this announcement was set to about 1.2 years.

As a publicity stunt, NetFlix made DVDs of Clinton's grand jury testimony in the Monica Lewinsky case available for two cents within a week of the broadcast. NetFlix received unexpected extra publicity when some customers were surprised to discover an X-rated movie in the package. In an unrelated story, Project X, the rumored game cum DVD player, was given the official moniker of Nuon. The company wisely recognized that it would do much better by positioning its technology as an enhanced DVD player rather than yet another game console. Nuon-based players were promised for 1999. Blockbuster, well-known for its refusal to rent any movie with a rating above R, began DVD rentals in 500 stores in September. Philips and Blockbuster partnered on a rental promotion program, as did Sony with NetFlix, and Warner Bros. and Columbia TriStar with West Coast Video and Hollywood Entertainment. This flurry of rental programs just happened to coincide with the nationwide expansion of Divx. By this time NetFlix, the online rental service, had become the largest DVD rental "store" with more than 2,000 titles. Online sales of DVDs by companies such as DVDExpress, Reel.com, Buy.com, and Amazon.com had become fiercely competitive, with many titles being sold at a 50-percent discount.

The UK's FACT made headlines again with a raid on Laser Enterprises in Essex, England, to seize imported US DVDs. A month later, in October 1998, the official launch of DVD in Western Europe occurred; sort of. At least more players and more discs became available. This was followed by another less-than-spectacular official launch, that of HDTV in the US on November 1, 1998. Approximately 40 TV stations began broadcasting bits and pieces of HDTV programming to a few thousand HDTV sets around the country. The next non-event was the release of the DVD-Audio spec, with promises of players in 1999 once the content protection issues were resolved. As the format that cried wolf, DVD-Audio was losing credibility among DVD technology analysts. The release date of new 4.7G DVD-R drives was bumped to the second quarter of 1999, to the dismay of developers desperate to use the larger-capacity format for testing works in progress.

In November, as Comdex rolled around, DVD-ROM drive speeds were notched up to 6X. Various combinations of player features and drive speeds were touted as "2nd generation" and "3rd generation," or even "DVD II" and "DVD III," again leading to much confusion among customers, many of whom thought that a generation or two of DVD had already become obsolete. Microsoft released *Encarta* on DVD, the one-millionth DVD player was sold to dealers in the US, and *Lost In Space*, the first major PC-enhanced movie, sold an incredible 200,000 copies in the first week. At the time there were fewer than 500,000 players in homes, so the obvious conclusion — given that such a lame movie couldn't possibly appeal to one of every two player owners — was that DVD PC owners were buying movies in much higher numbers than anyone anticipated. InterActual, the company that developed the PCFriendly software used on *Lost In Space*, began clinching deals with most of the major Hollywood studios for PC-enhanced titles.

Demolishing opinions that dual-layer, double-sided discs (DVD-18s) would never be achieved, WAMO announced a new process to make them. This was immediately followed by rumors that the 195-minute *Titanic* would be the first DVD-18 release, since director James Cameron was insisting that the DVD contain both fullscreen and widescreen versions. DVD fans had been miffed when the videotape version went on sale September 1 sans the DVD version. They had a long time to wait for the DVD, although it still came out too soon for DVD-18.

DVD had a very merry Christmas in 1998. Musicland sold $5 million worth of DVDs in the last week before Christmas, compared to sales of $50 million for all of 1998. Over five million DVD titles were sold in the last five weeks of the Christmas buying season. By the end of the year, 1.4 million players had been sold to dealers in the US with about 1 million installed in homes. Europe trailed with about 125,000 players. Warner Home Video, perennial barometer of DVD, made $170 million in US DVD sales, 17 percent of all sellthrough video rental. Shortages of standard DVD players apparently led to brisk sales of premium-priced Divx players. DVD-ROM , however, did not prosper comparably. Microsoft's predictions of 15 million DVD-ROM drives were off by 6 to 7 million.[10] More than 100,000 DVD-RAM drives had shipped since the middle of the year, which was nothing to sneer at, but rather anticlimactic for a technology with the potential to take over from CD-RW.

The Year of DVD

1999, by most reckonings, was the year of DVD. It came off a highly successful Christmas season and continued to exceed many expectations. Internet sales boomed — Image Entertainment reported that over 23 percent of its revenue came from Internet retailers. DVD fulfilled its role as the standard-bearer of quality video: nearly 75 percent of movie titles were enhanced for 16:9 widescreen, 10 percent were collector's editions, and 20 percent were on dual-layer DVD-9 discs. The winter CES show was filled with demonstrations of writable DVD products, both for computer use and for home video recording, although no products were expected before late 1999 or 2000. The bombshell at CES was Thomson's demonstration of a prototype high-definition Divx player. Thomson was counting on cautious movie studios being more likely to release high-definition movies on a secure platform such as Divx. What Thomson wasn't counting on was that Divx would be defunct six months later.

Pioneer's LD-V7200 player finally appeared, about eight months late. eMachines set a new entry level for DVD PCs with the announcement of a 333-MHz Celeron PC with a 5X DVD-ROM drive, without monitor, for $600. The first few DTS DVDs trickled out. Also trickling out were rumors that Sony was developing a new DVD-based version of its PlayStation game console.

---

[10]For those of you keeping track, yes, by this time Jim Taylor was contracting to Microsoft as a DVD Evangelist. He was quietly predicting 10 million DVD-ROM drives instead of the 15 million number thrown out by others at Microsoft.

In February, Philips, Sony, and Pioneer officially began their patent licensing program for DVD players and discs. Things were busy on the copyright front as well. Hitachi, IBM, NEC, Pioneer, and Sony (dubbed the Galaxy group) announced they had agreed on a digital water-marking technology to enhance DVD copyright protection. In the same month, apparently as a move to combat bootleggers, *Titanic* was released in China as a four-disc Video CD set. An estimated 3.5 million pirated copies had already been sold by the time the official version hit the streets. And, the *Digital Display Working Group* (DDWG) announced on February 23 the completion of the *Digital Visual Interface* (DVI) specification for final draft review. Studios were looking forward to this replacement for computer VGA output, since DVI included a content protection mechanism. On March 3, IBM, Intel, Matsushita, and Toshiba (dubbed the "4C"), announced a content protection framework for DVD-Audio, the fruit of more than 12 months of work with music labels BMG, EMI, Sony Music Entertainment, Universal Music Group, and Warner Music Group. This would later become CPSA (content protection system architecture) and cover both DVD-Video and DVD-Audio.

On March 1, 1999, Sony Computer Entertainment officially announced the PlayStation 2. "We have not come to a decision at this time whether we will put in the ability to play DVD [video] or not," said Kaz Hirai, executive vice president and chief operating officer.

DVD Gets Connected

On March 14, director John Frankenheimer hosted a live web event for owners of the *Ronin* DVD. As he answered questions over the Internet, he played back hidden content from the disc to illustrate behind-the-scenes events and to explain details of producing the movie. Since it was impossible to stream DVD video over the Internet, the trick was to use InterActual's PCFriendly software to remotely control the DVD-ROM drives of the tens of thousands of chat participants. WebDVD had reached a new milestone. Research indicated that more than half of DVD-Video player owners also owned DVD-ROM PCs. Time Warner Chairman and CEO Gerald Levin called WebDVD the "key to consumer entertainment and information in the next decade." The second European DVD Summit, held in Dublin, also witnessed growing interest in Web-connected and enhanced DVDs. Nuon, the forerunner in creating an enhanced DVD player for the consumer electronics market, announced at the DVD Summit that the first Nuon players would be available in spring 2000, later than originally projected.

On April 20, another DVD milestone was set with the release of *A Bug's Life*, the first feature film to be created and distributed from beginning to end using all-digital technology. The visual quality of the disc was stunning, especially when played back in high resolution on a computer. The producers of the film remarked that the DVD rendition, unlike the film release, finally achieved the look they had created on their computer graphics workstations.

DVD began to pick up speed in Europe, with many new titles and DVD-Video players selling for under £250 in UK stores such as Woolworth.

At the NAB conference in April 1999, Sonic Solutions re-announced DVDit, now reborn as a $500 authoring package, available in July. Apple released a public beta version of QuickTime 4.0, without the hoped-for support for MPEG-2 or DVD. Daikin announced that it was partnering with InterActual to produce an *Enhanced DVD Kit* (EDK) to integrate WebDVD production with its Scenarist authoring system. Matsushita revealed that it planned to ship two DVD-Audio players under the Panasonic and Technics brands, priced at $800 and $1700, with delivery date tentatively set for fall. Sony released its SACD player in Japan at the tear-inducing price of $4,500. Pioneer shipped the first samples of the 4.7G DVD-R drive, based on version 1.9 of the specification. Not surprisingly, the final 2.0 version was held up by work on content protection features. HP announced that its first DVD+RW drive would be out in June. The *Consumer Electronics Manufacturers Association* (CEMA) announced that digital television (DTV) sales over the previous seven months reached a total of 25,694.

At the E3 conference in May, Nintendo said that its upcoming DVD-based game console, code-named Dolphin, would play DVD movies. Later it was revealed that only the Matsushita (Panasonic) versions of the box would play video because Nintendo was concentrating on a bare-bones, lowest-cost version. At the same conference Nuon admitted that players using its chip would not debut in 1999. Nuon, which had once been a unique avant-garde technology, was losing its avantness and gaining competitors.

Patents and Protections

In June 1999 the other patent pool, comprising Hitachi, Matsushita, Mitsubishi, Time Warner, Toshiba, and JVC commenced worldwide joint licensing of patents essential for DVD-Video players, DVD-ROM drives, DVD decoders, and DVD-Video and DVD-ROM discs. Sony announced that it had developed a single-chip laser with dual wavelengths; 650 nm for reading DVDs and 780 nm for reading CDs. Oddly, Sony said that it intended to use this breakthrough technology only in their PlayStation 2.

The 4C group said that at its June 11 meeting it expected to pick a watermarking technology for its content-protection framework. The competing proposals were from the Galaxy group (Hitachi, IBM, NEC, Pioneer, and Sony) and the Millennium group (Macrovision, Digimarc, and Philips). June came and went with no decision. 1999 came and went with no decision. 2000 came and looked to end with no decision. Related efforts by the *Secure Digital Music Initiative* (SDMI) fared better, when at a June 23-25 meeting 100 companies from the music, consumer electronics, and information technology industries adopted a specification for portable devices for digital music.

On June 16, Divx announced its own obituary.

At the PC Expo, on June 22, Philips announced that its 3.0G DVD+RW drive would be out in September 1999 for $700. At the same conference a year earlier, Sony had made a similar announcement of DVD+RW availability in 1998.

About this time, 4.7G DVD-R drives finally began shipping to anxious developers, many of whom would have gladly paid the old $17,000 price rather than the new $5,000 price if it would have gotten the drives to them any sooner. The ability to quickly test full DVD-5 volumes on DVD-R was a critical step in the development of the DVD industry, where year-to-date sales of players had just surpassed 1 million.

Also in June, the *DVD Association* (DVDA) was formed. Many in the industry recognized the need for an association to support non-Hollywood DVD developers. The *Interactive Digital Media Association* (IDMA), long-time home to CD-i developers, stepped forward to host the new association.

On July 6, 1999, another obituary of sorts was reported by Pioneer Entertainment, which announced its completed transition out of the laserdisc business. After years of being the leading laserdisc supplier, Pioneer Entertainment had shifted to DVD and VHS only, which by then accounted for more than 90 percent of the company's title business. Although laserdisc players were still made by Pioneer New Media Technologies for education and industrial applications, the decision by Pioneer Corporation's entertainment group to abandon laserdiscs marked the end of an era.

Panasonic, hoping to create a new era of digital tape for consumers, announced again that it would begin to sell its HD-capable D-VHS VCR. NEC announced another potential competitor to DVD, its GigaStation video recorder based on *Multimedia Video Disc* (MVDisc), the latest incarnation of the *Multimedia Video File* (MMVF) technology it had announced more than a year before. The recorders were expected to appear in Japan in September, followed by worldwide release in 2000. Panasonic reiterated that its DVD-Audio players would ship in the US in October. As it turned out, all three of these long-delayed technologies had yet more postponements in store.

Hitachi, Sega, and Nippon Columbia announced that they had developed a conditional-access hybrid DVD that incorporated a programmable ROM chip to store information about what parts of the disc had been purchased. Sonopress, a leading disc manufacturer, announced a different kind of hybrid, christened DVD Plus, that combined DVD and CD layers in a single disc. American Airlines announced that in September it would become the first airline to offer DVD movies on scheduled flights.

Blockbusters and Logjams

At the end of August 1999, James Cameron's *Titanic* was finally released on DVD-9, in widescreen letterbox format. The long delay and non-anamorphic version dampened enthusiasm. The disc didn't do nearly as well as *The Matrix*, which came out three weeks later and quickly became the first DVD to sell a million copies.[11] *The Matrix* gained notoriety as buyers reported problems playing it on dozens of different player models. While there were a couple of errors on the disc itself, it was discovered that many players had not been properly engineered to handle a disc that aggressively exercised DVD features and included extra content for use on PCs. The success of *The Matrix* amplified the magnitude of the problem. Under pressure from disgruntled customers, manufacturers released firmware upgrades to correct the flaws in their players. In November, the first major Steven Spielberg movie, *Saving Private Ryan*, was released on DVD. Customers reported video problems on the disc, only to be told that the flaring, distorted video in the beach scene was an intentional effect in the film.

---

[11]It took two years for DVD to reach the first million-copy point for a title. Audio CD took four years with George Michael's *Faith*, and it took 11 years before a million VHS copies of *Top Gun* were shipped.

Toshiba shipped its progressive scan DVD player, which had been languishing in warehouses for a year. Panasonic also released a progressive scan player. The true potential of DVD video quality was finally being unlocked in places other than computers.

Digital Theater Systems moved more towards the mainstream when the company released DTS encoders in October. Previously, all DTS encoding had been directly done by the company. DVD took a giant step in mainstream consciousness when, on October 12, 1999, it received an Emmy award from the National Academy of Television Arts & Sciences. The award was presented to Dolby, Matsushita, Philips, Sony, Time Warner, and Toshiba.

DVD technology on PCs continued to improve with 10X drives. Another important landmark was the release of Stephen King's *The Stand*, the first commercial release on DVD-18, and on November 23, DVD officially became the hottest-selling home entertainment product in history. HDTV was not faring so well, with almost half the nation's 1,600 television stations supporting the Sinclair Broadcast Group's campaign to revise the broadcast standard. Still, factory-to-dealer sales of digital televisions continued to grow, reaching a year-to-date total of more than 97,000.

At the last minute, Sony, Philips, and HP cancelled their planned DVD+RW launch. The DVD+RW camp had decided to retrench, abandoning the 3.0G format, which would have been incompatible with every existing DVD-ROM drive and player, and focusing on the improved 4.7G version that promised backward compatibility with most DVD readers.

Pioneer announced that it would release a DVD-Audio player in Japan without content protection, since that was the only part that was unresolved. The player would only be able to play unprotected DVD-Audio discs until it was updated with final content protection support. This turned out to be a reasonable compromise because decisions on DVD-Audio encryption and watermarking were almost a year away.

Crackers

In November 1999 the SDMI approved selection of Verance's watermarking technology for the portable music device standard. Former competing companies Aris Technologies and Solana Technology Development had merged to form Verance, whose watermarking technology also would be later chosen for DVD-Audio.

In November and December, fallout began from a Windows software program called DeCSS that had spread across the Internet in late October. The program was designed to remove CSS encryption from discs and to copy the video files to a hard drive. DeCSS was written by 16-year-old Jon Johansen of Norway, based on code created by a German programmer who was a member of an anonymous group called *Masters of Reverse Engineering* (MoRE). The MoRE programmers reverse-engineered the CSS algorithm and discovered that the Xing software DVD player had not encrypted the key it used to unlock protected DVDs. Given the general weakness of the CSS design, additional keys were quickly generated by computer programs that guessed at values and tried them until they worked. Johansen claimed his intent in turning the MoRE code into an application was to be able to play movies using the Linux operating system, which had no licensed CSS implementation.

Anyone familiar with CSS was surprised that it had taken so long for the system to be cracked. After all, the first edition of this book had predicted three years earlier that CSS would be compromised. In spite of this, frenetic press reports portrayed a "shocked" movie industry, taken aback by the failure of the system that was supposed to protect its assets. DVD-Audio player manufacturers announced a six-month delay, presumably to counter the threat of DeCSS by reworking the planned CSS2 copy protection system. Those familiar with the lack of DVD-Audio titles and the incompleteness of CSS2 saw the announcement as a convenient excuse to delay products that would not have been ready in any case.

Other DVD "ripping" software had been available for years, but DeCSS was different because it directly decrypted CSS rather than intercepting video after it was decrypted and decompressed by a legitimate software DVD player. This direct implementation of the CSS protection method could be considered illegal circumvention under the DMCA and the WIPO treaties.

Rumors circulated that the key used by the Xing player had been revoked by removing it from the set of keys hidden on new discs, but the gossip proved to be false. Causing a legal player to stop working, even though it had not properly protected the CSS secrets, would not have been a good move, and by this time the entire set of 400 player keys had been guessed at, so revoking one key would have done little good.

On December 27, the *DVD Copy Control Association* (DVD CCA), the corporate entity responsible for licensing CSS, sued 21 individuals and 72 Websites, along with hundreds of "Does" — as in John Doe — to be filled in later. They were accused of posting or linking to DeCSS software as a misappropriated trade secret, a rather shaky argument. The *Electronic Frontier Foundation* (EFF) responded on behalf of the defendants with a shaky argument of their own, tied to free speech. "It is EFF's opinion that this lawsuit is an attempt to architect law to favor a particular business model at the expense of free expression. It is an affront to the First Amendment (and UN human rights accords) because the information the programmers posted is legal. EFF also objects to the DVD CCA's attempt to blur the distinction between posting material on one's own Website and merely linking to it (i.e., providing directions to it) elsewhere." On December 29, the California court enjoined the posting of DeCSS but denied the injunction request against linking to DeCSS, as the court rightly feared the side effects of banning the mere act of linking to information on the Internet, citing such an order as "overbroad and extremely burdensome."

Thus ended the year of DVD. Over 4 million new players had been sold in the US, bringing the installed base to around 5.5 million. Player sales had generated more than $300 million. About 2.5 million players operated in European homes, and the worldwide base of DVD PCs was estimated at somewhere around 40 million. Fifty million discs shipped in the US during the Christmas season alone. Total US interactive entertainment sales for the year topped $7 billion.

The Medium of the New Millennium

The DeCSS saga continued. On January 14, 2000, the seven top US movie studios — Disney, MGM, Paramount, Sony (Columbia/TriStar), Time Warner, Twentieth Century Fox, and Universal — backed by the MPAA, filed lawsuits in Connecticut and New York in a further attempt to stop the distribution of DeCSS on Websites in those states. These suits were based on the DMCA and alleged circumvention of DeCSS. On January 24, Jon Johansen, the Norwegian programmer who first distributed DeCSS, was questioned by local police, who raided his house and confiscated his computer equipment and cell phone.

This strengthened the viewpoint of many observers that Hollywood was twisting the law to intimidate the opposition in a losing battle. By this time the DeCSS source code was available on hundreds of Websites and on thousands of T-shirts, and had even been made publicly available by the DVD CCA itself in court records.

Around this time a new wrinkle appeared, under a confusingly familiar name. A major drawback in attempting to copy DVDs was that they quickly filled even huge hard drives and took literally a week to download using a 56K modem. Copies could be made on DVD-Rs, but blank discs cost more than the original DVD. As a result, a new DVD redistribution technology called DivX ;-) appeared. (Yes, the smiley face is part of the name.) DivX ;-) was a simple hack of Microsoft's MPEG-4 video codec coupled with MP3 audio, allowing DeCSSed video to be re-encoded at a lower data rate (and lower quality) so that it could be downloaded more easily and played using the Windows Media Player.

In another ironic twist, Sigma Designs, the leading producer of DVD playback add-in cards, announced at the beginning of February that it would add Linux support to its NetStream 2000 DVD playback card.

At the CES show in January, most DVD player manufacturers either showed or talked about progressive scan players, even though the number of potential customers with the necessary progressive scan displays was miniscule. DVD-Audio players were also on display, but few company representatives were willing to guess when they might finally be for sale.

On February 14, Jack Valenti, then the head of the MPAA, speaking of DeCSS and other threats to Hollywood's intellectual property, referred to the Internet, "where some obscure person sitting in a basement can throw up on the Internet a brand new motion picture, and with the click of a button have it go with the speed of light to 6 billion people around the world, instantaneously." This led people to wonder where they could get these new ISL modems[12] that 6 billion other people already had. This Chicken Little attitude, shared by so many in the motion picture industry, painted digital video as a dangerous new technology that opened the floodgates of piracy. What they seemed unable or unwilling to recognize was that any analog source such as laserdisc or even VHS tape can be digitized. Once digitized, the file can be copied and distributed the same as any other digital video file. And, given the loss of image quality caused by compression methods such as Divx ;-), there is no discernable difference.

Region coding, part of the CSS license, came under attack in London when British supermarket group Tesco wrote to Warner Home Video demanding an end to the policy.

---

[12]ISL = instantaneous speed of light, but, of course.

That's No Moon, That's a PlayStation!

George Lucas, under pressure from an increasingly vocal campaign to have *Star Wars* released on DVD, wrote a letter to fans on February 20 explaining why *Star Wars* would not be available on DVD any time soon. Essentially his excuse was that it had to be done right, which meant he had to do it, but that he was too busy at the moment.

In March, things improved slightly on the compatibility front. Toshiba announced a new combination CD-RW/DVD-ROM drive, and the Optical Storage Technology Association heralded the release of MultiRead 2, a specification requiring DVD units to read all CD formats as well as DVD-RAM discs.

Sony's PlayStation 2, which shipped more than 1 million units in its first 12 days, embarrassed the company with a flaw that allowed users to get around DVD region coding restrictions. Since a memory card glitch had also been discovered, Sony said, "We are asking buyers to return memory cards or consoles for checks and repairs while at the same time investigating the reasons for the glitches." Most owners declined to send their cards in to have the region code trick disabled. A week later yet another loophole was discovered: that the game console's analog RGB output could be used to get around Macrovision protection in order to copy DVDs to videotape. These glitches and loopholes, however, were insignificant compared to the overall success of PlayStation 2, which in less than a month doubled the number of DVD players in Japan. DVD industry analysts worldwide quickly updated their forecasts to account for the expected impact of this new kind of DVD player.

Throughout March, the MPAA kept busy, sending cease and desist letters to any Websites it found posting or linking to DeCSS.

In April 2000, Sonic Solution's DVDit finally made its debut as a $500 retail product for Windows. Two years late, and quite different from the original conception, it nevertheless heralded a new generation of DVD authoring tools that were affordable to just about anyone interested in doing professional DVD production.

JVC announced a consumer version of D-VHS. The digital tape format, originally designed only to record data, had been reworked with a standard way of recording and playing back digital video, along with the now requisite content protection. JVC had high hopes that movies would be released on prerecorded tapes. In a surprise move, Apple bought Astarte, an up-and-coming developer of DVD authoring software for Macintosh computers. Sonic, the main developer of Macintosh DVD authoring systems, announced hDVD, an unofficial variation of DVD-Video that incorporated high-definition video. The target market was high-end computers with fast DVD-ROM drives and powerful processors to decode the HD video. Ravisent, Sonic's partner in the initiative, was providing the software decoder.

In May, Circuit City recalled the Apex DVD player, one of its best-selling items. The recall and the strong sales had the same cause — loopholes in the player that allowed it to be easily modified to play discs from any region and to disable Macrovision content protection. The recall occurred primarily because of pressure from Macrovision.

Also in May, Sony announced plans for a third release of their SACD player models, dropping the price from over $3,000 to just over $700, and shifting focus from audiophiles to mainstream audio buyers. Philips moved in the opposite direction, releasing a $7,500 SACD player. Fewer than a hundred SACD titles were available by this time. Pioneer announced that its DVD video recorder had sold 25,000 units in the seven months since its release in Japan.

Macrovision was back in the news in June with the announcement that it had implemented content protection technology for 525-line progressive scan output. Constellation 3D, the company that had been busy for months issuing press releases about its *fluorescent multilayer disc* (FMD) technology, announced that it had adapted the design to be readable with standard red laser technology. This raised the interesting possibility that DVD players with minor modifications would be able to read 25 gigabyte FMDs. On the other hand, C3D, a relative newcomer to the field, did not explain how discs with six or ten or more layers could be reliably mass-produced, given how hard it had been just to get two-layer DVD-9s to work. Nevertheless, C3D confidently predicted that the new players would be available by the summer of 2001.

On June 12, Hollywood Entertainment announced that it was closing the e-commerce portion of its Website, Reel.com. The company had experienced losses in 1999, and Reel.com was blamed for much of its continuing losses in 2000. Given the cutthroat discounting on Internet DVD sites, this was no surprise.

Good news on compatibility came on June 27, 2000, when the DVD Forum announced a plan to deal with conflicts among recordable formats. A new specification called DVD Multi defined requirements of physical compatibility between all the Forum-sanctioned formats. Any player with a DVD Multi logo would read all discs, including DVD-RAM and DVD-RW, and any recorder with the DVD Multi logo would write to all formats.

In July 2000, more than three years after the introduction of DVD-Video and DVD-ROM in the US, DVD-Audio players belatedly shipped. It was not what anyone would call a grand coming out. There was no marketing or PR hoopla and the players were in short supply and hard to find. Verance watermarking technology, the same system chosen by the SDMI, was included. Meanwhile, the *Watermarking Review Panel* (WaRP), the successor to the *Data-Hiding Sub-Group* (DHSG) of the *Copy Protection Technical Working Group* (CPTWG), had still not chosen between Millennium and Galaxy watermarking proposals for DVD-Video. DVD-Audio production equipment and watermarking equipment was still so scarce that music studios did not expect to release more than a few titles for the Christmas 2000 season. Sonic Solutions, the only company working on a commercial DVD-Audio authoring system, had only pre-release versions of the software available for use.

Tony Faulkner, an audiophile's audiophile, had been leading a grassroots effort calling for open listening tests to prove that watermarking would not destroy the high-quality audio that the format had been so carefully designed to preserve. Invoking the specter of copycode, a disastrous earlier attempt to hide copyright data in audio streams, he pointed out that results of private tests on the new watermarking methods had not been released, and that almost no testing had been done by members of the audio production community. Verance itself admitted that it had done no testing on 192 KHz content. Soon after, watermarking "tests" for DVD-Audio were conducted in London, but since the decision had already been made to use Verance technology, it was more of a demo effort to woo the audio community. Participants complained that the samples were poorly chosen and the listening environment was terrible. Faulkner, who participated in the tests, claimed that he had been able to correctly identify the watermarked streams in six of eight cases. He worried that if he was able to detect watermarked samples with only 2-bit copy management payload, what would happen if a full 72-bit payload were used? The poor selection of "sub-Walkman standard" test pieces — analog recordings with tape hiss and "dreadfully recorded pop" — and the use of loudly whirring computer hard drives to play the music in the listening environment was alarming to many, who wondered if any of the engineers working on watermarking techniques truly understood high-fidelity audio. A few music industry executives were quick to point out that watermarking was optional.

On July 5, Sony clouded the optical disc waters further with the announcement of a new format that was halfway between CD and DVD. *Double-density CD* (DDCD) versions of CD, CD-R, and CD-RW reduced track pitch and pit length to increase capacity from 650 million bytes to 1.3 billion bytes. The CIRC error correction and addressing information (ATIP) were tweaked slightly to accommodate the higher density of data, and a content protection scheme was added. The new specification, "Purple Book," was supposed to be finalized by September 2000. The new DDCD discs were not compatible with existing players.

The convergence process took another step forward when EchoStar Communications demonstrated the first satellite television receiver combined with a DVD player. The entire system, complete with satellite dish and receiver/DVD player, was priced at only $400.

Pioneer released software to upgrade version 1.9 DVD-R drives to version 2.0. New version 2.0 discs, with CPRM copy-protection features, could only be written in 2.0 drives. Older 1.9 and 1.0 discs could still be written in 2.0 drives, although 1.9 media was no longer being produced.

On August 1, 2000, Warner Home Video announced that *The Matrix* DVD was the first disc to sell over three million copies in the US.

On August 17, Judge Kaplan, presiding over the DeCSS suit in New York, granted the requested injunction against the Website maintained by 2600: The Hacker Quarterly. 2600 had long since removed the DeCSS code, but it maintained that it had a right to link to Websites containing DeCSS or information about DeCSS, thus the suit. The district court granted a permanent injunction against (1) posting on any Internet site, or in any other way manufacturing, importing or offering to the public, providing, or otherwise trafficking in DeCSS or any other technology primarily designed to circumvent CSS, and (2) linking any Internet website, either directly or through a series of links, to any other Internet website containing DeCSS. In response to the injunction, 2600 revised its list of some 450 Websites so that the URLs were listed without embedded links. Users interested in going to any of the sites had to copy and paste the URLs into their Web browser. This was the sole tangible outcome of the MPAA's suit. In an inspired bit of irony, the 2600 Website also directed visitors to the Web search engine at Disney's Go.com, which provided hundreds of links to sites such as the DeCSS Distribution Center. In theory, all Internet search engines were in violation of DMCA as long as a single copy of DeCSS existed, even in countries where it was not illegal. Following the favorable outcome of the suit, the MPAA sent new rounds of threatening letters to Internet sites posting or linking to DeCSS. "If we have to file a thousand lawsuits a day, we'll do it," said Valenti. "It's less expensive than losing control of all your creative works."

DVD Turns Five

On October 29, 2001, the DVD format officially turned 5 years old. During this one year, over 12.7 million DVD-Video players shipped to form an installed base of more than 26 million players in the US alone, with over 14,000 titles to chose from.[13] There were more than 45 million DVD-ROM drives in the US, with approximately 90 million installed worldwide. And, for the first time, DVD player sales exceeded VCR sales. Not bad for a format with only five candles on its cake!

Even as DVD was reaching these milestones, changes were still happening in the industry. In early 2001, Hitachi and Panasonic released camcorders that use small DVD-RAM discs and offer random access and on-board editing of the video. Pioneer finally abandoned even small runs of laserdiscs in Japan (having halted US production in 1999) in order to focus its efforts on DVD. Sonic Solutions acquired Daikin US, Comtec's professional DVD authoring tool business, to augment its growing empire of DVD related hardware and software tools, and on February 7, 2001, NASA sent two multiregion DVD players to the International Space Station (amongst wild rumors that Antarctica and outer space would be considered for the eighth and final region code).

In March, Apple made a huge dent in consumer DVD production by releasing iDVD and DVD Studio Pro for the Macintosh, with support for the newly released Pioneer DVR-103 DVD-R(G) recorder. Later that month, Apple took a giant step backward by releasing OS X 10.0 (Cheetah), which was incompatible with iDVD, DVD Studio Pro, and their own DVD Player software.[14] A few months later, DVD-RW media made its US debut, which could be used with the DVR-103 DVD-R(G) drives as well as the generally released DVR-A03 (the same drive but sold separately).

On July 7th, Divx finally passed quietly into the void as the last of the active players were decommissioned when they signed in for the last time. They would continue to play standard DVD discs, but not Divx DVDs.[15]

Also in July, Apple re-entered the picture with a surprise move to acquire Spruce Technologies, manufacturer of the popular DVDMaestro authoring system. The engineering team and code of DVDMaestro formed the basis of what would be Apple's next release of DVD Studio Pro, one of the least expensive and most functional DVD authoring tools available to consumers and "pro-sumers" alike. Unfortunately, this left many users of the professional DVDMaestro authoring tool wondering where they would find future support for their product, as Apple had no plans to continue development of this Windows-platform software.

While the end of 2001 celebrated DVD's fifth birthday with the special limited edition release of Disney's seminal classic *Snow White and the Seven Dwarfs*, it also sounded the beginning of the end for several other formats and platforms. By the end of the year, just under 200 DVD-Audio titles had made it out onto the market, a sign of the slow start that would stick with the format as it limped through the next several years. Likewise, the D-

---

[13]This figure does not include adult video titles, which account for an additional 15%.

[14]This was later corrected with OS X version 10.1 (Puma).

[15]The Divx technology would appear again years later, however, in a slightly different form intended to protect limited release DVDs, such as screeners for the Academy Awards.

Theater format was released, which provided copy-protected high-definition movies at excellent quality on D-VHS tape. However, after a limited test run by Universal and others, the format never took hold. Finally, VM Labs, creators of the long awaited Nuon platform, filed for Chapter 11 backruptcy. A few months later, Genesis Microchip purchased the VM Labs assets and started the Nuon Semiconductor division to market Nuon chips under the name *Aries*. The following July, a mere four months later, Genesis laid off the entire Nuon division.

The next several years saw DVD flourish as sales of DVD players in the US continued to grow to form an installed base of over 73 million players by the fall of 2003. By that time, more than 27,000 DVD-Video titles were available, of which over 1.5 billion copies had been shipped. Within a year, 40,000+ DVD-Video titles were available as the industry topped $22 billion. (That's right, "billion" is the one with *nine* zeros after it.) By this point, home video releases on DVD began to overtake theatrical revenue for the same titles, marking the start of a new paradigm in the content community.

Meanwhile, in a small corner of the world, tensions were brewing — a new war was starting, a format war.

Format Wars — The Next Generation

Imagine, if you will, a meglomaniacal playwright who has taken the first parts of this book, rearranged the characters, twisted the plot, and is now putting the concoction into motion. "Dr. Alan Bell, you're now with Warner Brothers. Panasonic, you're paired up with your archenemy, Sony. Philips, you stay right where you are. Thomson...where's Thomson? Oh, there you are. You get to play both sides this time; you're not allowed to take either side...though at various times you will take one...and then the other...well, you know what I mean — just play the roll as a schizophrenic. Okay, places everyone! And...ACTION!"

In February 2002, the nine companies forming the Blu-ray Founders Group, led by Sony, Philips and Matsushita, announced the Blu-ray Disc format, their new high density recordable disc format with up to 50 GB capacity based on solid state blue laser technology. Seeing a storm on its horizon, the next month the DVD Forum (which includes the nine companies from Blu-ray) announced that it would investigate next-generation standards for high capacity, blue laser discs. In addition, the DVD Forum announced, at the request of Warner Bros., it would begin to investigate putting high definition video content on existing DVDs (which later became known as the HD DVD-9 format). Many in the industry predicted that Blu-ray would soon be adopted by the DVD Forum and there would be a smooth road to the next-generation of optical disc formats. Insiders quickly realized, however, that the beginning of a politically charged conflict was being staged that would create new friends and new enemies in a fight for the multi-billion dollar legacy from DVD.

By March 2003, at least five candidates vied to be anointed as the new high-definition DVD format — HD DVD-9, Blu-ray Disc, Advanced Optical Disc (AOD, jointly developed by Toshiba and NEC), Blue-HD-DVD-1 and Blue-HD-DVD-2 from the Advanced Optical Storage Research Alliance (AOSRA), an industry group in Taiwan. It quickly became clear that political lines had been drawn between the Blu-ray Disc Forum (BDF) members and the Toshiba/NEC team. In June and September, the AOD format was twice voted down in meetings of the DVD Forum Steering Committee, with Blu-ray supporters mostly voting against or abstaining (which had the effect of a "no" vote). Only after the voting rules were changed (possibly due to anti-trust allegations) was AOD finally approved by the DVD Forum Steering Committee in November 2003 and later renamed HD DVD.

Also that month, the Chinese government announced that *enhanced versatile disc* (EVD) would be launched by Christmas 2003. EVD was created as an alternative to DVD in part to decrease the flow of licensing royalties to Japan and other DVD patent holders. EVD uses its own optical disc format and originally planned on using proprietary video compression technologies (VP5 and VP6, developed by On2 in the US) but instead used HD MPEG-2. The EVD format supports HD video resolutions, including 1280×720 at 60 fps progressive (720p) and 1920×1080 at 30 fps interlaced (1080i). In April 2004, another group led by Amlogic and several major Chinese television manufacturers, announced development of the HVD format, and in May the DVD Forum began investigating a proposal from the Industrial Research Technology Institute (ITRI) of Taiwan for the FVD (forward versatile disc) format, intended specifically for the low-cost Chinese market.

By Spring 2005, as a result of strong industry pressure, talks had started between Toshiba, Sony and Matsushita in an attempt to reach agreement on format unification. Within weeks, however, it was reported that the talks had completely broken down with neither side willing to compromise. Toshiba and Memory Tech, an HD DVD supporter, announced a new triple-layer disc that would increase HD DVD's capacity from 30 Gbytes to 45 Gbytes, making it a viable alternative to Blu-ray's 50 Gbyte dual-layer disc. In a somewhat glib response, Blu-ray promoters announced a theoretical four-layer version that could hold up to 100 Gbytes (though many doubt the commercial feasibility of such a disc). Sony followed up with an announcement that their upcoming PlayStation® 3 game console would host a BD-ROM drive and would be able to play BD movies.

Those hoping for a single next-generation format are watching the situation go from bad to worse. With a format war raging in China between EVD, HVD and FVD (though not involving the DVD Forum), the desired convergence between HD DVD and BD is dead-ended with the two camps spinning off in separate directions. Multiple formats, all designed to perform essentially the same functions in nearly identical ways, and no one is willing to put the conflict to rest and work to merge the best parts of each into one.

Conspicuously, Sony launched the PlayStation® Portable (PSP™) and its new *universal media disc* (UMD™) format, which is capable of playing games, music and movies. Taking Hollywood by storm, UMD just might show us what a successful format should look like.

Chapter 3
Technology Primer

This chapter explains key technologies developed for or used by DVD, and what may be expected with the next-generation optical disc formats — UMD™, HD DVD and BD. Many of the fundamental technologies in use with DVD are being used by the upcoming disc formats, but there are a few key differences.

Gauges and Grids: Understanding Digital and Analog

We live in an analog world. Our perceptions are stimulated by information that is received in smooth, unbroken form, such as sound waves that apply varying pressure on our eardrums, a mercury thermometer showing infinitely measurable detail, or a speedometer dial that moves reflexively across its range. Digital information, on the other hand, is a series of snapshots depicting analog values coded as numbers, such as a digital thermometer that reads 71.5 degrees or a digital speedometer that reads 69 mph.[1]

The first recording techniques used analog methods — changes in physical material such as wavy grooves in plastic discs, silver halide crystals on film, or magnetic oxides on tape. After transistors and computers appeared, it was discovered that information signals could be isolated from their carriers if stored in digital form. One of the major advantages of digital information is that it is infinitely malleable. It can be processed, transformed, and copied without losing a single bit of information.

Analog recordings always contain noise (such as tape hiss) and random perturbations, so each successive generation of recoding or transmission decreases in quality. Digital information can pass through multiple generations, such as from a digital video master, through a studio network, over the Internet, into a computer bus, out to a recordable DVD, back into the computer, through the computer graphics chips, out over a FireWire connection, and into a digital monitor, all with no loss of quality. Digital signals representing audio and video can also be numerically processed. Digital signal processing is what allows AV receivers to simulate concert halls, surround sound headphones to simulate multiple speakers, and studio equipment to enhance video or correct colors.

---

[1]There are endless debates about whether the "true" nature of our world is analog or digital. Consider again the thermometer. At a minute level of detail, the readings can't be more accurate than a molecule of mercury. Physicists explain that the sound waves and photons that excite receptors in our ears and eyes can be treated as waves or as particles. Waves are analog, but particles are digital. Research shows that we perceive sound and video in discrete steps, which means our internal perception is actually a digital representation of the analog world around us. There are a finite number of cones in the retina, similar to the limited number of photoreceptors in the CCD of a digital camera. At the quantum level, all of reality is determined by discrete energy states that can be thought of as digital values. However, for the purposes of this discussion, referring to gross human perception, it's sufficiently accurate to say that sound and light, and our sensation of them, are analog.

When storing analog information in digital form, the trick is to produce a representation that is very close to the original. If the numbers are exact enough (such as a thermometer reading of 71.4329 degrees) and repeated often enough, they closely represent the original analog information.[2]

Digital video is a sheet of dots, called *pixels*, each holding a color value. It's similar to drawing a picture by coloring in a grid, where each square of the grid can only be filled in with a single color. If the squares are small enough and there is sufficient range of colors, the drawing becomes a reasonable facsimile of reality. For DVD, each grid of 720 squares across by 480 or 576 squares down represents a still image, called a *frame*. Thirty frames are shown each second to convey motion. (For PAL DVDs, 25 frames are shown each second.)

Digital audio is a series of numbers representing the intensity, or amplitude, of a sound wave at a given point. For DVD, these numbers are "sampled" over 48,000 times a second (sampling may be performed as frequently as 192,000 times a second for super-high–fidelity audio), providing a much more accurate recording than is possible with the rough analogues (pun intended) of vinyl records or magnetic tape. When a digital audio recording is played back, the stream of numerical values is converted into a series of voltage levels, creating an undulating electrical signal that drives a speaker.

Pits and Marks and Error Correction

Data is stored on optical discs in the form of microscopic *pits* (see Figure 3.1). The space between two pits is called a *land*. On writable discs, pits and lands are often referred to as *marks* and *spaces*. Read-only discs are stamped in a molding machine from a liquid plastic such as polycarbonate or acrylic and then coated with a reflective metallic layer. Writable discs are made of material designed to be physically changed by the heat of a laser, creating marks. As the disc spins, the pits (or marks) pass under a reading laser beam and are detected according to the change they cause in the intensity of the beam. These changes happen very fast (over 300,000 times per second) and create a stream of transitions spaced at varying intervals: an encoded digital signal.

Figure 3.1 Optical Disc Pits

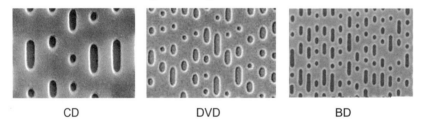

| CD | DVD | BD |

---

[2]Ironically, digital data is stored on analog media. The pits and lands on a DVD are not of a uniform depth and length, and they don't directly represent ones and zeros. They produce a waveform of reflected laser light that represents coded runs of zeros and transition points. Digital tape recordings use the same magnetic recording medium as analog tapes. Digital connections between AV components (digital audio cables, IEEE-1394/Firewire, etc.) encode data as square waves at analog voltage levels. However, in all cases, the digital signal threshold is kept far above the noise level of the analog medium so that variations don't cause errors when the data is retrieved.

NOTE

At current 16x DVD recording speeds, a disc makes 10,800 rotations per minute, corresponding to a linear velocity of 56 meters per second (over 200 km/h), while marks are burned with a precision of less than 0.05 micrometer!

Many people assume that the digital ones and zeros that comprise the data stored on the disc are encoded directly as pits and lands, but the reality is actually much more complicated. Pits and lands both represent strings of zeros of varying lengths, and each transition between them represents a one, but neither is a direct representation of the contents of the disc. Almost half of the information has been used to pad and rearrange (*modulate*) the data in sequences and patterns designed to be accurately readable as a string of pulses. Modulation makes sure strings of zeros aren't too long (no more than 10) or too short (no fewer than two) and that there is only a single one between them (since ones are represented by "edges" and it's not physically possible to have two edges together.)

About 13 percent of the digital signal before modulation is extra information for correcting errors. Errors can occur for many reasons, such as imperfections on the disc, dust, scratches, a dirty lens, and so on. A human hair is about as wide as 150 pits, so even a speck of dust or a minute air bubble can cover a large number of pits. However, the laser beam focuses past the surface of the disc so the spot size at the surface is much larger and is hardly affected by anything smaller than a few millimeters. This is similar to the way dust on a camera lens is not visible in the photographs because the dust is out of focus. As the data is read from the disc, the error correction information is separated and checked against the remaining information. If it doesn't match, the error correction codes are used to try to correct the error.

The error correction process is like a number square, where you add up columns and rows of numbers (see Figure 3.2). You could play a game with these squares where a friend randomly changes a number and challenges you to find and correct it. If the friend gives you the sums along with the numbers, you can add up the rows and columns and compare your totals against the originals. If they don't match, then you know that something is wrong — either a number has been changed or the sum has been changed.[3] If a number has been changed,

Figure 3.2 Number Squares

Original data	Sums calculated	Error (4 changed to 1), data and sums transmitted	New sums calculated, mismatches found	Corrected by adding difference (14 - 11 or 8 - 5)

3	5	8
4	1	3
7	7	2

3	5	8	16
4	1	3	8
7	7	2	16
14	13	13	40

3	5	8	16
(1)	1	3	8
7	7	2	16
14	13	13	40

3	5	8	16	16
1	1	3	8	5
7	7	2	16	16
14	13	13	40	37
11	13	13	37	

3	5	8
(4)	1	3
7	7	2

---

[3]It's possible for more than one number to be changed in such a way that the sum still comes out correct, but the DVD encoding format makes this an extremely rare occurrence.

then a corresponding sum in the other direction also will be wrong. The intersection of the incorrect row and incorrect column pinpoints the guilty number, and in fact, by knowing what the sums are supposed to be, the original number can be restored. The error correction scheme used by DVD is a bit more complicated than this but operates on the same general principle.

It's always possible that so much of the data is corrupted that error correction fails. In this case, the player must try reading the section of the disc again. In the very worst cases, such as an extremely damaged disc, the player will be unable to read the data correctly after multiple attempts. At this point, a movie player will continue on to the next section of the disc, causing a brief glitch in playback. A DVD-ROM drive, on the other hand, can't do this. Computers will not tolerate missing or incorrect data, so the DVD-ROM drive must signal the computer that an error has occurred so that the computer can request that the drive either try again or give up.

Layers

One of the clever innovations of DVD is to use layers to increase storage capacity. The laser that reads the disc can focus at two different levels so that it can look through the first layer to read the layer beneath. The outside layer is coated with a semireflective material that enables the laser to read through it when focused on the inner (bottom) layer. When the player reads a disc, it starts at the inside edge and moves toward the outer edge, following a spiral path. If unwound, this path would stretch 11.8 kilometers (7.3 miles), three times around the Indianapolis 500 Speedway. When the laser reaches the end of the first layer, it quickly refocuses onto the second layer. At that point, depending on the construction of the disc, it may start reading in the opposite direction, known as *opposite track path* (OTP) — from the outer edge toward the inner. Alternatively, the laser may follow a *parallel track path* (PTP) by jumping back to the inner edge and starting to read outward again, just as it did for the first layer. This, however, requires extra time for the laser pickup to move across the surface of the disc to pick up the new start point. Refocusing happens very quickly, but on most players the video and audio pause for a fraction of a second even with opposite track path configurations as the player searches for the resumption point on the second layer. If the player has a large enough buffer, or if the disc is carefully designed to lower the data rate at the layer switch point (so that the buffer will take longer to empty), the laser pickup has time to refocus and retrack without causing a visible break.

The DVD standard does not actually require compatibility with existing CDs, but manufacturers recognize the vital importance of backward compatibility. If the hardware were unable to read CDs, DVD wouldn't have a snowball's chance in Hollywood of surviving. The difficult part is that the pits on a CD are at a different level than those on a DVD (see Figure 3.3). In essence, a DVD player must be able to focus a laser at three different distances. This problem has various solutions, including using lenses that switch in and out, and holographic lenses that are actually focused at more than one distance simultaneously. An additional difficulty is that CD-R discs don't properly reflect the 635 to 650 nanometer wavelength laser required for DVD, so DVD players and DVD-ROM drives intended to read recordable CDs must include a second 780nanometer laser.

Figure 3.3 Optical Disc Layers

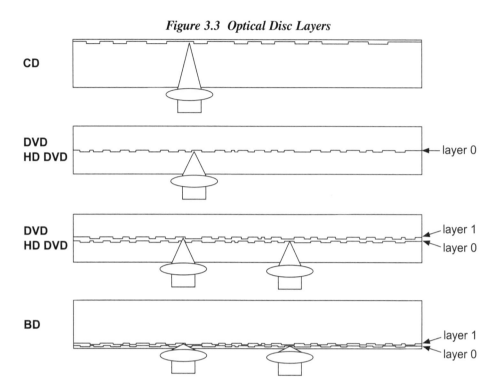

The remaining task to ensure CD compatibility requires an extra bit of circuitry and firmware for reading CD-format data. However, the CD family is quite large and includes some odd characters, not all of which fit well with DVD. The prominent members of the CD family are audio CD, Enhanced CD (or CD Plus), CD-ROM, CD-R, CD-RW, CD-i, Photo CD, CDV, and Video CD. It would be technically possible to support all these, and most of them require specialized hardware. Therefore, most manufacturers choose to support only the most common or easiest-to-support versions. Some, such as Enhanced CD and Video CD, are easy to support with existing hardware. Others, such as CD-i and Photo CD, require additional hardware and interfaces, so they are not commonly supported. Given that the data on a CD can be read by any DVD system, conceivably any CD format could be supported. DVD-ROM computers support more CD formats than DVD players partly because some are designed for computer applications and also because specialized CD systems can be simulated with computer software.

Two HDs: High Definition and High Density

With the upcoming disc formats, HD DVD and Blu-ray Disc, the acronym "HD" has a different meaning than one might expect. Although it could be shorthand for "high definition" because the disc can contain high definition video, the acronym actually means *high density* when refering to the physical disc. Both formats use the same basic technology as DVD and

CD, but one of the key differences is that the pits and lands are recorded at a much higher density. As a result, a shorter wavelength blue laser must be used to read the disc, just as DVD required a shorter wavelength (635-650 nm) laser relative to CD (780 nm).

Backward compatibility with DVD and CD will be a *de facto* requirement for HD DVD and BD, just as DVD players needed to support CD formats to succeed in the marketplace. This results in significant challenges for the player manufacturers, who must design pick-up heads that contain three lasers, one for each wavelength, all aiming down the same emission path while using the same lens. For Blu-ray Disc, this will be particularly challenging because BD uses a different type of lens than its fellow travellers. Instead of the 0.65 NA (*numeric aperture*) lens that CD and DVD both use, Blu-ray hosts a 0.85 NA lens, which corresponds to a much shorter focal length. As a result, the pickup head and drive mechanism have to be much more sophisticated in order to adjust for the difference in focal length between BD and the other disc formats.

Another effect of the Blu-ray 0.85 NA lens is that it results in a much smaller spot on the surface of the disc. As a result, dust, hair, fingerprints and scratches all have a much greater negative impact on the player's ability to read the disc. The size of a particle or a scratch is considerably larger compared to the laser spot diameter of a BD player than that of a DVD or CD. As a result, there is far less of the defocusing effect described earlier and the resulting signal is much more degraded.

The BD format has taken two steps to address this issue. First, disc caddies (BD-RE) and special hard surface coats (BD-ROM) have been adopted to help protect the surface of the discs from fingerprints, scratches, or anything that may threaten the readability of the disc. Second, BD uses much stronger error correction, allowing it to more easily detect and correct errors that may have occured while reading the disc. One interesting technique used was to almost double the size of the error control block such that a read error from a hair or finerprint has much less significance relative to the size of the error control block. In essence, the larger error control block in BD offers essentially the same increase in resiliency as the defocusing effect for DVD and CD.

Birds Over the Phone: Understanding Video Compression

After compact discs (CDs) appeared in 1982, digital audio became a commodity. It took many years before the same transformation began to work its magic on video. The step up from digital audio to digital video is a doozy, for in any segment of television there is about 250 times as much information as in the same-length segment of CD audio. Despite its larger capacity, DVD is not even close to 250 times more spacious than CD-ROM. The trick is to reduce the amount of video information without significantly reducing the quality of the picture. As we have seen, the solution is digital compression.

In a sense, you employ compression in daily conversations. Picture yourself talking on the phone to a friend. You are describing the antics of a particularly striking bird outside your window. You might begin by depicting the scene and then mentioning the size, shape, and color of the bird. But when you begin to describe the bird's actions, you naturally don't repeat your description of the background scene or the bird. You take it for granted that your friend remembers this information, so you only describe the action — the part that changes. If you had to continually refresh your friend's memory of every detail, you would have very high phone bills. The problem with TV is that it has no memory — the picture has to be continually refreshed. It's as if the TV were saying, "There's a patch of grass and a small tree with a 4-inch green and black bird with a yellow beak sitting on a branch. Now there's a patch of grass and a small tree with a 4-inch green and black bird with a yellow beak hanging upside down on a branch. Now there's a patch of grass and a small tree with a 4-inch green and black bird with a yellow beak hanging upside down on a branch trying to eat some fruit," and so on, only in much more detail, redescribing the scene 30 times a second. In addition, a TV individually describes each piece of the picture even when they are all the same. It would be as if you had to say, "The bird has a black breast and a green head and a green back and green wing feathers and green tail feathers and..." (again, in much more meticulous detail) rather than simply saying, "The bird has a black breast, and the rest is green." This kind of conversational compression is second nature to us, but for computers to do the same thing requires complex algorithms. Coding only the changes in a scene is called *conditional replenishment*.

The simplest form of digital video compression takes advantage of *spatial redundancy* — areas of a single picture that are the same. Computer pictures are comprised of a grid of dots, each one a specified color, but many of them are the same color. Therefore, rather than storing, say, a hundred red dots, you store one red dot and a count of 100. This reduces the amount of information from 100 pieces to 3 pieces (a marker indicating a run of similar colored dots, the color, and the count) or even 2 pieces (if all information is stored as pairs of color and count), as shown in Figure 3.4. This is called *run-length encoding* (RLE). It's a form of *lossless* compression, meaning that the original picture can be reconstructed perfectly with no missing detail. Run-length encoding is great for *synthetic images* — simple, computer-generated images containing relatively few colors. However, this method does not work well for most *natural images* (e.g., photographs) because in the natural world there are continuous, subtle changes in color that thwart the RLE process.

Figure 3.4 Run-length Compression Example

DVD-Video uses run-length encoding for subpictures, which contain captions and graphic overlays. The legibility of subtitles is critical and it's important that no detail be lost. DVD limits subpictures to four colors at a time, so there are lots of repeating runs of colors, making them perfect candidates for run-length compression. Compressed subpicture data takes up less than 1 percent of the data consumed by a typical DVD-Video program.

If you take the fundamental concept behind run-length encoding, which is to find redundancies by identifying correlations in the data, and then apply much more sophisticated pattern search algorithms and efficient symbol replacement (e.g., replace a 10-byte pattern with an 8-bit symbol each time that pattern appears), you will have an even more effective version of lossless compression. The PNG image format, which is commonly used in the upcoming next-generation formats, supports *compression filters*, which provide different search algorithms for identifying and reducing redundant patterns in the image data. With these formats, even natural images can start to see at least some benefit (though it is still more effective for synthetic images where there tends to be more and larger repeating patterns).

To reduce picture information even more, *lossy* compression is required. This results in information being removed permanently. The trick is to only remove detail that will not be noticed. Many such compression techniques, known as *psychovisual* encoding systems, take advantage of a number of aspects of the human vision system:

1. The eye is more sensitive to changes in brightness than changes in color.

2. The eye is unable to perceive brightness levels above or below certain thresholds.

3. The eye cannot distinguish minor changes in brightness or color. This perception is not linear, with certain ranges of brightness or color more important visually than others. For example, variegated shades of green such as leaves and plants in a forest are more easily discriminated than various shades of dark blue such as in the depths of a pool.

4. Gentle gradations of brightness or color (such as a sunset blending gradually into a blue sky) are more important to the eye and more readily perceived than abrupt changes (such as pinstriped suits or confetti).

The human retina has three types of color photoreceptor cells, called *cones*.[4] Each is sensitive to different wavelengths of light that roughly correspond to the colors red, green, and blue. Because the eye perceives color as a combination of these three stimuli, any color can be described as a combination of these primary colors.[5] Televisions work by using three electron beams to cause different phosphors on the face of the television tube to emit red, green, or blue light, abbreviated to RGB. Television cameras record images in RGB format, and computers generally store images in RGB format.

RGB values are a combination of brightness and color. Each triplet of numbers represents the intensity of each primary color. As just noted, however, the eye is more sensitive to brightness than to color. Therefore, if the RGB values are separated into a brightness component and a color component, the color information can be more heavily compressed. The brightness information is called *luminance* and is often denoted as Y.[6] Luminance is essentially what you see when you watch a black-and-white TV. Luminance is the range of intensity from black (0 percent) through gray (50 percent) to white (100 percent). A logical assumption is that each RGB value would contribute one-third of the intensity information, but the eye is most sensitive to green, less sensitive to red, and least sensitive to blue, so a

---

[4]Rods, another type of photoreceptor cell, are only useful in low-light environments to provide what is commonly called night vision.

[5]You may have learned that the primary "colors" are red, yellow, and blue. Technically, these are magenta, yellow, and cyan and usually refer to pigments rather than colors. A magenta ink absorbs green light, thus controlling the amount of green color perceived by the eye. Since white light is composed of equal amounts of all three colors, removing green leaves red and blue, which together form magenta. Likewise, yellow ink absorbs blue light, and cyan ink absorbs red light. Reflected light, such as that from a painting, is formed from the character of the illuminating light and the absorption of the pigments. Projected light, such as that from a television, is formed from the intensities of the three primary colors. Since video is projected, it deals with red, green, and blue colors.

[6]The use of Y for luminance comes from the XYZ color system defined by the Commission Internationale de L'Eclairage (CIE). The system uses three-dimensional space to represent colors, where the Y axis is luminance and X and Z axes represent color information.

uniform average would yield a yellowish green image instead of a gray image.[7] Consequently, it's necessary to use a weighted sum corresponding to the spectral sensitivity of the eye, which is about 70 percent green, 20 percent red, and 10 percent blue (see Figure 3.5).

Figure 3.5 Color and Luminance Sensitivity of the Eye

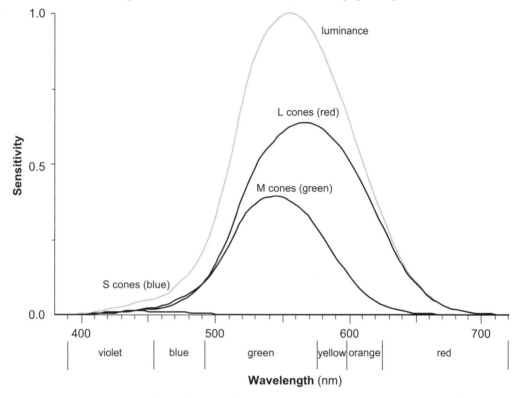

The remaining color information is called *chrominance* (denoted as C), which is made up of *hue* (the proportion of color: the redness, orangeness, greenness, etc.), and *saturation* (the purity of the color, from pastel to vivid). For the purposes of compression and converting from RGB, it's easier to use *color difference* information rather than hue and saturation.

The color information is what's left after the luminance is removed. By subtracting the luminance value from each RGB value, three color difference signals are created: R-Y, G-Y, and B-Y. Only three stimulus values are needed, so only two color difference signals need to be included with the luminance signal. Since green is the largest component of luminance, it has the smallest difference signal (G makes up the largest part of Y, so G-Y results in the smallest values). The smaller the signal, the more it is subject to errors caused by noise, so B-Y and R-Y are the best choice for use as the color difference values. The green color information can be recreated by subtracting the two difference signals from the Y signal (rough-

---

[7]Luminance from RGB can be a difficult concept to grasp. It may help to think of colored filters. If you look through a red filter, you will see a monochromatic image composed of shades of red. The image would look the same through the red filter if it were changed to a different color, such as gray. Since the red filter only passes red light, anything that's pure blue or pure green won't be visible. To get a balanced image, you would use three filters, change the image from each one to gray, and average them together.

ly speaking). Different weightings are used to derive Y and color differences from RGB, such as YUV, YIQ, and $Y'C_bC_r$. DVD uses $Y'C_bC_r$ as its native storage format. (Details of the variations are beyond the scope of this book.)

The sensitivity of the eye is not linear, and neither is the response of the phosphors used in television tubes. Therefore, video is usually represented with corresponding nonlinear values, and the terms *luma* and *chroma* are used. These are denoted with the prime symbol as Y′ and C′, as is the corresponding R′G′B′. (Details of nonlinear functions are also beyond the scope of this book.)

Compressing Single Pictures

An understanding of the nuances of human perception led to the development of compression techniques that take advantage of certain characteristics. Just such a development is *JPEG compression*, which was produced by the Joint Photographic Experts Group and is now a worldwide standard. JPEG separately compresses Y, B-Y, and R-Y information, with more compression done on the latter two, to which the eye is less sensitive.

To take advantage of another human vision characteristic — less sensitivity to complex detail — JPEG divides the image into small blocks and applies a *discrete cosine transform* (DCT) to the blocks, which is a mathematical function that changes spatial intensity values to spatial frequency values. This describes the block in terms of how much detail changes, and roughly arranges the values from lowest frequency (represented by large numbers) to highest frequency (represented by small numbers). For areas of smooth colors or low detail (low spatial frequency), the numbers will be large. For areas with varying colors and detail (high spatial frequency), most of the values will be close to zero. A DCT is an essentially *lossless transform*, meaning that an inverse DCT function can be performed on the resulting set of values to restore the original values. In practice, integer math and approximations are used, causing some loss at the DCT stage. Ironically, the numbers are bigger after the DCT transform. The solution is to *quantize* the DCT values so that they become smaller and repetitive.

Quantizing is a way of reducing information by grouping it into chunks. For example, if you had a set of numbers between 1 and 100, you could quantize them by 10. That is, you could divide them by 10 and round to the nearest integer. The numbers from 5 to 14 would all become 1s, the numbers from 15 to 24 would become 2s, and so on, with 1 representing 10, 2 representing 20, and so forth. Instead of individual numbers such as 8, 11, 12, 20, and 23, you end up with "3 numbers near 10" and "2 numbers near 20." Obviously, quantizing results in a loss of detail.

Quantizing the DCT values means that the result of the inverse DCT will not exactly reproduce the original intensity values, but the result is close and can be adjusted by varying the quantizing scale to make it finer or coarser. More importantly, as the DCT function includes a progressive weighting that puts bigger numbers near the top left corner and smaller numbers near the lower right corner, quantization and a special zigzag ordering result in runs of the same number, especially zero. This may sound familiar. Sure enough, the next step is to use run-length encoding to reduce the number of values that need to be stored. A variation of run-length coding is used, which stores a count of the number of zero values followed by the next nonzero value. The resulting numbers are used to look up symbols from a table. The symbol table was developed using Huffman coding to create shorter symbols for the most commonly appearing numbers. This is called *variable-length coding* (VLC). See Figures 3.6 and 3.8 for an example of DCT, quantization, and VLC.

Figure 3.6 Block Transforms and Quantization

Encoding

Macroblock (luma)

Pixel values

134	142	145	131	114	122	131	130
129	143	134	130	135	144	134	118
123	117	118	111	97	109	130	143
129	116	112	116	120	126	130	118
118	127	141	138	138	148	141	125
125	129	119	127	143	149	145	136
131	126	128	142	141	135	126	116
131	140	146	154	133	118	124	124

DCT coefficients

+1037	-1	-6	+1	-12	+8	-4	-4
-16	+1	+28	-6	-14	0	+4	0
+19	+32	-7	-19	+2	-1	-4	-3
+29	-9	-14	+13	-10	-6	+1	0
+4	+14	-6	-13	-2	+7	+1	+2
-26	+2	+16	+2	+11	+6	+1	+1
-10	-11	+27	-18	+4	+1	0	0
-2	+1	+1	-19	-1	+6	+6	0

Quantized coefficients
QÅ(DCT*16)/(QM*8)

+130	0	0	0	-1	0	0	0
-2	0	+3	0	-1	0	0	0
+2	+3	0	-1	0	0	0	0
+3	-1	-1	+1	0	0	0	0
0	+1	0	-1	0	0	0	0
-2	0	+1	0	0	0	0	0
-1	-1	+2	-1	0	0	0	0
0	0	0	-1	0	0	0	0

Decoding

Reconstructed coefficients (dequantize)

+1040	0	0	0	-9	0	0	0
-12	0	+24	0	-10	0	0	0
+14	+24	0	-10	0	0	0	0
+24	-8	-9	+10	0	0	0	0
0	+9	0	-10	0	0	0	0
-19	0	+10	0	0	0	0	0
-9	-10	+21	-12	0	0	0	0
0	0	0	-14	0	0	0	0

Reconstructed values (IDCT)

136	141	138	125	119	125	132	134
136	137	133	130	134	139	133	121
121	125	122	112	107	117	130	138
125	123	119	117	121	128	127	122
123	129	135	139	140	139	136	131
129	125	124	130	139	144	141	136
129	130	134	139	140	135	127	120
132	138	144	141	131	122	122	127

Difference

-2	+1	+7	+6	-5	-3	-1	-4
-7	+6	+1	+0	+1	+5	+1	-3
+2	-8	-4	-1	-10	-8	0	+5
+4	-7	-7	-1	-1	-2	+3	-4
-5	-2	+6	-1	-2	+9	+5	-6
-4	+4	-5	-3	+4	+5	+4	0
+2	-4	-6	+3	+1	0	-1	-4
-1	+2	+2	+13	+2	-4	+1	-3

Quantization matrix

8	16	19	22	26	27	29	34
16	16	22	24	27	29	34	37
19	22	26	27	29	34	34	38
22	22	26	27	29	34	37	40
22	26	27	29	32	35	40	48
26	27	29	32	35	40	48	58
26	27	29	34	38	46	56	69
27	29	35	38	46	56	69	83

The result of these transformation and manipulation steps is that the information that is thrown away is least perceptible. Since the eye is less sensitive to color than to brightness, transforming RGB values to luminance and chrominance values means that more chrominance data can be selectively thrown away. And, since the eye is less sensitive to high-frequency color or brightness changes, the DCT and quantization process removes mostly the high-frequency information. JPEG compression can reduce picture data to about one-fifth the original size with almost no discernible difference and to about one-tenth the original size with only slight degradation.

Compressing Moving Pictures

Motion video adds a *temporal* dimension to the spatial dimension of single pictures. Another worldwide compression standard, *MPEG*, from the Moving Pictures Expert Group, was designed with this in mind. MPEG is similar to JPEG but also reduces redundancy between successive pictures of a moving sequence.

Just as your friend's memory allows you to describe things once and then only talk about what's changing, digital memory allows video to be compressed in a similar manner by first storing a single picture and then only storing the changes. For example, if the bird moves to another tree, you can tell your friend that the bird has moved without needing to describe the bird over again.

MPEG compression uses a similar technique called *motion estimation* or *motion-compensated prediction*. As motion video is a sequence of still pictures, many of which are very similar, and each picture can be compared with the pictures next to it, the MPEG encoding process breaks each picture into blocks, called *macroblocks*, and then hunts around in neighboring pictures for similar blocks. If a match is found, instead of storing the entire block, the system stores a much smaller *vector* describing how far the block moved (or didn't move) between pictures. Vectors can be encoded in as little as 1 bit, so backgrounds and other elements that don't change over time are compressed most efficiently. Large groups of blocks that move together, such as large objects or the entire picture panning sideways, are also compressed efficiently.

MPEG uses three kinds of picture storage methods. *Intra* pictures are like JPEG pictures, in which the entire picture is compressed and stored with DCT quantization. This creates a reference frame from which successive pictures are built. These *I frames* also allow random access into a stream of video, and in practice occur about twice a second. *Predicted* pictures, or *P frames*, contain motion vectors describing the difference from the closest previous I frame or P frame. If the block has changed slightly in intensity or color (remember, frames are separated into three channels and compressed separately), then the difference (*error*) is also encoded. If something entirely new appears that doesn't match any previous blocks, such as a person walking into the scene, then a new block is stored in the same way as in an I frame. If the entire scene changes, as in a cut, the encoding system is usually smart enough to make a new I frame. The third storage method is a *bidirectional* picture, or *B frame*. The system looks both forward and backward to match blocks. In this way, if something new appears in a B frame, it can be matched to a block in the next I frame or P frame. Thus P and B frames are much smaller than I frames.

Experience has shown that two B frames between each I or P frame work well. A typical second of MPEG video at 30 frames per second looks like I B B P B B P B B P B B P B B I B B P B B P B B P B B P B B (see Figure 3.7). B frames are more complex to create than P frames, requiring time-consuming searches in both the previous and subsequent I or P frame. For this reason, some real-time or low-cost MPEG encoders only create I and P frames. Likewise, I frames are easier to create than P frames, which require searches in the subsequent I or P frame. Therefore, the simplest encoders only create I frames. This is less efficient but may be necessary for very inexpensive real-time encoders that must process 30 or more frames a second.

Figure 3.7 Typical MPEG Picture Sequence

Storage and decode order

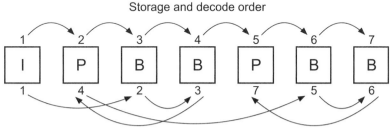

Display order (IBBPBBP)

MPEG-2 encoding can be done in real time (where the video stream enters and leaves the encoder at display speeds), but it's difficult to produce quality results, especially with *variable bit rate* (VBR). Variable bit rate allows varying numbers of bits to be allocated for each frame depending on the complexity of the frame. Less data is needed for simple scenes, whereas more data can be allocated for complex scenes. This results in a lower average data rate and longer playing times but provides room for data peaks to maintain quality. DVD encoding frequently is done with variable bit rate and is usually not done in real time, so the encoder has plenty of time for macroblock matching, resulting in much better quality at lower data rates. Good encoders make one pass to analyze the video and determine the complexity of each frame, forcing I frames at scene changes and creating a compression profile for each frame. They then make a second pass to do the actual compression, varying quantization parameters to match the profiles. The human operator often tweaks minor details between the two passes. Many low-cost MPEG encoding hardware or software for personal computers use only I frames, especially when capturing video in real time. This results in a simpler and cheaper encoder, since P and B frames require more computation and more memory to encode. Some of these systems can later reprocess the I frames to create P and B frames. MPEG also can encode still images as I frames. Still menus on a DVD, for example, are I frames.

The result of the encoding process is a set of data and instructions (see Figure 3.8). These are used by the decoder to recreate the video. The amount of compression (how coarse the quantizing steps are, how large a motion estimation error is allowed) determines how closely the reconstructed video resembles the original. MPEG decoding is *deterministic* — a given set of input data always should produce the same output data. Decoders that properly implement the complete MPEG decoding process will produce the same numerical picture even if they are built by different manufacturers.[8] This doesn't mean that all DVD players will produce the same video picture. Far from it, since many other factors are involved, such as conversion from digital to analog, connection type, cable quality, and display quality. Advanced decoders may include extra processing steps such as block filtering and edge enhancement. Also, many software MPEG decoders take shortcuts to achieve sufficient performance. Software decoders may skip frames and use mathematical approximations rather than the complete but time-consuming transformations. This results in lower-quality video than from a fully compliant decoder.

Figure 3.8 MPEG Video Compression Example

DC   **symbols**
130   01 0

run codes   **symbols**
(zeros, value)   (from lookup table)

1,-2	0011 01
0,+2	1100
3,+3	0000 0001 1100 0
0,+3	0111 0
0,+3	0111 0
1,-1	0101
2,-1	0010 11
1,-1	0101
0,-1	101
0,-1	101
0,+1	100
0,-2	1101
0,-1	101
2,+1	0010 10
7,-1	0000 1001
0,+1	100
0,-1	101
2,+2	0000 1110
9,-1	1111 0001
1,-1	0101
EOB	0110

bytestream (14 bytes: 22 symbols)
46 E0 0E 1C E5 2D 6D 9B 4A 09 94 3B C5 58

compression
4.6:1 (64:14)

134	142	145	131	114	122	131	130
129	143	134	130	135	144	134	118
123	117	118	111	97	109	130	143
129	116	112	116	120	126	130	118
118	127	141	138	138	148	141	125
125	129	119	127	143	149	145	136
131	126	128	142	141	135	126	116
131	140	146	154	133	118	124	124

pixel values

+130	0	0	0	-1	0	0	0
-2	0	+3	0	-1	0	0	0
+2	+3	0	-1	0	0	0	0
+3	-1	-1	+1	0	0	0	0
0	+1	0	-1	0	0	0	0
-2	0	+1	0	0	0	0	0
-1	-1	+2	-1	0	0	0	0
0	0	0	-1	0	0	0	0

quantized coefficients

	1	5	6	14	15	27	28
2	4	7	13	16	26	29	42
3	8	12	17	25	30	41	43
9	11	18	24	31	40	44	53
10	19	23	32	39	45	52	54
20	22	33	38	45	51	55	60
21	34	37	47	50	56	59	61
35	36	48	49	57	58	62	63

zig-zag scan sequence

---

[8]Technically, the inverse discrete cosine transform (IDCT) stage of the decoding process is not strictly prescribed, and is allowed to introduce small statistical variances. This should never account for more than an occasional least significant bit of discrepancy between decoders.

Encoders, on the other hand, can and do vary widely. The encoding process has the greatest effect on the final video quality. The MPEG standard prescribes a *syntax* defining what instructions can be included with the encoded data and how they are applied. This syntax is quite flexible and leaves considerable room for variation. The quality of the decoded video depends on how thoroughly the encoder examines the video and how cleverly it goes about applying the functions of MPEG to compress it. DVD video quality steadily improves as encoding techniques and equipment get better. The decoder chip in the player will not change, but the improvements in the encoded data will provide a better result. This can be likened to reading aloud from a book. The letters of the alphabet are like data organized according to the syntax of language. The person reading aloud from the book is similar to the decoder — the reader knows every letter and is familiar with the rules of pronunciation. The author is similar to the encoder — the writer applies the rules of spelling and usage to encode thoughts as written language. The better the author, the better the results. A poorly written book will come out sounding bad no matter who reads it, but a well-written book will produce eloquent spoken language.[9]

It should be recognized that random artifacts in video playback (aberrations that appear in different places or at different times when the same video is played over again) are not MPEG encoding artifacts. They may indicate a faulty decoder, errors in the signal, or something else independent of the MPEG encode-decode process. It's impossible for a fully compliant, properly functioning MPEG decoder core module to produce visually different results from the same encoded data stream, although in practice there are variations in implementation along with added video processing steps that make the output from every DVD player or decoding software application look different.

MPEG (and most other compression techniques) are *asymmetric*, meaning that the encoding process does not take the same amount of time as the decoding process. It's more effective and efficient to use a complex and time-consuming encoding process because video generally is encoded only once before being decoded hundreds or millions of times. High-quality MPEG encoding systems can cost tens of thousands dollars, but since most of the work is done during encoding, decoder chips cost less than $20, and decoding can even be done in software.

Some analyses indicate that a typical video signal contains over 95 percent redundant information. By encoding the changes between frames, rather than re-encoding each frame, MPEG achieves amazing compression ratios. The difference from the original generally is imperceptible even when compressed by a factor of 10 to 15. DVD-Video data typically is compressed to approximately one-twentieth of the original size (see Table 3.1).

---

[9]Obviously, it would sound better if read by James Earl Jones than by Ross Perot, but the analogy holds if you consider the vocal characteristics to be independent of the translation of words to sound. The brain of the reader is the decoder, the diction of the reader is the post-MPEG video processing, and the voice of the reader is the television.

Table 3.1 Compression Ratios for DVD Technologies

Native data	Native rate (kbps)	Compression	Compressed Rate (kbps)[a]	Ratio	Percent
Video					
$720 \times 576 \times 25$ fps $\times 12$ bits	124,416	MPEG-2	5,400	23:1	96
$720 \times 576 \times 24$ fps $\times 12$ bits	119,439	MPEG-2	5,200	23:1	96
$720 \times 480 \times 30$ fps $\times 12$ bits	124,416	MPEG-2	5,400	23:1	96
$720 \times 480 \times 24$ fps $\times 12$ bits	99,533	MPEG-2	4,400	23:1	96
$352 \times 288 \times 25$ fps $\times 12$ bits	30,413	MPEG-1	1,150	26:1	96
$352 \times 288 \times 24$ fps $\times 12$ bits	29,196	MPEG-1	1,150	25:1	96
$352 \times 240 \times 30$ fps $\times 12$ bits	30,413	MPEG-1	1,150	26:1	96
$352 \times 240 \times 24$ fps $\times 12$ bits	24,330	MPEG-1	1,150	21:1	95
Audio					
2 ch $\times$ 48 kHz $\times$ 16 bits	1,536	Dolby Digital 2.0	192	8:1	87
6 ch $\times$ 48 kHz $\times$ 16 bits	4,608	Dolby Digital 5.1	384	12:1	92
6 ch $\times$ 48 kHz $\times$ 16 bits	4,608	Dolby Digital 5.1	448	10:1	90
6 ch $\times$ 48 kHz $\times$ 24 bits	6,912	Dolby Digital 5.1[b]	448	15:1	94
6 ch $\times$ 48 kHz $\times$ 16 bits	4,608	DTS 5.1	768	6:1	83
6 ch $\times$ 48 kHz $\times$ 16 bits	4,608	DTS 5.1	1,536	3:1	67
6 ch $\times$ 96 kHz $\times$ 20 bits	11,520	MLP[c]	5,400	2:1	53
6 ch $\times$ 96 kHz $\times$ 24 bits	13,824	MLP[c]	7,600	2:1	45

[a]MPEG-2 and MLP compressed data rates are an average of a typical variable bit rate

[b]The source word length can be anywhere from 16 to 24 bits and the output rate does not change.

[c]Note that the compressed rate of MLP streams is heavily dependent on the source material.

Advanced Video Codecs

Among the key new technologies coming with the next-generation formats is a new set of advanced video codecs that are considered to be almost twice as efficient as those used for MPEG-2. The first of these codecs, *SMPTE VC-1*, is a video codec defined by the Society of Motion Picture Television Engineers (SMPTE) as "open standard" and is based on Microsoft's Windows Media Video 9 architecture. Development for Windows Media began years after the creation and standardization of MPEG-2 and has the benefit of building on the lessons learned there. Windows Media (and now VC-1) have demonstrated dramatic improvements in compression efficiency over MPEG-2 and MPEG-1 codecs. For example, prior to adoption of the codec in either HD DVD or BD, Microsoft began promoting a pro-prietary Windows Media Video HD format that showcased Windows Media Video by putting high definition content onto standard red-laser DVDs. The content ($1280 \times 720 \times 24$ fps and $1920 \times 1080 \times 24$ fps) was encoded at constant data rates from 6.0 to 8.0 Mbps for playback on Windows Media-capable PCs. Although most personal computers at the time did not have the processing power needed to present the full size HD video at a consistent frame rate, the WMV-HD format did go a long way to establish Windows Media, and later VC-1, as a serious video codec.

Taking a more traditional development path, *MPEG-4 Advanced Video Codec* (AVC), also known as H.264, was developed by the Joint Video Team (JVT), a co-operation between MPEG (under the ISO) and the ITU-T.[10] Designed with the intention of improving coding efficiency of video by a factor of two or more, uses for H.264 span multimedia applications ranging from video on cell phones to high definition televisions, utilizing a dramatic range of data rates and screen resolutions. In the evaluation of advanced codecs performed by both the DVD Forum and the Blu-ray Disc Founders group, MPEG-4 AVC quickly established itself as the leader and was the first advanced codec to be accepted into each format. The AVC codec includes in its toolset of compression algorithms a wide range of options, allowing AVC encoders to achieve high quality at lower data rates.

Both VC-1 and AVC push asymmetric encoding/decoding to extremes, with first generation implementations of the encoders requiring approximately 30 to 50 times more processing power than equivalent MPEG-2 encoders. Likewise, compared to MPEG-2, these advanced video codecs require significantly more processing power when decoding, but there the difference is perhaps only a factor of 4 to 1.

These advanced codecs achieve a 50 percent reduction in the size of a video stream, which has already played a critical factor in the launch of UMD, for example. Table 3.2, below, shows some comparative data rates for these next-generation compression technologies.

Table 3.2 Compression Ratios for Next-Generation Technologies

Native data	Native rate (kbps)	Compression	Compressed Rate (kbps)[a]	Ratio	Percent
Video					
$1920 \times 1080 \times 30$ fps $\times 12$ bits	746,496	MPEG-4 or VC-1	14,800	50:1	98
$1440 \times 1080 \times 30$ fps $\times 12$ bits	559,872	MPEG-4 or VC-1	11,200	50:1	98
$1280 \times 720 \times 60$ fps $\times 12$ bits	663,552	MPEG-4 or VC-1	13,400	50:1	98
$1920 \times 1080 \times 24$ fps $\times 12$ bits	597,197	MPEG-4 or VC-1	12,000	50:1	98
$1440 \times 1080 \times 24$ fps $\times 12$ bits	447,898	MPEG-4 or VC-1	9,000	50:1	98
$1280 \times 720 \times 24$ fps $\times 12$ bits	265,421	MPEG-4 or VC-1	5,300	50:1	98
$720 \times 480 \times 30$ fps $\times 12$ bits	124,416	MPEG-4 or VC-1	2,700	46:1	98
$720 \times 576 \times 25$ fps $\times 12$ bits	124,416	MPEG-4 or VC-1	2,700	46:1	98
$720 \times 576 \times 24$ fps $\times 12$ bits	119,439	MPEG-4 or VC-1	2,600	46:1	98
$720 \times 480 \times 24$ fps $\times 12$ bits	99,533	MPEG-4 or VC-1	2,200	45:1	98

continues

[10]ITU-T stands for the International Telecommunications Union. The Moving Pictures Experts Group (MPEG) falls under the International Organization for Standardization (ISO, whose name is not an acronym, but is derived from the Greek *isos*, meaning "equal"). Because the video codec was developed jointly by both groups, the resulting standard exists within both bodies. In the ITU-T, the codec is referred to as H.264. Under ISO/IEC, it is referred to as MPEG-4 Part 10 Advanced Video Codec (AVC).

Table 3.2 Compression Ratios for Next-Generation Technologies, continued

Native data	Native rate (kbps)	Compression	Compressed Rate (kbps)[a]	Ratio	Percent
Audio					
2 ch × 48 kHz × 16 bits	1,536	Dolby Digital Plus[b]	192	8:1	88
6 ch × 48 kHz × 16 bits	4,608	Dolby Digital Plus[b]	448	10:1	90
8 ch × 48 kHz × 16 bits	6,144	Dolby Digital Plus[b]	896	7:1	85
8 ch × 48 kHz × 24 bits	9,216	Dolby Digital Plus[b,c]	896	10:1	90
2 ch × 192 kHz × 24 bits	9,216	DTS-HD[d]	4,900	2:1	47
6 ch × 48 kHz × 16 bits	4,608	DTS-HD[e]	1,717	3:1	63
6 ch × 48 kHz × 16 bits	4,608	DTS-HD[f]	2,044	2:1	56
6 ch × 48 kHz × 24 bits	6,912	DTS-HD[f]	3,954	2:1	43
6 ch × 96 kHz × 24 bits	13,824	DTS-HD[f]	8,198	2:1	41
6 ch × 48 kHz × 16 bits	4,608	Dolby TrueHD[g]	1,450	3:1	69
6 ch × 48 kHz × 24 bits	6,912	Dolby TrueHD[g]	3,440	2:1	50
6 ch × 96 kHz × 24 bits	13,824	Dolby TrueHD[g]	5,300	3:1	62

[a]MPEG-2 and MLP compressed data rates are an average of a typical variable bit rate.

[b]Dolby Digital Plus offers improved performance over a wider range of data rates and channel configurations compared to Dolby Digital. However, recommended data rates remain unchanged for these 2- and 6-channel configurations

[c]As with Dolby Digital, source word lengths of 16, 20 and 24 bits all result in the same output rate.

[d]Losslessly encoded with an embedded DTS 2.0 core compressed at 384 kbps. Note that the compressed rate of DTS-HD streams is heavily dependent on the source material.

[e]Losslessly encoded with an embedded DTS 5.1 core compressed at 768 kbps. Note that the compressed rate of DTS-HD streams is heavily dependent on the source material.

[f]Losslessly encoded with an embedded DTS 5.1 core compressed at 1,509 kbps. Note that the compressed rate of DTS-HD streams is heavily dependent on the source material.

[g]Note that the compressed rate of Dolby TrueHD streams is heavily dependent on the source material. Values shown here represent movie source material, which can typically be more highly compressed than music.

Birds Revisited: Understanding Audio Compression

Audio takes up much less space than video, but uncompressed audio coupled with compressed video uses up a large percentage of the available bandwidth. Compressing the audio can result in a small loss of quality, but if the resulting space is used instead for video, it may improve the video quality significantly. In essence, reducing both the audio and the video is more effective. Usually video is compressed more than audio, since the ear is more sensitive to detail loss than the eye.

Just as MPEG compression takes advantage of characteristics of the human eye, modern audio compression relies on detailed understanding of the human ear. This is called *psychoacoustic* or *perceptual coding*.

Picture again your telephone conversation with a friend. Imagine that your friend lives near an airport, so that when a plane takes off, your friend cannot hear you over the sound of the airplane. In a situation like this, you quickly learn to stop talking when a plane is taking off, since your friend won't hear you. The airplane has *masked* the sound of your voice. At the opposite end of the loudness spectrum from airplane noise is background noise, such as a ticking clock. While you are speaking, your friend can't hear the clock, but if you stop, then the background noise is no longer masked.

The hairs in your inner ear are sensitive to sound pressure at different frequencies (pitches). When stimulated by a loud sound, they are incapable of sensing softer sounds at the same pitch. Because the hairs for similar frequencies are near each other, a stimulated audio receptor nerve will interfere with nearby receptors and cause them to be less sensitive. This is called *frequency masking*.

Human hearing ranges roughly from low frequencies of 20 Hz to high frequencies of 20,000 Hz (20 kHz). The ear is most sensitive to the frequency range from about 2 to 5 kHz, which corresponds to the range of the human voice. Because aural sensitivity varies in a nonlinear fashion, sounds at some frequencies mask more neighboring sounds than at other frequencies. Experiments have established certain *critical bands* of varying size that correspond to the masking function of human hearing (see Figure 3.9).

Figure 3.9 Frequency Masking and Hearing Threshold

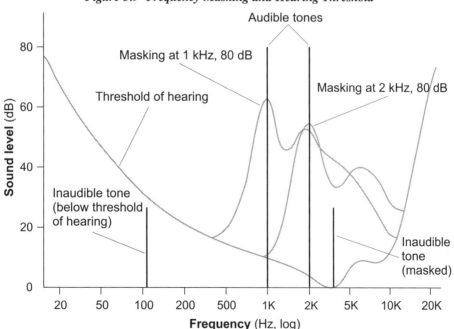

Another characteristic of the human audio sensory system is that sounds can't be sensed when they fall below a certain loudness (or amplitude). This sensitivity threshold is not linear. In other words, the threshold is at louder or softer points at different frequencies. The overall threshold varies a little from person to person — some people have better hearing than others. The threshold of hearing is adaptive; the ear can adjust its sensitivity in order to pick up soft sounds when not overloaded by loud sounds. This characteristic causes the effect of *temporal masking*, in which you are unable to hear soft sounds for up to 200 milliseconds after a loud sound and for 2 or 3 milliseconds before a loud sound.[11]

Perceptual Coding

DVD uses three audio data reduction systems: Dolby Digital (AC-3) coding, MPEG audio coding, and DTS (Coherent Acoustics) coding. All use mathematical models of human hearing based on sensitivity thresholds, frequency masking, and temporal masking to remove sounds that you can't hear. The resulting information is compressed to about one-third to one-twelfth the original size with little to no perceptible loss in quality (see Table 3.1).

Digital audio is *sampled* by taking snapshots of an analog signal thousands of times a second. Each sample is a number that represents the amplitude (strength) of the waveform at that instance in time. Perceptual audio compression takes a block of samples and divides them into frequency bands of equal or varying widths. Bands of different widths are designed to match the sensitivity ranges of the human ear. The intensity of sound in each band is analyzed to determine two things: (1) how much masking it causes in nearby frequencies and (2) how much noise the sound can mask within the band. Analyzing the masking of nearby bands means that the signal in bands that are completely masked can be ignored. Calculating how much noise can be masked in each band determines how much compression can be applied to the signal within the band. Compression uses quantization, which involves dividing and rounding, and this can create errors known as *quantization noise*. For example, the number 32 quantized by 10 gives 3.2 rounded to 3. When re-expanded, the number is reconstructed as 30, creating an error of 2. These errors can manifest themselves as audible noise. After masked sounds are ignored, remaining sounds are quantized as coarsely as possible so that quantization noise is either masked or is below the threshold of hearing. The technique of noise masking is related to *noise shaping* and is sometimes called *frequency-domain error confinement*.

Another technique of audio compression is to compare each block of samples with the preceding and following blocks to see if any can be ignored on account of temporal masking — soft sounds near loud sounds — and how much quantization noise will be temporally masked. This is sometimes called *temporal-domain error confinement*.

Digital audio compression also can take advantage of the redundancies and relationships between channels, especially when there are six or eight channels. A strong sound in one channel can mask weak sounds in other channels, information that is the same in more than

---

[11]How can masking work backward in time? The signal presented by the ear to the brain is a composite built up from stimuli received over a period of about 200 milliseconds. A loud noise effectively overrides a small portion of the earlier stimuli before it can be accumulated and sent to the brain.

one channel need only be stored once, and extra bandwidth can be temporarily allocated to deal with a complex signal in one channel by slightly sacrificing the sound of other channels.

MPEG-1 Audio Coding

MPEG-1 digital audio compression carries either monophonic or stereophonic audio. It divides the signal into frequency bands (typically 32) of equal widths. This is easier to implement than the slightly more accurate variable widths.

MPEG-1 has three *layers*, or compression techniques, each more efficient but more complicated than the last. Layer II is the most common and is the only one allowed by DVD. Layer II compression typically uses a sample block size of 23 milliseconds (1152 samples). Layer III, commonly called MP3, is a popular format for compressed music on the Internet and the iPod, but it is not directly supported by the DVD-Video or DVD-Audio formats. Some DVD players can in fact play MP3 files from CD-ROM or DVD-ROM discs.

MPEG-2 Audio Coding

MPEG-2 digital audio compression adds multiple channels and provides a mode for backward compatibility with MPEG-1 decoders. This backward-compatible mode is required for DVD. Five channels of audio are *matrixed* into the standard left/right stereo signal, which is encoded in the normal MPEG-1 Layer II format. *Phase matrix encoding* is the process of mixing multiple audio channels into two channels according to a defined mathematical relationship relying on different audio signal phases. This relationship can then be used to later reconstruct a total of four channels (left, right, center, and surround).[12] The advantage of matrixing is that the two-channel audio can still be played on standard stereo audio systems. MPEG provides predetermined matrixing formulas depending on the intended audience for the stereo audio signal. One formula is a conventional stereo signal, and another is designed to deliver a signal that is compatible with Dolby Surround decoding to recreate a center channel with left/right surround channels. Additional discrete channel separation information, plus the low-frequency channel, is put in an *extension stream* so that an MPEG-2 audio decoder can recreate six separate signals. For eight channels, an additional layer is provided in another extension stream. The extension streams are compressed in a way that reduces redundant information shared by more than one channel. Because of the need to be backward compatible with MPEG-1 decoders, the center and surround channels that are matrixed into the two MPEG-1 channels are duplicated in the extension stream. This allows the MPEG-2 decoder to subtract these signals from the matrixed signals, leaving the original left and right channels. This duplication of information adds an overhead of approximately 32 kbps, making the backward-compatible mode somewhat inefficient.

The hierarchical structure of MPEG-1 plus extension streams means that a two-channel MPEG decoder need only decode a two-channel data stream, and a six-channel MPEG

---

[12]A signal containing two main channels with additional channels matrixed onto them is often referred to as L_t/R_t, with the t standing for "total." A pure stereo signal that does not carry phase-shifted audio intended for a decoder is sometimes identified as L_o/R_o, with the o standing for "only."

decoder need only process a six-channel data stream. The eight-channel MPEG decoder is the only one that must decode the entire contents of an eight-channel stream. Thus, an MPEG-1 decoder "sees" only the MPEG-1–compatible data to produce stereo audio, whereas an MPEG-2 decoder combines it with the first layer of extension data to produce 5.1-channel audio or, with both layers of extension data, to produce 7.1-channel audio. This clever technique provides an advantage to MPEG-2 over other encoding schemes, since cheap MPEG-1 decoders can be used when necessary. Nevertheless, there are several drawbacks to the backward-compatibility scheme, in addition to the inefficiency of matrixed channel duplication. The original two-channel MPEG-1 encoding process was not developed with matrixed audio in mind and therefore may remove surround-sound detail. The matrix-canceling process tends to expose coding artifacts; that is, when the signal from a matrixed channel is removed, the remaining signal may no longer mask the same neighboring frequencies or noise as the original signal, thus unmasking the noise introduced by formerly appropriate levels of quantization. This problem can be mitigated in part with special processing by the encoder, but this makes the encoding process less efficient with a resulting loss in quality at a given bit rate.

MPEG-2 includes a non-backward-compatible process, originally labeled NBC but now known as *advanced audio coding* (AAC). This coding method deals with all channels simultaneously and is thereby more efficient, but it was not developed in time to be supported by DVD.

MPEG-2 allows a variable bit rate in order to handle momentary increases in signal complexity. Unfortunately, this becomes difficult to deal with in practice because audio and video are usually processed separately, and with two variable rates there is a danger of simultaneous peaks pushing the combined rate past the limit. This can be controlled by limiting the peak rate, but this may reduce audio quality in difficult passages.

Dolby Digital Audio Coding

Dolby Digital audio compression (known as AC-3 in standards documents) provides for up to 5.1 channels of discrete audio. One of the advantages of Dolby Digital is that it analyzes the audio signal to differentiate short, transient signals from long, continuous signals. Short sample blocks are then used for short sounds and long sample blocks are used for longer sounds. This results in smoother encoding without transient suppression and block boundary effects that can occur with fixed block sizes. For compatibility with existing audio/video systems, Dolby Digital decoders can downmix multichannel programs to ensure that all the channels are present in their proper proportions for mono, stereo, or Dolby Pro Logic reproduction.

Dolby Digital uses a frequency transform — somewhat like the DCT transform of JPEG and MPEG — and groups the resulting values into frequency bands of varying widths to match the critical bands of human hearing. Each transformed block is converted to a floating-point frequency representation that is allocated a varying number of bits from a common pool, according to the importance of the frequency band. The result is a *constant bit rate* (CBR) data stream.

Dolby Digital also includes dynamic range information so that different listening environments can be compensated for. Original audio mixes, such as movie sound tracks, which are designed for the wide dynamic range of a theater, can be encoded to maintain the clarity of the dialogue and to enable emphasis of soft passages when played at low volume in the home.

Dolby Digital was developed from the ground up as a multichannel coder designed to meet the diverse and often contradictory needs of consumer delivery. It also has a significant lead over other multichannel systems in both marketing and standards adoption. Over 50 million Dolby Digital decoders are entrenched in living rooms, compared with almost no MPEG-2 audio decoders, no SDDS decoders, and relatively few DTS decoders. Dolby Digital has been chosen for the U.S. DTV standard and is being used for digital satellite systems and most other digital television systems.

DTS Audio Coding

Digital Theater Systems (DTS) Digital Surround uses the Coherent Acoustics differential subband perceptual audio transform coder, which is similar to Dolby Digital and MPEG audio coders. It uses polyphase filters to break the audio into subbands (usually 32) of varying bandwidths and then uses ADPCM to compress each subband. The ADPCM step is a linear predictive coding that "guesses" at the next value in the sequence and then encodes only the difference. The prediction coefficients are quantized based on psychoacoustic and transient analysis and then are variable-length coded using entropy tables.

DTS decoders include downmixing features using either preset downmix coefficients or custom coefficients embedded in the stream. Dynamic range control and other user data can be included in the stream to be used in post-decoder processes.

The DTS Coherent Acoustics format used on DVDs is different from the format used in theaters, which is Audio Processing Technology's apt-X, a straight ADPCM coder with no psychoacoustic modeling.

MLP & Dolby TrueHD Audio Encoding

In addition to the lossy perceptual coding formats from DVD-Video, the DVD-Audio format includes a mathematical encoding technique called *Meridian Lossless Packing* (MLP), which was functionally expanded and renamed *Dolby TrueHD* in 2005 for inclusion into the HD DVD and BD specifications. MLP compresses audio data bit for bit, removing redundancy without discarding any data. In addition to storing more data in the same amount of space, MLP reduces the maximum peak data rate. Since six channels of 96-kHz 24-bit audio have an uncompressed data rate of 13.824 Mbps, reducing the peak data rate to fit into the DVD-Audio maximum of 9.6 Mbps is important. MLP produces variable data rates, thus providing longer playing times than a fixed-rate scheme.

MLP achieves a typical compression ratio of 2:1 by using a combination of three techniques: lossless matrixing to compress interchannel redundancy, lossless waveform prediction to compress intersample correlation, and entropy coding of the remaining values. MLP also incorporates stream buffering to help deal with transients and hard-to-compress segments so as to limit the peak data rate. Unlike lossy encoding methods, MLP cannot throw away data to stay within a specified data rate, so in some cases it will be unable to sufficiently reduce the data rate. In such cases, the encoder operator must use other options such as reducing the bit size of one or more channels or filtering out high-frequency data. For this reason, MLP is known as a *statistically lossless* codec, which means that to compress the audio data effectively, the operator may need to manipulate the audio data somewhat prior to encoding. However, the data that comes back out the decoder will match bit-for-bit with the (manipulated) audio that the operator fed into the system.

MLP can encode a 2-channel downmix along with a multichannel mix, and can carry additional data such as dynamic range control profiles, copy control information, time codes, and descriptive text. The format also has built-in error detection to recover from transmission errors.

DTS-HD

A recent addition to the statistically lossless group of codecs is the DTS-HD variable bit rate (VBR) codec. Built on the existing DTS foundation, the DTS-HD codec takes a backward compatible, hierarchical approach to implementing lossless compression. The encoded data is composed of a normal DTS core plus an additional extension to achieve lossless compression. The mechanism is conceptually simple: The original audio data is first passed through the optional DTS core with a fixed data rate set at 768 kbps, 1024 kbps, or 1509 kbps. Next, difference coding is applied to encode the differences between the original audio data and the DTS core, and the resulting data is stored in the DTS extension area. If the DTS core is not utilized, DTS-HD compression efficiency tends to be very similar to that achieved by MLP. However, when the DTS core is added, because it is able to very accurately reconstruct a close approximation to the original stream, the encoder is left only having to encode relatively small differences, which compress well. Another benefit to DTS-HD is that it can maintain backward compatibility by stripping out the DTS core, which can be decoded by a great majority of the A/V receivers on the market.

Effects of Audio Encoding

Aside from lossless MLP or DTS-HD encoding, which has no effect on the audio signal, lossy perceptual audio compression can affect the quality of the audio signal. Audio compression techniques result in a set of data that is processed in a specific way by the decoder. The form of the data is flexible, so improvements in the encoder can result in improved quality or efficiency without changing the decoder. As understanding of psychoacoustic models improves, perceptual encoding systems can be improved.

At minimal levels of compression, perceptual encoding removes only the imperceptible information and provides decoded audio that is virtually indistinguishable from the original. At higher levels of compression, and depending on the nature of the audio, *bit starvation* may produce identifiable effects of compression. These include a slightly harsh or gritty sound, poor reproduction of transients, loss of detail, and less pronounced separation and spaciousness.

Both Dolby Digital and MPEG-2 audio on DVD are usually compressed at a factor of about 10:1.

DTS on DVD is usually compressed at a factor of 6:1, or sometimes 3:1. Many listeners claim that DTS audio quality is better than Dolby Digital, but such claims are rarely based on accurate comparisons, and may be another form of perceptual influence — that of assuming higher bitrates means higher quality. Within the confines of any given codec, that is indeed true, but it is not able to be assumed when comparing codecs of differing efficiencies — something that has recently become abundantly clear when comparing MP3s to iTunes AAC recordings. DTS tracks are usually encoded at a reference volume level that is 4 dB higher, and most DTS soundtracks are mixed differently than their Dolby Digital counterparts, including different volume levels in the surround and LFE tracks. This makes it nigh impossible to compare them objectively, even using a disc that contains a soundtrack in both formats. The limited semi-scientific comparisons that have been done indicate that there is little perceptible difference between the two and that any difference is noticeable only on high-end audio systems.

Speakers Everywhere

One of the key selling points for DVD is its support for discrete multichannel audio. While most VCR users usually listened to mono or stereo audio through their television's built-in speakers, DVD players paired with A/V receivers are able to deliver a much more compelling experience — 5.1 channel surround sound with a quality equivalent to (or better than) CD audio. Interestingly, while a higher quality of video had usually been assumed to be the most compelling feature by the creators of DVD, it seems that a higher quality audio experience may have been an even more significant motivating factor. The fact that more than 39 million Dolby Digital and DTS capable surround sound receivers have been sold tends to support this hypothesis.

Most consumers, it seems, are familiar with the common 5.1 channel speaker setup, with left (L), center (C), right (R), left surround (LS), and right surround (RS) speakers and low-frequency effects subwoofer (LFE) arranged around the room. With the next-generation disc formats, new speaker configurations may be on their way. For example, the new 7.1-channel audio codecs allow for multiple different speaker arrangements, including the standard 5.1 plus either two additional "center surround" speakers, or two additional surround back speakers, or height speakers to provide altitude to the exprience. In the meantime, even releases to the theaters rarely use a 7.1-channel mix, and many more standard practices must be defined before these new configurations have much meaning or impact.

A Few Timely Words about Jitter

Apart from aspect ratios and anamorphic conversions, *jitter* is one of the most confusing aspects of DVD,[13] because even the experts disagree about its effects. Part of the problem is that jitter means many things, most of them quite technical. Modern episodes of *Star Trek* come closest to providing a comprehensive definition. Since it's not very dramatic to say, "Captain, we have detected jitter!" crew members instead say, "We have encountered a temporal anomaly!" In general terms, jitter is inconsistency over time.

When most people speak of jitter, they mean *time jitter*, also called *phase noise*, which is a time-base error in a clock signal — deviation from the perfectly spaced intervals of a reference signal. Figure 3.10 compares the simplified square wave of a perfect digital signal to the same signal after being affected by factors such as poor-quality components or poorly designed components, mismatched impedance in cables, logic-level mismatches between *integrated circuits* (ICs), interference and fluctuations in power supply voltage, *radiofrequency* (RF) interference, and reflections in the signal path. The resulting signal contains aberrations such as phase shift, high-frequency noise, triangle waves, clipping, rounding, slow rise/fall, and ringing. The binary values of the signal are encoded in the transition from positive voltage to negative voltage, and vice versa. In the distorted signal, the transitions no longer occur at regularly spaced intervals.

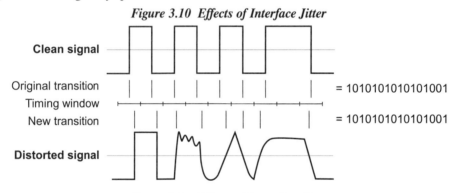

Figure 3.10 Effects of Interface Jitter

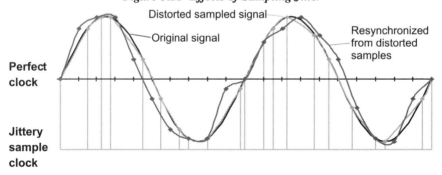

Figure 3.11 Effects of Sampling Jitter

---

[13]When the first edition of this book was written, it was thought that the most difficult task would be covering all the technical details of the format. It was soon discovered that the hardest job was explaining DVD's aspect ratio features in a way that was easy to understand without taking up an entire chapter. Jitter could likewise easily take an entire chapter to explain.

However, looking closely at Figure 3.10 reveals something interesting. Even though the second signal is misshapen to the point of displacing the transition points, the sequence of ones and zeros is still reconstructed correctly, since each transition is within the interval timing window. In other words, there is no data loss, and there is no error. The timing information is distorted, but it can be fixed. This is the key to understanding the difference between correctable and uncorrectable jitter. In the digital domain, jitter is almost always inconsequential. Minor phase errors are easily corrected by resynchronizing the data. Of course, large amounts of jitter can cause data errors, but most systems specify jitter tolerances at levels far below the error threshold.[14] Jitter is an interface phenomenon — it only becomes a problem when moving from the analog to the digital world or from the digital world to the analog world. For example, jitter in the sampling clock of an analog-to-digital converter causes uneven spacing of the samples, which results in a distorted measurement of the waveform (Figure 3.11). On the other end of the chain, jitter in a digital-to-analog converter causes voltage levels to be generated at incorrect moments in time, resulting in audio waveform distortions such as spurious tones and added noise, producing what is often described as a "harsh" sound. Jitter that passes into an analog speaker signal degrades spatial image, ambience, and dynamic range. Actual data errors produce clicks or pops or periods of silence.

In some cases there is nothing you can do about jitter (other than buy better equipment). In other cases, power conditioners and high-quality cables with good shielding reduce certain kinds of jitter. Before you do anything, however, it's important to understand the various types of jitter and which ones are worth worrying about. Many a shrewd marketer has capitalized on the fears of consumers worried about jitter and sonic quality, bestowing on the world such products as colored ink that supposedly reduces reflections from the edge of the disc, disc stabilizer rings that claim to reduce rotational variations, foil stickers alleged to produce "morphic resonance" to rebalance human perception, highly damped rubber feet or hardwood stabilizer cones for players, cryogenic treatments, disc polarizing devices, and other technological nostrums that are intimate descendents of Dr. Feelgood's Amazing Curative Elixir.

Basically, five types of jitter are relevant to DVD:[15]

■ **Oscillator jitter**   Oscillating quartz crystals are used to generate clock signals for digital circuitry. The quality of the crystal and the purity of the voltage driving it determine the stability of the clock signal. Oscillator jitter is a factor in other types of jitter, since all clocks are "fuzzy" to some degree.

---

[14]The AES/EBU standard for serial digital audio specifies a 163 ns clock rate with ±20 ns of jitter. The full 40 ns range is 24 percent of the unit interval. Testing has shown that correct data values are received with bandwidths as low as 400 kHz. Jitter in the recovered clock is reduced with wider bandwidths up to 5 MHz. The CD Orange Book specifies a maximum of 35 ns of jitter, but also recommends that total jitter in the readout system be less then 10 percent of the unit interval (that is, 23 ns out of 230 ns). The DVD-ROM specification states that jitter must be less than 8 percent of the channel bit clock period (8 percent of 38 ns comes to approximately 3 ns of jitter).

[15]One particular phenomenon is incorrectly referred to as jitter. When DVD-ROM drives and CD-ROM drives perform digital audio extraction (DAE) from audio CDs, they can run into problems if the destination drive can't keep up with the data flow. Most drives don't have block-accurate seeking, so they may miss or duplicate a small amount of data after a pause. These data errors cause clicks when the audio is played back. This is colloquially referred to as "jitter," and there are software packages that perform "jitter correction" by comparing successive read passes during DAE, but technically this is not jitter. It's a data error, not a phase error.

- **Sampling jitter** (*recording jitter*)   This is the most critical type of jitter. When the analog signal is being digitized, instability in the clock results in the wrong samples being taken at the wrong time (see Figure 3.11). Reclocking at a later point can fix the time errors but not the amplitude sampling errors. There is nothing the consumer can do about jitter that happens at recording time or during production, because it becomes a permanent part of the recording. Sampling jitter also occurs when the analog signal from a DVD player is sent to a digital processor (such as an AV receiver with DSP features or a video line multiplier). The quality of the DAC in the receiving equipment determines the amount of sampling jitter. Using a digital connection instead of an analog connection avoids the problem altogether.

- **Media jitter** (*pit jitter*)   This type of jitter is not critical. During disc replication, a laser beam is used to cut the pattern of pits in the glass master. Any jitter in the clock or physical vibration in the mechanism used to drive the laser will be transmitted to the master and thus to every disc that is molded from it. Variations in the physical replication process also can contribute to pits being longer or shorter than they should be. These variations are usually never large enough to cause data errors, and each disc is tested for data integrity at the end of the production line. Media jitter can be worse with recorded discs because they are subject to surface contamination, dust, and vibration during recording. Strange as it may seem, however, recorded media usually have cleaner and more accurate pit geometry than pressed media. However, with both pressed and recorded discs, the minor effects of jitter have no effect on the actual data.

- **Readout jitter** (*transport jitter*)   This type of jitter has little or no effect on the final signal. As the disc spins, phase-locked circuits monitor the modulations of the laser beam to maintain proper tracking, focus, and disc velocity. As these parameters are adjusted, the timing of the incoming signal fluctuates. Media jitter adds additional perturbations. Despite readout jitter, error correction circuitry verifies that the data is read correctly. Actual data errors are extremely rare. The data is buffered into RAM, where it is clocked out by an internal crystal. In theory, the rest of the system should be unaffected by readout jitter because an entirely new clock is used to regenerate the signal.

- **Interface jitter** (*data-link jitter*)   This type of jitter may be critical or it may be harmless, depending on the destination component. When data is transmitted to another device, it must be modulated onto an electrical or optical carrier signal. Many factors in the transmission path (such as cable quality) can induce random timing deviations in the interface signal (see Figure 3.10). There is also *signal-correlated jitter*, where the characteristics of the signal itself cause distortions. As a result of interface jitter, values are still correct (as long as the jitter is not severe enough to cause data errors), but they are received at the wrong time. If the receiving component is a digital recorder that simply stores the data, interface jitter has no effect. If the receiving component uses the signal directly to generate audio and does not sufficiently attenuate the jitter, it may cause audible distortion.[16] A partial solution is to use shorter cables or cables with more bandwidth and to properly match impedance.

To restate the key point, when the component that receives a signal is designed only to transfer or store the data, it need only recover the data, not the clock, so jitter below the error threshold has no effect. This is why digital copies can be made with no errors. However, when the component receiving a signal must reconstruct the analog waveform, then it must recover the clock as well as the data. In this case, the equipment should reduce jitter as much as possible before regenerating the signal. The problem is that there is a tradeoff between data accuracy and jitter reduction. Receiver circuitry designed to minimize data errors sacrifices jitter attenuation.[17] The ultimate solution is to decouple the data from the clock. Some manufacturers have approached this goal by putting the master clock in the DAC (which is probably the best place for it) and having it drive the servo mechanism and readout speed of the drive. RAM-buffered time-base correction in the receiver is another option. This *reclocks* incoming bits by letting them pile up in a line behind a little digital gate that opens and closes to release them in a retimed sequence. This technique removes all incoming jitter but introduces a delay in the signal. The accuracy of the gate determines how much new jitter is created.

The quality of digital interconnect cables makes a difference, but only to a point. The more bandwidth in the cable, the less jitter is introduced. Note that a "digital audio" cable is actually transmitting an analog electrical or optical signal. There's a digital to analog conversion step at the transmitter and an analog to digital conversion step at the receiver. That's the reason interface jitter can be a problem. However, the problem is less serious than when an analog interface cable is used because no resampling of analog signal values occurs.

Much ado is made about high-quality transports — disc readers that minimize jitter to improve audio and video quality. High-end systems often separate the transport unit from other units, which ironically introduces a new source of jitter in the interface cable. Jitter from the transport is a function of the oscillator, the internal circuitry, and the signal output transmitter. In theory, media jitter and transport jitter should be irrelevant, but in reality, the oscillator circuitry is often integrated into a larger chip, so leakage can occur between circuits. It's also possible for the servo motors to cause fluctuations in the power supply that affect the crystal oscillator, especially if they are working extra hard to read a suboptimal disc. Other factors such as instability in the oscillator crystal, temperature, and physical vibration may introduce jitter. A jitter-free receiver changes everything. If the receiver reclocks the signal, you can use the world's cheapest transport and get better quality than with an outrageously expensive, vacuum-sealed, hydraulically cushioned transport with a titanium-lead chassis. The reason better transports produce perceptibly better results is that most receivers

---

[16]For example, a string of ones or a string of zeros may travel faster or slower than a varying sequence because the transmission characteristics of the cable are not uniform across the signal frequency range.

[17]The jitter tolerance characteristic of a PLL circuit is inversely proportional to its jitter attenuation characteristic. That is, the more "slack" the circuit allows in signal transition timing, the more jitter passes through. This situation can be improved by using two PLLs to create a two-stage clock recovery circuit.

don't reclock or otherwise sufficiently attenuate interface jitter. Even digital receivers with DSP circuitry usually operate directly on the bit stream without reclocking it.

Interface jitter affects all digital signals coming from a DVD player: PCM audio, Dolby Digital, DTS, MPEG-2 audio, and so on. In order to stay in sync with the video, the receiver must lock the decoder to the clock in the incoming signal. Since the receiver depends on the timing information recovered from the incoming digital audio signal, it's susceptible to timing jitter.

There's an ongoing tug of war between engineers and critical listeners. The engineers claim to have produced a jitterless system, but golden ears hear a difference. After enough tests, the engineers discover that jitter is getting through somewhere or being added somewhere, and they go back to the drawing board. Eventually, the engineers will win the game. Until then, it's important to recognize that most sources of jitter have little or no perceptible effect on the audio or video.

Pegs and Holes: Understanding Aspect Ratios

The standard television picture is a specific rectangular shape: a third again wider than it is high. This aspect ratio is designated as 4:3, or 4 units wide by 3 units high, also expressed as 1.33.[18] This rectangular shape is a fundamental part of the NTSC and PAL television systems — it cannot be changed without redefining the standards and reengineering the equipment.[19]

The problem is that movies are generally wider than television screens. Most movies are 1.85 (about 5.5:3). Extra wide movies in the Panavision or Cinemascope format are around 2.35 (about 7:3). The conundrum is how to fit a wide movie shape into a not-so-wide television shape (see Figure 3.12).

Fitting a movie into television is like the old puzzle of putting a square peg in a round hole, but in this case it's a rectangular peg and a rectangular hole. Consider a peg that's twice as wide as the hole (see Figure 3.13). There are essentially three ways to get the peg in the hole:

1. Shrink the peg to half its original size or make the hole twice as big (see Figure 3.14).

2. Slice off part of the peg (see Figure 3.15).

3. Squeeze the peg from the sides until it's the same shape as the hole (see Figure 3.16).

---

[18]There's no special meaning to the numbers 4 and 3. They are simply the smallest whole numbers that can be used to represent the ratio of width to height. An aspect ratio of 12:9 is the same as 4:3. The ratio can be normalized to a height of 1, but the width becomes the repeating fraction 1.33333..., which is why the 4:3 notation is generally used. For comparison purposes it's useful to use the normalized format of 1.33:1 or 1.33 for short. On occasion, people will mention the "Academy Aperture" ratio of 1.37. In 1927, the Academy of Motion Picture Arts and Sciences officially chose 1.33 as the industry standard. In 1931 it was changed to 1.37 to allow room on the film for a soundtrack. From the 1920s to the early 1950s almost all films were shot at 1.37, providing a picture aspect ratio of 1.33.

[19]The next generation of television — known as HDTV, ATV, DTV, etc. — has a 1.78 (16:9) picture that's much wider than standard television. However, the new digital format is incompatible with the old standard recording and display equipment.

Figure 3.12 TV Shape vs. Movie Shape

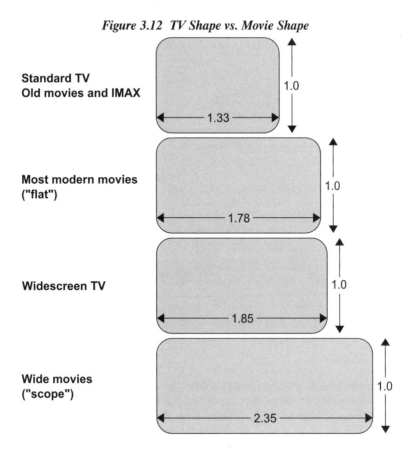

**Standard TV
Old movies and IMAX**

1.0
1.33

**Most modern movies
("flat")**

1.0
1.78

Widescreen TV

1.0
1.85

**Wide movies
("scope")**

1.0
2.35

Figure 3.13 Peg and Hole

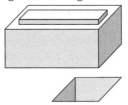

Figure 3.14 Shrink the Peg

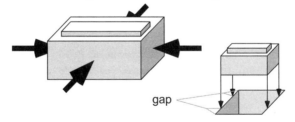

gap

Figure 3.15 Slice the Peg

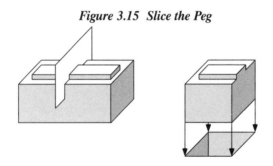

Figure 3.16 Squeeze the Peg

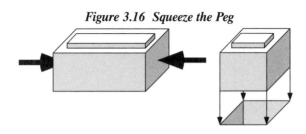

Now, think of the peg as a movie and the hole as a TV. The first two peg-and-hole solutions are used commonly to show movies on television. Quite often you'll see horizontal black bars at the top and bottom of the picture. This means that the width of the movie shape has been matched to the width of the TV shape, leaving a gap at the top and the bottom. This is called *letterboxing*. It doesn't refer to postal pugilism but rather to the process of putting the movie in a black box with a hole the shape of a standard paper envelope. The black bars are called *mattes*.

At other times you might see the words "This presentation has been formatted for television" at the beginning of a movie. This indicates that a *pan and scan* process has been used, where a TV-shaped window over the film image is panned from side to side, scanned up and down, or zoomed in and out (see Figures 3.17 and 3.18). This process is more complicated than just chopping off a little from each side; sometimes the important part of the picture is entirely or primarily on one side, and sometimes there is more picture on the film above or below what is shown in the theater, so the person who transfers the movie to video must determine for every scene how much of each side should be chopped off or how much additional picture from above or below should be included in order to preserve the action and story line. For the past 20 years or so, most films have been shot *flat*, sometimes called *soft matte*. The cinematographer has two rectangles in the viewfinder, one for 1.85 (or wider) and one for 4:3. He or she composes the shots to look good in the 1.85 rectangle while making sure that no crew, equipment, or raw set edges are visible above or below in the 4:3 area.

Figure 3.17 Soft Matte Filming

1.33
(1.78)
1.85

Figure 3.18 Pan and Scan Transfer

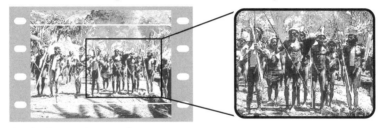

Flat (soft matte): zoom in

Flat (soft matte): full frame

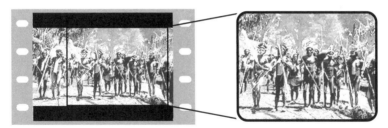

Scope (hard matte): extract

Figure 3.19 The Anamorphic Process

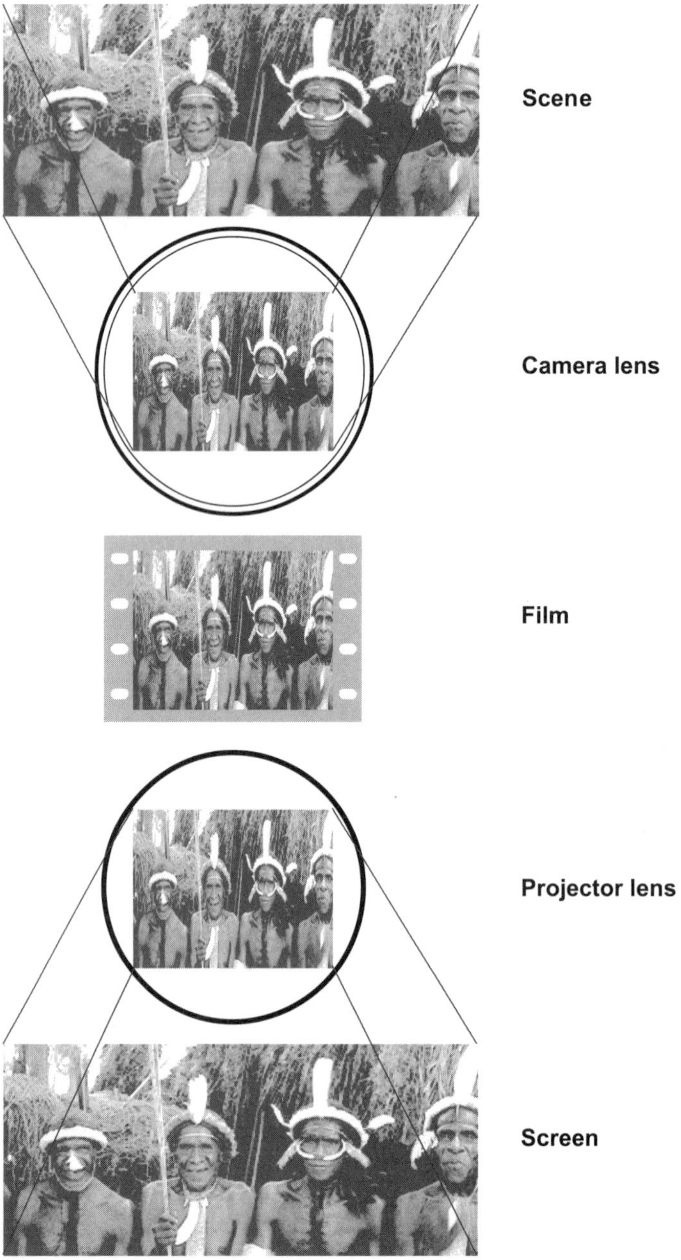

Scene

Camera lens

Film

Projector lens

Screen

Figure 3.20a-o Aspect Ratios, Conversions, and Displays

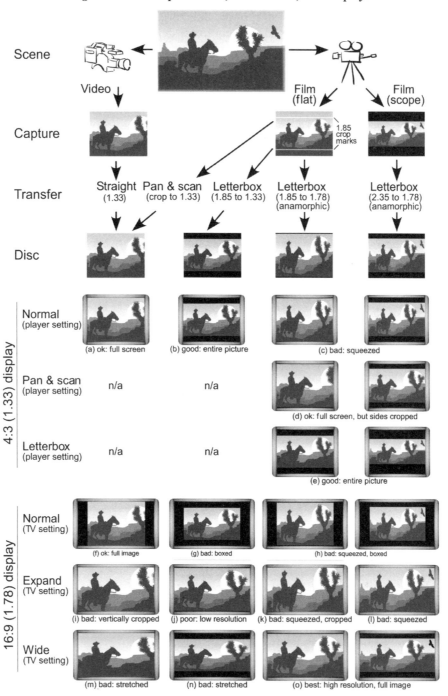

For presentation in the theater, a *theatrical matte* or *aperture plate* is used to mask off the top and bottom when the film is printed or projected. When the movie is transferred to video for 4:3 presentation, the full frame is available for the pan and scan (and zoom) process.[20] In many cases, the director of photography or the director approves the transfer to ensure that the integrity of the film is maintained. Full control over how the picture is reframed is very important. For example, when the mattes are removed, close-up shots become medium shots, and the frame may need to be zoomed in to recreate the intimacy of the original shot. In a sense, the film is being composed anew for the new aspect. The pan and scan process has the disadvantage of losing some of the original picture but is able to make the most of the 4:3 television screen and is able to enlarge the picture to compensate for the smaller size and lower resolution as compared with a theater screen.

There is another option that applies only to computer-generated films. The shots can be recomposed for 4:3 presentation by moving characters and objects closer together. The story, action, and sound all remain the same, but the movie is re-rendered to fit the dimensions of standard TVs.

The third peg-and-hole solution has been used for years to fit widescreen movies onto standard 35-mm film. As filmmakers tried to enhance the theater experience with ever wider screens, they needed some way to get the image on the film without requiring new wider film and new projectors in every theater. They came up with the *anamorphic* process, where the camera is fitted with an anamorphic lens that squeezes the picture horizontally, changing its shape so that it fits into a standard film frame. The projector is fitted with a lens that unsqueezes the image when it is projected (see Figure 3.19). It's as if the peg were accordion-shaped so that it can be squeezed into the square hole and then pop back into shape after it is removed. You may have seen this distortion effect at the end of a Western movie where John Wayne suddenly becomes tall and skinny so that the credits will fit between the edges of the television screen.

How It's Done with DVD

DVD mixes and matches all the preceding techniques. Three methods are provided for standard 4:3 displays, and a fourth targets widescreen displays. Together, the four techniques used for DVD are defined as follows (see Figure 3.20):

■ **Full Frame** ("the peg fits the hole")   Most material shot for television and some older movies such as *The Wizard of Oz* and *Casablanca* originated in 4:3 aspect ratio, so no special processing is required. The content can be captured straight and encoded as 4:3 without any extra mattes. A DVD player would then output the material as 4:3, which a standard television would display as is, while a widescreen television can display the content in normal, expanded or wide modes. Normal mode, of course, would provide the proper 4:3 aspect ratio, while the other two modes would distort the picture.

■ **Pan and Scan** ("chop off the sides")   This is the traditional "fill the frame" way of showing video on a standard TV, and is usually labeled as "full frame" or

---

[20]Contrast this to hard matte filming, where the top and bottom are physically — and permanently — blacked out to create a wide aspect ratio. Movies filmed with anamorphic lenses also have a permanently wide aspect ratio, with no extra picture at the top or bottom.

"full screen" on DVD packaging. When the film is converted to video, the transfer artist (also called the colorist or the telecine artist) may use a variety of techniques to make the picture fill the screen. This can include zooming in and out, moving up, down, left, and right, or simply choosing to "open the matte" and apply the full 4:3 film frame. (Many 1.85 films are transferred this way, which preserves the full width but shows additional material at the top and bottom that was not visible in the theater.) All of this work is typically done to the feature master prior to capture and encode, at which point it is treated simply as 4:3 material.

- **Letterbox** ("shrink and matte") This is the alternative way of showing widescreen video on a standard TV, preferred by videophiles and popularized by laserdiscs and DVD. The original theatrical image is boxed into the 4:3 frame by adding black mattes to the top and bottom of the picture. Again, this is done prior to capture and encode in order to create a 4:3 "source letterboxed" master, at which point it can be treated simply as 4:3 material as in item #1, above. The down side of this approach is that only a fraction of the total vertical resolution is being used for actual picture, and a relatively large portion of the video that must be encoded is nothing but black mattes. A better alternative for providing letterboxed output to a 4:3 television is available via the Widescreen method.

- **Widescreen** ("accordion squeeze") One of the advantages of DVD-Video for home theater systems is native widescreen support. DVD supports wide images by using a 16:9 (1.78) aspect ratio that is anamorphically squeezed into a 4:3 TV shape before being stored on the disc.[21] (The 4:3 ratio can be expressed as 12:9, so in order to get from 16:9 to 12:9, the width needs to be reduced by 25 percent, from 16 to 12.) Widescreen televisions have a mode for stretching the picture back out to the full 16:9 shape during playback. What makes this a particularly attractive option is that DVD players also can adapt anamorphic widescreen video for standard 4:3 displays, using essentially the same techniques described above, but doing it at playback.[22]

There are three ways to do this, each performed by the player:

- **Automatic letterbox** All DVD players can automatically reduce the vertical size of the anamorphic picture by 25% and add letterbox mattes in order to achieve proper aspect ratio on a 4:3 display while still showing the

[21]Some people prefer to think of anamorphic video as being stretched vertically rather than squeezed horizontally. The difference is largely a matter of semantics. If anamorphic video were stretched vertically from a letterboxed source after being transferred from film to video, it would lose resolution, but in practice, it happens during the transfer process, so it's largely a matter of perspective whether you think of the video as being squeezed horizontally or stretched vertically. Matching the source to the height of the 4:3 shape and squeezing horizontally comes out the same as matching the width of the 4:3 shape and stretching vertically. A widescreen TV increases horizontal sweep to stretch anamorphic video from a standard NTSC signal, whereas some 4:3 TVs decrease vertical scan pitch to achieve a 16:9 aspect ratio, so either point of view is valid.

[22]Anamorphic video is not unique to DVD. There are a few anamorphic laserdiscs and even rare anamorphic videotapes. The problem is that they can only be viewed properly on a widescreen TV. Unlike DVD players, standard laserdisc players and VCRs are unable to adapt anamorphic video for standard TVs.

full image. By taking this approach, the full vertical resolution of the capture and encode steps is used for actual picture (which gives a much better picture for the widescreen television), and the player itself does the processing to create the letterboxed output.[23]

■ **Automatic pan and scan** Center-of-interest information can be included with the widescreen video to tell the player which part of the image to extract and which part to truncate. The player chops off the indicated amount from the sides and then unsqueezes the remaining picture to create a 4:3 image for the TV. On each frame, the center-of-interest viewport can be moved to the left or right to smoothly track the action on screen. However, because DVD supports only left/right movement (*panning*) and none of the other functionality, such as zooming, cropping and squeezing, that is traditionally used in professional telecine transfers, essentially no DVDs use this approach on feature. However, it is often used for menu backgrounds (stills and video) to provide a simple center-cut view of the active area on 4:3 displays, while also allowing the menu to fill the screen of a widescreen television.

■ **Lie to the player** Your DVD player doesn't know what kind of TV you have, so you can tell it to send a 16:9 picture to a 4:3 TV. You will see the unchanged anamorphic picture, making Hardy look like Laurel. You're not supposed to do this, but just as there are no "mattress tag police," there are no "aspect ratio police" to come and take your player away.

How It Wasn't Done with DVD

There are other possible solutions for dealing with widescreen video that could have been used. DVD could have stored the full-width, undistorted image, but this would have used up more storage space[24] and would have required reducing the amount of video on a disc or reducing the video quality.

Another option would have been to always letterbox the video before storing it on DVD. The problems with this approach are that vertical information would have been lost, and storage space that could have held picture information would have been used to hold black mattes instead. A few widescreen films are put on DVD this way, usually because a letterboxed transfer from film to video is available and either the original elements are no longer available or the studio doesn't want to pay for a new anamorphic transfer.

Variable anamorphic squeeze also could have been used. The wider the video, the more it would be squeezed. The advantage to this approach is that every pixel of video storage space would be used to hold video, no matter its shape. The problem is that more expensive circuitry would be required to handle multiple squeeze ratios.

---

[23]Unfortunately, some players do a better job of letterboxing anamorphic content than others. See Chapter 9, Application Details, for more information about the letterbox conversion process in DVD-Video.

[24]Thirty-three percent more, to be exact, since 1.78 is 33 percent larger than 1.33. Even more data would be needed to store movies in their original aspect ratio. Most movies have an aspect ratio of 1.85, which would require 39 percent more data. Panavision and Cinemascope movies with a 2.35 ratio would require 76 percent more data.

The designers of DVD chose the reasonable compromise of using the anamorphic technique to fit the most amount of information into the standard television image space, and they settled on the standard 16:9 wide aspect ratio (see "Why 16:9?" following). Having the DVD player shrink the anamorphic picture vertically for letterbox display on a 4:3 TV gives the same result (and the same information loss) as shrinking and letterboxing the picture before storing it on the disc, yet preserves more picture information for widescreen TVs.

Unfortunately there's no standardized package labeling for anamorphic DVDs. The following terms are all used to mean the same thing: *enhanced for widescreen TVs*, *enhanced for 16:9 TVs*, *16:9*, *16:9 fullscreen version*, *widescreen 16×9*, *anamorphic video*, *anamorphic widescreen*, *1.78 edge-to-edge*, and *widescreen*. In general, look for the word *"enhanced"* or *"anamorphic."*

Some people complain that the term anamorphic should apply only to the optical process used in films, or even that the term as applied to DVD is incorrect. Frankly, this is like complaining that the computer industry should not describe regions of a computer screen as "windows" since the term means something else. In any case, "anamorphic" already has a very clear meaning for DVD in practice, independent of the aspect ratio of the film or the TV, and is the clearest and most unambiguous way of specifying the form of the video on the disc. The recommended label for DVD packaging is *anamorphic widescreen*.

Widescreen TVs

Widescreen displays are quite flexible in the way they deal with different input formats. They can display 4:3 video with black bars on the sides — a kind of sideways letterbox that is sometimes called a *windowbox* or *pillarbox* — and they also have display modes that enlarge the 4:3 video to fill the entire screen.

Wide mode stretches the picture horizontally (see Figure 3.21). This is sometimes called *full* mode. This is the proper mode to use with anamorphic video, but it makes everything look short and fat when applied to 4:3 video. Some widescreen TVs have a *parabolic* or *panorama* version of wide mode, which uses nonlinear distortion to stretch the sides more and the center less, thus minimizing the apparent distortion. This mode should not be used with anamorphic DVD output or very strange fun-house-mirror effects will occur.

Figure 3.21 Wide (full) Mode on a Widescreen TV

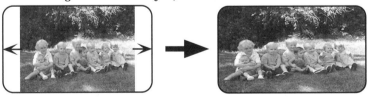

Expand mode proportionally enlarges the picture to fill the width of the screen, thus losing the top and bottom (see Figure 3.22). This is sometimes called *theater* mode. Expand mode is for use with letterboxed video because it effectively removes the mattes. If used with standard 4:3 picture, this mode causes a Henry VIII "off with their heads" effect.

Figure 3.22 Expand (theater) Mode on Widescreen TV

 NOTE
Expand mode should only be used for video that is already letterboxed into a 4:3 picture, as on a laserdisc, a letterbox-only DVD, or other nonanamorphic source. Letting the DVD player autoletterbox anamorphic video causes a loss of resolution, which is amplified when the TV expands the picture (see Figure 3.20j).

Most widescreen TVs also have other display modes that are variations of the three basic modes. Some standard 4:3 TVs, especially in Europe, can adjust vertical scan size to create a letterboxed display from anamorphic pictures. The advantage is that the picture is full resolution because no pixels are lost by having the player do the letterboxing.

When DVD contains widescreen video, the different output modes of the player can be combined with different widescreen TV display modes to create a confusing array of options. Figure 3.20 shows how the different DVD output modes look on a regular TV and on a widescreen TV. Note that there is one "good" way to view widescreen video on a standard TV (Figure 3.20e) but that a very large TV or a widescreen TV is required to do it justice. Also note that there is only one good way to view widescreen video on a widescreen TV, and this is with widescreen (anamorphic) output to wide mode (Figure 3.20o).

Clearly, it's easy to display the wrong picture in the wrong way. In some cases, the equipment is smart enough to assist. The player can send a special signal embedded in the video blanking area or via the s-video connector to the widescreen TV, but everything must be set up properly:

1. Connect the DVD player to the widescreen TV with an s-video cable.

2. Set the widescreen TV to s-video in (using the remote control or the front-panel input selector).

3. Set the TV to automatic or normal mode.

4. Set the DVD player to 16:9 output mode (using the on-screen setup feature with the remote control or with a switch on the back of the player).

If everything is working correctly and the TV is equipped to recognize widescreen signaling, it will automatically switch modes to match the format of the video.

Aspect Ratios Revisited

To review, video comes out of a DVD-Video player in basically four ways:

1. Full frame (4:3 original)

2. Pan and scan (widescreen original)

3. Letterbox (widescreen original)

4. Anamorphic (widescreen original)

All four can be displayed on any TV, but the fourth is specifically intended for widescreen TVs. This may seem straightforward, but it becomes much more complicated. The problem is that very few movies are in 16:9 (1.78) format. The preceding discussions dealt with 16:9 widescreen in a general case. A small proportion of video is being created in 16:9 format as 16:9 HD cameras are being more widely used, and inexpensive home video cameras may have an anamorphic 16:9 mode.

Most movies are usually 1.85 or wider, although European movies are often 1.66. DVD only supports aspect ratios of 1.33 (4:3) and 1.78 (16:9) because they are the two most common television shapes. Movies that are a different shape must be made to fit, which brings us back to pegs and holes. In this case, the hole is DVD's 16:9 shape (which is either shown in full on a widescreen TV or formatted to letterbox or pan and scan for a regular TV). There are essentially four ways to fit a 1.85 or wider movie peg into a 1.78 hole (see Table 3.3):

- **Letterbox to 16:9** Perhaps the most common approach, when the movie is transferred from film, black mattes are added to box it into the 16:9 shape. These mattes become a permanent part of the picture. The position and thickness of the mattes depend on the shape of the original.

  - For a 1.85 movie, the mattes are very small. On a widescreen TV or in automatic pan and scan on a regular TV (where the player is extracting a vertical slice from the letterboxed picture), the mattes are hidden in the overscan area.[25] On a standard TV in automatic letterbox mode (where the player is letterboxing an already letterboxed picture), the thin permanent mattes merge imperceptibly with the thick player-generated mattes.

  - For a 2.35 movie, the permanent mattes are much thicker. In this case, the picture has visible mattes no matter how it is displayed. When the picture is letterboxed by the player, the permanent mattes merge with the player-generated mattes to form extra thick mattes on the television. These mattes are the same size as if the movie had been letterboxed originally for 4:3 display (as with a laserdisc).

  - For a 1.66 movie, thin mattes are placed on the sides instead of at the top and bottom and generally will be hidden in the overscan area.

---

[25]Overscan refers to covering the edges of the picture with a mask around the screen. Overscan was originally implemented to hide distortion at the edges. Television technology has improved to the point where overscan isn't usually necessary, but it is still used. Most televisions have an overscan of about 4 to 5 percent. Anyone producing video intended for television display must be mindful of overscan, making sure that nothing important is at the edge of the picture. It should be noted that when DVD-Video is displayed on a computer there is no overscan and the entire picture is visible.

■ **Crop to 16:9**   For 1.85 movies, slicing 2 percent from each side is sufficient and probably will not be noticeable. The same applies to 1.66 movies, except that about 3 percent is sliced off the top and off the bottom, but to fit a 2.35 movie requires slicing 12 percent from each side. This procrustean approach throws away a quarter of the picture and is not as likely to be a popular solution.

■ **Pan and scan to 16:9**   The standard pan and scan technique can be used with a 16:9 window (as opposed to a 4:3 window) when transferring from film to DVD. For 1.85 movies, the result is essentially the same as cropping and is hardly worth the extra work. For wider movies, pan and scan is more useful, but the original aspect ratio is lost, which goes against the spirit of a widescreen format. When going to the trouble of supporting DVD's widescreen format, it seems silly to pan and scan inside it, but if the option is there, someone is bound to use it.

■ **Open the soft matte to 16:9**   When going from 1.85 to 16:9, a small amount of picture from the top and bottom of the full-frame film area can be included. This stays close to the original aspect ratio without requiring a matte, and the extra picture will be hidden in the overscan area on a widescreen TV or when panned and scanned by the DVD player. Even wider movies are usually shot full frame, so the soft matte area can be included in the transfer (see Figure 3.23).

Table 3.3 Combinations of Output Formats and Video Transfers

| | 4:3 Original | 16:9 Original | | Non-16:9 Original | |
	Stored in 4:3	Stored in 4:3	Stored in 16:9	Stored in 4:3	Stored in 16:9
Full frame (1)	Direct transfer(a)	n/a	n/a	n/a	n/a
Pan and scan (2)	P&S transfer from full frame (a)	P&S transfer(a)	Auto P&S by player(b1)	P&S transfer(a)	Auto P&S by player on P&S transfer (b2) or on LB transfer (b3)
Letterbox (3)	LB transfer from inside soft matte (a)	LB transfer(a)	Auto LB by player (b1)	LB transfer(a)	Auto LB by player on P&S transfer (b2) or on LB transfer (b3)
Anamorphic (4)	Anamorphic transfer from16:9 soft matte (a)	n/a	Anamorphic transfer (a)	n/a	P&S transfer to anamorphic (b) or LB transfer to anamorphic(c)

Note: Labels in parentheses refer to the outline in text.

Figure 3.23 *Opening the Frame from 1.85 to 1.78*

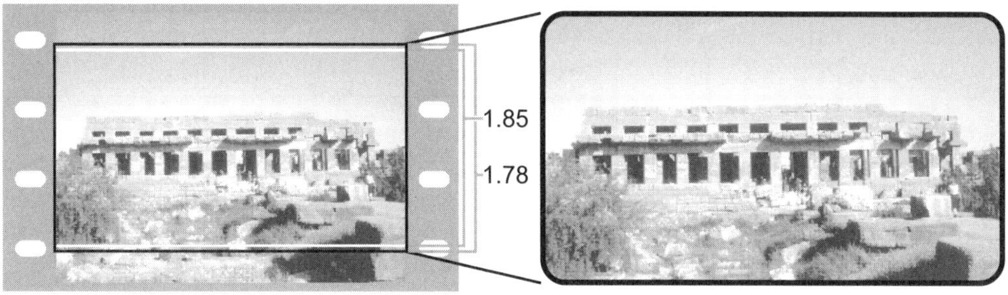

Most movies are converted to anamorphic DVD using either the Letterbox to 16:9 method or the Open the soft matte to 16:9 method. Many directors are opposed for artistic reasons to the pan and scan process, especially if it's done mechanically by the player. They may choose to make their movies on DVD viewable only in widescreen or letterbox format, or they may choose to do a full-frame transfer in 4:3 format. This brings up the issue of different transfers, which is discussed after the following brief digression into why 16:9 is the widescreen aspect ratio of choice.

Why 16:9?

The 16:9 ratio has become the standard for widescreen. Most widescreen televisions are this shape, it's the aspect ratio used by almost all high-definition television standards, and it's the widescreen aspect ratio used by DVD. You may be wondering why this ratio was chosen, since it does not match television, movies, computers, or any other format. But this is exactly the problem: There is no standard aspect ratio (see Figure 3.24).

Figure 3.24 *Common Aspect Ratios*

	7:9	2.3:3	0.77:1	Paper (8.5 x 11)
	12:9	4:3	1.33:1	Television
	15:9	5:3	1.66:1	Photographs, many European movies
	16:9	5.3:3	1.78:1	DVD, HDTV, most widescreen TV
	16.7:9	5.6:3	1.85:1	Most movies
	19.9:9	6.6:3	2.21:1	Wide movies (70 mm)
	21.2:9	7.1:3	2.35:1	Wide movies (Panavision, Cinemascope)
	24.3:9	8.1:3	2.7:1	Extra-wide movies (Ultra Panavision)

Current display technology is limited to fixed physical sizes. A television picture tube must be built in a certain shape. A flat-panel *liquid-crystal display* (LCD) screen must be made with a certain number of pixels. Even a video projection system is limited by electronics and optics to project a certain shape. These constraints will remain with us for decades until we progress to new technologies such as scanning lasers or amorphous holographic projectors. Until then, a single aspect ratio must be chosen for a given display. The cost of a television tube is based roughly on diagonal measurement (taking into account the glass bulb, the display surface, and the electron beam deflection circuitry), but the wider a tube is, the harder it is to maintain uniformity (consistent intensity across the display) and convergence (straight horizontal and vertical lines). Therefore, too wide a tube is not desirable.

The 16:9 aspect ratio was chosen in part because it's an exact multiple of 4:3. Going from 4:3 to 16:9 merely entails adding one horizontal pixel for every three (3⇒4) and going from 16:9 to 4:3 requires simply removing one pixel from every four (4⇒3).[26] This makes the scaling circuitry for letterbox and pan and scan functions much simpler and cheaper. It also makes the resulting picture cleaner.

The 16:9 aspect ratio is also a reasonable compromise between television and movies. It's very close to 1.85 and it's close to the mean of 1.33 and 2.35. That is, $4/3 \times 4/3 \times 4/3 \sim= 2.35$. Choosing a wider display aspect ratio, such as 2:1, would have made 2.35 movies look wonderful but would have required huge mattes on the side when showing 4:3 video (as in Figure 3.20f, but even wider).

Admittedly, the extra space could be used for picture outside picture (POP, the converse of PIP), but it would be very expensive extra space. To make a 2:1 display the same height as a 35-inch television (21 inches) requires a width of 42 inches, giving a diagonal measure of 47 inches. In other words, to keep the equivalent 4:3 image size of 35-inch television, you must get a 47-inch 2:1 television. Figure 3.25 shows additional widescreen display sizes required to maintain the same height of common television sizes.

Figure 3.26 demonstrates the area of the display used when different image shapes are letterboxed to fit it (that is, the dimensions are equalized in the largest direction to make the smaller box fit inside the larger box).[27] The 1.33 row makes it clear how much smaller a letterboxed 2.35:1 movie is: only 57 percent of the screen is used for the picture. On the other hand, the 2:1 row makes it clear how much expensive screen space goes unused by a 4:3 video program or even a 1.85:1 movie. The two middle rows are quite similar, so the mathematical relationship of 16:9 to 4:3 gives it the edge.

---

[26]In each case a weighted scaling function generally is used. For example, when going from 4 to 3 pixels, 3/4 of the first pixel is combined with 1/4 of the second to make the new first, 1/2 of the second is combined with 1/2 of the third to make the new second, and 1/4 of the third is combined with 3/4 of the fourth to make the new third (see Figure 9.20). This kind of scaling causes the picture to become slightly softer but is generally preferable to the cheap alternative of simply throwing away every fourth line. Similar weighted averages can be used when going from 3 to 4 (see Figure 9.22).

[27]If you wanted to get the most for your money when selecting a display aspect ratio, you would need to equalize the diagonal measurement of each display because the cost is roughly proportional to the diagonal size. This approach is sometimes used when comparing display aspect ratios and letterboxed images, and it has the effect of emphasizing the differences. The problem is that wider displays end up being shorter (for example, a 2:1 display normalized to the same diagonal as a 4:3 display would be 4.47:2.25, which is 12 percent wider but 25 percent shorter). In reality, no one would be happy with a new widescreen TV that was shorter than their existing TV. Therefore, it's expected that a widescreen TV will have a larger diagonal measurement and will cost more.

Figure 3.25 Display Sizes at Equal Heights
Diagonal size (width x height)

1.33 (4:3)	1.78 (16:9)	2.0
27" (22 x 16)	33" (29 x 16)	36" (32 x 16)
32" (26 x 19)	39" (34 x 19)	43" (38 x 19)
34" (27 x 20)	42" (36 x 20)	46" (41 x 20)
36" (29 x 22)	44" (38 x 22)	48" (43 x 22)
42" (34 x 25)	51" (45 x 25)	56" (50 x 25)
46" (37 x 28)	56" (49 x 28)	62" (55 x 28)
50" (40 x 30)	61" (53 x 30)	67" (60 x 30)
53" (42 x 32)	65" (57 x 32)	71" (64 x 32)
60" (48 x 36)	73" (64 x 36)	80" (72 x 36)
65" (52 x 39)	80" (69 x 39)	87" (78 x 39)
84" (67 x 50)	103" (90 x 50)	113" (101 x 50)
100" (80 x 60)	122" (107 x 60)	134" (120 x 60)
120" (96 x 72)	147" (128 x 72)	161" (144 x 72)

Figure 3.26 Relative Display Sizes for Letterbox Display

Image shape

Display shape	1.33 (4:3)	1.85	2.35
(4:3) 1.33	100%	72%	57%
(16:9) 1.78	75%	96%	76%
1.85	72%	100%	79%
2.0	67%	92%	85%

In summary, the only way to support multiple aspect ratios without mattes would be to use a display that can physically change shape — a "mighty morphin' television." Since this is currently impossible (or, if possible, outrageously expensive), 16:9 is the most reasonable compromise. That said, the designers of DVD could have improved things slightly by allowing more than one anamorphic distortion ratio. If a 2.35 movie were stored using a 2.35 anamorphic squeeze, then 24 percent of the internal picture would not be wasted on the black mattes, and the player could automatically generate the mattes for either 4:3 or 16:9 displays. This was not done, probably because of the extra cost and complexity it would add to the player. The limited set of aspect ratios presently supported by the MPEG-2 standard (4:3, 16:9, and 2.21:1) also may have had something to do with it.

The Transfer Tango

Of course, the option still remains to transfer the movie to DVD's 4:3 aspect ratio instead of 16:9. At first glance, there may seem to be no advantage in doing this because 1.85 movies are so close to 16:9 (1.78). It seems simpler to do a 16:9 transfer and let the player create a letterbox or pan and scan version. However, there are disadvantages to having the player automatically format a widescreen movie for 4:3 display: the vertical resolution suffers by 25 percent, the letterbox mattes are visible on movies wider than 1.85, and the player is limited to lateral motion. In addition, many movie people are averse to what they consider as surrendering creative control to the player. Therefore, almost every pan and scan DVD is done in the studio and not enabled in the player. During the transfer from film to video, the engineer has the freedom to use the full frame or to zoom in for closer shots, which is especially handy when a microphone or a piece of the set is visible at the edge of the shot.

Many directors are violently opposed to pan and scan disfigurement of their films. Director Sydney Pollack sued a Danish television station for airing a pan and scan version of his *Three Days of the Condor*, which was filmed in 2.35 Cinemascope. Pollack feels strongly that the pan and scan version infringed his artistic copyright. He believes that "The director's job is to tell the film story, and the basis for doing this is to choose what the audience is supposed to see, and not just generally but exactly what they are to see." Years later, when talking about DVD, Pollack said, "DVD hasn't changed my approach to filmmaking, but what it has changed radically is my emotional reaction to the afterlife of the films that I do. I was always terribly disturbed by the fact that the overwhelming majority of people who see the work that you do as a filmmaker do not ever see it in its original form. … More and more people were watching videos, which were more and more often being altered by panning and scanning. Those pictures were getting butchered on video. … You have superb quality with DVD. Plus, you can see the movies in their original widescreen format, framed as they were intended."[26] Some directors, such as Stanley Kubrick, accept only the original aspect ratio. Others, such as James Cameron, who closely supervise the transfer process from full-frame film, feel that the director is responsible for making the pan and scan transfer a viable option by recomposing the movie to make the most of the 4:3 TV shape.

About two-thirds of widescreen movies are filmed at 1.85 (flat) aspect ratio. When a 1.85 film is transferred directly to full-frame 4:3 by including the extra picture at the top and bottom, the actual size of the images on the TV are the same as for a letterbox version. In other words, letterboxing only covers the part of the picture that also was covered in the theater.

A pan and scan transfer to 4:3 makes the "I didn't pay good money for my 30-inch TV just to watch black bars" crowd happy, but a letterbox transfer to anamorphic 16:9 is still needed to appease the videophiles who demand the theatrical aspect ratio and to keep the "I paid good money for my widescreen TV" crowd from revolting. Ordinarily, this would call for two separate products, but not with DVD. The producer of the disc can put the 4:3 version on one side (or one layer) and the letterboxed 16:9 version on the other. This more or less doubles the premastering cost and slightly increases the mastering and replication costs, but with

[28]Interview in *Widescreen Review*, Issue 37, 2000.

production runs of 100,000 copies or more, this adds less than a dollar to the unit cost. On the other hand, the widescreen letterbox transfer is sometimes reserved for a special edition and sold as a separate product at a higher price.

As widescreen TVs and HDTVs slowly replace traditional TVs, 4:3 transfers will become less common, and even letterboxed 4:3 transfers will become more appreciated. In Japan and Europe, where widescreen TVs already outsell standard TVs, letterboxed video is more popular.

The Pin-Striped TV: Interlaced vs. Progressive Scanning

One of the biggest problems facing early television designers was displaying images fast enough to achieve a smooth illusion of motion. Early video hardware was simply not fast enough to provide the required flicker fusion frequency of around 50 or 60 frames per second. The ingenious expedient solution was to cut the amount of information in half by alternately transmitting every other line of the picture (see Figure 3.27). The engineers counted on the persistence of the phosphors in the television tube to make the two pictures blur into one.[29] For a 525-line signal, first the 262 (and a half) odd lines are sent and displayed, followed by the 262 (and a half) even lines. This is called *interlaced scanning*. Each half of a frame is called a *field*. For the NTSC system, 60 fields are displayed per second, resulting in a rate of 30 frames per second. There are 480 active lines of video, meaning that only 240 lines are visible at a time. For the PAL and SECAM systems, there are 50 fields per second, resulting in a rate of 25 frames per second, and there are 576 active lines out of a total of 625, giving 288 lines per field.

Figure 3.27 Interlaced Scan and Progressive Scan

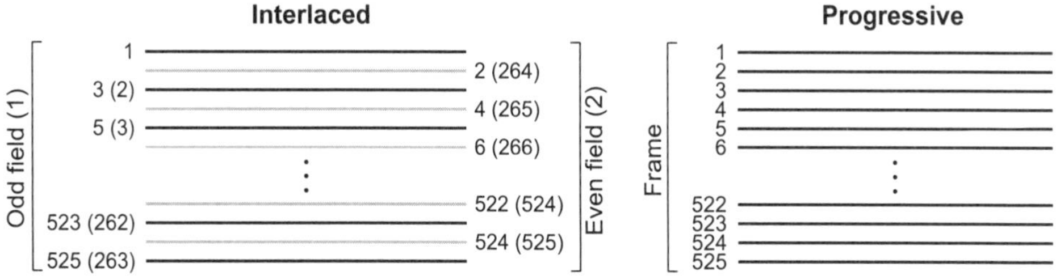

The alternative approach, *progressive scanning*, displays every line of a complete frame in one sweep. Progressive scanning requires twice the frequency in order to achieve the same refresh rate. Progressive scan monitors are more expensive and generally are used for computers. *High-definition television* (HDTV) also includes progressive scanning.

Progressive scan provides a superior picture, overcoming many disadvantages of interlaced scanning. In interlaced scanning, small details, especially thin horizontal lines, appear

---

[29]Many texts refer to "persistence of vision" as the phenomenon that allows interlaced video and motion pictures in general to create a seemingly continuous moving image. This is largely incorrect (see "A Brief History of Video Technology" in Chapter 2).

only in every other field. This causes a disturbing flicker effect, which you can see when someone on TV is wearing stripes. A common practice in video production is to filter the video to eliminate vertical detail smaller than two scan lines. This improves the stability of the picture but cuts the already poor resolution in half. NTSC video frames must be reduced to 200 lines of detail before interline flicker disappears. The flicker problem is especially noticeable when computer video signals are converted and displayed on a standard TV. The alternating black and white horizontal lines in Macintosh window titles were especially problematic. In addition to flicker, line crawl occurs when vertical motion matches the scanning rate. Interlaced scanning also causes problems when the picture is paused. If objects in the video are moving, they end up in a different position in each field. When two fields are shown together in a freeze-frame, the picture appears to shake. One solution to this problem is to show only one field, but this cuts the picture resolution in half. You may have seen this effect on a VCR: when the tape is paused, much of the detail disappears.

Since film runs at 24 progressive frames per second, displaying it at NTSC rates of 60 video fields per second requires a process called *2-3 pulldown*, where one film frame is shown as two fields, and the following film frame is shown as three fields (see Figure 3.28).

Figure 3.28 Converting Film to Video

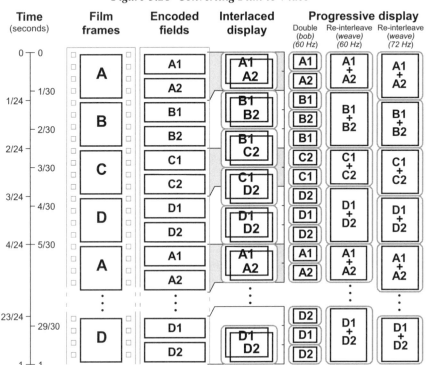

This pattern results in pairs of 24-per-second film frames converted to 60-per-second TV fields ($[2+3] \times 12 = 60$). Unfortunately, this causes side effects. One is that film frame display times alternate between 2/60 of a second and 3/60 of a second, causing a *motion judder* artifact — a jerkiness that's especially visible when the camera pans slowly. Another side effect is that two of every five television frames contain fields derived from two different film frames, which doesn't cause problems during normal playback but can cause problems when pausing or playing in slow motion. A minor problem is that NTSC video actually plays at 59.94 Hz, so the film runs 0.1 percent slow and the audio must be adjusted to match. Displaying film at PAL rates of 50 video fields per second is simpler and usually is achieved by showing each film frame as two fields and playing it 4 percent faster.[30] This is sometimes called *2-2 pulldown*.

Most video is encoded from videotape. The videotape is created by a *telecine* machine, which performs 2-3 pulldown when making an NTSC tape. Since it would be inefficient to encode the extra fields, they are not duplicated in the MPEG-2 stream. A good encoder recognizes and removes the duplicate fields. This is called *inverse telecine* (no, it's not called 3-2 pushup). There are flags in the MPEG-2 stream that indicate which fields to show when and which fields to repeat when. The encoder sets the repeat_first_field flag on every fifth field, which instructs the decoder to repeat the field, thus recreating the 2-3 pulldown sequence needed to display the video on an interlaced TV. In other words, the decoder in the player performs 2-3 pulldown, but only by following the instructions in the MPEG-2 stream. Therefore, a film on DVD must be encoded for the intended display rate, either NTSC or PAL, but not both. Some players can convert PAL to NTSC or NTSC to PAL; this is covered in Chapter 9, Application Details.

Technically, DVD-Video can only be stored in interlaced format.[31] Signals from standard video cameras are already in interlaced format. Film, which is inherently progressive, is encoded into MPEG-2 as paired fields. Even though the encoding is field-based, the frame-based nature of the source can be preserved. This allows progressive-scan players to put Humpty Dumpty back together again.

Progressive DVD Players

When DVD was developed, there were about 1 billion interlaced TV sets in the world and fewer than 100,000 progressive TVs (not counting computer monitors). Not surprisingly, DVD is biased toward encoding and displaying in interlaced format. This doesn't mean that DVDs can't be displayed in progressive-scan format, but it does mean that it's not a trivial process. Nevertheless, it's worth doing. A major advantage of DVD is that computers and

[30]Since the video is sped up 4 percent when played, the audio must be adjusted before it is encoded. In many cases the audio speedup causes a semitone pitch shift that the average viewer will not notice. A better solution is to digitally shift the pitch back to the proper level during the speedup process.

[31]In MPEG-2 encoding, the decision between progressive and interlaced format can be made all the way down at the macroblock level. The DVD-Video specification limits MPEG-2 video to nonprogressive sequences, which can include both progressive and interlaced frames. Progressive frames are still encoded for display as two fields, but they are identified as progressive. Interlaced frames can further include both progressive and interlaced macroblocks. Since progressive macroblocks are more efficient (using one motion vector instead of two), even interlaced source is often encoded with more than 50 percent progressive macroblocks. However, each frame is represented as two fields of 720×240 pixels each for NTSC, or 720×288 pixels each for PAL/SECAM.

progressive players can deinterlace the MPEG-2 video and display it progressively with considerably better quality than on standard interlaced displays. Progressive players work with all standard DVD titles but look best with video encoded from a progressive source such as film. The result is a significant increase in perceivable vertical resolution for a more detailed and filmlike picture.

A progressive-scan DVD player converts the interlaced (480i) video from DVD into progressive (480p) format for connection to a progressive display at 31.5 kHz or higher. It's also possible to buy an external *line multiplier* to convert the output of a standard DVD player to progressive scanning. All DVD computers are progressive players because the video is displayed on a progressive monitor. However, quality of deinterlacing and video playback varies wildly from computer to computer.

Converting interlaced DVD video to progressive video involves much more than putting film frames back together. There are essentially four methods of converting from interlaced video to progressive video:

1. Reinterleave (also called *weave*; see Figure 3.29). If the original video is from a progressive source, the two fields can be recombined into a single frame.

2. Line-double or line-multiply (also called *bob*; see Figure 3.30). If the original video is from an interlaced source, simply combining two fields will cause motion artifacts (the effect is reminiscent of a zipper), so instead, each line of a single field is repeated twice to form a frame. Better line doublers use interpolation to produce new lines that are a combination of the lines above and below. The term line doubler is vague and misleading because cheap line doublers only bob, whereas expensive line doublers (those that contain digital signal processors) also can weave, and in many cases, the number of lines is more than doubled.

3. Field-adaptive deinterlacing. This method examines individual pixels across three or more fields and selectively weaves or bobs regions of the picture as appropriate. Regions of pixels that do not change across frames can be reinterleaved without creating motion artifacts. Sections where there is motion can be bobbed or can be averaged across fields to create a motion blur effect. Only a few years ago, the cost of field-adaptive deinterlacing was $10,000 and up, but it already can be found in some consumer DVD players.

4. Motion-predictive deinterlacing.[32] This method uses massive image processing to identify moving objects in order to selectively weave or bob regions of the picture as appropriate. These systems decompose the picture into two-dimensional or three-dimensional representations that are used to generate a progressive version of the picture. For example, converting interlaced video of a basketball game to progressive format would require that a model be generat-

---

[32]The term motion-adaptive originally applied only to systems that perform object motion analysis, but it now tends to be used for both field-adaptive and motion-predictive deinterlacing.

ed for the frame of reference, for the ball, and for each player (each of which might be moving in simple motion across the screen or might be moving toward or away from the camera, growing and shrinking in apparent size). Some systems use MPEG-2 motion vectors as clues to guide the analysis process, but since motion vectors also can be used to replicate similar areas that appear anywhere in the frame, they can't be relied on exclusively for motion detection. High-quality motion-predictive systems tend to be quite costly, even today.

Figure 3.29 Weave

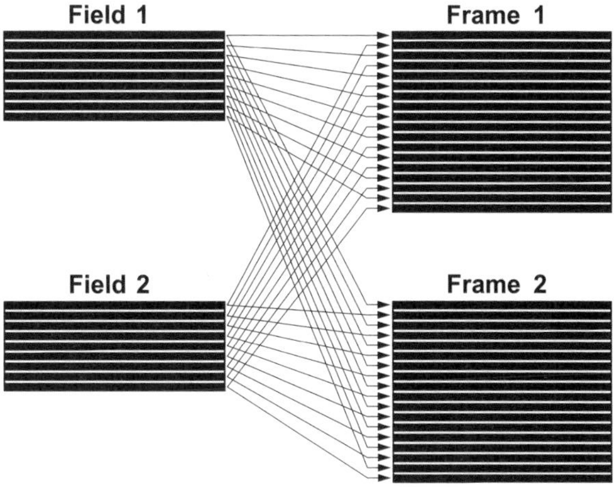

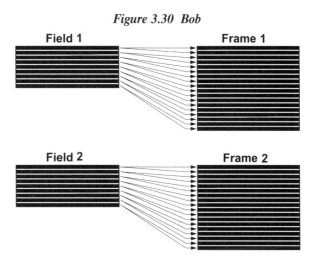

Figure 3.30 Bob

The three common categories of deinterlacing systems are:

1. **Integrated** This is usually best, where the deinterlacer is integrated with the MPEG-2 decoder so that it can read MPEG-2 flags and analyze the encoded video to determine when to bob and when to weave. Most DVD computers use this method.

2. **Internal** The decoded digital video is passed from the MPEG-2 decoder to a separate deinterlacing chip. A potential disadvantage is that flags, motion vectors, and other information in the MPEG-2 stream may no longer be available to help the deinterlacer determine the original format and cadence of the progressive source.

3. **External** Analog video from the DVD player is passed to a separate line multiplier or to a display with a built-in line doubler. In this case, the video quality is slightly degraded as it is converted to analog, back to digital, and often back again to analog. In addition, ancillary data from the MPEG-2 stream are not available. For high-end projection systems, a separate line multiplier (which bobs, weaves, and interpolates to a variety of scanning rates) may achieve the best results.

As deinterlacers become better, using field-adaptive and motion-predictive techniques, the differences in quality between the three categories, or at least the first two, will mostly disappear.

For better quality, any deinterlacing process other than simple bobbing also must undo the 2-3 pulldown process on a film source. Since the MPEG-2 encoder already did this, the deinterlacer simply must ignore the repeat field flags. This could be called "synthetic inverse telecine." An external deinterlacer has a harder time because it receives the signal after the player has repeated the fields.

A progressive DVD player has to determine whether the video should be line-doubled or reinterleaved. When reinterleaving film-source video, the player also has to deal with the difference between film frame rate (24 Hz) and TV frame rate (30 Hz). Since the 2-3 pulldown trick can't be used to spread film frames across progressive video frames, there are worse motion artifacts than with interleaved video. Progressive video is commonly displayed at 60 Hz, twice the normal rate, so frames are repeated in a 2-3 sequence, which means that the smoothing effect of interlaced fields is lost. However, the increase in resolution more than makes up for it. Advanced progressive players and DVD computers can get around the problem by displaying at multiples of 24 Hz, such as 72 Hz, 96 Hz, etc.

A progressive player also has to deal with problems such as video that doesn't have clean cadence (such as when it is edited after being converted to interlaced video, when bad fields are removed during encoding, or when the video is sped up or slowed down to match the audio track). Figure 3.28 shows how film frames cross video frames (frames B and D). If the video is cut on any of these frames, the frame coherency is lost, and the deinterlacer no longer has a clean sequence of paired fields to weave back together. The MPEG encoder is also affected because it becomes harder to do inverse telecine.

Another problem is that many DVDs are encoded with incorrect MPEG-2 flags, so a reinterleaver that uses these flags has to recognize and deal with pathological cases. In some instances it's practically impossible to determine if a sequence is 30-frame interlaced video or 30-frame progressive video.

A related problem is that many TVs with progressive input don't allow the aspect ratio to be changed. When a non-anamorphic signal is sent to these TVs, they stretch it horizontally instead of properly windowboxing or proportionally enlarging it.

Just as early DVD computers did a poor job of progressive-scan display of DVDs, the first generation of progressive consumer players also were a bit disappointing. However, as techniques improve, and as DVD producers become more aware of the steps they must take to ensure that their content looks good on progressive displays, and as more progressive displays appear in homes, the experience will undoubtedly improve, bringing home theaters closer to real theaters.

New Display Technologies

In recent years, there has been a dramatic expansion in the area of new display technologies. We have gone from a world dominated by *cathode ray tubes* (CRTs) to one in which it will soon be difficult to find one. With advances in these new display technologies, it will be important to understand how DVD and the next-generation disc formats will be affected.

What's Wrong with CRT?

If you go down to your local consumer electronics store today, you will see a very interesting trend: a large majority of the standard definition televisions on display are CRTs, while almost all of the high definition displays are not. One of the leading reasons for this dicotomy is size. It is very difficult to manufacture a glass tube CRT at very large sizes. Given that the CRT is a type of vacuum tube technology, the tube itself must be able to structurally withstand the pressure of containing a vacuum without imploding. As the tube grows larger, this becomes more difficult without any sort of internal support. Additionally, it's just simply a lot of glass, making it expensive and, more importantly, HEAVY.

Does that mean CRT is going away? Probably, but not completely. It's true that a 500+ pound CRT television will have a difficult time competing against the convenience of a relatively light, wall-mounted plasma screen or LCD. Even in situations where taking up a large space is not a problem, the convenience of a lightweight DLP system may outweigh (pun intended) the CRT heavyweight. However, it would be a shame to lose CRTs. For one thing, there has been over 60 years of research applied to developing the color display technologies that CRT uses, including phosphors, screen masks and color dot distribution to name a few.

Types of New Displays

In order to really see the benefit of CRTs, all you have to do is put a properly calibrated CRT television next to any of the other display types and just try to get them to look as good as the CRT. The current limitations of these other display types quickly become evident. Below is a quick summary of some of the new display technologies and the problems that tend to come with them.

Digital light processing (DLP) This type of display is based on the *digital micromirror device* (DMD), a silicon chip whose surface is covered by a matrix of tiny mirrors. Through electrical signals, these mirrors can toggle from an "off" position to an "on" position and back. Basically, each micromirror represents a pixel; by reflecting a light source off the mirror, one can project a black and white image. Flickering the mirror at variable high rates creates the impression of varying amounts of light, thereby displaying a grayscale. Passing the image through a rotating color wheel that is properly synchronized with the mirror flashes generates a color display. This is the basis for most consumer rear-projection DLP televisions, which can offer a bright, crisp picture. However, depending on the color wheels used and the sophistication of the display logic, DLP displays tend to have difficulty reproducing some of the darker color ranges. Likewise, being a rear-projection device, there may be some *light bleed*, which can wash out the image, making black areas appear gray. Some DLP displays may exhibit color shift problems, exaggerated sharpness, and excessive brightness. The latter two characteristics, in particular, can overemphasize encoding artifacts in MPEG video.

Plasma display device (PDP) Plasma displays work by applying a charge to a small gas-filled cell. The gas becomes ionized and, in turn, interacts with a phosphor on the surface of the cell, which glows a given color. Because plasma displays work with phosphors, they are able to utilize much of the research that has gone into phosphor research for CRTs and reproduce more natural colors. However, it also means that they may burn in quickly, just like a CRT, leaving a permanent after-image where something remained on screen for a length of time. For example, channel logos that occupy a corner of the screen can begin to burn in after a period of time. Plasma displays can consume large amounts of power compared to other types of displays, and tend to suffer from bleed-over effects in which the plasma from one cell bleeds over to other cells.

Liquid-crystal display (LCD) Liquid crystal displays operate by sandwiching color filters and polarizing filters on either side of a liquid crystal cell matrix, and placing a white backlight behind the sandwich. The backlight remains lit while the display is active. The liquid crystal in each cell, by default, stays "curled" up such that it blocks the polarized illumination from the backlight. When a charge is applied to a cell, the crystal "unwinds," allowing the illumination from the backlight to escape through the color filter. One of the most noticeable problems with LCD is the limited viewing angle, although with the latest updates to these technologies, the viewing angle has been extended dramatically (up to as much as 178 degrees). Increased viewing angle tends to lead to increases in *light bleed* — white light that escapes to wash out the picture and raise black levels. In addition, the "unwinding" process takes time at a molecular level. As a result, the time it takes to fully refresh an LCD display is currently limited to approximately 24 ms. This is fine for a progressively displayed frame of film, which usually lasts for about 40 ms, but interlaced video (59.94 fields per second) requires an update to the screen every 16.7 ms; LCD is not currently able to refresh that rapidly.

• **_Liquid crystal on silicon (LCoS)_** Similar to DLP rear projection displays, this type of television operates by projecting colored light through a translucent matrix of liquid crystals, in which each cell can be individually made opaque, transparent or somewhere in between. Unfortunately, because LCoS uses liquid crystal, these displays tend to suffer from the same performance issues as LCD and often require three separate light sources (or one light source with three pathways through three separate color filters) in order to provide full color reproduction.

With the rapid acceptance of these new display technologies and with new ones on the way, such as _organic light emitting diodes_ (OLED) and _high dynamic range_ (HDR) displays, the DVD content producer's job becomes much more complicated. It used to be that a producer only had to worry about NTSC and PAL video formats. Now they have that plus a half dozen new display technologies that bring their own problems to the production mix.

The New Video Formats

As the next-generation disc formats enter the marketplace, they will meet the collection of recently introduced video formats, with a new set of matching considerations. For example, there is an ongoing battle between two high definition video formats — 720p (1280×720 resolution at 59.94 progressive frames per second) and 1080i (1920×1080 resolution at 29.97 interlaced frames per second). Both formats require a similar pixel bandwidth in terms of the total number of pixels that must be drawn each second, but they each look different to viewers. What is particularly challenging is that as more HD displays accommodate both formats, what they do to the video internally before it is displayed may be extremely different.

Most high-definition displays today have a native resolution of approximately 1280×720. Some are slightly smaller (e.g., 1024×768 for most DLPs) while others may be slightly higher (e.g., 1368×848 for some plasma displays). One might expect that the 720p format would be more widely adopted. However, there are a significant number of existing displays (rear projection or standard CRTs) that prefer 1080i. The problem is that some of these displays, when they receive a 720p signal down-convert the image to 480p (720×480 at 59.94 progressive frames per second) to maintain the progressive frame rate. That defeats the purpose of creating high definition content, especially if it begins to interfere with the readability of onscreen text. Similarly, if material created at 1080i is going to be down-converted to 720p resolutions, because the displays have a lower native resolution and are progressive devices, what is the point of designing for 1080i resolution?

Perhaps the greatest concern is that a display may state that it supports both 720p and 1080i video formats, but they rarely indicate what scaling, frame rate conversion, deinterlacing, or other signal processing these formats may undergo prior to being displayed. Finding a good match between video format and display device may be just a matter of luck for some time to come.

Chapter 4
Features

Bells and Whistles

The creators of DVD realized that, to succeed, DVD had to be more than just a roomier CD or a more convenient laserdisc. Hollywood had started the ball rolling by requesting a digital video consumer standard that would hold a full-length feature film, had better picture quality than existing high-end consumer video plus widescreen aspect ratio support, contained multiple versions of a program with parental control, supported high-quality surround-formatted audio with soundtracks for at least three languages, and had built-in content protection. The computer industry added their requirements of a single format for computers and video entertainment with a common cross-platform file system, high performance for both movies and computer data, compatibility with CDs and CD-ROMs, compatible recordable and rewritable versions, no mandatory caddy or cartridge, and high data capacity with reliability equal to or better than CD-ROM. Then, Hollywood agreed that they too wanted a content protection system, with the added requirement that the new standard has to include a locking system to control release across different geographic regions of the world.

Next, the designers threw in a few more features, such as multiple camera angles and graphic overlays for subtitling or karaoke, and DVD was born. Unlike CD, where the computer data format was cobbled on top of the digital music format, the digital data storage system of DVD-ROM is the base format for the new standard. DVD-Video is built on top of DVD-ROM, using a specific set of file types and data types. A DVD-ROM may contain digital data in almost any conceivable format, as long as a computer or other device can make use of it. DVD-Video, on the other hand, requires simple and inexpensive video players, so its capabilities and features are strictly defined.

From the beginning, the plan was to create a separate DVD-Audio format based on input from the music industry. Their requirements included copyright identification and content protection, compatibility with DVD-ROM and DVD-Video, CD playback (including an optional hybrid DVD/CD format that could play in CD players), navigation with random access similar to DVD-Video but also usable on players without an attached video display, slideshow features and, of course, superior sound quality. The result is that DVD-Audio supports a subset of DVD-Video features and, apart from portable devices and car audio systems, eventually every new DVD-Audio player will also play DVD-Video discs. The converse, however, may not be true. DVD player manufacturers are not required, let alone obligated, to support DVD-Audio.

The following sections present the features of DVD-Video. The term *player* also applies to software players on computers, as well as other devices such as video game consoles that have the ability to play DVD-Video discs. DVD-Audio features are mentioned when relevant.

Over 2 Hours of High-Quality Digital Video and Audio

More than 95 percent of Hollywood movies are shorter than 2 hours and 15 minutes, so 135 minutes was chosen as the goal for the digital video disc. Uncompressed, this much video would require more than 255 gigabytes.[1] DVD uses MPEG-2 compression to fit high-resolution digital video onto a single disc. The MPEG-2 encoding system compresses video in two ways: spatially, by reducing areas of repetitive detail and removing information that is not perceptible, and temporally, by reducing information that does not change over time. Reducing the video information by a factor of almost 50 enables it to be less than 5 gigabytes. Unfortunately, compression can cause unwanted effects such as blockiness, fuzziness, and video noise. However, the variable data rate compression method used for DVD enables extra data to be allocated for more complex scenes. Carefully encoded video is almost indistinguishable from the original studio master.

The 135 minute length (or the absurdly precise 133 minute length, see "Myths") is a rough guideline based on estimates of average video compression and number of audio tracks. The length of a movie that can fit on a standard DVD depends almost entirely on how many audio tracks are available and how heavily the video is compressed. Other factors come into play, such as the frame rate of the source video (24, 25, or 30 frames per second), the quality of the original (soft video is easier to compress than sharp video or video with high levels of film grain, and clean video is easier to compress than noisy or dirty video), and the complexity (slow, simple scenes are easier to compress than scenes with rapid motion, frequent changes, and intricate detail).

In any case, the average Hollywood movie easily fits on one side of a DVD. This overcomes one of the big objections to laserdisc — that you had to flip the disc over or wait for the player to flip it over after each hour of playing time.

For DVD-Audio, quality is significantly improved by doubling the PCM sampling rates of DVD-Video and using lossless packing to increase disc playing times. A single-layer DVD-Audio disc can play 74 minutes of super-fidelity multichannel audio or over seven hours of CD-quality stereo audio.

Widescreen Movies

Television and movies shared the same rectangular shape until the early 1950s when movies began to get wider. Television stayed unchanged until recently. Widescreen TVs are appearing in greater numbers each year, and DVD has been a factor in the increase in demand. Movies can be stored on DVD in widescreen format to be shown on widescreen TVs close to the width envisioned by the director. DVD includes techniques to show these widescreen movies on regular televisions, while also addressing the new widescreen television display formats. The different aspect ratios are discussed in Chapter 3.

---

[1]Digital studio masters for standard definition video generally use 4:2:2 10-bit sampling, which at 270 Mbps eats up over 32 megabytes every second.

Multiple Surround Audio Tracks

The DVD-Video standard provides for up to eight soundtracks to support multiple languages and supplemental audio. Each of these tracks may include surround sound with 5.1 channels of discrete audio.[2] DVD surround sound audio may use either Dolby Digital (AC-3), DTS Digital Surround, or MPEG-2 audio encoding. The 5.1-channel digital tracks can be downmixed by the player for compatibility with legacy stereo systems and Dolby Pro Logic audio systems. An option is also available for better-than-CD-quality linear PCM audio. Almost all DVD players include digital audio connections for high-quality output.

The usefulness of multiple audio tracks was discovered when digital audio was added to laserdiscs, leaving the analog tracks free. Visionary publishers, such as Criterion, used the analog tracks to include audio commentary from directors and actors, musical sound tracks without lyrics, foreign-language audio dubs, and other fascinating or obscure audio tidbits.

DVD-Audio improves on the audio features of DVD-Video with higher sampling rates for PCM and improved support for multichannel PCM audio tracks and audio downmixing.

Most DVD players allow the user to select a preferred language so that the appropriate menus, language track, and subtitle track are selected automatically when available. In many cases, the selection also determines the language used for the player's on-screen display.

Karaoke

DVD keeps karaoke fans singing because it includes special karaoke audio modes to play music without vocals or to add vocal backup tracks. More importantly, DVD's subtitle feature breaks the language barrier with up to 32 different sets of lyrics in any language, complete with bouncing balls or word-by-word (or ideogram-by-ideogram) highlighting.

Karaoke support is optional for players. Karaoke players have the ability to mix karaoke audio tracks (guide/vocal tracks and melody tracks) into the base stereo tracks. Provisions are also included for identifying the music and singing, such as male vocalist, female soloist, chorus, and so on.

Subtitles

Video can be supplemented with one of 32 subpicture tracks for subtitles, captions, and more. Unlike existing closed captioning or teletext systems, DVD subpictures are graphics that can fill the screen. The graphics can appear anywhere on the screen and can create text in any alphabet or symbology. Subpictures can be Klingon characters, karaoke song lyrics,

---

[2]Discrete means that each channel is stored and reproduced separately rather than being mixed together (as in Dolby Surround) or simulated. The ".1" refers to a low-frequency effects (LFE) channel that adds bass impact to the presentation. MPEG-2 and SDDS audio allow 7.1 channels, but this feature is unlikely to be used for home products.

Monday Night Football-style motion diagrams, pointers and arrows, highlights and overlays, and much more. Subpictures are limited to a few colors at a time, but the graphics and colors can change with every frame, which means subpictures may be used for simple animation and special effects. Some DVDs use subpictures to show silhouettes of the people speaking on a commentary track, in the style of the *Mystery Science Theater 3000* TV show. Being able to see the gestures and finger pointing of the commentators enhances the audio commentary.

A transparency effect can be used to dim areas of the picture or make picture areas stand out. This technique can be used to great effect for educational video and documentaries. Other options include covering parts of a picture for quizzes, drawing circles or arrows, and even creating overlay graphics to simulate a camcorder, night-vision goggles, or a jet fighter cockpit.

Different Camera Angles

One of the most innovative features of DVD is the option to view scenes from different angles. A movie or television event may be filmed with multiple cameras, which could then be authored so that the DVD viewer can switch at will between up to nine different viewpoints. In essence, camera angles are multiple simultaneous video streams. As you watch, you can select one of the nine video tracks just as you can select one of the eight audio tracks.

This feature presents a paradigm shift that could be as significant as the way sound changed motion pictures. The storytelling opportunities are fascinating to contemplate. Imagine a movie about a love triangle that can be watched from the point of view of each main character; or a murder mystery with multiple solutions; or a scene that can be played at different times of the day, different seasons, or different points in time. Music videos can include shots of each performer, enabling viewers to focus on their current favorite or to pick up instrumental techniques. Classic sports videos can be designed so that armchair quarterbacks have control over camera angles and instant replay shots. Exercise videos may allow viewers to choose their preferred viewpoints. Instructional videos can provide close-ups, detail shots, and picture insets containing supplemental information. The options are endless, and they enable new tools for new approaches to filmmaking and video production.

The disadvantage of this feature is that each angle requires additional footage to be created and stored on the disc. A program with three angles available the entire time can only be one-third as long if it has to fit in the same amount of space as a single angle DVD.

Multistory Seamless Branching

A major drawback of almost every previous video format, including laserdiscs, Video CD, and even computer-based video formats, such as QuickTime, is that any attempt to switch to another part of the video causes a break in play. DVD-Video achieves completely seamless branching. For example, a DVD may contain additional "director's cut" scenes for a movie but can jump over them without a break to recreate the original theatrical version.

This creates endless possibilities of the mix-and-match variety. At the start of a movie, the viewer could choose to see the extended director's cut, with alternate ending number four, and the punk rock club scene rather than the jazz club scene, and the player would jump around the disc, unobtrusively stitching the selected scenes together. Of course, this requires significant additional work by the director or producer, and most mass-market releases skip this option, leaving it to small, independent producers with more creative energy.

Parental Lock

DVD includes parental management features that can be set to block playback and to only allow viewing of a particular rating level if multiple versions of a movie are on a single disc. Players can be set to a specific parental level using password-protected onscreen menus. If a disc with a rating above this level is put in the player, it will not play. In some cases, different programs on the disc could have different ratings.

A disc also can be designed so that it plays a different version of the movie depending on the parental level that has been set in the player. By taking advantage of the branching feature of DVD, scenes can be skipped over automatically or substituted during playback, usually without a visible pause or break. For example, a PG-rated scene can be substituted for an R-rated scene, along with the appropriate dialog containing less profanity. This requires that the disc be carefully authored with alternate scenes and branch points that do not cause interruptions or discontinuities in the soundtrack.

Unfortunately, fewer than 1 percent of DVDs use the multirating feature. Hollywood studios are not convinced that the demand merits the extra work involved, which includes shooting extra footage, recording extra audio, editing new sequences, creating seamless branch points, synchronizing the soundtrack across jumps, submitting new versions for MPAA rating, dealing with players that do not implement parental branching properly, having video store chains refuse to carry discs with unrated content, and much more. The few discs that have multirated content do not have standard package labeling or other ways to be easily identified.

Another option is to use a software player or a specialized settop DVD player that can read a "playlist" telling it where to skip scenes or mute the audio. Playlists can be used to "retrofit" the thousands of DVD movies that have been produced without parental control features.

Menus

To provide access to many of the advanced features, the DVD-Video standard includes onscreen menus. Menus are used to select from multiple programs, choose different versions of program content, navigate through multilevel or interactive programs, activate features of the player or the current disc, and more. The video can stop at any point for interaction with the viewer, or selectable hot spots can be "hidden in plain sight" on live video.

For example, a movie disc may have a main menu from which you can choose to watch the movie, view a trailer, watch a "making of" featurette, or peruse production stills. Another menu may also be available from which you can choose to hear the regular sound track, foreign-language sound track, or director's commentary. Selecting the supplemental information option from the main menu may bring up another menu with options such as production stills, script pages, storyboards, and outtakes.

Interactivity

In addition to menus, DVD can be even more interactive if the creator of the program takes advantage of the rudimentary command language that is built into all DVD players. DVD-Video can be programmed for simple games, quizzes, branching adventures, and so on. DVD brings a new level of personal control to video programs. While it is not apparent just how much control the average couch potato desires, directing the path and form of a presentation is definitely an appealing option. "Choose your own ending" stories have graduated from paper to video. The creative community could embrace an entire new genre of nonlinear cinema.

For example, a music video disc could provide an editing environment where the viewer can choose music, scenes, and so on to create their own custom version. An instructional video can include comprehension check questions. If a wrong answer is selected, a special remedial segment can be played to further explain the topic.

On-Screen Lyrics and Slideshows

The DVD-Audio format includes features for displaying lyrics on the screen and highlighting parts of the lyrics in time with the audio. This is also possible using the subtitle feature of the DVD-Video format, but it is not as straightforward.

The DVD-Audio format also includes a slideshow feature for showing pictures as the audio plays. The pictures may appear automatically at preselected points in the program, or the viewer can choose to browse them at will, independent of the audio. Although the DVD-Video format also supports programmed slideshows, it does not support browsable pictures that don't interrupt the audio.

Customization

As mentioned, DVD players may be customized with a parental lock. Other options can be set on a DVD player to also customize the viewing experience. Most DVD players can be set for the preferred soundtrack language and subtitle language and even menus in the chosen language, when available. Preferred aspect ratio — widescreen, letterbox, or pan and scan — also can be set.

For example, if you were studying French, you could set your preferences to watch movies with French dialog and English subtitles. If these are available on the disc, they would be selected automatically.

Instant Access

Consumer surveys have indicated that one of the most appealing features of DVD is that one never has to rewind or fast forward it. It cannot be underestimated how important time and convenience can be to consumers — just consider our penchant for microwave ovens, electric pencil sharpeners, and escalators. A DVD player can obligingly jump to any part of a disc — program, chapter, or time position — in less than a second.

Trick Play

In addition to near-instantaneous search, most DVD players include features such as freeze-frame, frame-by-frame advance, slow motion, double-speed play, and high-speed scan. Most DVD players scan backward at high speed, but due to the nature of MPEG-2 video compression, most cannot play at normal speed in reverse or step a frame at a time in reverse. This is only possible on advanced players that have more sophisticated video processors.

Access Restrictions

DVDs include a feature that enables the author of the disc to restrict *user operations* (UOPs) such as fast forward, chapter search, and menu access. Almost every button on the remote control can be blocked at any point on the disc. This is not always a benefit to the viewer (as when you are locked into the FBI warning or advertisements at the beginning of a disc), but it is helpful in complicated discs to keep button-happy viewers from going to the wrong place at the wrong time.

Durability

Unlike tape, DVDs are impervious to magnetic fields. A DVD left on a speaker or placed too close to a motor will remain unharmed. Discs are also less sensitive to extremes of heat and cold. Because they are read by a laser that never touches the surface, the discs will never wear out — even your favorite one that you play six times a week or the kids' favorite one that they play six times a day. DVDs are susceptible to scratching, but their sophisticated error-correction technology can recover from minor damage.

Programmability

Some DVD players are viewer-programmable, similar to CD players. Chapters may be selected for playback in a specified order. Just as you can rearrange the sequence of tracks in a music video to your own taste, you can drive your friends crazy by having the catchiest song reappear at strategically annoying points. To impress visitors, multidisc players may be programmed to show a demo of your favorite scenes from different discs. Note that the access restrictions mentioned previously may make it difficult to program jumps to arbitrarily selected parts of a disc.

Interoperability with ROM Players

DVDs can have more content than what may be available from the menu selections shown when using the disc in a settop DVD player. Computer software such as screen savers, games, and interactive enhancements can be included on a disc. DVD computers, home video game systems and WebTV boxes can support enhanced DVD features. A single DVD could contain a movie, a video game based on the movie, an annotated screenplay with "hot" links to related scenes or storyboards, as well as the searchable text of the screenplay's novelization complete with illustrations and hyperlinks. The disc could contain links to Internet Websites with fan discussion forums, related merchandise, and special promotions.

DVD is becoming the most common component of multipurpose settop boxes, such as a box that combines a digital video recorder, cable TV receiver, with a DVD player/recorder, or a video game console that is also a DVD player and a Web browser.

Availability of Features

Many of the enhanced DVD features entail extra work by the producer of the disc. Adding additional scenes, multiple language tracks, subtitles, ratings information, supplemental menus, authoring branch points, and more requires additional effort and expense. The extent to which movie producers support these features largely depends on two factors: the level of customer demand for them and how much customers are willing to pay for them.

In the laserdisc market, a thriving special-edition industry emerged, titillating videophiles with restored footage, outtakes, director's commentaries, production photos, and documentaries. These special editions required hundreds of hours of extra work by dedicated or obsessed professionals, and they generally sold for over $100 — three times the cost of a regular edition. Whereas, special editions of DVDs are even more common and more popular than those on laserdisc, but the content rarely rises to the level previously seen for laserdiscs, as the price can typically be increased by only $10 or so, if at all.

Beyond DVD-Video and DVD-Audio

Several new optical disc formats are currently under development, and all of them are contending for the coveted status as the successor to DVD. These new formats expand on the current features of DVD-Video and DVD-Audio by adding support for high definition video and higher-quality lossy and lossless audio compression systems. These formats hope to establish new paradigms for user interactivity and delivery of content. The following sections describe some of the new or impending features.

Navigation & Interactivity

The new formats accentuate and expand on the navigation principles of DVD-Video and DVD-Audio, bringing new and exciting levels of user interactivity. While continuing to offer the "Standard Content" level of currently available DVDs, an entirely new level of functionality with tremendous programming capabilities is a major feature of the next-generation of optical disc, and is referred to as "Advanced Content."

Standard Navigation (DVD-like)

The next-generation formats establish a standard content mode that is based largely on the navigation structure of DVD-Video. This mode retains support of essentially every feature noted above, along with additional extensions for high definition video and higher fidelity audio. Improvements have been made in defining the relationship between the content and the user, along with fixes to problems that were found in the previous DVD versions.

Advanced Navigation (PC-like)

In addition, each format offers a level of navigation and programmability that previously was only available on computers. For example, instead of limiting user control to operations evident on an infrared remote, software can be written that uses alternative controllers, such as a keyboard, mouse, or game pad. More sophisticated applications can be created that go well beyond the simple menus and games that standard navigation offers. For example, imagine having an application that allows the viewer to chose a frame of video from a movie, offload a coloring book style version of that frame, color it with a variety of painting tools, and then upload the resulting image to a computer for printing.

Picture in Picture

Picture-in-Picture (PIP) support is an added feature of the upcoming formats, in which a second video stream can be decoded and overlayed on the primary, full-resolution video. The PIP feature supports the *luma keying* (luminance keying) technique, where any pixel darker than a certain level becomes transparent so that the primary video remains visible. For example, a film's director could appear superimposed on the film while they explain the different scenes as they play. The visual portion of this commentary could be turned on or off while the commentary audio and feature video proceed uninterrupted.

Given the difficulties of scaling video well, predetermined size ratios have been defined for PIP — full resolution (720×480 for standard definition NTSC picture-in-picture video), 1/2 resolution (360×240), 1/4 resolution (180×120), etc. Initial implementations of this feature are expected to provide only standard definition PIP video. However, high definition PIP may be available in some high-end players as well. In addition, it is expected that it will be possible to scale the standard definition secondary video to full screen, allowing for a wide range of overlay effects.

Each of the next-generation disc specifications differ on how picture-in-picture will be implemented and on what functionality may be mandatory. Most likely, the PIP feature will be available for network-streamed content, for video from the persistent storage of the player, and for content that is multiplexed on the disc with the primary video. Depending on the origin of the PIP content, the producer can decide whether the PIP video should be tightly synchronized to the primary video or left to be freely started and stopped.

Menu Noises

Added to the new formats is the ability to incorporate sound effects with the disc content, such as click sounds for menus. Current DVD specifications only allow for a single audio stream to be decoded and played back at any one time. In the new formats, audio mixers have been added so that authored sound effects may be triggered and mixed with the feature audio or played independently. The feature may also be useful, for example, by adding a "pop" sound when popup information appears during movie playback, or providing comedic "heckling" of scenes in a movie, or creating a censor's "bleep" sounds to block certain words in the dialog (though the timing of the effect may not be terribly accurate).

One challenge that will come with menu sounds will be how to ensure that the effects can be mixed with the feature audio, while allowing for the feature audio to be experienced in its full multi-channel format, on legacy audio equipment.

Subtitles & Captions

Enhancements have been added that provide for text-based versions of subtitles and captions that can be rendered on-the-fly during DVD playback. Based on very small text files, such subtitles and captions can potentially be delivered via a network after the disc has already been released.

Network Capabilities

Each of the new formats supports various levels of network support, allowing the player to be connected to a home network or to the Internet. Extra content can be delivered via this network connection, such as updated information about the cast and crew members or previews of their latest films. The potential exists for supporting e-commerce transactions, allowing the viewer to browse a collection of products and perform a purchase online. The networking support does not go so far as to implement general web browsing capabilities, but it does provide for content streaming, uploading and downloading of files, and in some cases opening up individual socket connections for more sophisticated network capabilities.

Persistent Storage

The next-generation discs and players will include a level of persistent storage, allowing a title to store information like game high scores or chapter and title information for later access. This storage may be small, fixed amounts, such as compact flash memory, or it may be composed of an internal hard disk capable of storing tens or hundreds of gigabytes of information. This data can even include content files downloaded via the network connection or transferred from disc for later recall and playback.

Chapter 5
Content Protection

Before Hollywood would embrace DVD, it had to be assured that DVD would not put Hollywood's bread and butter out on the open market for anyone to make perfect digital copies. Thus were born various schemes intended to reduce consumer copying of video and audio. Four major components comprise a complete content protection ecosystem: conditional access, protected distribution, protected transmission and protected storage (see Figure 5.1).

Figure 5.1 Copy Protection

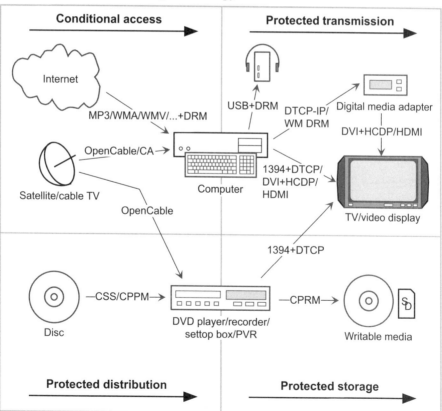

Conditional access. Conditional access ensures that only those users who are intended to receive content actually do. It primarily applies to broadcast, cable and Internet delivery of content, in which "open roads" are used to distribute content to an end user. With packaged media like DVD, the condition to access it is that you paid for the disc at the store.

Protected distribution. In order to ensure that the content remains secure in delivery to the end user, the distribution method must be protected. For DVD, this protection is provided by the *content scrambling system* (CSS) and by a method called *content protection for prerecorded media* (CPPM).

Protected transmission. The transmission of the content, usually in decoded *baseband* form to the presentation device, either a video display or an audio receiver, must be protected to prevent unauthorized duplication. Transmission potentially opens the "analog hole" in a secure content protection scheme. When digital video is decoded and transmitted as a baseband analog signal, it can easily be copied by a VCR or other recording device. Solutions exist to "plug the hole" for standard definition video, but not high definition, which is odd considering that the problem has been identified and understood for nearly a decade. Yet, early high definition televisions (HDTVs) were released with unprotected HD analog inputs which, in fact, were the only means to connect and view the high definition video!

Protected storage. The final component of a complete content protection system is to ensure that copies, when authorized, are properly protected. Fair use rules generally suggest that one backup copy of the content is permissable, but no more than one. Therefore, most protection systems provide a means of securing the initial copy and preventing further copies, unless subsequently authorized by the content.

In the purist sense, true content protection is a futile exercise, as a completely foolproof protection method would make it impossible to use the disc — if you can see it or hear it, you can copy it. Dr. Alan Bell, chairman of the *Copy Protection Technical Working Group* (CPTWG), points out that "really strong digital encryption is always ultimately defeatable by analog output." He elaborates that watermarking is the best solution, as long as players recognize watermarked analog copies. What many proponents of content protection apparently fail to recognize is that a digital copy of an analog output is only slightly degraded from the original digital source, and the newly created digital copy can be distributed and recopied as easily and endlessly as a digital copy of a digital source. Nevertheless, millions of dollars and hundreds of thousands of person-hours have been spent creating technical measures that make it harder to create digital and analog copies from DVD. The result is that the average DVD buyer cannot simply make a videotape copy of a DVD or copy DVD files to a hard drive or a writable DVD. A determined consumer, on the other hand, will find many ways to circumvent content protection.

Implementation of effective content protection must address three fronts: technology (the protection method of encryption in the digital domain and watermarking in the analog domain), business (using license guidelines that require compliant devices from manufacturers, maintaining reasonably priced content for consumers), and legislation (enforcement). These must be balanced according to the needs of content owners, manufacturers, and consumers.

Requirements of content owners are:

- No effect on the quality of the content
- Effective against unauthorized consumer use
- Robust and tamper-resistant
- Renewable (to recover from a breach of the system)
- Applicable to all forms of distribution or media
- Suitable for implementation on CE devices and PCs
- Low cost

Requirements of system manufacturers are:

- No effect on normal use of system
- Low additional resource requirement
- Tamper-resistant
- Voluntary
- Low cost

Requirements of consumers are:

- No effect on normal use of system
- No loss of quality
- Fair-use copying
- No additional cost
- No limitations on playback equipment or environment
- No artificial barriers (inconvenience) to access the content

Because the goals of each group sometimes conflict, the resulting protection methods are a compromise. Each group, however, seems reasonably happy with its ability to use DVD and the associated content protection measures. Manufacturers and studios are busy making products, and consumers are busy buying them.

When DVD was first being developed, content protection was intended to be part of the specification. After numerous lengthy delays, the DVD Forum recognized the need to separate the legal and technical aspects of content protection from the rest of the DVD specification. Content protection features are covered under separate licenses, where the makers of DVD playback systems essentially agree to implement content protection features in return for being granted access to the decryption keys and algorithms needed to play back encrypted content. The DVD Forum does not specify content protection technologies. The industry's CPTWG solicits proposals and makes recommendations. The DVD Forum Working Group 9 (WG9) is responsible for coordinating with the CPTWG. WG9 reviews content protection systems and submits them to the DVD Forum for approval. Once a content protection system is approved, the various working groups of the DVD Forum amend specifications as needed to support the requirements of the content protection system.

Over time, it was recognized that a framework was needed for security and to control access across the entire DVD family and beyond. The 4C entity (Intel, IBM, Matsushita, and Toshiba), in cooperation with the CPTWG and the *Secure Digital Music Initiative* (SDMI), developed the *Content Protection System Architecture* (CPSA). CPSA covers encryption, watermarking, playback control, protection of analog and digital outputs, etc. It is broadly defined to include physical and electronic distribution of analog and digital audio and video in consumer electronics and in computer systems (see Figure 5.2).

Figure 5.2 Content Protection System Architecture

	Prerecorded media, recordable media, and electronic distribution

Content Protection System Architecture (CPSA)

CSS	VEIL	CPRM VCPS	DTCP	Verance	CPPM
Protected storage (DVD-Video)	Content identification (video watermark)	Protected recording (writable DVD)	Protected transmission (IEEE 1394)	Content identification (audio watermark)	Protected storage (DVD-Audio)

Video

Audio

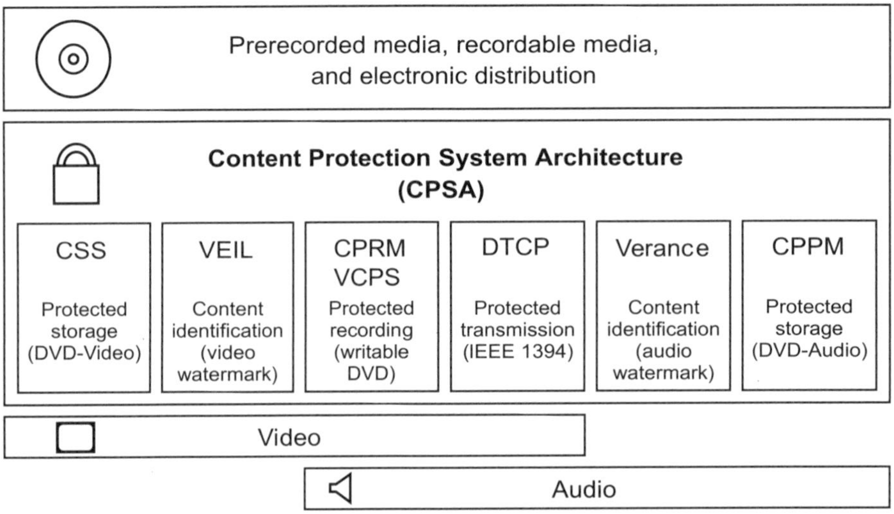

The CPSA creates a structure that defines the content protection obligations of compliant modules (see Figure 5.3). It defines how *content management information* (CMI) and *copy control information* (CCI) are carried and verified throughout the playback chain. CMI, also known as *usage rules*, specifies the conditions under which the content can be used. It also may contain other information such as *triggers* telling the player when and how to protect audio and video outputs. CCI, a subset of CMI, constrains how the content can be copied.

Eight forms of content protection apply to DVD. Each is explained below.

Content Scrambling System (CSS)

Because of the potential for perfect digital copies, worried movie studios forced a deeper content protection requirement into the DVD-Video standard. The *content scrambling system* (CSS) is a data encryption and authentication scheme intended to prevent copying video files directly from the disc. Occasional sectors containing A/V data (audio, video, or subpicture) are scrambled in such a way that the data cannot be used to recreate a valid signal. Scrambled sectors are encrypted with a combination of a *title key* and a *disc key*. The title key is stored in the sector header, which is normally not readable from a DVD-ROM drive or other DVD reader. Each *video title set* (VTS) on the disc has a separate key. The disc key is hidden in the control area of the disc, which is also not directly accessible.

Figure 5.3 Copy Protection System Architecture

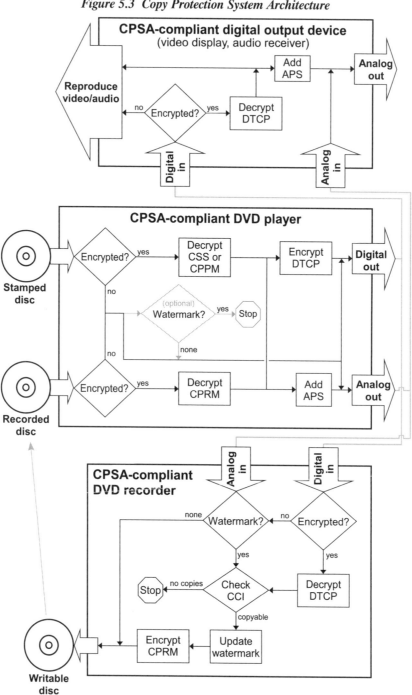

Use of CSS is strictly controlled by licensing. Each CSS licensee is given a *player key* from a master set of 400 keys that are stored on every CSS-encrypted disc. This allows a license to be revoked by removing its key from future discs. The CSS algorithm exchanges player keys with the drive unit to generate an encryption key that is then used to obfuscate the exchange of disc keys and title keys that are needed to decrypt data from the disc.

All standard DVD players have a decryption circuit that decrypts the data before displaying it. The process is similar to scrambled cable channels, except that the average consumer will never see the scrambled video and will have no idea that it has gone through an encryption/decryption process. The process does not degrade the data; it merely shifts the data around and alters it so that the original values are unrecognizable and difficult to decipher. The decryption process completely restores the data. The only case in which someone is likely to see a scrambled video signal is if they attempt to play the disc on a player or computer that does not support CSS or if they attempt to play a copy of the data or a copy of the disc. Since the copy does not include the key, the video signal cannot be decrypted and appears garbled or blank.

On the computer side, DVD decoder hardware and software must include a CSS decryption module. The encrypted data is sent from the drive to the decoder to be decrypted and then MPEG decoded before being displayed. DVD-ROM drives have extra firmware to exchange authentication and decryption keys with the CSS module in the computer in order to protect movies. Beginning in 2000, new DVD-ROM drives were required to support regional management in conjunction with CSS. Makers of equipment used to display DVD-Video (drives, decoder chips, decoder software, display adapters, and so on) must license CSS. CSS is not required of video players or DVD computers; systems without it will not be able to play scrambled movies.

NOTE
In the context of CSS, scrambling and encryption mean the same thing.

CSS Technical Overview

CSS encryption is applied at the sector level. If a disc contains encrypted data, the corresponding decryption keys are stored in the sector headers and in the control area of the disc in the lead-in, which is not directly readable by a PC. The keys can be read only by the drive in response to certain I/O commands that are controlled by authentication procedures. When an encrypted sector is encountered, the drive and the decoder exchange a set of keys, further encrypted by bus obfuscation keys to prevent eavesdropping by other programs. This authentication process eventually produces the key used by the decryptor to decrypt the data. Until an authentication success flag is set in the drive, it does not read encrypted sectors. The drive itself does not decrypt the data; it merely participates in the two-way process of establishing an authenticated key and then sends the encrypted data to the decryptor.

All components participating in this process must be licensed to use CSS. Although regional management is technically independent of content protection, it's included as a requirement of CSS-compliant components. See the Regional Management section at the end of this chapter for a few other details.

CSS requires that the analog output of decrypted video be covered by an *analog protection system* (APS), such as Macrovision or similar. That is, display adapters with TV outputs are required by license to include Macrovision circuitry, which is also described later in this chapter. The decoder or operating system queries the card. If the card reports that it has no TV output or that it has Macrovision-protected output, the system will send the card decrypted, decoded video. Otherwise, the card will not receive video and will show a black screen when playing encrypted DVDs.

A digital bus output, such as USB or IEEE 1394 (FireWire) must be protected by a *digital protection system* (DPS). The standards for these systems are still under development, although DTCP (*digital transmission content protection*) is the forerunner, as described later in this chapter. Digital video outputs, such as DVI and HDMI, must also be protected (see the DHCP section, also later in this chapter).

Video, audio, and subpictures are stored in MPEG packs. Each pack is one sector, so encryption can be selectively applied. About 50 percent of sectors are encrypted on a disc because 100 percent encryption can be a burden on software decoders. Encrypted content must be decrypted before the MPEG video decoder, audio decoder, and subpicture decoder can process it. The decryptor can be part of the same circuitry as the decoder, or it can be independent. The decryptor and the decoder may be implemented in hardware or software.

The system is expected to protect the decrypted data from being copied. This is the state in which the content is considered to be most valuable. Once it has been decoded, requirements to protect the content are less stringent, particularly because no mechanism protects RGB VGA output. After the video and audio is decrypted and decoded, the video is sent to the display and the audio is sent to the audio hardware.

CSS Authentication and Decryption

CSS is essentially a cryptographic method for the distribution and management of cryptographic keys. The decoder or the OS (referred to as the player in this discussion) and the DVD drive engage in a handshaking protocol in which all the communication between them is encrypted. After verifying that the decryption module is registered and not compromised (which is done by using a specific hash algorithm to match the decoder key to a key in the key block on the disc), the DVD drive passes the content key, copy control information, and encrypted content to the decryption module. The decrypted content is then sent on a secure channel to the decoder. The decoded content, along with the copy control information, is communicated to the video controller where it is converted to video signals. The control information tells the video controller if an analog protection scheme must be applied prior to delivering the video signals to the display.

The authentication process works roughly as follows (see Figure 5.4): the player and the drive generate random numbers and send them to each other. Each encrypts the number using a CSS hash function (which turns a 40-bit number into an 80-bit number) and sends back the result, called a *challenge key*. If the recipient can decode the number using the same function, then it knows that the other device is a bona fide member of the CSS club. Once the authentication process is complete, the drive sets a flag to provide access to encrypted sectors. All sectors on the disc can then be read (by any program), but encrypted sectors must be decrypted to be of any use.

Figure 5.4 The CSS Authentication Process

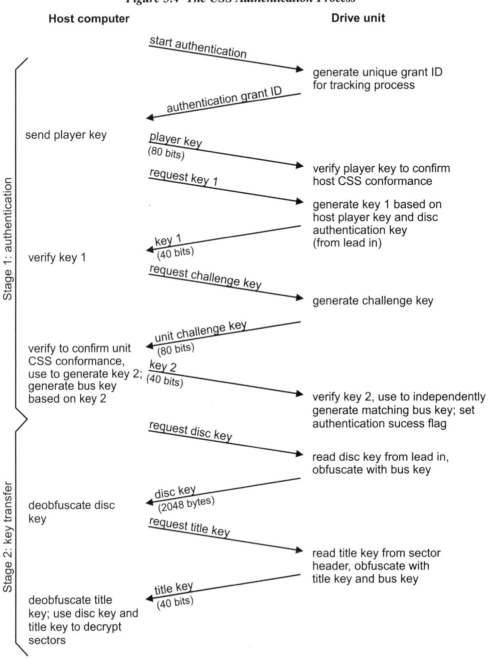

The second stage of CSS is to decrypt protected content. Each device uses challenge keys from the authentication step to generate a bus key. The bus key, which is never sent over the bus, is used to encrypt further key exchanges using a simple XOR operation. Each player has a key (or set of keys) assigned to it by the licensing authority. Each disc has a 2048-byte key block stored in the control area. The key block contains the disc key encrypted with 409 player keys. The purpose of this arrangement is that if a player key is known to be compromised, future discs can omit the encrypted key for that player, rendering the player unable to retrieve the disc key. Once a player retrieves and decrypts the disc key from its slot in the key block, it can use the disc key to decrypt the title key stored in the sector header of each encrypted sector. Each VTS on the disc uses a different title key. The decrypted title key is used to decrypt the video or audio contents of sectors that are encrypted with the CSS stream cipher.

CSS Licensing and Compliance for PCs

CSS-compliant computers are required to have the following:

- A CSS-licensed DVD-ROM drive with authentication hardware
- CSS-licensed decoding hardware/software, including an authenticator and decryptor
- CSS-licensed video components
- Regional management support
- Protection of digital output
- Protection of analog video output (other than RGB)
- Prevention against copying decrypted files
- Rejection of "no copies allowed" material on a recordable disc (may be provided by the drive or system)

Any computer industry segment involved in the production or integration of hardware or software components that are affected by CSS may be required to obtain a license and abide by its restrictions, as follows:

Operating System Makers. A CSS license is not officially required, but the operating system should protect decrypted data so that it cannot be copied either as files or as data streams. For example, a video *T-filter* must not be allowed to split off a stream of decrypted video during playback. The OS must also not enable complete bit-image copies of DVD-Video discs.

The OS must support the regional management for DVD-Video discs if not handled elsewhere. The OS must provide for, or not interfere with, DPS and APS information in the video output.

Application Developers. A CSS license may be required if the application directly manipulates DVD-Video files. The application is subject to the same restrictions as an OS.

Disc Manufacturers. Companies that master and replicate encrypted discs must have a CSS license in order to apply disc keys and encrypt sectors. They are responsible for maintaining the secrecy of the keys and the encryption algorithms.

Drive Makers. A CSS license is required only if the drive supports CSS authentication. If so, the drive must perform authentication handshaking and key obfuscation. The drive must also include firmware that stores a region code and enables it to be changed a limited number of times by the user.

Component Makers. A CSS license is required for any hardware or software component, such as a decoder or decryption module, which directly deals with the decrypted data stream. Decrypting components are strictly licensed because they use the "secret" descrambling algorithm and the "secret" keys. The components must safeguard the algorithm, the keys, and the decrypted data. Licensed components can only be sold to other CSS-licensed integrators or OEMs.

A CSS license is also required for DVD add-in cards that incorporate DVD-Video playback features. The hardware must protect decrypted digital and analog outputs.

A CSS license is not required for graphic cards and other video display components, but other licensed components are not allowed to be connected to display components that include television video output unless analog protection is provided. If a method is devised for protecting RGB outputs, it may be required as well.

System Integrators and OEMs. A CSS license is required for companies that produce or assemble DVD-Video-enabled computers. The final system must comply with all requirements, including video output protection, digital output protection, decrypted file protection, and regional management. No hardware or software can be added that enables the circumvention of any license requirements. Partly assembled subsystems must be distributed only to licensed integrators or resellers.

Retailers and Resellers. A CSS license is required only for retailers who do an additional assembly or integration of CSS-related components. The restrictions are then the same as for system integrators.

Users. No license is required, although users who desire to assemble their own systems might not be able to legally purchase licensed components. The goal is to make DVD-Video a plug-and-play system, with the end user unaffected by CSS entanglements.

A Recipe for Resistance

The DVD Copy Protection Technical Working Group (CPTWG) worked hard to make the licensing process as simple and as unrestrictive as possible, but any system with this many rules and requirements can easily go awry. Strenuous objections from computer hardware and software companies had to be overcome, but the golden glow of Hollywood movies on computers provided a strong incentive to compromise.

Most of the computer industry had no say in these negotiations. This is an industry notorious for its lack of regard for regulations and artificial barriers. Computer software developers long ago gave up the fight for copy-protected software because it did little to slow down those determined to make copies. However, it sorely inconvenienced honest users who were unable to make legal backup copies or had problems simply installing or uninstalling the software they had purchased. Microsoft, for instance, loses billions of dollars to illegal software copying, yet it still makes plenty of money. The company educates users, employs sophisticated anti-counterfeiting techniques such as holograms and thermographic ink, and aggressively pursues counterfeiters and commercial pirates, but it does not usually encrypt its software applications or use other technical protection methods. Hollywood, however, is not this sanguine about its bread-and-butter assets.

A worse problem is the potentially onerous restrictions on computer retailers and owners. Many mom and pop computer stores assemble low-cost systems from diverse components. How happy will they be with the licensing requirements demanded of them? How many of them will simply turn to the inevitable gray market for supplies? And what about the computer owner who simply wants to upgrade to a new video card? Even a lowly computer owner who purchases a system assembled from licensed components by a licensed integrator is presumably barred from buying a new licensed component because these components can only be sold to licensed integrators. Of course, nothing can prevent end users from swapping for different video boards without APS and making copies to their hearts' content. Clearly, however, the same factors of cost and convenience that keep consumers from engaging in mass duplication of audio CDs and videotapes apply to DVD as well, with or without complex technical content protection schemes.

The CSS algorithm and keys were supposed to remain a big secret, but the algorithm was reverse engineered and all the keys were derived by computer hackers. Security experts who analyzed the CSS implementation noted that its 40-bit key length made it easily compromised through brute force attacks, especially because only 25 bits of the key were uniquely employed. Still, CSS prevents the average user from using a computer to copy a DVD movie, which is what it was intended to do.

Content Protection for Prerecorded Media (CPPM)

CPPM replaces CSS for use with DVD-Audio.[1] It is also known as *4C* for the group of four companies that developed it: IBM, Intel, Matsushita, and Toshiba. CPPM is a method for encrypting and protecting content on prerecorded (read-only) discs. Copyright information is stored in every sector of the disc indicating whether or not the sector is allowed to be copied. Some sectors of the disc are encrypted. Licensing and compliance rules are similar to those described in the earlier section on CSS, as is the authentication mechanism, so no changes are required to existing drives. A disc with both DVD-Video and DVD-Audio content may use both CSS and CPPM. CPPM is planned for use on other prerecorded media such as SD memory cards.

CPPM is generally more robust and sophisticated than CSS. It has no title keys, and the disc key is replaced by an *album identifier*. The role of the album identifier is to provide a key that cannot be duplicated on recordable media, since it is stored in the control area of the lead-in, which is not accessible on writable discs. A 56-bit key is applied to 64-bit chunks of data for the encryption/decryption process. The final 1,920 bytes of an MPEG pack (a sector on the disc) are encrypted. DVD-Audio data is interleaved with five 64-bit bit key conversion values used to vary the CPPM encryption after each pack.

Each player has a set of 16 *device keys*. Key sets may be either unique to each device or shared by multiple devices. Device keys are highly confidential. Rather than secretly storing

---

[1]A slightly improved system called CSS2, originally intended for use on DVD-Audio discs, was abandoned and replaced by the more sophisticated CPPM when CSS was cracked in 1999.

the set of all known device keys on the disc, CPPM stores a *media key block* (MKB) in the DVDAUDIO.MKB file on the disc. The media key block is provided by the 4C entity to disc replicators. The player performs a series of decryptions and mathematical transforms on the media key block's logically ordered rows and columns of data with its device key. The resulting *media key* is used with the album identifier to decode the encrypted portions of the disc. If a device key is revoked, the media key block is changed on future discs. Revoked players will then generate an invalid media key that will not decrypt the disc.[2]

Each column of the media key block may contain up to 65,536 rows. The first generation of CPPM for DVD-Audio defines 16 device key columns, resulting in a maximum file size of 3,145,728 bytes (96 ECC blocks). Someone designing CPPM has a sense of humor, because a successful media key verification step returns the value DEADBEEF in hex digits. If the device key is revoked, the media key block-processing step results in a key value of 0. As with CSS, the media key block can be updated to revoke the use of compromised player keys.

The CPPM authentication process roughly works as follows:

1. **Authentication and key sharing.** The drive and computer (or player) carry out a CSS-style authentication and key exchange process. If successful, each calculates a shared bus key.

2. **Encrypted transfer of control area data.** The computer requests that the drive read sector 2 from the control data area of the disc. The drive encrypts (obfuscates) the data using the bus key. This is also the same as in CSS.

3. **Decryption of the album identifier.** The computer decrypts sector 2 using the bus key and extracts the album identifier from bytes 80 through 87. The album identifier is then used to decrypt content on the disc. Unlike CSS, CPPM does not use an intermediate title key cryptographic step.

As with CSS, only audio, video, subpicture, and still picture sectors are encrypted. Other sectors containing navigation, highlight, and real-time information are not encrypted.

Content Protection for Recordable Media (CPRM)

CPRM is a mechanism that ties a recording to the medium on which it is recorded. It was developed by the same 4C group that created CPPM and shares many of its features. CPRM is supported by all DVD recorders released after 1999. The goal of CPRM is to enable a recording to be made and played on different devices while ensuring that copies of the recording will not be playable. CPRM is defined for writable DVD formats and for SD memory cards, and is intended to be used for other recordable media such as compact flash cards and microdrives.

Each blank recordable DVD disc contains a unique, 64-bit *media identifier* (media ID) permanently etched into the *burst cutting area* (BCA, see Chapter 7), allowing each disc to be

---

[2]This feature is called renewability by the creators of the process, but it does not renew anything. On the contrary, it can make formerly functioning devices cease to work. Perhaps they were thinking of "renew" as used in *Logan's Run*. A problem with revocable devices is that if keys are stolen or the system is cracked so that legitimate keys can be used in unauthorixzed devices, the keys cannot be revoked without causing hundreds or thousands of genuine players to stop working.

uniquely distinguished from all other recordable discs. The first four bits are reserved, the second four bits identify the media type, the next two bytes are a manufacturer ID (assigned by the licensing authority), and the last five bytes are the unique serial number. The media ID does not have to be secret but must be an unalterable value tied to the medium. In addition, the disc contains a *media key block* (MKB), which is prerecorded in the control area of the disc, where it cannot be changed. The recorder uses its set of 16 device keys to process the media key block and produce a *media key*. The authentication process uses the existing CSS mechanism to read the media key block, and a cryptomeria (C2) hash function is used to verify the integrity of the media key block and the media ID. The media key block is stored during disc manufacturing in the embossed control data area of the lead-in. Sectors 4 through 15, a 24,576-byte extent, hold the media key block pack, which is repeated 12 times.

The actual content stored on the disc is encrypted using a *title key*, which is itself encrypted with a *media unique key* generated from the media ID and from the media key block prior to storing on the disc. If the content is copied to another disc, the media ID will be absent or changed. As a result, the wrong media unique key will be generated, which will not properly decrypt the title key, and therefore prevents the content on the disc from being properly decrypted. Due to how this system operates, neither the media key block nor the media ID need to be secret.

When recording, the recorder generates the title key for encrypting the content on the disc. Each disc includes exactly one title key. If something has been recorded, then the existing title key is used to encrypt new content as well. The recorder uses the media key (derived from the media key block) and the media ID to encrypt the title key before recording it on the disc. The media key and media ID are not used to directly encrypt recorded content, which is why they do not need to remain a secret. Only the title key is used (along with CCI) to encrypt the actual A/V sectors.

Each audio, video, subpicture and still image pack contains a 56-bit *title key conversion value* that varies the encryption from pack to pack. Navigation data, such as the *real-time data information* (RDI) packs used by DVD-VR, are not encrypted. Each RDI pack indicates the CGM and APS status (see below) of subsequent A/V packs. Because this information is not protected from potential tampering, devices often control the copying of the recorded content solely based on whether or not it is encrypted. Unencrypted content can be copied without restriction, whereas CPRM-encrypted content is not permitted to be copied.

The encryption process, performed by a DVD recorder when writing new protected content to a disc, works roughly as follows.

1. **Derive the media unique key.** The recording device reads the media key block from the disc and uses its 16 device keys to calculate the media key. It also reads the media ID from the disc. Using the two values, it calculates a media unique key.

2. **Generate a title key.** A single 56-bit title key is used for all protected video and audio data on the disc. The recording device examines the title key status field (in the VMGI_MAT of a DVD-VR disc, or in the AMGI_MAT of a DVD-AR disc) to determine if a title key is already recorded on the disc. If not, the recording device generates a random title key and records it on the disc.

3. **Encrypt the data.** For each AV pack to be encrypted, the recording device uses the pack's title key conversion data (generated as the pack is formatted) to calculate a 56-bit *content key*. The content key is then used to encrypt the last 1,920 bytes of the pack.

The decryption process, performed by a CPRM-compliant DVD reader, works roughly as follows.

1. **Derive the media unique key.** The player reads the media key block from the disc and uses its 16 device keys to calculate the media key. The player reads the media ID from the disc and combines it with the media key to calculate the media unique key.

2. **Decrypt the title key.** The player reads the encrypted title key from the disc and uses the media unique key to decrypt it.

3. **Decrypt the data.** Each pack on the disc contains title key conversion data, which is combined with the title key to calculate a 56-bit *content key* to decrypt the last 1,920 bytes of the pack.

CPRM improves on the revocation features of CPPM by allowing recorders to update the media key block by writing a *media key block extension* to a file on the disc. This allows new revocation information to be disseminated quickly. Since the media key block is not a secret, it can be sent over the Internet to CPRM devices. Only one media key block extension is included on a disc. It is replaced by newer versions as they become available. Writing a new media key block extension changes the media key, so the title key must be re-encrypted with the new key. Media key block extensions can only alter a media key generated by the static media key block (and thus revoke device keys); they cannot generate a media key, so they cannot be hacked to enable new device key sets. Writing media key block extensions is optional for recorders; reading is mandatory.

Analog Protection System (APS)

Copying from DVD to VHS or other analog recording systems is prevented by a Macrovision or similar circuit in the player. The general term is *analog protection system* (APS). Computer video cards with composite or s-video TV output also must use APS. DVD players or DVD computers can be built without APS, but they will not be licensed to play CSS-protected video.

The Macrovision 7.0 process provides two separate antitaping processes: *automatic gain control* (AGC) and *Colorstripe*. Macrovision AGC technology has been in use since 1985 to protect prerecorded videotapes. It works by adding pulses to the vertical blanking sync signal to confuse the automatic-recording-level circuitry of a VCR, causing it to record a noisy, unstable picture. The Colorstripe technology was developed in 1994 for digital settop boxes and digital video networks. (It cannot be applied to prerecorded tapes.) The Colorstripe process produces a rapidly modulated colorburst signal that confuses the chroma processing circuitry in VCRs, resulting in horizontal stripes when the recording is played back.

AGC works on approximately 85 percent of consumer VCRs, including NTSC, PAL and SECAM units. Colorstripe, on the other hand, works on approximately 95 percent, but only on NTSC models. Macrovision is intended to affect only VCRs, but unfortunately, it may degrade the picture, especially with old or nonstandard television equipment. Macrovision makes DVD players unusable when used with some line doublers. Effects of Macrovision may appear as stripes of color, distortion, repeated darkening and brightening, rolling or tearing, or a black-and-white picture.

Just as with videotapes, some DVDs are Macrovision-protected and some are not. The discs themselves tell the player whether to enable Macrovision AGC or Colorstripe. The producer of the disc decides what amount of content protection to allow and pays Macrovision royalties accordingly. Each video object unit of the disc contains "trigger bits" telling the player whether or not to enable Macrovision AGC, with the optional addition of two- or four-line Colorstripe. This finely detailed selective control, occurring about twice a second, enables the disc producer to disable content protection for scenes that may be adversely affected by the process.

Macrovision protection is provided on the composite and s-video output of all but a few commercial DVD players. Macrovision protection was not present on the interlaced component video outputs of early DVD players but is required by the CSS license for all players (the AGC process only, because no colorburst exists in a component signal). A version of AGC for progressive scan component output was developed in 1999 to be required in 2001 by the CSS license and incorporated into progressive scan players (settop only, not computers).

Macrovision protection can be defeated with inexpensive video processing boxes that clean up a video signal. The Macrovision Corporation has been very aggressive in buying the patents for these technologies in order to remove them from the market, but some remain and are always available, although only a few work with the new Colorstripe feature. These devices go under names such as Video Clarifier, Image Stabilizer, and Color Corrector. They are a necessary accessory for people who have combination VCR-TVs, which usually route the video signal through the VCR, thus preventing protected DVDs from being watched on the TV. Many newer digital devices, such as digital camcorders and computer video capture cards, recognize the Macrovision process on incoming video signals and refuse to copy them.

Recently, an alternative to Macrovision has been considered for inclusion in the DVD standards and new formats. Dwight Cavendish Systems (DCS) offers a more technologically up-to-date approach to securing analog signals and would provide player manufacturers and content providers an alternative choice for APS. At the time of this writing, however, it was not yet clear which specifications would approve DCS for the analog protection system.

Copy Generation Management System (CGMS)

Digital video copying — and some analog video copying — is controlled by information on each disc specifying if the data can be copied. This is a serial *copy generation management system* (CGMS), designed to prevent copies or copies of copies. The information is embedded in the outgoing analog and digital video signals. Obviously, the equipment making the copy has to recognize the CGMS information and abide by the rules.

The analog (CGMS/A) information is encoded into the XDS service of line 21 of the NTSC television signal. The digital standards (CGMS/D) are incorporated into DTCP and

HDCP (see the following). Digital recording devices generally check for CGMS/A information in analog inputs.

The CGMS information indicates whether no copies, one copy, or unlimited copies can be made. If no copies are allowed, the recording device will not make a copy. If one copy is allowed, the recording device will make one copy and change the CGMS information to indicate that no copies can be made from the copy.

As with APS information, CGMS flags are present for each sector on a disc. However, since CGMS information could be tampered with, devices are required to establish the protection status of content by the presence or absence of encryption. If the content is encrypted with CSS, CPPM, or CPRM, the "no copies" status is assumed. If no encryption is present, the "copy freely" status is assumed. Some DVD devices are designed to check for no-copy flags on recordable media. If a no-copy flag is found, the disc is presumed to be an illegal copy and will not be played. As a result of these two behaviors, the "copy once" option is essentially rendered useless.

Obviously, CGMS does not prevent multiple copies from being made from the original. However, it is the most fair and reasonable form of content protection, in that it allows fair-use copies by consumers for their own personal use (or at least it would if the "copy once" option were effective).

DVD-Audio extends the concepts of CGMS for recordings on legacy media such as CD-R, Minidisc, and DAT. Unlike digital copies of protected content on writable DVD media, which must be encrypted, unencrypted authorized copies can be made at sound quality no better than audio CD. Specifically, the copies can only have one or two channels of no greater than 48 kHz sampling frequency at no more than 16 bits per sample. The recorder must watermark the copies (see the following section) and keep track of what copies have been made so that only one copy of original audio content can be made per recorder unless otherwise authorized by the content owner by setting CCI parameters for number of allowed copies. An ISRC must be included with the content so that the recorder can track the number of copies it makes of any title. The content owner also can specify the allowed sound quality for copies (CD-Audio, two-channel full-quality, multichannel full-quality). Aside from analog and CD-quality digital audio (IEC-958) outputs, all other outputs from a copy-protected DVD-Audio disc must be encrypted by a method such as DTCP.

Watermarking

Watermarking is a technical process of embedding information into content in a way that is intended to be transparent to the user of the content, but cannot be removed or altered easily. Each digital video frame or segment of digital audio is permanently marked with noise that is supposedly undetectable by human ears or eyes. (As discussed in Chapter 2, there is much debate about how undetectable watermarking is in practice. The amount of watermarking can be varied so that an especially dynamic piece could have a lower level of watermarking to reduce its impact.) The noise carries a digital signature that can be recognized by recording and playback equipment. The signature stays connected to the content regardless of digital or analog transformations. Watermarking does not directly protect the content, it only identifies the status of the content. When used with a content protection system, watermarking usually carries CMI. When the content is played on compliant devices, they recognize the CMI carried in the watermark and abide by its constraints. This only works if a "hook" is present that compels devices to be compliant. Encryption is the carrot to which watermarking is attached. To get the keys and secrets needed to play encrypted content, manufacturers must sign a license that may require that watermark detection be implemented.

Another use of watermarking is to detect if an analog copy has been made or if the content has been digitally re-encoded. A *fragile watermark*, as used by the Secure Digital Music Initiative (SDMI), is designed to be destroyed by any processing of the content, thus indicating that it is not the original version.

DVD-Audio uses watermarking technology developed by Verance. All DVD-Audio players licensed to play CPPM or CPRM discs are required to include circuitry to recognize the Verance watermark. Watermarking will be added to DVD-Video at some point, as a requirement for new players only. It will not render existing DVD-Video players obsolete.

DVD-Audio recorders include *remarking* encoders that change the watermark for copy-generation management. The CCI embedded in the watermark from a "copy once" source is changed in the recording to "copy never" or "no more copies." The inclusion of remarking encoders in millions of consumer devices may make the system more vulnerable to being compromised.

Digital Transmission Content Protection (DTCP)

A digital medium such as DVD deserves to be connected digitally to other digital devices such as digital televisions or digital video recorders. Unfortunately, DVD is far ahead of any digital interconnect standards, which are perennially held back by content protection concerns. Thus the pristine digital content from DVD is usually converted to analog by the player and then converted back to digital by the receiving device — kind of like sending a photocopy of a work of art instead of the real thing.

Digital transmission content protection (DTCP) is the leading technology for protecting content sent over digital connections. Often called *5C* for the five companies that developed it (Intel, Sony, Hitachi, Matsushita, and Toshiba), DTCP focuses on IEEE 1394/FireWire but can be applied to other transmission protocols. Under DTCP, devices that are digitally connected, such as a DVD player and a digital TV or a digital VCR, exchange keys and authentication certificates to establish a secure channel. The DVD player encrypts the encoded audio/video signal as it sends it to the receiving device, which must decrypt it. This keeps other connected but unauthenticated devices from hijacking the signal. No encryption is needed for content that is not protected on the disc. Security can be "renewed" by new content (such as new discs or new broadcasts) and by new devices that carry updated key blocks and revocation lists that identify unauthorized or compromised devices. Digital devices that do nothing more than reproduce audio and video are able to receive all data, as long as they can authenticate that they are playback-only devices. Digital recording devices are only able to receive data that is marked as copyable, and the recorders must change the flag to "do not copy" or "no more copies" if the source is marked "copy once." DTCP requires new DVD players with digital connectors (such as those on DV equipment). Since the encryption is done by the player, no changes are needed to existing discs.

High-bandwidth Digital Content Protection (HDCP)

In 1998, the *Digital Display Working Group* (DDWG) was formed to create a universal interface standard between computers and displays, a new digital replacement for the venerable analog "VGA" connection standard. Founding group members included Silicon Image, Intel, Compaq, Fujitsu, Hewlett-Packard, IBM, and NEC. The resulting *Digital Visual Interface* (DVI) specification, released in April 1999, was based on Silicon Image's panelLink technology. DVI quickly gained wide acceptance as the new industry standard for low-cost, high-speed digital links to video displays. DVI supports 1600×1200 (UXGA) resolution, which covers all of the HDTV resolutions. Even higher resolution can be supported with dual links. Contemporary IEEE 1394 implementations are limited to 40 Mbps, which is plenty for compressed MPEG video but not in the same league with DVI's 4.95 Gbps. Many new HDTV displays are likely to have both IEEE 1394 and DVI connections.

Intel proposed a security component for DVI: *High-bandwidth digital content protection* (HDCP). Used with digital video monitor interfaces such as DVI and HDMI, HDCP has three components: authentication and key exchange, encryption, and revocation. The HDCP ciphers and related circuitry are implemented in DVI digital transmitters and receivers and add approximately 10,000 gates to each device. Each transmitter and receiver has *programmable read-only memory* (PROM) to hold device keys. Receivers are integrated into display devices, and transmitters are integrated into computers and playback devices. (HDCP software drivers may also be required in host computers.)

HDCP authentication is a cryptographic process that verifies that the display device is authorized to receive protected content. A licensed computer or playback device (transmitter) and a licensed display (receiver) each have secret keys, supplied by the HDCP licensing authority and stored in PROM circuitry. HDCP is not mandatory, and early DVI monitors were released before HDCP was ready. When an HDCP-equipped DVI card senses that the connected monitor does not support HDCP, it lowers the image quality of protected content.

Each device has an array of 40 secret keys, 56 bits each, and a corresponding 40-bit binary *key-selection vector* (KSV), all provided by the HDCP licensing entity. The transmitter begins the authentication process by sending its KSV and a random 64-bit value. The receiver then sends back its own KSV, and the transmitter checks that the received KSV has not been revoked. The two devices use the exchanged data to calculate a shared value that will be equal if both devices have a valid set of keys. This shared value is then used to encrypt and decrypt protected data. A block cipher is used during the authentication process. Reauthentication occurs approximately every two seconds (128 data frames) to confirm that the link is still secure. That is, the computer checks that it has not been unplugged from an authorized device and plugged into an unauthorized device.

Content is encrypted at the transmitter so that devices tapped into the connection cannot make any unauthorized copies. A stream cipher is XORed with the video data to provide pixel-by-pixel content protection. Because the receiving device has calculated the same cipher key, it can decrypt the incoming stream. If the encrypted content is viewed on a display device without decryption, it appears as random noise.

If a display device is compromised and its secret keys are exposed, the licensing administrator places the device's KSV on a key-revocation list, which is carried by *system renewability messages* (SRMs). The player updates its key-revocation list whenever it receives a new SRM, which can come from prerecorded or broadcast sources, or from a connected device.

HDCP was key to winning Hollywood's support for high definition movies on computers and other digital display systems.

Summary of Content Protection Schemes

All of the various content protection implementations are optional for the producer of a disc. If someone wishes to protect content on a disc, he or she can choose to apply APS, CSS, CPPM, or watermarking; a combination of more than one is usually applied. It is possible to use APS (Macrovision) without CSS, but it is easy to circumvent the APS trigger bits if the content is not protected with CSS. Therefore, Macrovision is almost always combined with CSS.

CPRM encryption is performed automatically by DVD recorders. DTCP and HDCP are handled by the DVD player or computer, not by the disc developer. As long as the developer has included CMI with the content, it will be honored or passed along by the digital output protection system.

CSS, CPPM, and CPRM decryption are optional for hardware and software player manufacturers; a player or computer without decryption capability can play only unencrypted discs.

Watermarking is being retrofitted into the CSS and CPPM licenses so that future compliant devices will be required to check for watermarking in unencrypted content.

Regional Management (RMA)

It's important to some motion picture studios to control the geographic distribution of DVD-Video titles. This is partly due to the timing for release of the home video version of a title. A movie may come out on video in the United States when it is just hitting theatrical screens in Europe, although this is much less common now than when DVD was being developed. The primary reason for regional management is to preserve exclusive distribution arrangements with local distributors. For example, one studio may have distribution rights for a particular film in North America, but a different studio may have distribution rights for the same film elsewhere.

The DVD-Video standard includes codes that can be used to prevent playback of certain discs in certain geographic regions. Each disc contains a set of region flags.[3] If a flag is cleared, the disc is allowed to be played in the corresponding region; if the flag is set, the disc

---

[3]Regions apply to disc sides. Thus, it is possible to have a disc that is one region on one side and a different region on the other.

is not allowed to be played. Players are branded with the code of the region where they are intended to be sold. The player puts up a message and refuses to play a disc that is not flagged to play in the player's region. This means that discs bought in one country may not play on players bought in another country.

The use of regional codes is entirely optional. Discs with no region locks will play on any player in any country. The codes are not an encryption system, just one bit of information on the disc that the player checks. In many cases, discs are released without any region locks, but in other cases regional control is very important to the business model of movie distribution. Many studios sell exclusive foreign release rights to other distributors. If the foreign distributor can be assured that discs from other distributors will not be competing in its region, then the movie studios can sell the rights for a better price. The foreign distributors are free to focus on their region of expertise, where they may better understand the cultural and commercial environment.

Regions apply only to DVD-Video titles, not DVD-Audio titles, DVD-ROM titles, or other applications of DVD. Region coding of players is enforced only by the CSS license, not by the DVD specification, so region codes officially apply only to CSS encrypted discs. However, the reality is that it's technically possible to regionally code non-encrypted discs, and players don't check for CSS along with the region codes.

The DVD standard specifies eight regions, also called locales.[4] Players and discs are identified by a region number that is usually superimposed on a world globe icon. If a disc plays in more than one region, it will have more than one number on the globe or sometimes the word "all" or the number 0 on the globe. See Figure 5.5 for a map of the DVD regions and Table 5.1 for a list of countries in each region. Also see Table A.1 for the region of every country.

Figure 5.5 Map of DVD Regions

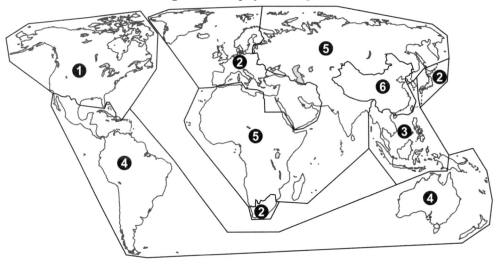

---

[4]Since each region is represented by a bit, a single byte can hold 8 region flags. The neighboring byte is reserved, so it would be possible for the DVD Forum to designate a total of 16 regions in the future.

Table 5.1 DVD Regions

1 Canada, United States, Puerto Rico, Bermuda, the Virgin Islands, and some islands in the Pacific

2 Japan, Europe (including Poland, Romania, Bulgaria, and the Balkans), South Africa, Turkey, and the Middle East (including Iran and Egypt)

3 Southeast Asia (including Indonesia, South Korea, Hong Kong, and Macau)

4 Australia, New Zealand, South America, most of Central America, western New Guinea, and most of the South Pacific

5 Most of Africa, Russia (and former Russian states), Mongolia, Afghanistan, Pakistan, India, Bangladesh, Nepal, Bhutan, and North Korea

6 China and Tibet

7 Reserved

8 Special nontheatrical venues (airplanes, cruise ships, hotels)

Whenever a deterrent is artificially imposed, a way around it is inevitably found. For example, many video game systems introduced since 1995 include regional restrictions. Workarounds quickly appeared for buyers who were interested in games from other countries. Not surprisingly, as soon as DVD players were released, numerous ways were found to defeat the regional coding. Some early players could be set to "region 0"[5] with a switch on the circuit board or sequence of keys on the remote control. Movie studios quickly complained, and manufacturers made it more difficult for users to modify the region setting. Nevertheless, code-free players can be purchased, even from legitimate manufacturer outlets outside the United States, and after-market "region mod" chips are available for many players. The Internet is replete with details on how certain players can be made region-free.

The second salvo from movie studios, particularly Fox, Buena Vista/Touchstone/Miramax, MGM/Universal, and Polygram, was to add program code to some of their discs to check for the proper region in the player. These "smart discs" (*There's Something About Mary* and *Psycho* are examples) query the player for its region code and refuse to work if the player is not set to the single correct region. These discs prevent code-free players from working, so the response from player modification designers was code-switchable players, which enable the region code to be changed by using the remote control. Autoswitching players also check the region on the disc when it is inserted and then set the player region to match. These players do not always work with smart discs, since a disc can have all its region flags set so that the player does not know which region to switch to.

Some people believe that region codes are an illegal restraint of trade, but no legal cases have occurred to establish this. Conversely, rumors have evolved that the major movie studios are pursuing legislation to make region-modification devices illegal in the United States.

---

[5]*Region 0* is a common but misleading term. There is no region 0. Region-free players and all-region discs exist, but region 0 players or region 0 discs are nonexistent. A player modified to work in all regions may have all the bits in the region mask set, which means that it is technically a region 65535 or region FFFF (hex) player.

The only requirement for manufacturers to make region-coded players is the CSS license. Physically modifying a player will void the warranty but is not illegal.[6]

The average consumer who buys DVDs from local stores need not worry about regions. However, those who buy imported discs from other regions or move to other countries will encounter problems.

Regional Management on PCs

Before 2000, regional management was handled independent of the drive and most DVD-ROM drives had no built-in region code. These drives were designed according to *region playback control phase 1* (RPC1). Because RPC1 drives did not contain hardware support for region management, the software player application, the decoder, or the operating system was responsible for maintaining the region code. The region code could be set once by the user on initial use. Some older DVD decoders were preset for a specific region. Though, generally speaking, the user could not change a decoder's region.

To make it easier for drive manufacturers to ship their products anywhere in the world and for computer makers to allow their customers to set the proper region code for where they live, as well as for users to change the region code if they move to a different region, *region playback control phase 2* (RPC2) was implemented. As of January 1, 2000, all drive manufacturers are required to make only RPC2 drives. These drives maintain region code and region change count information in firmware. The user can change the region of the drive up to five times. After that, it can't be changed again unless the vendor or manufacturer resets the drive. RPC2 moves region management into the drive hardware where it's more secure and more easily controlled.

An RPC2 drive has four region states that can be queried and set by the host computer:

- **None:** The drive region has not been set. The computer must set the initial region before playing discs with region information. When the region is first set, the region-setting counter goes from 5 to 4.

- **Set:** The drive region has been set. The region-setting counter will decrement on each change until it reaches 1.

- **Last chance:** The region-setting counter is 1, and the drive region can be set one more time. In order to change to a new region, it must be the same as an inserted single-region disc.

- **Permanent:** The region-setting counter is 0, and the region can no longer be changed. At this point, the drive region can be reinitialized only by the vendor to the None state.

---

[6]At least this is the general consensus. Anyone relying on this book, rather than a lawyer, for legal advice deserves whatever happens to them.

Next-Generation Copy Protection

With the cracking of the content protection scheme for DVDs less than two years after the formats' launch, imagine how the studios feel about releasing pristine digital copies in high definition with quality and resolution equal to or better than what may be seen in a theater. Content protection is one of the leading issues for the new formats, and one of the driving forces to get the formats released. To address their concerns, a group of eight companies (IBM, Intel, Microsoft, Matsushita, Sony, Toshiba, Walt Disney and Warner Bros.) came together to define the *Advanced Access Content System* (AACS) for managing content on both prerecorded and recordable optical media. The Blu-ray Disc format will further augment AACS with *self-protecting digital content* (SPDC), which provides an extended level of renewability by making it possible for each disc to have its own unique security software. Combined with additional transmission protection schemes such as DTCP and HDCP, it is hoped that the next generation of content protection will provide security long enough for the new disc formats to reach success.

Advanced Access Content System(AACS)

AACS uses the *advanced encryption standard* (AES) with 128-bit keys that provide secure encryption. However, as with most content protection systems, the encryption itself is almost never attacked directly because of the brute force power needed to defeat it. Instead, the focus of attacks tends to be the key handling methodology. In the case of AACS, a next-generation media key block is used along with unique device identifiers for each and every device. The combination allows for precise revocation, making it possible to eliminate specific devices that are known to have been compromised. AACS implements an enhanced drive authentication process to further ensure that all of the devices in the content display chain meet the AACS security requirements. One of the most interesting attributes of AACS is that it has been designed to allow content to be securely "moved" from device to device. For example, content may arrive on an optical disc, but it may be downloaded to an automobile-based player or a home server prior to playback. As long as each of the devices is AACS compliant, this would be possible.[7]

Self-Protecting Digital Content (SPDC)

As Sony recently discovered (again) with their content protection for PlayStation® Portable games, it helps to be able to rewrite the security software when someone cracks it. That is essentially what Cryptography Research proposed to do with *self-protecting digital content* (SPDC). In short, each playback device implements a *security virtual machine* onto which software is loaded from the disc at startup. This software provides the key handling routines necessary to properly manage the content on the disc. In the event that the security

---

[7]For those interested in knowing more about AACS, the specifications are available for public review on the web at http://www.aacsla.com.

software is circumvented, new software can be created for managing future discs that simply operates differently, thus creating a moving target for those trying to defeat content protection on these discs. SPDC represents an innovative approach to security that goes beyond static systems like CSS which, once broken, remain wide open.

Ramifications of Copy Protection

All of these content protection schemes are designed to guard against casual copying, which the studios claim causes billions of dollars in lost revenue. The goal is to "keep the honest people honest." The people who developed the content protection measures are the first to admit that they will not stop well-equipped pirates or even determined consumers. Video pirates have equipment that can easily make complete bit-by-bit copies of discs or create identical master copies for mass replication. Bit-by-bit copiers are available to computer owners who know where to look.

Movie studios began promoting legislation in 1994 that would make it illegal to defeat technical content protection measures. The result is the *World Intellectual Property Organization* (WIPO) Copyright Treaty, the WIPO Performances and Phonograms Treaty (December 1996), and the compliant US *Digital Millennium Copyright Act* (DMCA), passed into law in October 1998. Processes or devices intended specifically to circumvent content protection are now illegal in the United States and many other countries. As the legislation was being developed, a cochair of the CPTWG stated, ". . . in the video context, the contemplated legislation should also provide some specific assurances that certain reasonable and customary home recording practices will be permitted, in addition to providing penalties for circumvention." It is not at all clear how this may be "permitted" by a player or by studios that set the "no copies" flag on all their discs.

Although the Content Protection System Architecture promotes compatibility and consistency between various content protection schemes, presumably improving the customer experience, many people regard it as an Orwellian nightmare of big brother intrusion. It can be argued that CPSA and associated content protection measures do away with the time-honored (and court-honored) doctrines of first sale rights and fair use. You can buy a DVD, but unless you have a playback device that is sanctioned by a licensing entity, you may not be able to play it. If the copyright owner chooses to not allow any copying of the disc, then reasonable copying is denied for personal use, such as compiling a disc of favorite songs or making a copy to play in the car. Access to content is being tied to a monopolistic cabal of copyright holders and player manufacturers because an unbreakable system that protects the content creation while affording unfettered lawful private use has yet to be developed.

Effect of Copy Protection on Computers

DVD-ROM drives and computers, including DVD-ROM upgrade kits, are required to support Macrovision, CGMS, and CSS. Computers that play DVD-Audio discs or record video onto writable DVDs must support CPPM and CPRM, respectively. Computer video cards with TV outputs that do not support Macrovision will not work with CSS-encrypted movies. Computers with IEEE 1394/FireWire connections must support the DTCP standard in order to work with other DTCP devices. New computers with DVI outputs must support HDCP. Every DVD-ROM drive includes CSS/CPPM circuitry to establish a secure conntection to the decoder hardware or software in the computer, although CSS can only be used on DVD-Video content and CPPM can only be used on DVD-Audio content. Writable DVD drives must include support for CPRM. The various protection systems are only used for audio and video data, not for other types of computer data. Of course, since DVD-ROM can hold any form of data, any desired encryption scheme can be implemented beyond those which are part of CPSA.

Operating systems such as Windows and Mac OS are beginning to integrate security and content protection measures into the core of the system. More robust hardware-based security, such as the encryption/decryption chips proposed by Intel, is also on the horizon. This kind of support from the computer industry makes content owners more willing to trust their content to computer and Internet environments, but it also makes playback much more complicated and often interferes with the use and creation of nonprotected content.

The Analog Hole

One key problem for the next-generation high definition disc formats that has yet to be adequately addressed is the "Analog Hole." This term refers to the fact that high definition display manufacturers, in their hurry to get their product to market, failed to consider content protection in their designs. As a result, millions of "HD-ready" displays were sold that are capable of presenting high definition video, but have no built-in HD tuner or digital inputs. These devices can only display HD video that is delivered via analog inputs. Unfortunately, these displays predate any sort of analog content protection system for high definition video signals. As a result, such a signal would be "in the clear" and easily copied.

As one might imagine, Hollywood has railed against this reality, stating that high definition studio content will not be allowed into the analog realm unprotected. When confronted by the number of HD-ready televisions that can only take unprotected analog input, the studio's response to the display manufacturers is usually along the lines of "that's your problem." While it is true that the TV set manufacturers should probably have known better, given that Macrovision protection for analog standard definition video had been around for over a decade, it doesn't necessarily make sense to block those early adopters who purchased the initial HDTVs, especially since it is likely that these same early adopters will buy the first HD DVD and BD discs and players.

Owners of HD-ready televisions may have to settle for displaying broadcast-flag protected high definition television programming, while also having to accept the inability to display either the digital or the analog high definition output from their HD DVD and BD players.

After all is said and done, the analog hole may not be plugged to the satisfaction of all the parties.

Chapter 6
Overview of the Formats

This chapter deals with the family, fundamental formats, and features of DVD technology, with a focus on the most important format, DVD-Video. It also introduces the new generations of optical disc technology that have been spawned by DVD.

The DVD Family

The DVD family started off as promisingly as the Brady Bunch. Mr. Laserdisc and Mrs. CD-ROM produced rotund twins that everyone loved: DVD-ROM and DVD-Video. Little brother DVD-R came next and got along well enough despite a case of split personality when he was 2 years old. Like the sibling squabbles of the Bradys, however, rivalries and friction soon popped up. DVD-RAM followed DVD-R but refused to play with the others. DVD-Audio, after an interminable gestation, was so different from DVD-Video that at first the two could not play together. Cousin DVD+RW was ostracized by the rest of the family, even though she tried hard to fit in. And the young triplets, DVD-Video Recording, DVD-Audio Recording, and DVD-Stream Recording, were odd enough that it would take the rest of the family awhile to be able to handle them. The happy ending, where differences are resolved and everyone gets along, which always came at the end of every Brady Bunch episode, is much longer in coming for the DVD family. Figure 6.1 shows the relationships between the members of the DVD family. Then come the grandkids — HD DVD, BD, UMD™, EVD, and the rest — determined to put their aging parents out to pasture. We'll get to them later in this chapter.

Figure 6.1 The DVD Family

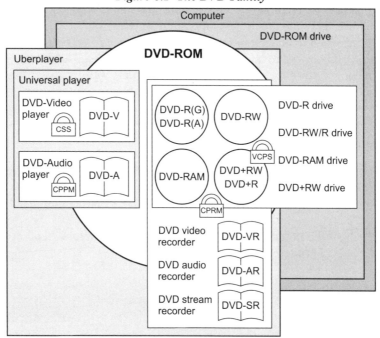

DVD-ROM is the base format for the others. The writable formats are all variations of DVD-ROM, each with its particular good and bad points. DVD-R can record data once, whereas DVD-RAM, DVD-RW, and DVD+RW can be rewritten thousands of times. When first released, DVD-R and DVD-RAM were available for computers only, but by 2001 all of the recordable formats were being used in home video recorders. DVD-R, based on organic dye media similar to CD-R, is generally compatible with other DVD drives and players. The related DVD-RW format, based on phase-change technology similar to CD-RW, is also generally compatible with other drives and players. Just as with CD-R, problems reading DVD-R and DVD-RW will soon disappear and be forgotten. On the other hand, DVD-RAM, a concoction of magneto-optical and phase-change technologies, was not compatible with anything when it was released. It took more than a year before DVD-ROM drives that could read DVD-RAM discs were released. It took more than two years before compatible DVD-Video players began to trickle out. DVD+RW, championed by Philips, Sony, and Hewlett Packard, is not an official member of the DVD family. Similar to DVD-RW, it has about the same level of compatibility.

Manufacturers of DVD-RAM, DVD-RW, and DVD+RW claim that specific features make their format better than the others or more suited for particular uses, but the reality is that the technical distinctions make little difference, especially as drives get faster and buffers get bigger. They all record data on a writable disc.

Beyond the various physical formats of DVD, there are logical formats, or application formats, that define how data gets organized on the disc for a specific purpose. DVD-Video was the first application format, designed for video and audio. DVD-Audio, with specific features aimed at extra-high-fidelity audio, was launched as a separate format, but after a few years it will have mostly merged with DVD-Video. In 2000 and 2001, additional application formats for recording video (DVD-VR), audio (DVD-AR), and streaming data such as from a camcorder or digital satellite receiver (DVD-SR) were rolled out. These application formats are designed to handle real-time recording and custom playlists.

Just as it is important to understand the difference between DVD-ROM and DVD-Video (or DVD-Audio), it is crucial to understand the difference between DVD data recorders and DVD video recorders (or audio recorders). Data recorders were released for computers in 1997, while audio/video (A/V) recorders were not available for another three years or so. Of course, video and audio are just another kind of data; so data recorders connected to computers may be used to write discs with audio and video on them, but it has to be processed in the computer. A/V recorders, like VCRs, have TV tuners and external inputs for analog audio and video, as well as built-in real-time encoders for Dolby Digital audio and MPEG-2 video. Because of the incompatibility of the new DVD-VR and DVD-AR recording formats with existing drives and players, some recorders provide the option to write the standard DVD-Video or DVD-Audio formats even though they were never optimized for real-time recording.

Further details of the physical formats are presented in Chapters 7 and 8; details of the application formats are covered in Chapter 9.

The DVD Format Specification

The DVD Forum is the voluntary association of manufacturers, content developers, and other interested companies that establishes the DVD formats and promotes their acceptance. The official specification for DVD is documented in a series of books published by the DVD Forum. The books are divided into various parts for physical specification, file system specification, and application specifications (see Tables 6.1 and 6.2). Forum members participate in working groups (WGs) that are assigned to develop and maintain various areas of the specification (see Table 6.3). The books are available under NDA and license from the DVD Format and Logo Licensing Corporation (FLLC). The books do not include information about copy protection schemes, which are licensed and documented separately (see "Licensing" in Chapter 10). The DVD formats incorporate various standards defined in other documents (see Appendix B).

Table 6.1 DVD Specification Books

Book	Physical	File System③	Application	First Published
DVD-ROM	Part 1 (Ver. 1.0)②	Part 2 (Ver. 1.0)		August 1996
DVD-Video			Part 3 (Ver. 1.1)①	August 1996
DVD-Audio			Part 4 (Ver. 1.2)④	March 1999
DVD-ENAV			Part 5 (Ver. 0.9)①	March 2004
DVD-R (3.9G) (recordable)	Part 1 (Ver. 1.0)⑥	Part 2 (Ver. 1.0)		July 1997
DVD-R(G) (recordable)	Part 1 (Ver. 2.1)⑥	Part 2 (Ver. 2.1)		March 2000
DVD-R(A) (recordable)	Part 1 (Ver. 2.0)⑥	Part 2 (Ver. 2.0)		July 2000
DVD-R DL (recordable, dual layer)	Part 1 (Ver. 3.0)⑥			Early 2005
DVD-RAM (2.6G) (rewritable)	Part 1 (Ver. 1.0)⑤	Part 2 (Ver. 1.1)		July 1997
DVD-RAM (4.7G) (rewritable)	Part 1 (Ver. 2.2)⑤	Part 2 (Ver. 2.0)		September 1999
DVD-RW (rerecordable)	Part 1 (Ver. 1.2)⑥	Part 2 (Ver. 1.0)		November 1999
DVD-VR (video recording)			Part 3 (Ver. 1.1)①	October 1999
DVD-AR (audio recording)			Part 4 (Ver. 1.1)④	Late 2000
DVD-SR (stream recording)			Part 5 (Ver. 1.0)①	Mid 2001

① through ⑩ indicate responsibility of DVD Forum working groups (Table 6.3).

Table 6.2 DVD Specifications Timeline

Book	Version	Published	Product Available
DVD-ROM	0.9	April 1996	
	1.0	August 1996	Q1 1996 (Q3 1996 in Japan)
	1.01	December 1997	
	1.02	September 1999	
DVD-Video	0.9	April 1996	
	1.0	August 1996	Q1 1996 (Q3 1996 in Japan)
	1.1	December 1997	
	1.11	May 1999	
DVD-R	0.9	April 1997	
	1.0	July 1997	Q4 1997
	1.9	November 1998	Q2 1999
	1.01	April 1999	
DVD-R(A)	2.0	July 2000	Q3 2000 (firmware upgrade)
DVD-R(G)	2.0	October 2000	Q4 2000
DVD-RAM	0.9	April 1997	
	1.0	July 1997	Q2 1998
	1.9	October 1998	
	2.0	September 1999	Q3 2000
	2.1	February 2000	
DVD-Audio	0.9	May 1998	
	1.0	March 1999	Q3 2000
	1.1	May 1999	
	1.2	May 1999	
DVD-RW	0.9	October 1999	
	1.0	November 1999	Q4 2000 (Q4 1999 in Japan)
	1.1	October 2000	
DVD-VR	0.9	January 1999	
	1.0	October 1999	Q4 2000 (Q4 1999 in Japan)
	1.1	February 2000	
DVD-AR	0.9	Mid 2000	
	1.0	Early 2001	
DVD-SR	0.9	Mid 2000	
	1.0	Mid 2001	
DVD-ENAV	0.9	March 2004	

Note: File system specifications are omitted for simplicity

Table 6.3 DVD Forum Working Groups

WG-1	DVD-Video applications
WG-2	Physical specifications for DVD-ROM
WG-3	File system specifications for all DVD variations
WG-4	DVD-Audio applications
WG-5	Physical specifications for DVD-RAM
WG-6	Physical specifications for DVD-R and DVD-RW
WG-9	Content protection liaison
WG-10	Professional applications of DVD
WG-11	Physical specifications for blue laser DVD

The DVD format specification defines the discs and their content. It does not define players. It provides guidelines on player features and basic design, but manufacturers generally are free to implement players as they desire. This encourages innovation but also leads to inconsistencies and incompatibilities among players.

Variations and Capacities of DVD

Believe it or not, more than 144 possible variations of DVD exist. DVDs come in about 24 physical incarnations (see Table 6.4) and more than six data-format variations, such as general data (aka DVD-ROM)[1], DVD-Video, DVD-Audio, DVD-Video Recording, DVD-Audio Recording, and DVD-Stream Recording. The combinations of these formats make for quite a variety of discs. See Table A.3, DVD/CD Capacities for further information and additional disc details.

Two sizes of discs are available: 12 centimeters (4.7 inches) and 8 centimeters (3.1 inches), both 1.2 millimeters thick. These are the same diameters and thickness as CD, but DVDs are made of two 0.6 millimeter substrates glued together. This makes them more rigid than CDs so that they spin with less wobble and can be tracked more reliably by the laser. The thinner substrate reduces birefringence and improves tilt margins for more accurate data readout. A DVD can be single-sided or double-sided. A single-sided disc is a stamped substrate bonded to a blank, or dummy, substrate. A double-sided disc is two stamped substrates bonded back to back. To complicate matters, each side can have one or two layers of data. This is part of what gives DVD its enormous storage capacity. A double-sided, dual-layer disc has data stored on four separate planes. See Chapter 7 for more details on dual-layer construction.

---

[1]Although ROM stands for read-only memory and originally referred to read-only DVDs such as the ones that most movies come on, the term DVD-ROM later became commonly used for recordable DVDs containing general data files, as distinguished from those holding video. This oxymoronic terminology, as in "I burned a DVD-ROM of all my computer files," persists because no good alternative exists; calling them data DVDs isn't much better because video and audio are just other forms of data.

Table 6.4 Physical Format Variations

Type	Size	Sides and Layers[a]	Billions of Bytes[b]	Giga-bytes[b]	Approx. Playing Time[c]
DVD-ROM (DVD-5)	12 cm	1 side (1 layer)	4.70	4.37	2.25 h
DVD-R(G) 2.0					
DVD-R(A) 2.0					
DVD-RAM 2.0					
DVD-RW 1.0					
DVD+RW 2.0					
DVD-ROM (DVD-9)	12 cm	1 side (2 layers)	8.54	7.95	4 h
DVD-R DL (DVD-R9)					
DVD+R DL (DVD+R9)					
DVD-ROM (DVD-10)	12 cm	2 sides (1 layer each)	9.40	8.75	4.5 h
DVD-R(G) 2.0					
DVD-RAM 2.0					
DVD-RW 1.0					
DVD+RW 2.0					
DVD-ROM (DVD-14)	12 cm	2 sides (1 and 2 layers)	13.24	12.33	6.25 h
DVD-ROM (DVD-18)	12 cm	2 sides (2 layers each)	17.08	15.91	8 h
DVD-ROM	8 cm	1 side (1 layer)	1.46	1.36	0.75 h
DVD-RAM 2.0					
DVD-R					
DVD+R					
DVD-ROM	8 cm	1 side (2 layers)	2.65	2.47	1.25 h
DVD-ROM	8 cm	2 sides (1 layer each)	2.92	2.72	1.5 h
DVD-RAM 2.0					
DVD-ROM	8 cm	2 sides (1 and 2 layers)	4.12	3.83	2 h
DVD-ROM	8 cm	2 sides (2 layers each)	5.31	4.95	2.5 h
DVD-R 1.0	12 cm	1 side	3.95	3.67	1.75 h
DVD-RAM 1.0	12 cm	1 side	2.58	2.40	1.25 h
DVD-RAM 1.0	12 cm	2 sides	5.16	4.80	2.5 h
CD-ROM	12 cm	1 side	0.68	0.64	0.25 h[d]
DDCD-ROM	12 cm	1 side	1.36	1.28	0.5 h[d]

[a]DVD-14 (and corresponding 8 centimeter size) has one layer on one side and two layers on the other side.

[b]Reference capacities in billions of bytes (10^9) and gigabytes (2^{30}). Actual capacities can be slightly larger if the track pitch is reduced.

[c]With an average aggregate data rate near 4.7 Mbps. Actual playing times can be much longer or shorter.

[d]Assuming that the data from the CD is transferred at typical DVD video data rate, about four times faster than a single-speed CD-ROM drive.

Six configurations of layers and substrates are possible (see Figure 6.2 and refer to Table 6.4):

Figure 6.2 Layers and Sides

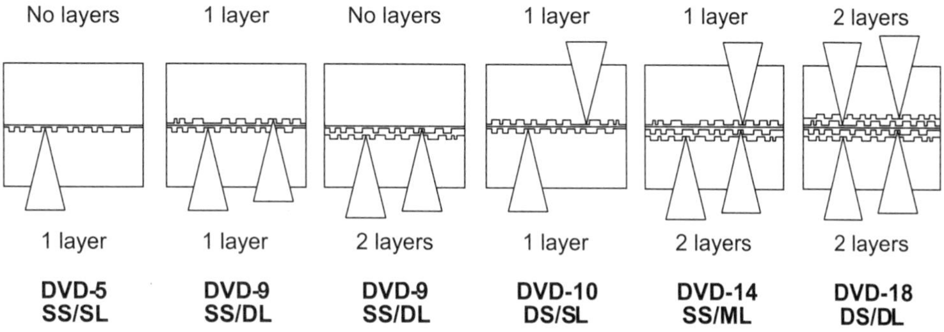

No layers	1 layer	No layers	1 layer	1 layer	2 layers
1 layer	1 layer	2 layers	1 layer	2 layers	2 layers
DVD-5 **SS/SL**	**DVD-9** **SS/DL**	**DVD-9** **SS/DL**	**DVD-10** **DS/SL**	**DVD-14** **SS/ML**	**DVD-18** **DS/DL**

- One single-layer substrate bonded to a blank substrate (one side, one layer; DVD-5)

- Two single-layer substrates with a transparent bond (one side, two layers; DVD-9)

- One dual-layer substrate bonded to a blank substrate (one side, two layers; DVD-9 - uncommon variation)

- Two single-layer substrates bonded together (two sides, one layer each; DVD-10)

- One dual-layer substrate bonded to a single-layer substrate (two sides, one and two layers; DVD-14)

- Two dual-layer substrates bonded together (two sides, two layers each; DVD-18)

 TIP
RULE OF THUMB: It takes about 2 gigabytes to store 1 hour of average video.

A single-sided, single-layer DVD holds 4.7 billion bytes of data (4.37 gigabytes), seven times more than a CD-ROM, which holds more than 650 megabytes.[2] A double-sided, dual-layer DVD holds just more than 17 billion bytes (15.9 gigabytes), which is 25 times what a CD-ROM holds. See the "Units and Notation" in the Introduction's "About This Book" for a discussion of the difference between billions of bytes and gigabytes.

For the first few years of DVD production (1998-2000), approximately 78 percent of all DVD-Video titles were released on DVD-5, with about 14 percent on DVD-9 and 8 percent on DVD-10. The share of DVD-14 and DVD-18 discs was insignificant. By 2005 DVD-5 titles had dropped to approximately 75 percent, with about 17 percent on DVD-9 and less than 7 percent on other configurations. In terms of actual disc production, the percentage of DVD-9 discs is much higher, especially since they are commonly used for blockbuster movies that sell tens of millions of DVDs.

---

[2]The loose tolerance of the CD standard enables tracks to be placed more tightly together, so CDs actually can hold 750 megabytes or more. An example is the so-called 84-minute CD.

Hybrids

A hybrid disc is one that combines the features of one or more disc formats. Anyone who speaks in general terms of a hybrid disc is being dangerously vague because as many potential hybrids exist as there are physical and logical formats. Table 6.5 lists some of the more common DVD hybrids.

Hybrid players are also available: video-capable audio players (DVD-Audio players with video output) and universal players (players that can play both DVD-Video and DVD-Audio). Devices such as PVRs with DVD players built in are also known as hybrids.

Table 6.5 DVD Hybrids

Enhanced DVD	A disc that works in both DVD-Video players and DVD-ROM PCs. This is the most common use of the term hybrid.
Cross-platform DVD	A DVD-ROM disc that runs on Windows and Mac OS computers.
WebDVD	A DVD-ROM or DVD-Video disc that also contains HTML content, usually designed to work with a connection to the Internet. Also called a connected DVD.
Universal DVD	A disc that contains both DVD-Video and DVD-Audio content. Also called a DVD-AV.
CD-compatible DVD	A disc with two layers, one that can be read in DVD players and one that can be read in CD players. Also called a legacy disc. Three variations of this hybrid are:
	A CD substrate is bonded to the back of a 0.6 millimeter DVD substrate. The CD substrate is usually thinner than normal, around 0.9 millimeters, which causes problems with some CD readers. The CD substrate can be read by CD players and the other side of the disc can be read by DVD players. The resulting disc is 0.3 to 0.6 millimeters thicker than a standard CD or DVD, which can cause problems in players with tight tolerances, such as portables. Sonopress, the first company to announce this type of disc, calls it DVDPlus. It is colloquially known as a "fat" disc. A variation is known as *DualDisc™*.
	A 0.6 millimeter CD substrate is bonded to a semitransparent 0.6 millimeter DVD substrate. Both layers are read from the same side, with the CD player being required to read through the semitransparent DVD layer, causing problems with some CD players.
	A 0.6 millimeter CD substrate is given a special refractive coating to create a 1.2 millimeter focal depth. The CD substrate is bonded to the back of a 0.6 millimeter DVD substrate. One side can be read by CD readers and the other side can be read by DVD readers.
DVD-PROM	A disc with two layers, one containing pressed (DVD-ROM) data and one containing writable (DVD-RAM, DVD-RW, and so on) media for recording. Also called a mixed-media or rewritable sandwich disc. (PROM comes from the computer term *programmable read-only memory*.)
Chipped DVD	A disc with an embedded memory chip for storing custom usage data and access codes.

New Formats

Even before DVD debuted in 1997, research groups across the industry were working on its replacement. The goal was "HD" in both senses of the acronym: high definition (better quality video) and high density (more data on a disc). The consensus was that blue or violet lasers, with a wavelength about half that of DVD, would yield the smaller pits and thinner tracks needed to provide the added capacity for storing high-definition video, which requires roughly three times more space than standard-definition video.

An interesting variation, however, was that improvements in video coding technology also made it possible to fit two hours of high-definition video on a standard dual-layer DVD. Ironically, computers supported HD video long before settop players because 2x and faster DVD-ROM drives could provide the requisite data rates, and computer monitors could display the high-definition video. This led to various "720p DVD" projects, which used the existing DVD format to store video in 1280×720 or 1920×1080 resolution at 24 progressive frames per second. Even as new physical formats were being hammered out in laboratories, interim solutions such as Microsoft's WMV HD appeared for PCs.

Different groups proposed variations on the theme of DVD, and the dance of technopolitics has continued. The major Japanese consumer electronics companies eventually consolidated into two camps: the DVD Forum with HD DVD, and the Blu-ray Disc Association with BD. Meanwhile, three wannabe-HD formats battled for supremacy in China, while Taiwan also created its own HD formats.

The following sections present brief overviews of various spin-offs and progeny of DVD. Table 6.6 summarizes the new formats, and Table 6.7 details the physical format variations. For more details on the key formats, see Chapters 8 and 9.

Table 6.6 Next Generation DVD Formats

	Physical Format	Laser	Capacity (single/dual layers)	Video Audio	Copy Protection	Key Backers
HD DVD	0.6 mm cover, smaller pits	Blue	15G / 30G (ROM) 20G / 40G (RW)	MPEG-2, VC-1, MPEG-4 AVC (H.264) PCM, Dolby Digital Plus, DTS HD, MLP	AACS	DVD Forum (primarily Toshiba and NEC), Paramount, Universal, Warner/New Line
Blu-ray	0.1 mm cover, smaller pits	Blue	25G / 50G	MPEG-2, VC-1, MPEG-4 AVC (H.264) PCM, Dolby Digital Plus, DTS HD	AACS	BDA (primarily Panasonic, Pioneer, Sony, Dell, HP, and Apple), Disney/Buena Vista, Fox, MGM, Sony Pictures

continues

Table 6.6 Next Generation DVD Formats (continued)

	Physical Format	Laser	Capacity (single/dual layers)	Video Audio	Copy Protection	Key Backers
WMV HD	Dual-layer DVD	Red	4.7G / 8.5 G	WMV9 WMA9	Windows Media DRM	Microsoft
DivX HD	Dual-layer DVD	Red	4.7G / 8.5 G	DivX HD MP3, Dolby Digital	DivX, DRM	DivX
EVD	0.6 mm cover	Red, Blue	4.7G / 8.5 G, 24G / 48G	MPEG-2 HD, AVS, On2 PCM, EAC	Proprietary	E-World, Chinese government
FVD	0.6 mm cover, tighter tracks	Red	6G / 11G	WMV9 WMA9	Windows Media DRM	ITRI, AOSRA[a]

[a]Taiwan's Industrial Technology Research Institute and Advanced Optical Storage Research Alliance

Table 6.7 Next-Generation Disc Physical Format Variations

Type	Size	Sides and Layers[a]	Billions of Bytes[b]	Giga-bytes[b]	Approx. Playing Time[c]
UMD	6 cm	1 Side (1 layer)	0.90	0.84	1.1 h[c]
UMD	6 cm	1 Side (2 layer)	1.80	1.68	2.2 h[c]
3X DVD-ROM (DVD-9)[d]	12 cm	1 Side (2 layer)	8.54	7.95	1.5 h[e]
HD DVD-ROM	12 cm	1 Side (1 layer)	15.00	13.97	2.5 h[e]
HD DVD -R HD DVD-RW	12 cm	1 Side (1 layer)	20.00	18.63	
HD DVD-ROM HD DVD-R DL	12 cm	1 Side (2 layer)	30.00	27.94	5.1 h[e]
HD DVD-ROM	12 cm	1 Side (3 layer)[f]	45.00	41.91	7.7 h[e]
HD DVD-ROM	12 cm	2 Side (1 layer)	30.00	27.94	
HD DVD -R HD DVD-RW	12 cm	2 Side (1 layer)	40.00	37.25	
HD DVD-ROM	12 cm	2 Side (2 layer)	60.00	55.88	
HD DVD Twin	12 cm	1 Side, 1 HD DVD layer and 1 DVD layer	15.00 +4.70		
BD-ROM BD-R BD-RE	12 cm	1 Side (1 layer)	25.00	23.28	4.3 h[e]

continues

Table 6.7 Next-Generation Disc Physical Format Variations (continued)

Type	Size	Sides and Layers[a]	Billions of Bytes[b]	Giga-bytes[b]	Approx. Playing Time[c]
BD-ROM	12 cm	1 Side (2 layer)	50.00	46.57	8.5 h[e]
BD-R					
BD-RE					
BD-ROM	12 cm	1 Side (3 layer)[g]	75.00	69.85	
BD-ROM	12 cm	1 Side (4 layer)[h]	100.00	93.13	
BD-ROM	12 cm	2 Side (1 layer)	25.00	23.28	
BD-R					
BD-RE					
BD-ROM	12 cm	2 Side (2 layer)	100.00	93.13	
BD-R					
BD-RE					

[a]DVD-14 (and corresponding 8-centimeter size) has one layer on one side and two layers on the other side.

[b]Reference capacities in billions of bytes (10^9) and gigabytes (2^{30}). Actual capacities can be slightly larger if the track pitch is reduced.

[c]With an average aggregate data rate near 1.8 Mbps, sufficient for MPEG-4 AVC video at SD resolution. Actual playing times vary with data rate.

[d]3X DVD-ROM has the same physical characteristics of DVD-ROM, except that it is designed to operate at three times the data rate (33.21 Mbps) to support HD DVD-Video content.

[e]With an average aggregate data rate near 13 Mbps for HD resolution video for comparison. HD DVD-ROM and BD-ROM discs will typically use higher data rates to achieve higher quality. Actual playing times will vary with data rate.

[f]Recently introduced to the HD DVD-ROM family, this triple-layer approach provides nearly the capacity of a dual-layer BD-ROM disc.

[g]Late development, not in initial specification.

[h]Introduced in response to the triple-layer HD DVD-ROM, this *hypothetical* disc format is unlikely to ever reach market.

Please note that this book uses the term "next-generation," not HD. There are two reasons for this. First, the DVD Forum cleverly, if confusingly, snapped up the "HD DVD" moniker, so talking about "HD DVD" is ambiguous. No one's sure if you're talking about one format or all formats. Second, "HD" can stand for high density or high definition, and because it's possible to put HD video on standard-density DVDs, it's even more ambiguous.

General Similarities & Differences of HD DVD and BD

In actuality, HD DVD and BD are much more alike than they are different. They both have the same form factors as CD and DVD, they both reflect your face when you look into them, they both have high enough capacities to hold a few hours of high-definition video, and they both add much more interactivity than DVD has. In short, if you want to watch HD movies, record your own HD movies, or store computer data; either serves nicely. In October 2003 the Hollywood studios came up with a wish list of features that both formats support almost equally (see Table 6.8).

Table 6.8 Hollywood Studio Requirements for Next-Generation Formats

Proper launch timing	No cartridge
Best quality	Play CDs and DVDs
Up to 1920 × 1080 video	Advanced content protection
Multistream mixing in player	Prevent casual copying
Adequate capacity	Deter professional piracy
2-hour movie	Protect all outputs
Bonus content, often in SD	Watermarking
Advanced features	Regions
Interactivity (WebDVD)	Internet connection
Single format for PC and CE, including games	
Uncompressed and compressed multichannel audio	

The formats can be sliced into eight architectual layers. Some of the layers are the same for both formats, while others differ. See Table A.8.

HD DVD

HD DVD is an extension of the original DVD format using a blue laser for smaller pits and tighter tracks without changing the data layer depth. Prerecorded (ROM) discs hold 15 billion bytes on a single layer, 30 billion bytes per side in a dual-layer configuration. A triple-layer variation, assuming it proves possible, may hold 45 gigabytes on one side. (It cannot easily be made in a two-sided version.) A hybrid disc is also possible, with standard DVD on one side and HD DVD on the other. Recordable HD DVD discs can hold more than prerecorded HD DVD discs: 20 billion bytes on one layer, 40 billion on two.

The familiar DVD-Video specification has been updated for higher resolution video, better subpictures (up from 4 colors to 256 colors!), 64 GPRMs (up from 16), fewer restrictions on jumping between places on the disc, browseable slideshows (that allow you to step through still images while the audio plays uninterrupted), and other enhancements, but it's not a major overhaul.

The official specifications for the various HD DVD formats are documented in books published by the DVD Forum. As with DVD, these books are divided into various parts for physical specification, file system specification, and application specifications (see Tables 6.9 and 6.10).

Table 6.9 HD DVD Specification Books

Book	Physical	File System③	Application	First Published
HD DVD-ROM	Part 1 (Ver. 1.2)⑪	Part 2 (Ver. 0.9)		June 2004
HD DVD-R (recordable)	Part 1 (Ver. 1.0)⑪	Part 2 (Ver. 0.9)		February 2005
HD DVD-R DL (recordable, dual layer)	Part 1 (Ver. 1.9)⑪			Late 2005
HD DVD-RW (rewritable)	Part 1 (Ver. 1.0)⑪	Part 2 (Ver. 0.9)		September 2004
HD DVD-Video			Part ?[a] (Ver. 1.0)①	Late 2005
HD DVD-VR (video recording)			Part ?[a] (Ver. 1.0)①	Late 2005

[a]Book part numbers for HD DVD-Video and HD DVD-VR are not yet defined at the time of this writing.

① through ⑪ indicate responsibility of DVD Forum working groups (Table 6.3).

Table 6.10 HD DVD Specifications Timeline

Book	Version	Published
HD DVD-ROM	0.9	November 2003
	1.0	June 2004
	1.1	February 2005
	1.2	May 2005
HD DVD-R (recordable)	0.9	September 2004
	1.0	February 2005
HD DVD-R DL (recordable, dual layer)	1.9	Late 2005
HD DVD-RW (rewritable)	0.9	February 2004
	1.0	September 2004
HD DVD-Video	0.9	March 2005
	1.0	Late 2005
HD DVD-VR (video recording)	0.9	March 2005
	1.0	Late 2005

Note: File system specifications are omitted for simplicity

The updated DVD-Video format is called *standard content* to differentiate it from *advanced content*, a much more interactive variation based on the WebDVD work done over the past several years by InterActual Technologies, Microsoft, and others. Advanced content uses subsets of XHTML, ECMAScript (aka JavaScript), and SMIL for a combination of synchronized layout markup and scripted programming. In other words, a lot of what can be done on Web pages today can be done on HD DVD titles and will work in all players, not just PCs.

The HD DVD application formats are intended to work on standard DVD media as well as new HD DVD media. That means shorter programs can be put on inexpensive red-laser discs instead of more expensive blue-laser discs. See Chapter 8 for details of the physical format and Chapter 9 for information on the application formats.

Blu-ray Disc (BD)

BD is a significantly changed version of DVD that uses smaller pits and tracks to fit about 25 billion bytes (more than five times standard DVD capacity) on a single layer, and about 50 billion bytes on two layers. Prototypes of four-layer discs have been demonstrated, holding 100 gigabytes on a single side. Blu-ray achieves this higher density by using a 0.1-mm cover layer to move the data closer to the lens (standard DVD uses a 0.6-mm cover, CD uses a 1.2-mm cover). A blue laser at 405-nm wavelength is required to read the smaller pits.

Recordable BD technology has been on the market in Japan since April 2003 in the form of BD-RE home recorders designed to capture HD video broadcast. The recorders were designed for home recording only (not for playing pre-recorded HD movies), and only work with Japan's digital HD broadcast system. BD-RE was updated to a 2.0 version called BDAV for general video recording, based primarily on MPEG transport streams.

The Blu-ray Disc Association is responsible for developing the format specifications, and the association members are organized in Technical Expert Groups (see Tables 6.11 and 6.12).

Table 6.11 Blu-ray Disc Association Specification Books

Book	Physical	File System⑤	Application	First Published
BD-RE (rewritable)	Part 1 (Ver. 2.0)①	Part 2 (Ver. 2.0)	Part 3 (Ver. 2.0)①	June 2002
BD-R (recordable)	Part 1 (Ver. 1.0)④	Part 2 (Ver. 1.0)		Not yet published[a]
BD-ROM	Part 1 (Ver. 1.2)③	Part 2 (Ver. 1.0)	Part 3 (Ver. 0.9)②	Not yet published[a]

① through ⑤ indicate responsibility of Blu-ray Disc Association

[a]These specification books have not yet been published at the time of this writing.

Table 6.12 Blu-ray Disc Technical Expert Groups

TEG1	Physical specifications for BD RE (Blu-ray Disc Rewritable)
TEG2	AV application
TEG3	Physical specifications for BD ROM (Blu-ray Disc ROM)
TEG4	Physical specifications for BD R (Blu-ray Disc Recordable)
TEG5	File system and command set

BD developers, after years of experience with DVD-Video, engineered a format called HDMV that provides similar functionality with a cleaner implementation. HDMV adds a few tweaks such as two high-definition graphics planes, animated menu buttons and pop-up menus over video, and sound effects for menu button selection (which, like fonts on laser printers, will be mercilessly abused in early titles). Like HD DVD, BD has an advanced interactive format called BD-J, based on a subset of Java similar to the Multimedia Home Platform (MHP) developed for interactive TV. The BD application formats are intended to work only on BD media.

Microsoft WMV HD

WMV HD uses the Windows Media 9 coding format to fit high-definition video and multichannel audio on dual-layer DVD-9 discs. Menus and interactivity are provided using HTML and JavaScript. The quality can be very good, but in some respects WMV HD is inferior to current DVD because it doesn't manage features such as subtitles, trick play, and random access well. Some titles require a live Internet connection for playback because of the Windows DRM (digital rights management) content protection. WMV HD demands a 2-3 GHz Windows PC for playback. A few consumer DVD players can play WMV HD files, but don't support the interactive menus. Despite its limitations, WMV HD was the best interim option for releasing HD titles into the marketplace. WMV HD titles, primarily IMAX movies, were first released in 2003; more mainstream features were released on WMV HD in Europe and have done relatively well.

DivX HD

DivX HD is another application of DVD-9 discs for HD format, using DivX HD video and MP3 or Dolby Digital multichannel audio and DivX DRM for content protection. You might think of it as the bohemian alternative to WMV HD. As of 2005 the only DivX HD content available was movie trailers downloadable from divx.com, which could be played on PCs and a few DivX HD-compatible settop players.

UMD

In a way, UMD (universal media disc) doesn't quite belong in this list because it employs red laser discs and standard-definition video. On the other hand it's an interesting and successful format resulting from the marriage of DVD with Sony MiniDisc. UMDs hold videogames or video playable on Sony PlayStation® Portable (PSP™) handheld game players, released in Japan at the end of 2004, in the US in March 2005, Asia in May 2005, and Europe in September 2005. Yet, studios saw enough potential in video on 4.3-inch screens that in less than six months more than 90 UMD video titles were released in the US with another 140 scheduled to debut before the end of the year. The 60-mm discs in hard-shell cartridges hold 1.8 billion bytes. UMD uses MPEG-4 AVC for video and Sony's own ATRAC3Plus or PCM for audio, with 128-bit AES encryption. Interactive menus are created with UMD Video Resource XML and UMD Video Script (similar to the combination of HTML and JavaScript). See Chapter 7 for details of the physical format and Chapter 9 for more on the application format.

FVD

FVD (forward versatile disc) is a tweaked version of red-laser DVD that achieves slightly higher capacities of 6 billion bytes on a single layer and 11 billion bytes on two layers. FVD uses Microsoft Windows Media Video (WMV9) and Audio (WMA9). FVD is backed by Taiwan's government-sponsored Industrial Technology Research Institute (ITRI) and Advanced Optical Storage Research Alliance (AOSRA), but drive and player manufacturers in Taiwan seem much more interested in BD and HD DVD.

EVD, HVD, and HDV

EVD (enhanced versatile disc) is a domestic format developed in China, partly in an attempt to break free from expensive patent royalties charged by the incumbent format owners. Ambitious plans to release a highly advanced format using domestic technology were soon revised into phases, most of which have never materialized. The homegrown AVS video encoding format wasn't ready, and an attempt to license On2 video from the U.S. failed.

The first phase, with players released in China in December 2003, was based on standard, prerecorded red-laser DVDs using MPEG-2 HD video and China's own EAC multichannel audio format. Future phases are planned to add stronger content protection, recordability, AVS video, multilevel recording (using Calimetrix's volumetric recording technique of variable pit depth to achieve higher capacity), and eventually graduate to blue laser.

EVD was promoted by E-World, a consortium of Chinese consumer electronics (CE) manufacturers and government entities. The format received government endorsement in July 2004, but many of the promises of a strong content protection system were never realized, so Hollywood was in no hurry to "endorse" the format with movies. In the meantime, two competing HD formats in China, HVD (high-definition versatile disc) and HDV (high-definition digital video), both variations of red-laser DVD , began bundling illegally copied titles with their players to attract buyers, undermining EVD's half-hearted attempts to play by the rules. FVD also began to make inroads in China. In 2005, about two years after its launch with a goal of selling 20 million players in the first year, EVD had sold about 200,000 players and was not exactly taking the country by storm.

Format Compatibility

Attempting to understand optical disc compatibility could drive a granite plinth insane. It does not help that compatibility has many facets and meanings that all become increasingly complex as formats are added for DVD and beyond. There's *backward compatibility*, such as DVD players being able to play CDs, and *forward compatibility*, such as DVD players being able (or not being able) to play high-definition discs. There's also compatibility within a format or format family.

The specifications have very little to say about players. They almost exclusively define physical discs, file systems, and applications. Player makers are free to implement the specifications as they see fit. Unfortunately, failures of compatibility occur at various levels: physical, file system, application, and implementation (see Table 6.13). The end result is that consumers cannot assume that any disc with "DVD" in its name will work in any player or computer with "DVD" in its name. If someone records video onto a DVD to send to someone else, it cannot be assumed that it will play on the recipient's DVD hardware.

 NOTE
DVD compatibility breakdowns occur at the physical level, logical level, and implementation level.

Table 6.13 Examples of Compatibility Problems

Problem	Incompatibility	Explanation
A DVD player cannot play a DVD-Audio disc.	Application	Unless the player was designed to read the DVD-Audio data format, it will not recognize the contents of the disc. Luckily, many DVD-Audio discs include audio in DVD-Video format so that they will play in DVD-Video players and DVD computers.
A DVD player cannot play a recorded DVD	Physical	The disc may be a type that the player cannot physically read, such as DVD-RAM. Or, the player may not have been designed to recognize recordable discs.
	Application	The disc may be recorded using an application format that the player does not recognize, such as DVD-VR or DVD-SR.
A DVD player cannot play a PC-enhanced disc.	Application	The settop DVD player does not recognize the computer applications and HTML pages on the disc. It can play the contents of the DVD-Video zone, and perhaps, the DVD-Audio zone.
	Implementation	Some players were not designed to properly deal with extra files and extra directories on the disc. Even though this is allowed by the DVD specification, it confuses certain players to the point that they cannot even play the DVD-Video content.
A player cannot play a CD-R but plays a CD-RW.	Physical	The player does not have a second laser at the wavelength needed to read a CD-R. CD-RW reflectively is different, so the laser for reading DVDs can read a CD-RW (and a CD).
A player cannot play a CD.	Physical	If the disc is recorded on CD-R media, the player may not be able to physically read it.
	Application	Every DVD player has been designed to read CDs, but the CD may have computer data on it, not CD-Audio.

continues

Table 6.13 Examples of Compatibility Problems (continued)

Problem	Incompatibility	Explanation
A DVD player cannot play a Video CD or a Super VCD.	Physical	If the disc is recorded on CD-R media, the player may not be able to physically read it.
	Application	Even if the player can read the disc, the manufacturer may have chosen not to add firmware and circuitry needed to play other video formats.
A DVD computer cannot play a Sony PlayStation DVD.	Application	Although the computer can read the data, it does not know how to execute the application or interpret the proprietary file formats.
A computer cannot play or copy some movies.	Application	Most movies are encrypted with CSS. This prevents direct copying of files without an authentication process between the decoder and the drive. If the computer does not have a software player and decoder that implements CSS, it is not able to decrypt and play the files.
	Implementation	Bugs may be present in the computer driver or player software that prevent proper playback, or the data may have been incorrectly formatted on the disc.
A DVD player cannot play certain movies.	Application	A DVD-Audio-only player cannot play movies. It was not designed to read the DVD-Video files.
	Implementation	The disc may have been authored or encoded in a way that is not compliant with the DVD specification, or the information on the disc may be compliant with the specification but not handled properly by the player.
A DVD player cannot read a disc recorded in a computer.	Physical	The disc may be a type that is not physically readable in the DVD player. Or, the player may not have been designed to recognize recordable discs.
	File system	DVD players use the UDF file system to find and read the files containing audio and video. The computer may not have used UDF. DVD players also expect the files to be physically contiguous. The computer may not have ordered the video files in contiguous order from the beginning of the disc.
	Application	The data may be recorded as a set of MPEG-2 files or MP3 files, which the player is not designed to recognize and process.

continues

Table 6.13 Examples of Compatibility Problems (continued)

Problem	Incompatibility	Explanation
One computer cannot read or play the disc written on another computer.	Physical	The disc may be a type that is not physically readable in the computer drive, such as DVD-RAM or DVD+RW.
	File system	The second computer may not recognize the file system used by the first. For example, newer versions of Windows natively use FAT32, which many other operating systems cannot read.
	Application	The second computer may not have the software needed to play the disc.
	Implementation	Bugs may be present in the drivers or formatting software of the first or second computer.

Physical Compatibility

Physical compatibility is not an issue with the base DVD-ROM format. Every player, drive, and recorder is physically able to read data from DVD-ROM discs. Players may not know what to do with the data or may not even attempt to read certain sections of the disc, but they are capable of reading the bits. Physical compatibility only becomes a problem with the writable formats, as illustrated in Table 6.13. Physical compatibility also applies to other formats such as CD and CD-R, where it is up to the manufacturer to decide whether or not to support a particular physical medium.

File System Compatibility

File system compatibility is generally not a problem. The file system determines how the data is organized and accessed. UDF and ISO 9660 are the two common file systems on DVD; in fact, most discs contain both. Because DVDs are simply storage media, other file systems such as Microsoft FAT, NTFS, Macintosh HFS, Unix, and so on can be used to write data files to the disc. Compatibility problems generally occur only with these specialized file systems, such as when a disc formatted with Windows FAT32 is placed in the drive of a Mac.

Application Compatibility

Application compatibility is the most confusing area. It is not always clear why a disc does not work in a specific player, because it may not be obvious which application formats the player supports. For example, a DVD-Video player can read the data and UDF files on a DVD-Audio disc, but if it is not built to read and process data from the DVD-Audio files in the AUDIO_TS directory, it will not play the audio. Both the player and the disc have "DVD" on them, but it may not be clear why they do not work together. The proliferation of

DVD Forum application formats such as DVD-VR and DVD-AR, along with other custom formats for computers and game consoles, places a burden on the consumer to understand which discs use which application formats and which players can play them.

Implementation Compatibility

Implementation compatibility has to do with flaws or omissions in players (including computers and any other device that can play DVDs). Some players are poorly designed, whereas others behave in unexpected ways with unanticipated content. Each player implements the DVD specification in a slightly different way, complicated by the fact that the specification is ambiguous and confusing. The result is that discs may play differently or not at all in different players. Implementation errors also can occur on the production side as well. That is, a bug might exist in the encoder or in the system used to author the disc, or the person who authored the disc may have violated the DVD specification and thus a disc will not work on some or all players. Implementation problems are discussed in more detail under "Playback Incompatibilities" in Chapter 12.

New Wine in Old Bottles: DVD on CD

Another compatibility problem, which falls between application and implementation, results from DVD content being placed on media other than DVD. Many people would like to put short DVD-Video programs onto CDs or inexpensive CD-Rs. The idea of a "mini-DVD" is quite appealing, particularly for testing titles during development and viewing short programs such as music video singles, home movies, or corporate marketing clips. However, each would-be clever inventor is disappointed to learn that such CDs do not play on settop players. Only a few odd player models designed around DVD-ROM drives can play the discs. All other DVD-Video players fail to play DVD-Video content from CD media. This failure is due to several reasons:

1. The player does not expect DVD content on CD media. The first thing most players do after a disc is inserted is check focus depth and reflectivity. If nothing can be read at DVD focus, the player switches focus (and usually switches lens and laser as well) to see if it can read data at CD focus depth. If it determines that a CD is in the drive, it goes to CD mode. It checks for audio CD content and it might check for Video CD content or MP3 files, but it does not look for DVD files.

2. The drive unit is not fast enough. It is simpler and cheaper for players to spin CDs at 1X speed rather than the 9X speed needed to provide the 11 Mbps data rate required by DVD-Video content.

3. Many players cannot read CD-R discs. A player without a laser at CD wavelength cannot read CD-R media.

Hollywood may be concerned about movies being copied easily and cheaply to CD-Rs that would play in DVD players, although the quality would be poor even if the video was spread across more than one CD. Of course, player manufacturers could deal with all three of these obstacles, but they do not believe that the demand justifies the expense. Critics also could accuse them of not wanting to sustain the CD market when they can make more money (or pay fewer royalties) with DVD.

Computers are more forgiving. Most of them are media agnostic when it comes to DVD content. DVD-Video files from any source with fast enough data rates, including CD-R or CD-RW, or even hard drives or Jaz drives, with or without UDF formatting, will play back on almost any DVD computer as long as the drive can read the media. In fact, many DVD authoring systems include the option to write small volumes to CD-R or CD-RW.

An alternative is to put Video CD or Super Video CD content on CD-R or CD-RW media for playback in a DVD player. About half the settop player models can play VCDs, although very few can play SVCDs. The limitations of VCD apply (MPEG-1 video and audio, 1.152 Mbps, 74 minutes of playing time).[7] All DVD-ROM computers able to read recordable CD media can play recorded VCD discs if they have the necessary playback software. An MPEG-2 decoder and SVCD player software are needed to play SVCDs.

Compatibility Initiatives

Many efforts have been made to deal with compatibility problems at various levels. In October 1997, the Optical Storage Technology Association (OSTA) produced the MultiRead specification, which provides a logo for devices — including DVD drives and players — that read CD audio, CD-ROM, CD-R, and CD-RW. The followup MultiRead 2 specification from December 1999 added DVD-ROM and DVD-RAM to the list. The DVD Forum manages a certification program for discs and players (see Appendix C). In 2000, the DVD Forum began finalizing the DVD Multi program to promote physical compatibility and limited application compatibility for reading DVD-ROM, DVD-R, DVD-RW, and DVD-RAM discs and writing DVD-R, DVD-RW, and DVD-RAM discs. A player or drive with the DVD Multi logo is guaranteed to read a certain set of disc and application formats (see Table 6.14 at the end of this chapter).

This state of affairs is far from satisfactory, but at least the industry is working toward read compatibility across formats, if not write compatibility.

Backward Compatibility

Player manufacturers have it rough in the digital age. No one ever expected cassette tape players to play old reel-to-reel tapes or even eight-track tapes, but today it would be suicidal for a manufacturer to produce a DVD player that can't play CDs. And, blue-laser players have to play both CDs and DVDs. The new formats will slowly supersede the original DVD format, but new players will play old DVD discs and will often make them look even better with progressive scan video, upscaling, and image processing. So, anyone who owns a large

---

[7]These limitations can be stretched with the so-called DVCD and DSVCD formats, which squeeze more tracks on a disc to extend the capacity. In the case of DVCD, playing time is increased from 74 to 100 minutes at the expense of minor physical compatibility problems on some VCD and DVD players.

collection of DVDs can count on backwards compatibility of future players. DVDs will only become "obsolete" in that you might want to replace them with new high-definition versions. However, next-generation discs won't be playable in older DVD players unless they are special hybrid discs in both HD and SD format.

Sideways Compatibility

The warring blue-laser formats are different enough that one type of player will not be able to play the other type of disc, at least not initially. This raises the specter of VHS vs. Betamax, with consumers checking every movie before they buy or rent to make sure it will work in their player. Fortunately, if competing blue-laser formats come to market, it probably won't take more than a year before the first dual-format players make their appearance, forcing the other player manufacturers to match.

Table 6.14 DVD Multi Compatibility

| | DVD Multi Consumer Electronics Device | | | | DVD Multi Computer | |
	Player	Audio-Only Player	Video-Only Player	Recorder	Drive	Record
DVD-ROM disc	Will not play	Will not play	Will not play	Will not play, cannot record	Reads	Reads, cannot write
DVD-Video disc	Plays	Will not play	Plays	Plays, cannot record	Plays[a]	Plays[a], cannot write
DVD-Audio disc	Plays	Plays	Will not play	Plays, cannot record	Plays[b]	Plays[b], cannot write
DVD-R disc	Plays	Plays (audio)	Plays (video)	Plays, records	Reads	Reads, writes
DVD-RW disc	Plays	Plays (audio)	Plays (video)	Plays, records	Reads	Reads, writes
DVD-RAM disc	Plays	Plays (audio)	Plays (video)	Plays, records	Reads	Reads, writes

[a]Can play if computer has DVD-Video decoder.

[b]Can play if computer has DVD-Audio decoder.

Chapter 7
Red Laser Physical Disc Formats

This chapter details the original DVD disc and data recording formats. Although akin to a primer, this chapter assumes that the reader has a basic understanding of the terms and concepts described in Chapters 3 and 4.

DVD-ROM is the foundation of DVD: the physical layer and the file system layer. DVD-Video and DVD-Audio are *applications* of DVD-ROM as well as MPEG-2, Dolby Digital, MLP, and other standardized formats. The DVD-Video and DVD-Audio formats define subsets of other standards to be applied in practice for audio and video playback. DVD-ROM may contain any desired digital information, but DVD-Video and DVD-Audio are limited to certain data types designed for video and audio reproduction. The DVD-Video data types are matched to the NTSC and PAL television formats, as NTSC and PAL televisions are the primary target systems.

The DVD family includes recordable versions of DVD-ROM. Application specifications for disc recording are built on top of the physical recordable format specifications. These are DVD Video Recording (DVD-VR), DVD Audio Recording (DVD-AR), and DVD Stream Recording (DVD-SR). There are also provisions whereby the DVD-Video format may be used on recordable discs. See Chapter 9 for information on the DVD application specifications, particularly DVD-Video.

Physical Composition

Table 7.1 lists the physical characteristics of DVD. Most of these characteristics are shared by all the physical formats (read-only, writable, 12-centimeter and 8-centimeter).

Table 7.1 Physical Characteristics of DVD

Thickness	1.2 mm ($+0.30/-20.06$) (two bonded substrates)
Substrate thickness	0.6 mm ($+0.043/-0.030$)
Spacing layer thickness	55 mm (±15)
Mass	13 to 20 g (12-cm disc) or 6 to 9 g (8-cm disc)
Diameter	120 or 80 mm (±0.30)
Spindle hole diameter	15 mm ($+0.15/-0.00$)
Clamping area diameter	22 to 33 mm
Inner guardband diameter	33 to 44 mm
Burst cutting area diameter	44.6 mm ($+0.0/-0.8$) to 47 (±0.10) mm
Lead-in diameter	45.2 to 48 mm ($+0.0/-0.4$)

continues

Table 7.1 Physical Characteristics of DVD (continued)

Data diameter	48 mm ($+0.0/-0.4$) to 116 mm (12-cm disc) or 76 mm (8-cm disc)
Lead-out diameter	Data $+2$ mm (70 mm min. to 117 mm max. or 77 mm max.)
Outer guardband diameter	117 to 120 mm or 77 to 80 mm
Radial runout (disc)	<0.3 mm, peak to peak
Radial runout (tracks)	<100 μm, peak to peak
Index of refraction	1.55 (±0.10)
Birefringence	0.10 μm max.
Reflectivity	45 to 85% (SL), 18 to 30% (DL)[a]
Readout wavelength	650 or 635 nm (640 ±15) (red laser)
Polarization	Circular
Numerical aperture	0.60 (±0.01) (objective lens)
Beam diameter	1.0 mm (±0.2)
Optical spot diameter	0.58 to 0.60 μm
Refractive index	1.55 (±0.10)
Tilt margin (radial)	$\pm0.8°$
Track spiral (outer layer)	Clockwise
Track spiral (inner layer)	Clockwise or counterclockwise
Track pitch	0.74 μm (±0.01 avg., ±0.03 max.)
Pit length	0.400 to 1.866 μm (SL), 0.440 to 2.054 μm (DL) (3T to 14T)
Data bit length (avg.)	0.2667 μm (SL), 0.2934 μm (DL)
Channel bit length (avg.)	0.1333 (±0.0014) μm (SL), 0.1467 (±0.0015) μm (DL)
Jitter	<8% of channel bit clock period
Correctable burst error	6.0 mm (SL), 6.5 mm (DL)
Maximum local defects	100 μm (air bubble), 300 μm (black spot), no more than six defects between 30 and 300 μm in an 80-mm scanning distance
Rotation	Counterclockwise to readout surface
Rotational velocity[b]	570 to 1630 rpm (574 to 1528 rpm in data area)
Scanning velocity[b]	3.49 m/s (SL), 3.84 m/s (DL) (±0.03)
Storage Temperature	-20 to 50°C (-4 to 112°F), $\leq$ 15°C/h variation (59°F/h)
Storage humidity	-5 to 90% relative, 1 to 30 g/m^3 absolute, $\leq$ 10%/h variation
Operating temperature	-25 to 70°C (-13 to 158°F), $\leq$ 50°C sudden change (122°F)
Operating humidity	-3% to 95% relative, 0.5 to 60 g/m^3 absolute, $\leq$ 30% sudden change

[a]SL = single layer; DL = dual layer.

[b]Reference value for a single-speed drive.

Substrates and Layers

A DVD disc is composed of two 0.6 mm bonded plastic substrates (see Figure 7.1).

Figure 7.1 DVD/HD DVD Disc Structure

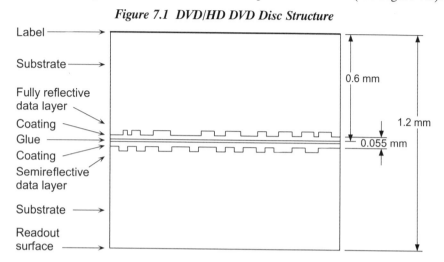

Each substrate may contain one or two layers of data. The first layer, called *layer 0*, is closest to the side of the disc from which the data is being read. The second layer, called *layer 1*, is further from the readout surface. A disc is read from the bottom, so layer 0 is below layer 1. The layers are spaced very close together, 55 microns apart, which is less than one-tenth of the thickness of the substrate (less than one-twentieth of the thickness of the disc).

Mastering and Stamping

DVDs are usually made from polycarbonate, but they also may be made from acrylic, polyolefin, or similar transparent material that can be stamped by a mold while in molten form. Material with lower birefringence, such as acrylic (PMMA) or lexan, is preferable but often prohibitively expensive.

The physical format of a DVD disc resembles that of a CD. In fact, most of the production process matches that for a CD, but more care needs to be taken to properly replicate tinier pits in thinner substrates, and two substrates need to be bonded together. In many cases the same replication equipment can alternate between making CDs and DVDs.

Mastering refers to the process of creating the physical stamping molds used in the replication process (see Figure 7.2). The formatted data signal is used to modulate the cutting laser of a laser beam recorder machine (LBR), which creates pits in a glass disc that has been covered with a thin photoresist coating. The laser does not actually cut the glass but rather exposes sections of the photosensitive polymer. The coating is then developed and etched to remove the exposed areas. A vacuum evaporation or sputtering process applies an electrically conductive layer that is then thickened by an electrochemical plating process to create the stamping master, or *father*. The stamping master, usually made of nickel, contains a mirror image of the pits that are to be created on the disc. The raised bumps on the stamper are pressed into the liquid plastic during the injection molding process to create the pits. In some cases, such as when millions of discs need to be produced, the father is used to create *mother* discs, which can then be used to create multiple stampers called *sons*. Masters also may be made using a dye polymer and metalization process.

Figure 7.2 Mastering and Stamping (single layer)

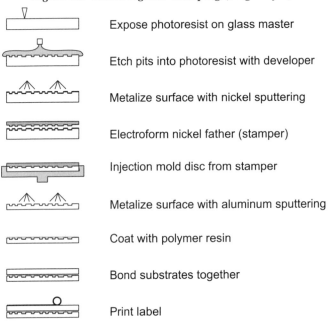

Expose photoresist on glass master

Etch pits into photoresist with developer

Metalize surface with nickel sputtering

Electroform nickel father (stamper)

Injection mold disc from stamper

Metalize surface with aluminum sputtering

Coat with polymer resin

Bond substrates together

Print label

NOTE
The term *mastering* is often misused to refer to the authoring or premastering process.

In the case of a single layer, the stamped plastic surface is covered with a thin reflective layer of aluminum, silver, or other reflective metal by a sputtering process. This creates a metallic coating between 60 and 100 angstroms thick. Silicon can also be used for the reflective layer. For two layers, the laser must be able to reflect from the first layer when reading it, but also focus through it when reading the second layer thus the first layer has a semi-transparent coating. In many cases, gold is used for the semireflective layer, giving dual-layer discs their characteristic gold tint. The first layer is about 20 percent reflective, the second layer about 70 percent. The top reflective layer is optionally given a protective overcoat before being bonded to the other substrate.

Adding to the tangle of layers and sides, is confusion over how to make a single-sided, dual-layer disc. Two processes exist: the first, developed by Matsushita, puts one layer on each substrate, separated by a very thin, transparent adhesive film. The second method, developed by 3M (now Imation), puts one layer on a substrate and then adds a second layer to the same substrate using a photopolymer (2P) material. Figure 7.3 depicts the two methods for making a dual-layer disc.

Figure 7.3 DVD/HD DVD Dual-layer Construction

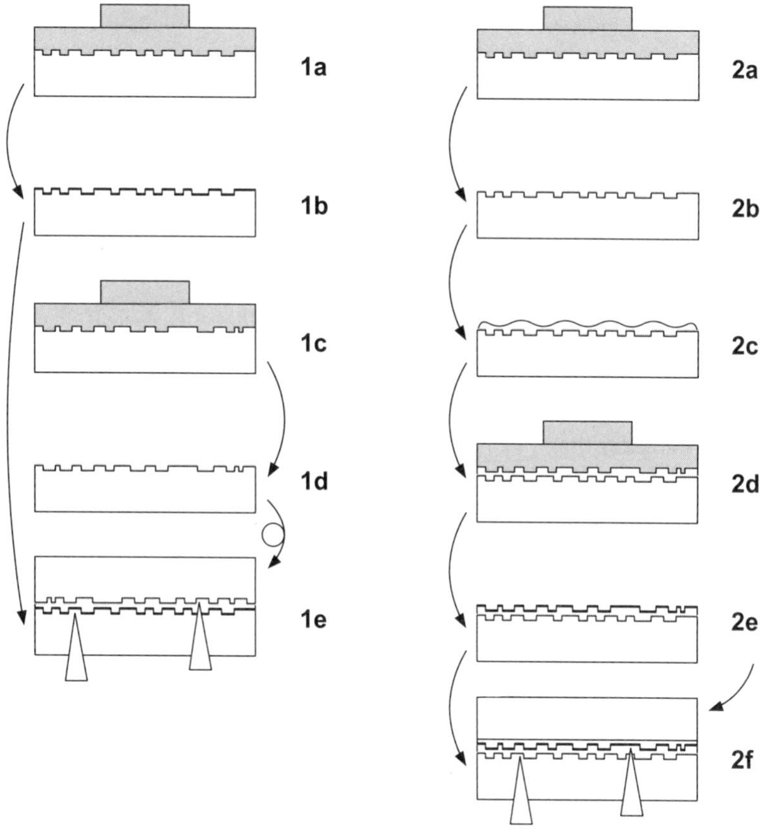

Method 1 — One layer on each side:
   a. Stamp the outer data layer into the first substrate,
   b. Add a semireflective coating,
   c. Stamp the inner layer "upside down" on the second substrate,
   d. Add fully reflective coating,
   e. Flip the second substrate over and glue with transparent adhesive to the first.

Method 2 — Two layers on one side:
   a. Stamp the outer data layer into first substrate,
   b. Add a semireflective coating,
   c. Cover with molten transparent material (ultraviolet resin),
   d. Stamp the inner layer into the transparent material[1],
   e. Add a fully reflective coating,
   f. Glue to the second substrate.

---

[1]The second layer may be transferred to the adhesive from an intermediate surface stamped into a temporary substrate made of a more "slippery" material such as PMMA or polyolefin.

Clearly, the first method is simpler. The second method is required for double-sided, dual-layer discs, which have four layers, making them slightly thicker than other discs.

In either case, the result is essentially the same, because the data layers end up at the proper distance from the outer surface. In the first case, the laser focuses through the first side and through the transparent glue onto the layer at the inner surface of the second side. In the second case, the laser focuses through the added transparent material onto the layer at the inner surface of the first side.

Hybrid Disc Construction

There are various approaches to mixing physical disc formats, such as an SACD/CD disc or a DVD-Audio DualDisc, usually to make backward compatible discs that will play on CD players (See Table 6.5 for common hybrids). CD audio pits can be molded on the outside of the otherwise blank DVD substrate, and a semi-reflective DVD data layer on the other substrate allows a conventional CD player to read through it. Alternatively the non-DVD substrate can be made thicker than normal — somewhere between 1.2 mm CD thickness and 0.6 mm DVD thickness — so that one side can be read by a CD player and the other side by a DVD player.

Bonding

A variety of methods of bonding substrates together also exist. Some roll the adhesive on, some "print" the adhesive with a screen, some spin the disc to spread the adhesive across the surface, and some rely on capillary flow. The most popular adhesives are hot-melt glue and ultraviolet (UV)-cured photopolymers. Tolerances are extremely tight (on the order of 30 microns), so this is one of the most challenging aspects of the DVD replication process and the one that sets it apart from CD replication.

The hot-melt method rolls a thin coat of melted thermoplastic resin onto each substrate and then presses the substrates together with a hydraulic ram. Once the glue has set, a strong bond is formed that remains stable at temperatures up to 70°C (150°F). The hot-melt bonding technique is cheaper and easier than the UV technique.

Variations of UV bonding use a photopolymer (2P) lacquer that is spread across the bonding surface by spinning the disc (radical adhesive), applying a silk-screened layer (cationic adhesive), or attaching a UV film in a vacuum. The lacquer is hardened by UV light, forming an extremely strong bond that is unaffected by temperature. A transparent adhesive such as 2P is required for dual-layer discs, since the laser must be able to read through it. The silk-screening method is more appropriate for writable discs because the cationic adhesive curing process generates less heat, making it less likely to affect the dye or phase-change layers.

Similar bonding techniques were used with both magneto-optical (MO) discs and laserdiscs. Laserdiscs sometimes suffered from deterioration of the aluminum layer, which was colloquially referred to as "laser rot" even though it had nothing to do with the laser beam or rotting. Laserdiscs were molded from acrylic (PMMA), which absorbed about 10 times more moisture than the polycarbonate used in DVDs and could lead to oxidation of the aluminum. The flexibility of laserdiscs allowed movement along the bond, which also could damage the aluminum layer. Laser rot problems were more prevalent when material purity was lower and the interaction between the materials that comprise a disc was not as well understood as it is today. Because DVDs are molded from improved materials and are more rigid than laserdiscs, they are not as susceptible to laser rot. Some discs have "cloudy" discolorations, usually as a result of improper molding, which sometimes causes problems and sometimes not. You can't tell by looking at the disc. Occasionally, there are manufacturing defects that are not visible but can cause progressive deterioration of the disc.

Burst Cutting Area

A section near the hub of the disc, called the *burst cutting area* (BCA), can be used for individualizing discs during the replication process. A strong laser is used to cut a series of low-reflectance stripes, somewhat like a bar code, that stores up to 188 bytes of information such as ID codes or serial numbers. The BCA sits between 22.3 and 22.5 millimeters from the center of the disc (see Figure 7.4).

Figure 7.4 Burst Cutting Area

A variation called *narrow burst cutting area* (NBCA) is used by recordable DVDs. NBCA bar codes are cut in the 22.7- to 23.5-millimeter-radius section. The BCA can only be used on single-layer discs (one- or two-sided). BCA data is recorded using a system similar to but simpler than DVD data recording: sync byte, preamble, data, error detection code (EDC), Reed-Solomon error correction code (ECC), postamble, and resync bytes, which are 2:1 phase encoded to represent 0 with 10 and 1 with 01. The resulting channel bits are modulated with the return-to-zero (RZ) method, where each 1 bit is represented by a pulse, created by a low-reflectance stripe.

Inscribing the BCA is optional for disc makers, but reading it is mandatory for drive makers, although not all drives reliably read it. The BCA can be read by the same laser pickup head that reads the disc. BCA information is usually unique to each disc so it may be used for identifying individual copies of a disc or by storage systems, such as disc jukeboxes, to quickly identify discs. The BCA is also used by content protection systems to hold a key used to encrypt the data on a recordable disc, tying the content to the unique key, or media ID, on that particular disc. Manufacturers of recordable media write a unique media ID in the BCA of each disc. During the development of this system, the DVD Forum realized that data in the BCA needed to be organized, otherwise the existing free-for-all would cause problems. They defined a BCA pack format, where the first two bytes of each pack hold an application ID to identify the chunk of data stored in the pack. The third byte is a version number, the fourth byte indicates the length of the data in the pack, and remaining bytes of the pack are the data. This allows multiple packs to be placed in the BCA. Various applications can read each pack until they find one that has the desired application ID.

Optical Pickups

A critical component of DVD units is the optical pickup, often called a pickup head (PUH), which houses one or more lasers, optical elements, and associated electronic circuitry. The optical pickup is mounted on an arm that physically positions the pickup under the disc, moving in and out to read different tracks. DVD readers use semiconductor red laser diodes at a wavelength of 635 to 650 nanometers.

Because of the requirement to read DVDs (which have data layers recorded at two different distances from the surface of the disc) as well as CDs (which have data recorded at yet a third level), a variety of techniques have been developed for controlling focus depth. The technology is further complicated by the fact that CD-ROM discs can be read with the 650 nanometer DVD laser but CD-R discs can only be reliably read using a 780 nanometer laser.

Holographic pickups use aspherical lenses with annular rings that focus reflected laser light from two different depths, one for DVDs and one for CDs. These pickups are unable to read CD-R discs unless used with dual lasers. Twin-lens pickups use two lenses to focus at DVD or CD depth. The appropriate lens is physically moved into the path of the laser beam by a magnetic actuator. These pickups are unable to read CD-R discs unless a second laser is included. Twin-laser pickups use two separate laser and lens assemblies, with one laser wavelength tuned for CD and CD-R and the other for DVD. In some cases these pickups include holographic lenses. Other approaches, such as twin-wavelength laser diodes that generate two wavelengths of light, are also used.

The same focusing control that compensates for movement of the disc as it spins is also used to adjust the focus depth for reading dual data layers. The DVD data layers are approximately 0.55 millimeter from the surface of the disc, and the CD data layer is approximately 1.15 millimeters deep.

NOTE

Pickups become even more complicated with blue-laser formats, because a third, 405-nanometer wavelength is required. Blu-ray introduces a 0.1-millimeter focus depth in addition to 0.6 and 1.2 millimeters.

Media Storage and Longevity

DVDs can be stored and used in a surprisingly wide range of environments (see Table 7.2). You can keep them in your refrigerator or in your hot water heater, although neither of these storage methods is a recommended alternative to a sturdy shelf or cabinet. A cool, dry storage environment is best for long-term data protection. If a disc has been kept in an environment significantly different from the operating environment, it should be conditioned in the operating environment for at least two hours before use.

Table 7.2 Environmental Conditions

	Ambient Temperature	Maximum Variation	Relative Humidity	Maximum Variation
Storage				
DVD-ROM, DVD-R	−20 to 50°C (−4 to 122°F)	15°C (59°F) per hour	5 to 90%	10% per hour
DVD-RAM	−10 to 50°C (14 to 122 °F)	10°C (50°F) per hour	3 to 85%	10% per hour
DVD+R	5 to 55°C (41 to 131°F)	10°C (50°F) per hour	3 to 85%	10% per hour
DVD+RW	−10 to 55°C (14 to 131°F)	15°C (59°F) per hour	3 to 90%	10% per hour
Operation				
DVD-ROM, DVD-R	−25 to 70°C (−13 to 158°F)	15°C (59°F) per hour	3 to 95%	10% per hour
DVD-RAM	5 to 60°C (41 to 140°F)	10°C (50°F) per hour	3 to 85%	10% per hour
DVD+R	−10 to 55°C (14 to 131°F)	15°C (59°F) per hour	3 to 90%	10% per hour
DVD+RW	−5 to 55°C (23 to 131°F)	10°C (50°F) per hour	3 to 85%	10% per hour

DVDs are quite stable, and if treated well, they usually will last longer than the person who owns them. Estimating the lifetime of a storage medium is a complex process that relies on simulated aging and statistical extrapolations. Based on accelerated aging tests and past experience with optical media, the consensus is that replicated discs will last from 50 to 300 years.

DVD-R and DVD+R discs are expected to last from 20 to 250 years, about as long as CD-R discs.[2] Shelf life before recording is about 10 years. The primary factor in the lifespan of DVD-R discs is aging of the organic dye material, which can change its absorbance properties. Long-term storage of DVD-Rs should be in a relatively dark environment, because the dyes are photosensitive, especially to blue or UV light (both of which are present in sunlight). Cyanine media are more susceptible than phthalocyanine media. Anecdotal reports have circulated of CD-R and DVD-R discs "wearing out" after being played for long periods of time, ostensibly because of alteration in the dye caused by reading it with a laser, but there are counter reports of discs that play in kiosks 24 hours a day for years without a hitch.

The erasable formats (DVD-RAM, DVD-RW, and DVD+RW) are expected to last from 25 to 100 years. The primary factor in the lifespan of these discs is the chemical tendency of phase-change alloys to separate and aggregate, thus reducing their ability to hold state. This also limits the number of times a disc can be rewritten.

In all cases, longevity can be reduced by materials of poor quality or a shoddy manufacturing process. Pressed DVDs of inferior quality may deteriorate within a few years, and cheap recordable DVDs may produce errors during recording or become unreadable soon after recording.

For comparison, magnetic media (tapes and disks) last 10 to 30 years; high-quality, acid-neutral paper can last a hundred years or longer, and archival-quality microfilm is projected to last 300 years or more.

Remember that computer storage media usually become technically obsolete within 20 to 30 years, long before they physically deteriorate. In other words, long before the discs become nonviable, it will become difficult or impossible to find the equipment to read them.

As discussed earlier under "Bonding," DVDs may be subject to "laser rot" — oxidation of the reflective layer. Media types such as dual-layer discs and DVD-Rs that use gold for the reflective layer are not susceptible to oxidation because gold is a stable element.

Writable DVD

There are six recordable versions of DVD-ROM: DVD-R for General, DVD-R for Authoring, DVD-RAM, DVD-RW, DVD+RW, and DVD+R.[3] The drive units of each format can read DVD-ROM discs, but each uses a different type of disc for recording. DVD-R and DVD+R record data only once (sequentially), whereas DVD-RAM, DVD-RW, and DVD+RW can be rewritten thousands of times. All writable DVD formats can be used for computer data storage as well as for recording video or audio in consumer devices.

---

[2]Kodak officially states that its CD-R media will last 100 years if stored properly. Testing by Kodak engineers indicates that 95 percent of the discs should last 217 years.

[3]R stands for "recordable." RAM stands for "random-access memory." Officially, DVD-RAM is "DVD rewritable," whereas DVD-RW, which ought to be "rewritable," is officially "DVD rerecordable."

It is possible to create a hybrid disc, sometimes called DVD-PROM,[4] where a portion of the disc is read-only (stamped in a replication line) and another portion is writable. This could be a disc with two layers, one prerecorded and one writable, or it could be a disc where part of one layer is stamped and the rest is writable.

The three erasable formats (DVD-RAM, DVD-RW, and DVD+RW) are essentially in competition with each other. DVD-RAM had a head start of about two years, but because the discs are not readable by most standard DVD drives or players, it has become more of a subformat that is supported along with one or both of the other formats. A major problem with writable formats is that none are fully compatible with each other or even with existing drives and players, although they have become more compatible and intermixed in the years since their introduction. See Table 7.3 for compatibility details.

Each writable DVD format is detailed in the next sections, followed by a summary and comparison.

Table 7.3 Compatibility of Writable DVD Formats

Disc Type	DVD Unit	DVD-R(G) Unit	DVD-R(A) Unit	DVD-RW Unit	DVD±RW Unit	DVD+RW Unit	DVD-RAM Unit
DVD-ROM	Reads	Reads	Reads	Reads	Reads	Reads	Reads
DVD-R(G)	May read	Reads, writes	Reads, does not write	Reads, writes	Reads, writes	Reads, does not write	Reads, does not write
DVD-R(A)	May read	Reads, does not write	Reads, writes	Reads, does not write	Reads, does not write	Reads, does not write	Reads, does not write
DVD-RW	May read	Reads, does not write	Reads, does not write	Reads, writes	Reads, writes	Usually reads, does not write	May read, does not write
DVD+R	May read	May read, does not write	Usually reads, does not write	May read, does not write	Reads, writes	Reads, writes	May read, does not write
DVD+RW	May read	Usually reads, does not write	May read, does not write	May read, does not write	Reads, writes	Reads, writes	May read, does not write
DVD-RAM	Rarely reads	Does not read, does not write	Does not read, does not write	Does not read, does not write	Does not read, does not write	Does not read, does not write	Reads, writes

DVD-R

DVD-R is the record-once version of DVD-ROM, similar in function to CD-R. Data can only be written permanently to the disc, which makes DVD-R well suited for archiving, especially for data that requires an audit trail. DVD-R is compatible with most DVD drives and players.

---

[4]PROM stands for "programmable read-only memory." The acronyms ROM, RAM, and PROM are all borrowed from roughly analogous solid-state memory devices.

First-generation DVD-R capacity was 3.95 billion bytes (3.68 gigabytes), later extended to 4.7 billion bytes (4.37 gigabytes) in version 2.0. Matching the full capacity of DVD-ROM was crucial for title development and testing.

DVD-R originally was a single format, but in early 2000 it was split into "DVD-R for Authoring", and "DVD-R for General". The general version uses a 650-nanometer laser (instead of 635 nanometers) for the ability to also write DVD-RAM. DVD-R(G) is intended for home use and has become the predominant variation, whereas DVD-R(A) is intended for professional development and is becoming rare. DVD-R(A) media are not writable in DVD-R(G) recorders, and vice versa, though both kinds of media are readable in most DVD players and drives. DVD-R(A) retains the characteristics of the original DVD-R format. The changes made to DVD-R(G), in addition to recording wavelength, are decrementing prepit addresses, a prestamped or prerecorded control area, and double-sided discs (see Table 7.4). Wobble frequency is the same for both. A major addition to DVD-R(A) 2.0 is the cutting master format (CMF) feature, which specifies how a disc description protocol (DDP) file is to be written in the lead-in area. The inclusion of a DDP file enables a DVD-R to be used in place of a DLT for submission to a replicator.

Table 7.4 Differences Between DVD-R Subformats

	DVD-R General (G)	DVD-R Authoring (A)
Recording wavelength	650 nanometers	635 nanometers
Prepit addressing	Decrementing (from sector FFCFFFh)	Incrementing (from sector 003000h)
Sides	One or two	One[a]
Control area content protection	Pre-recorded with 0s	0s written by drive
Content protection	CPRM (added in 2004)	None

[a]Two-sided DVD-R(A) discs are possible but were never made.

DVD-R uses organic dye technology. A dye material such as cyanine, phthalocyanine, or azo coats a wobbled groove molded into the polycarbonate substrate. The dye is backed with a reflective metallic layer — usually gold for high reflectivity. The wobbled pregroove provides a self-regulating clock signal to guide the laser beam of the DVD-R recorder as it "burns" pits into the photosensitive dye layer. The recorder writes data by pulsing the laser at high power (6 to 12 milliwatts) to heat the dye layer. The series of pulses avoids an over-accumulation of heat, which would create oversized marks. DVD-R uses two levels of laser power, while DVD-RW uses three levels to enable overwriting. The heated area of the dye becomes dark and less transparent, causing less reflection from the metallic layer underneath. The disc can be read by standard DVD readers, which ignore the wobble. Wobble frequency is eight times the size of a sync frame. The wobble is not modulated — that is, it does not contain addressing information — and is primarily used to generate a spindle motor control signal. Addressing and synchronization information is prestamped into the *land prepit* area that is positioned between the recording tracks at the beginning of each sector of the disc. As the recorder's laser beam follows the pregroove, the land prepits are contacted peripherally and create a second pattern of light reflected back to the photodetector. Since the land prepits generate a different signal frequency than the pregroove wobble, the encoded information can be extracted to tell the recorder what sector it is writing to. The land prepits are offset in alternating directions to avoid conflict when they coincide on each side of the groove.

Dual-layer DVD-R(G) discs are similar in structure, with two recording layers separated by a heat-diffusing spacer layer and a semi-transparent reflective layer that also acts as a heat sink. Refraction in the dye recording layer causes the recording beam spot to narrow to about 0.5 micrometers at the focal point, but be spread to about 40 millimeters on the neighboring layer, allowing control over writing to each layer.

Because sectors are already physically allocated on the disc, no physical formatting is required before recording. DVD-Rs usually are recorded in a single session, called *disc at once* (DAO). This is widely supported by DVD-R formatting tools and provides the most compatibility with readers.

Incremental writing, where data is appended to the end of a DVD-R in multiple sessions, can be done in two ways. The first method uses the multisession features of UDF, similar to the packet writing method developed for CD-R. This method is supported by only a few formatting tools. The second method is called *border zone recording*, where a small buffer zone called a *border-out* area is written at the end of each session. Border zone recording was adapted from multisession CD-R designs. The border-out acts as a temporary lead-out area that a reader can ignore once more data is written and a new border zone is added at the end. Reading a disc written with border zone recording requires drives or players that fully support the DVD-R specification. Many, however, do not support it, although the DVD Multi program requires it. As with CD-R, the price of DVD-R will drop to the point where it is easier to burn a new disc rather than go through the complications of incremental writing.

The first area on the disc is the power calibration area (PCA), which is used by the recorder to perform tests to optimize the laser power for the inserted disc. This process is called *optimal power calibration* (OPC) and is done automatically by the drive before recording. Power calibration can be performed about 7088 times on a single disc. The PCA extends from physical sectors 20800h to 223AFh in version 1.0 and from sectors 1E800h to 203AFh in version 2.0.

The PCA is followed by the recording management area (RMA), where the recorder stores recording management data (RMD) such as power calibration information, recorder ID, recording history, incremental recording details, etc. Beyond the RMA is the information area, which contains the lead-in, the data recording area, and the lead-out. Only the data recording area is user-accessible. In addition to recording user data, the drive can record copyright management information (CPM and CGMS) in sector headers of the data area. DVD-R(A) recorders can write to the control data in the lead-in area using special commands, although only 0s can be written in the part used for storing content protection keys. DVD-R(G) media are required to have prestamped or prerecorded control areas so that content protection keys from read-only media cannot be written. Content protection is provided by CPRM (see Chapter 5).

The DVD-R 1.0 format is standardized in ECMA-279.

DVD-RW

DVD-RW, officially known as *DVD rerecordable*, is a rewritable version of DVD-R. The name is pronounced as DVD "dash" RW, although many in the opposing camps pejoratively refer to it as DVD "minus" RW. It was briefly called DVD-ER and then DVD-R/W before being dubbed DVD-RW. The format is intended for use in authoring and consumer video and audio recording. Because of the physical similarities between DVD-R and DVD-RW, most DVD-RW recorders also write to DVD-R(G) media.

DVD-RW uses phase-change media. Compatibility of DVD-RW with most DVD drives and players is slightly less than that of DVD-R. The media has a low reflectivity, which confuses some readers into thinking it is dual-layer media. As with DVD-R, DVD-RW uses wobbled-groove recording with address info on land prepit areas for synchronization at write time. DVD-RW capacity is 4.7 billion bytes (4.37 gigabytes). DVD-RW media can be rewritten about 1000 times. DVD-RW media typically comes bare, but it can use the same cartridges as DVD-RAM.

Extra information, called a linking sector, is recorded between each ECC block on the disc. This extra padding allows for "write splices" to occur at the start and end of each write to the disc. Linking sectors have the same length and format as a normal sector, but they contain information that identifies them as not containing user data. A reader must allow for these gaps in the data by resynchronizing the data clock and otherwise ignoring them.

Unlike DVD-RAM, DVD-RW is designed for sequential access, using constant-linear-velocity (CLV) rotation control. Sequential access improves storage capacity and compatibility. DVD-RW uses a restricted overwrite mode, which reduces unrecorded gaps on the media. This is designed to limit extra seeks during recording (other than defect management sparing) in order to improve real-time recording performance. The alternative to restricted overwrite mode is sequential recording mode, which is similar to that of DVD-R. As with DVD-R, border zone recording can be done to provide compatibility with standard DVD-ROM drives and players. Defect management is performed in software, using features of UDF 2.0.

For restricted overwrite mode, ECC blocks must be formatted in advance. Since formatting the entire disc can take a long time (usually more than an hour), special quick format operations are specified for formatting small areas at a time and for adding or enlarging border zones.

The power management area (PMA) and recording management area (RMA) are the same as for DVD-R. The control area is pre-embossed to prevent content protection keys being written. Content protection is provided by CPRM (see Chapter 5).

Dual-layer DVD-RW discs are possible, but as of this writing, the technology had not yet made it out of the laboratory.

DVD-RAM

DVD-RAM, officially known as *DVD rewritable*, is an erasable, rerecordable version of DVD-ROM. Version 1.0 of DVD-RAM had a storage capacity of 2.58 billion bytes (2.4 gigabytes). It was increased in version 2.0 to 4.7 billion bytes (4.37 gigabytes). DVD-RAM media may be rewritten more than 100,000 times. DVD-RAM version 2.0 doubled the recording data rate to 22.16 Mbps (equivalent to 18X CD recording). Version 2.1 added an 80-millimeter disc size designed for use in portable devices such as digital camcorders.

DVD-RAM uses *phase-change dual* (PD) technology with some CD-RW and MO features mixed in. A wobbled groove provides clocking data with a 1.33-micron period. Marks are written in both the groove and the land between grooves. The alternating groove and land recording inverts the tracking signal once per revolution. The grooves and pre-embossed sector headers are molded into the disc during manufacturing.

Sectors on DVD-RAM are arranged in a *zoned CLV* layout, a hybrid of CLV and constant angular velocity (CAV). Rather than using the same angular velocity for the entire disc, which makes for fast access but is an inefficient use of recording surface, the disc is divided into multiple concentric rings, called *zones* (see Figure 7.5). The first sector of each revolution in the zones is always aligned. This allows for faster access than a traditional CLV layout. The speed of rotation remains constant within a zone. That is, CAV recording is used so that data density in each zone is the same. A 120 millimeter version 2.0 disc is partitioned into 35 zones. Each zone has a fixed radius in width so that each successive zone, moving out from the hub, contains more sectors. One sector per revolution is added to each zone out from the center of the disc, starting with 17 sectors per track, or a total of 39,200 sectors (2450 ECC blocks) in zone 0 and 105,728 sectors (6608 ECC blocks) in zone 34. Version 1.0 discs have 24 zones. A version 2.1 80-millimeter disc has 14 zones.

Figure 7.5 Zoned CLV

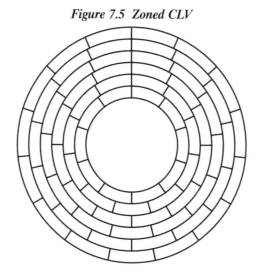

Of the writable DVD formats, DVD-RAM is the best suited for use in computers because of its hardware-based defect management, zoned CLV format for rapid access, and high number of rewrites. DVD-RAM is also intended for consumer audio and video recording. A major difference between DVD-RAM and all the other physical DVD formats is that embossed headers are used to identify the physical sectors. The address used by the drive to read or write sectors is the physical address, not the logical sector number. Because of this difference, along with low reflectivity, tracking changes for land and groove recording, modified access control for zoned CLV, defect management, and a different starting location of the data area, DVD-RAM discs cannot be read by many DVD drives and players.

Defect management is handled automatically by the drive. Two replacement methods are used. *Slipping replacement* is used first, where a defective physical sector is replaced by the first following nondefective physical sector. The primary defect list (PDL) records information about defective sectors found during manufacturing, as well as those found during later formatting. The PDL can only be changed by re-formatting. The number of sectors in a group to be listed in the PDL shall not exceed the number of sectors in the spare area in that group. *Linear replacement* is the second defect-management method, where a defective physical sector causes the containing ECC block (16 sectors) to be replaced by a block of sectors from a spare area. Defective sector blocks are listed in the secondary defect list (SDL) recorded on the disc.

DVD-RAM defines two types of spare areas from which to allocate replacement sectors. The *primary spare area* (PSA), consisting of 12,800 sectors, is preassigned during formatting. The *supplementary spare area* (SSA) is also assigned during formatting but can be expanded later. It can range from 0 to 97,792 sectors. Once the primary spare area is used up, the supplementary spare area is used and enlarged as needed.

Slipping replacement is quick because defective sectors are replaced by nearby sectors. Linear replacement is slower because it requires a head seek operation to get to the spare area to write or read the replacement block, but it is the only way to handle defects during recording without formatting the media.

Each time a data sector is overwritten, the recording start position is randomly shifted by 0 to 15 channel bits, and the lengths of the guard fields are changed by 0 to 7 bytes to accommodate. This random placement of the data keeps the recording surface from being stressed when the same data is written multiple times.

DVD-RAM media comes preformatted. A disc can be re-formatted to erase the contents and to reorganize defect-management information for better performance. A full format — which takes about an hour — verifies and initializes each sector, clears all but the manufacturer defect information from the primary defect list, and creates new spare areas. A quick improvement format — which takes only seconds — changes linear replacement sectors to slipped sectors, no data is erased. A quick clearing format — which also takes only seconds — initializes the media for use by clearing the PDL, as with full formatting, but without testing any sectors. All data on the disc is erased. Version 1.0 defined zoned formatting, which performs the equivalent of a quick clearing format to a single zone, erasing only the data in the zone.

In order to protect the media and make recording more reliable, DVD-RAM media usually comes in cartridges. Once the disc is written, it can be removed from the cartridge to be used in standard drives and players. There are three types of cartridges: type 1 is sealed, type 2 allows the disc to be removed and reinserted, and type 3 is an empty cartridge that a disc can be inserted into for more writing. There are also additional "type" labels for combinations of single/double sides and 120/80 mm sizes. Single-sided DVD-RAM discs come with or without cartridges. Double-sided DVD-RAM discs are usually available only in sealed type 1 cartridges. Special cartridges are available for 80-millimeter DVD-RAM; the two-part cartridge includes a disc holder that makes it easy to put the disc in and take it out without touching the recording surface. Adapter cartridges let 80-millimeter cartridges be used in recorders that only accept 120-millimeter cartridges. Standard cartridge dimensions are $124.6 \times 135.5 \times 8.0$ millimeters (see Figure 7.6).

Figure 7.6 Disc Cartridges

Images courtesy Panasonic and Marken Communications

Information placed in the embossed lead-in area on the disc specifies if it can be written without a cartridge. If the field is set to 00h, the recorder will not write to the disc unless it is mounted in a cartridge. Cartridges have write-inhibit holes. If the hole is open, the recorder will not write on the disc. A write-inhibit flag (in version 2.0 only) can be set or cleared by the recorder. If the flag is set, the recorder will not write on the disc. Each cartridge also has a sensor hole that indicates if the disc has been taken out at least once. If the sensor hole indicates that the disc has been removed, the recorder may reject certain operations, such as writing with verification turned off.

The DVD-RAM 1.0 format is standardized in ECMA-272 and ECMA-273. DVD-RAM 1.0 drives were able to read older 650 MB PD discs. Compatibility with PD media was dropped in version 2.0 drives, DVD-RAM 2.0 drives can read and write 1.0 media.

DVD+R and DVD+RW

The write-once DVD+R and erasable DVD+RW formats compete with DVD-R and DVD-RW and are not supported by the DVD Forum, even though the companies backing it (primarily Hewlett Packard, Philips, and Sony) are members of the DVD Forum.

The DVD+RW format is based on CD-RW technology and uses phase-change media with a high-frequency wobbled groove that provides highly accurate sector alignment during recording, thus eliminating linking sectors as used by DVD-RW. Marks are written in the groove only. Unlike the other writable formats, there is no embossed addressing information. The groove wobble is frequency modulated to carry addressing information called *address in pregroove* (ADIP). The DVD+RW specification provides for CLV formatting for sequential video access or CAV formatting for random access to computer data, but in practice no DVD+RW drives can write in CAV mode, largely because CAV-formatted discs are not compatible with standard DVD readers. CLV-formatted discs hold 4.7 billion bytes (4.37 giga-bytes). DVD+RW media may be rewritten about 1000 times. Most DVD+RW drives also write to CD-R and CD-RW media.

An early 1.0 version of DVD+RW, which held 3 billion bytes (2.8 gigabytes), was not com-patible with anything else and was abandoned in late 1999 in favor of the improved 2.0 ver-sion.

Optional defect management for DVD+RW is similar to that of DVD-RAM, with prima-ry and secondary defect lists and slipping and linear replacement. DVD+RW features, such as background formatting and hardware defect management, were later embodied in the Mount Rainier specification.

DVD+R is similar to DVD-R, but uses the same high-frequency wobble addressing as DVD+RW.

Phase-Change Recording

Phase-change recording technology, used for CD-RW, DVD-RW, DVD-RAM, DVD+RW, and other optical technologies, depends on changes in reflectivity between the amorphous state and the crystalline state of special alloys. When the alloy is heated by a laser at low power (*bias*) to reach a temperature around 200°C (400°F), it melts and crystallizes into a state of high reflectivity. When the alloy is heated by the laser at high power (*peak*) to a temperature between 500 and 700°C (900 and 1300°F), it melts and then cools rapidly to an amorphous state in which the randomized atom placement causes low reflectivity. As the disc rotates under the laser, it writes marks with high-power pulses and "erases" between the marks with low-power pulses. The marks can then be read with a much lower power setting to sense the difference in reflectivity (see Figure 7.7).

Figure 7.7 Phase-Change Recording

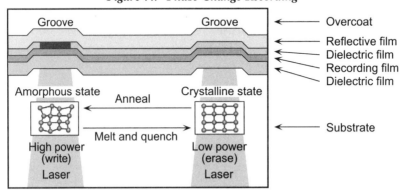

Phase-change discs are made of a polymer substrate to which a recording layer and a pro-tective lacquer are applied. The recording layer is a sandwich of four thin films: the lower dielectric film, the recording film, the upper dielectric film, and the reflective film. The upper and lower dielectric films rapidly draw heat away from the recording layer to create the 10 second supercooling effect needed to keep it from returning to a crystalline state. The recording film is composed of an alloy such as germanium, antimony, and tellurium (Ge-Sb-Te) or indium, silver, antimony, and tellurium (In-Ag-Sb-Te). The dielectric layers are made of a material such as zinc sulfide and silicon dioxide ($ZnS-SiO_2$). The reflective film is usually aluminum or gold.

Real-Time Recording Mode

The version 2.0 file system specifications for the various rewritable formats define a special real-time recording mode. Real-time stream recording and playback are critical applications, yet the drive units, especially those used in consumer players, have weak performance compared with hard-disk drives. In addition to slow seek times and lower data rates, dispersion of data across the medium may degrade performance. A practical solution to the problem is to make sure that streaming data is arranged in continuous order on the disc so that no head seeks are required during recording or playback. This helps guarantee a minimum data rate. The real-time stream model specifies new methods to handle defective sectors on writable discs.

When a player or operating system specifies that a file containing real-time streaming content is being recorded, data is placed in the largest unallocated extents that provide a physically contiguous space. Since linear replacement causes discontinuities in the data, it is disabled for streaming write operations (slipping, which is performed at formatting time, remains in effect). Even if the recorder encounters a defective block, it will not replace it, since the highest priority is given to continuity of data. Using real-time stream operations may result in erroneous data. Minor errors in an MPEG-2 video stream, for example, are often undetectable or not objectionable during playback and are preferable to loss of recording or playback continuity.

Summary of Writable DVD

Despite the many differences among writable DVD formats, they do have much in common (see Tables 7.5 and 7.6). Although some of the writable formats are slightly more suited for one application than for another, ultimately they all do a similar job of recording data. Some have speed advantages, yet faster drives and the use of CAV speeds when reading CLV discs makes most of the write/record speeds moot. Endless arguments percolate about which technology is superior, just as there are endless arguments about the benefits and drawbacks of cartridges. The bottom line, however, is that people prefer bare discs. The question of whether to use caddyless media was a point of some contention when deciding on the DVD read-only specs, but the consensus was to use bare discs like CDs. Early CD-ROM drives used caddies, but they were soon dropped in favor of drives that could accept bare discs. The customers had spoken. The recommendations of the DVD Forum working groups charged with designing rewritable DVD formats also were for bare discs. DVD-RAM uses optional cartridges (not caddies), as the engineers put more emphasis on protecting the media.

Table 7.5 Writable DVD Overview

	DVD-RAM		DVD-R			DVD+R	DVD-RW	DVD+RW
	1.0	2.0	1.0	2.0	DL	1.0	1.0	2.0
First available in U.S.	Q3 1998	Q2 2000	Q4 1997	Q2 1999			Q4 2000	Q1 2001
Capacity	2.6 GB/side	4.7 GB/side	3.95 GB/side	4.7 GB/side	8.5 GB/side	4.7 GB/side	4.7 GB/side	4.7 GB/side
Rewrites	~ 100,000		None	None		None	~1,000	~1,000
Write method	Wobbled land and groove		Wobbled groove			High-frequency wobbled groove	Wobbled groove	High-frequency wobbled groove
Recording technology (similar to)	Phase change (PD & MO & CD-RW)		Organic dye (CD-R)			Organic dye (CD-R)	Phase change (CD-RW)	Phase change (CD-RW)
Formatting	Zoned CLV		CLV			CLV	CLV	CLV
Applications	Data storage, backup, audio/video recording		Development testing and premastering short-run duplication, copying, archiving				Audio/video recording	Data storage, backup, audio/video recording
Backers	Hitachi, Matsushita, Samsung, Toshiba		Pioneer			Dell, Hewlett Packard, Philips, Sony, Thomson, Mitsubishi, Ricoh, Yamaha	Pioneer, Ricoh, Sharp, Sony, Yamaha	Dell, Hewlett Packard, Philips, Sony, Thomson, Mitsubishi, Ricoh, Yamaha

Table 7.6 Comparison of Physical Formats

	Capacity (12 cm, billion bytes)	Capacity (8 cm, billion bytes)	Recording Wavelength (nm)	Reflectivity (%)	Data Bit Length (µm)	Channel Pit Length (µm)	Min. Pit Length (µm)	Max. Pit Length (µm)	Track Pitch (µm)	Scanning Velocity[a] (m/s)
DVD-ROM SL	4.70	1.46	n/a	45-85	0.267	0.133	0.400	1.866	0.74	3.49
DVD-ROM DL	8.54	2.66	n/a	18-30	0.293	0.147	0.440	2.054	0.74	3.84
DVD-R 1.0	3.95	1.23	635	45-85	0.293	0.147	0.440	2.054	0.80	3.84
DVD-R 2.0	4.70	1.46	635/650[b]	45-85	0.267	0.133	0.400	1.866	0.74	3.49
DVD-R DL										
DVD-RW 1.0	4.70	1.46	650	18-30	0.267	0.133	0.400	1.866	0.74	3.49
DVD+R										
DVD+R DL										
DVD+RW	4.70	1.46	650	10-20		0.176			0.80	4.9-6.25[e] 3.02-7.35[f]
DVD-RAM 1.0	2.6	n/a	650	15-35	0.409-0.435[c]	0.205-0.218[c]	0.614-0.653[c]	2.863-3.045[c]	0.74	5.96-6.35[c]
DVD-RAM 2.0	4.70	1.46	650	15-35	0.280-0.291[c] (0.280-0.295[d])	0.140-0.146[c] (0.140-0.148[d])	0.420-0.437[c] (0.420-0.443[d])	1.960-2.037[c] (1.960-2.065[d])	0.615	8.16-8.49[c] (8.16-8.61[d])
CD-R	0.65	0.19								
CD-RW	0.65	0.19								

aReference rate. Many drives run faster.

b650 nanometers for DVD-R General, 635 nanometers for DVD-R Authoring.

cRange for zoned CLV format.

dValues for 8-centimeter media.

eRange for CLV format.

fRange for CAV format.

Data Format

On top of the physical formats sit the data and file formats. They are generally similar, with minor differences for writable discs to deal with error correction and rewriting of data (see Table 7.7).

Table 7.7 Data Storage Characteristics of DVD

Modulation	8/16 (EFMPlus)
Sector size (user data)	2048 bytes
Logical sector size (data unit 1)	2064 bytes (2048 + 12 header + 4 EDC)
Recording sector size (data unit 2)	2366 bytes (2064 + 302 ECC)
Unmodulated physical sector (data unit 3)	2418 bytes (2366 + 52 sync)
Physical sector size	4836 (2418 × 2 modulation)
Error correction	Reed-Solomon product code (208,192,17) × (182,172,11)
Error correction overhead	15% (13% of recording sector: 308/2366)
ECC block size	16 sectors (32678 bytes user data, 37856 bytes total)
Format overhead	16% (37856/32678)
Maximum random error	≤ 280 in 8 ECC blocks
Channel data rate[a]	26.16 Mbps
User data rate[a]	11.08 Mbps
Capacity (per side, 12 cm)	4.38 to 7.95GB (4.70 to 8.54 billion bytes)
Capacity (per side, 8 cm)	1.36 to 2.48GB (1.46 to 2.66 billion bytes)

[a]Reference value for a single-speed drive.

Sector Makeup and Error Correction

Each user sector of 2048 bytes is scrambled with a bit-shifting process to help spread the data around for error correction. Sixteen extra bytes are added to the beginning: 4 bytes for the sector ID (1 byte of sector information, 3 bytes for the sector number), 2 bytes for ID error detection, and 6 bytes of copyright management information (CPR_MAI). Four bytes of payload error detection code are also added to the end. This makes a data sector of 2064 bytes, called a data unit 1 (Figure 7.8). Each data sector is arranged into 12 rows of 172 bytes, and the rows of 16 data sectors are interleaved together (spreading them apart to help with burst errors) into error correction code blocks (192 rows of 172 bytes) (see Figure 7.9).

For each of the 172 columns of the ECC block, a 16-byte outer-parity Reed-Solomon code is calculated, forming 16 new rows at the bottom. For each of the 208 (192 + 16) rows of the ECC block, a 10-byte inner-parity Reed-Solomon code is calculated and appended. The ECC block is then broken up into recording sectors by taking a group of 12 rows and adding 1 row of parity codes. This spreads the parity codes apart for further error resilience. Each recording sector is 13 rows (12 + 1) of 182 bytes (172 + 10).

Figure 7.8 DVD Data Sector

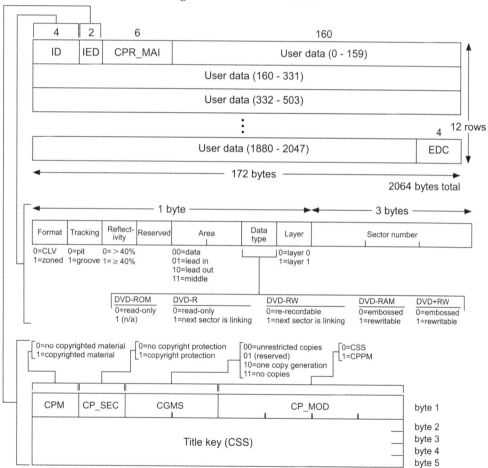

For DVD-RW 1.1 and DVD-R for General 2.0 media the CP_MOD field is replaced with a two-bit ADP_TY field (and a 2-bit reserved field). If ADP_TY is set to 01b it indicates that the sector is recorded in DVD-Video format.

Figure 7.9 DVD ECC Block and Recording Sectors

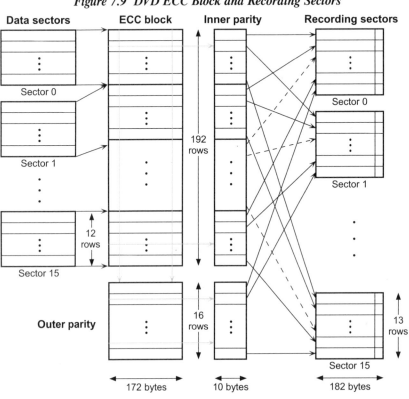

Each recording sector is split down the middle, and 1-byte sync codes are inserted in front of each half-row (Figure 7.10). This results in a recording data unit of 2418 bytes, called a data unit 3, which is identical for all DVD physical formats (DVD-ROM, DVD-R, DVD-RW, DVD-RAM, and DVD+RW).

Figure 7.10 DVD Physical Sector

Inner parity

sync 0	row 1, bytes 0-90	sync 5	row 1, bytes 91-181	
sync 1	row 2, bytes 182-272	sync 5	row 2, bytes 273-363	
sync 2		sync 5		
sync 3		sync 5		
sync 4		sync 5		
sync 1		sync 6		
sync 2		sync 6		
sync 3		sync 6		
sync 4		sync 6		
sync 1		sync 7		
sync 2		sync 7		
sync 3	row 12, bytes 2002-2092	sync 7	row 12, bytes 2093-2183	
sync 4	row 13, bytes 2184-2274	sync 7	row 13, bytes 2275-2365	

◄— 2 —►◄——— 91 bytes ———► ►◄— 2 —►◄——— 91 bytes ———►

Outer parity

(186 x 13 = 2418 bytes)

Channel bits:

sync 0	0-90	sync 5	91-181	sync 1	182-272	. . .	sync 7	2275-2365
32	1456	32	1456	32	1456		32	1456 = 38688

(4836 bytes)

DVD-RAM media adds data between each data unit 3 before recording (see Figure 7.11). The data unit is processed with 8/16 modulation (doubling each 8-bit byte to 16 bits). This creates a physical sector of 4836 bytes, which is written out row by row to the disc as channel data. Data is written using NRZI format (nonreturn to zero, inverted), where each transition from pit to land represents a one, and the lack of a transition represents a string of 2 to 10 zeroes.

Figure 7.11 DVD-RAM Physical Sector

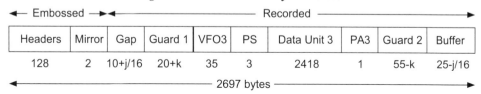

Embossed		Recorded							
Headers	Mirror	Gap	Guard 1	VFO3	PS	Data Unit 3	PA3	Guard 2	Buffer
128	2	10+j/16	20+k	35	3	2418	1	55-k	25-j/16

2697 bytes

j is varied randomly from 0 to 15 to shift data position a bit at a time
k is varied randomly from 0 to 7 to shift data position a byte at a time

The 8/16 modulation process replaces each byte with a 16-bit code selected from a set of tables (or a four-state machine). These codes have been chosen carefully to minimize low-frequency (DC) energy, and they also incorporate sync and merge characteristics. The codes guarantee at least two and at most ten zeroes between each pair of ones. 8/16 modulation is sometimes called EFMPlus. This is an odd name because EFM is the method used for channel data modulation on CDs and stands for "eight-to-fourteen modulation." Apparently "plus" means to add two to the second number.

The additional error detection and correction information takes up approximately 13 percent of the total data: 308 bytes out of 2366 bytes for each sector. Six error detection bytes are included with the sector data (2 for the ID and 4 for the complete sector). There are 302 bytes for RS parity codes (RS-PC) (120 for the 12 rows, 172 for the 172 columns, and 10 for the thirteenth row of column codes). The product code part of the error correction system refers to calculating parity codes on top of parity codes, where the rows of outer parity cross the columns of inner parity codes. This creates an extra level of detection and correction. RS-PC can reduce a random error rate from 2×10^{-2} (1 error in 200) to less than 1×10^{-15} (1 error in 1 quadrillion, or 1 million-billion). This is an error correction efficiency approximately 10 times that of CD.

Unlike one-dimensional error correction systems such as CIRC, which treat the data as a continuous stream, RS-PC works on relatively small blocks that are more appropriate for computer data. A disadvantage of RS-PC is that the matrix structure requires twice the buffer size, but given the ever-falling cost of memory, this is no longer a critical issue.

The maximum correctable burst error length for DVD-ROM is approximately 2800 bytes, corresponding to physical damage of 6 millimeters in length (6.5 millimeters for dual layers). In comparison, the CIRC structure of CDs can only correct burst errors of approximately 500 bytes, corresponding to 2.4 millimeters of physical damage. The writable formats each define slightly different correctable burst error lengths.

Writable media sector size is 2048 bytes. The writable formats cannot record encryption keys in the extra 6 bytes of copyright management information that is present on DVD-ROM discs. The MPAA has asked for a version of DVD-R that would be able to include CSS and CPPM information. The DVD Forum has been considering how to implement this request (which would create a third DVD-R subformat), but it may never see the light of day.

Data Flow and Buffering

Raw channel data is read off the disc at a constant 26.16 Mbps. The 16/8 demodulation process reduces the data by half to 13.08 Mbps. After error correction, the user data stream runs at a constant 11.08 Mbps (see Figure 7.12).

Figure 7.12 DVD-Video Data Flow Block Diagram

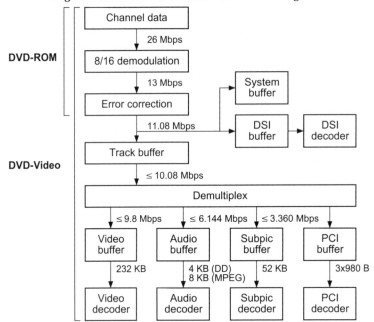

The track is broken into sectors of 2048 bytes (2 kilobytes). A 37,856-byte ECC block is made up of 16 sectors; 5088 bytes of this is overhead: 4832 bytes of RS-PC error correction, 96 bytes of sector error detection and correction, and 160 bytes of ID and content protection information. Because error correction is applied to an entire block, data must be read and written in complete blocks. Each ECC block contains 32,767 (32 kilobytes) of user data.

DVD drives, sometimes referred to as DVD logical units, transfer data in 2048-byte chunks — one sector's worth. An internal cache parcels out these units from an ECC block. This presents a special problem for rewritable drives that receive data from the host in 2-kilobyte chunks but must write it to the drive in 32-kilobyte chunks. Drives must implement an internal process for maintaining an ECC block that can be read, modified to add a new sector's worth of data, and then rewritten. This read-modify-write process may use an internal cache rather than writing to the disc.

If the data is not cached, more than one rotation is needed to write the data. First, the ECC block must be read from the drive, new data must be written and new ECC information calculated, and then the new ECC block must be written back to the same place on the disc. A technique to provide better performance with rewritable media is to write data in sizes that are a multiple of 32,768 bytes starting at logical block addresses that are a multiple of 16, which results in a one-pass direct overwrite operation.

Disc Organization

A DVD contains data written in a continuous spiral track. The track in layer 0 travels from the inner to the outer part of the disc. If layer 1 exists, its track can travel in either direction. Opposite track path (OTP) is designed for continuous data from layer to layer. The laser pickup assembly begins reading layer 0 at the center of the disc, starting with the lead-in area; when it reaches the outside of the disc, it enters the middle area and refocuses to layer 1 and then reads back toward the center of the disc until it reaches the lead-out area. Parallel track path (PTP) is designed for special applications where it is advantageous to switch between layers. In this case, each layer is treated separately, and each has its own lead-in and lead-out area, leaving no middle area (see Figure 7.13).

Figure 7.13 Dual Layer Track Path

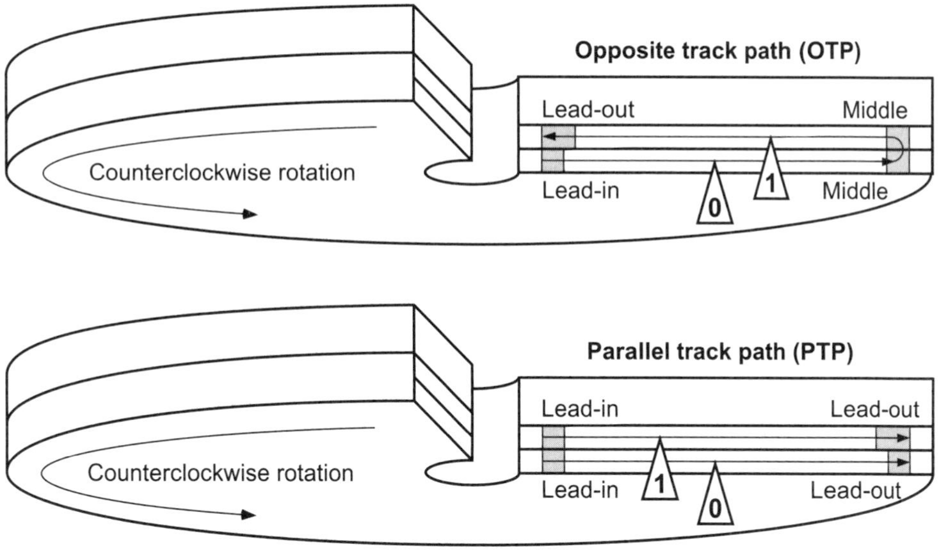

The lead-in area contains information about the disc (see Table 7.8 and Figure 7.14). content protection information and disc keys or media key blocks are also contained in the lead-in area.

The middle area and lead-out area each provide a guide to the readout system that it has reached the middle of a two-layer extent or the end of a layer. They are simply sectors full of zeros.

Table 7.8 Contents of Lead-in Area

Format version	4 bits (B0:b_0-b_3)	0 = 0.9x 1 = 1.0x 2 = 1.1x 4 = 1.9x 5 = 2.0x 6 = < 2.0 (specified at B27)
Disc type	4 bits (B0:b_4-b_7)	0 = DVD-ROM 1 = DVD-RAM 2 = DVD-R 3 = DVD-RW 9 = DVD+RW
Maximum data rate	4 bits (B1:b_0-b_3)	0 = 2.52 Mbps[a] 1 = 5.04 Mbps[a] 2 = 10.08 Mbps 15 = none specified
Disc size	4 bits (B1:b_4-b_7)	0 = 120 mm, 1 = 80 mm
Layer type	4 bits (B2:b_0-b_3)	b_0: 1 = read-only b_1: 1 = write-once b_2: 1 = write-many
Track path	1 bit (B2:b_4)	0 = PTP, 1 = OTP
Number of layers	2 bits (B2:b_5-b_6)	1, 2
Track density	4 bits (B3:b_0-b_3)	0 = 0.74 μm/track 1 = 0.80 μm/track 2 = 0.615 μm/track
Linear density	4 bits (B3:b_4-b_7)	0 = 0.267 μm/data bit 1 = 0.293 μm/data bit 2 = 0.409-0.435 μm/data bit 4 = 0.280-0.291 μm/data bit 8 = 0.353 μm/data bit
Starting sector number	3 bytes (B5-B7)	030000h (031000h for DVD-RAM and DVD+RW)
Ending sector number (main)	3 bytes (B9-B11)	
Ending sector number (layer 0)	3 bytes (B13-B15)	
BCA flag	1 bit (B16:b_7)	Does not exist, exists
Copyright protection system	1 byte	0 = none 1 = CSS or CPPM 2 = CPRM
Region management flags	1 byte	1 bit per region (8)
Encryption Data	2048 Bytes	
Manufacturing Data	2048 Bytes	1 per layer
Content Provider Information	28672 Bytes	

[a]Slow rotation for battery-operated drives. Not appropriate for video or audio playback.

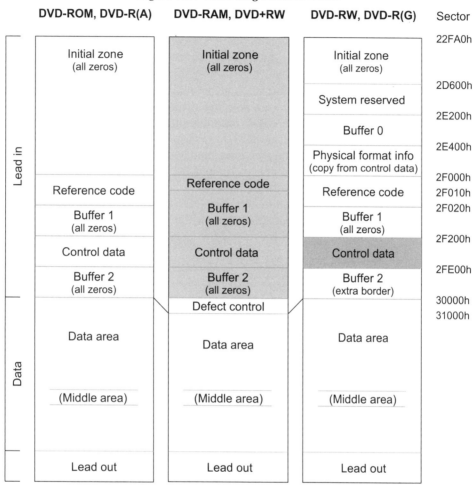

Figure 7.14 Recording Data Structure

Shaded areas are pre-embossed or pre-recorded and can't be changed

Unlike CD, DVD has no specially defined low-level data formats for audio or video. In addition, no TOC or subchannel data exist. There is only the concept of general data.

On most units, the rotational velocity of the disc varies; the farther from the center the laser pickup is, the slower the disc spins. This creates a CLV, keeping the current track surface moving at a constant speed above the pickup head so that the data density remains the same across the entire surface. Accordingly, there are more data sectors per revolution further from the center of the disc. Some units spin the disc at a constant angular velocity, keeping the revolutions per minute the same regardless of the position of the pickup head. They use more advanced circuitry to deal with the difference in data readout rates. These units only achieve their maximum rated speed when reading the outermost section of the disc.

File Format

Data sectors on a DVD-ROM may contain essentially any type of data in any format. Officially, the OSTA UDF file format standard is mandatory. UDF defines specific ways in which the ISO 13346 volume and file structure standard is applied for specific operating systems.

UDF limits ISO 13346 by making multivolume and multipartition support optional, defining filename translation algorithms, and defining extended attributes such as Mac OS file/creator types and resource forks. UDF provides information specific to DOS, OS/2, Macintosh, Windows 95, Windows NT, and UNIX. On top of the UDF limitations, the DVD-ROM standard requires that the logical sector size and logical block size be 2048 bytes. UDF includes an appendix defining additional restrictions for DVD-Video. See the "DVD File Format" section in Chapter 9 for details.

UDF also defines an extended attribute for storing DVD copyright management information. It is an implementation use-extended attribute with the implementation identifier set to "*UDF DVD CGMS Info."

DVD-ROM Premastering

The premastering process for a DVD-ROM is very similar to the premastering process for a CD. The computer directory structure and data files are written to a UDF, a UDF Bridge (UDF/ISO 9660), or other format disc image. Some premastering systems allow the disc image to be used to simulate an actual disc. The disc image is copied to DLT or DVD-R for creating a one-off test disc or for mastering and replication. The DVD-Video premastering process is described in Chapter 6.

Improvement over CD

The storage capacity of a single-layer DVD is seven times higher than that of a CD, due to a combination of smaller physical bit size, slightly larger data area, and improved logical channel coding (see Table 7.9).

Table 7.9 Incremental Improvements of DVD over CD-ROM

Factor	DVD	CD	Gain
Smaller pit length	$0.400 \ \mu m$	$0.833 \ \mu m$	2.08x
Tighter tracks	$0.74 \ \mu m$	$1.6 \ \mu m$	2.16x
Slightly larger data area	$8605 \ mm^2$	$8759 \ mm^2$	1.02x
More efficient modulation	16/8	17/8	1.06x
More efficient error correction	13% (308/2366)	34% (1064/3124)	1.32x
Less sector overhead	2.6% (62/2418)	8.2% (278/3390)	1.06x
Total increase			~7x

Chapter 8
Blue Laser Physical Disc Formats

This chapter covers the details of blue laser physical formats and data recording formats that are the basis for the next generation of optical discs. As with chapter 7, this chapter assumes that the reader has a basic understanding of the terms and concepts discussed in Chapters 3 and 4, as well as Chapters 5 and 7. The application details for the blue laser disc formats are covered in Chapter 9.

Physical Composition

The physical format and production process of an HD DVD disc are close to those of a DVD or CD. More care needs to be taken to properly replicate the tinier pits (compared to DVD) and to bond the substrates together. Blu-ray discs (BD) are even more complicated to produce because of their thinner data layer.

Table 8.1 lists the physical characteristics of blue laser discs. Most of these characteristics are shared by all the physical formats (read-only, writable, 12 centimeter and 8 centimeter).

Table 8.1 Physical Characteristics of HD DVD and BD

	HD DVD	BD
Thickness	1.2 mm (+0.30/-0.06) (two bonded substrates)	1.2 mm (±0.3) (two bonded substrates)
Substrate Thickness	0.6 mm (+0.043/-0.030)	0.1 (SL) or 0.75 mm (DL) plus 1.1 mm (±0.003)[a]
Spacing Layer Thickness	0.055 mm (±0.015)	0.025 mm (±0.005)
Hard coat thickness	n/a	2 μm
Diameter	120 (±0.3) or 80 mm	120 (±0.3) or 80 mm
Center hole diameter	15 mm (+0.15/-0.00)	15 mm (+0.10/-0.00)
Clamping area diameter	22.0 to 33.0 mm	23.0 to 33.0 mm
Burst cutting area (BCA) diameter	44.6 (+0.0/-0.8) to 46.6 (+0.10/-0.0) mm	42 (+0.0/-0.6) to 44.4 (+0.4/-0.0) mm
Lead-in diameter	46.6 to 48.2 mm	44.0 to 44.4 mm (+0.4/-0)
Data diameter (12-cm disc)	48 (+0.0/-0.4) to 115.78 mm	48.0 (+0, -0.2) to 116.2 mm
Data diameter (8-cm disc)	-	48.0 (+0/-0.2) to 76.2 mm

continues

Table 8.1 *Physical Characteristics of HD DVD and BD (continued)*

	HD DVD	BD
Lead-out outer diameter (12-cm disc)	-	117 mm
Lead-out outer diameter (8-cm disc)	-	77 mm
Mass (12-cm disc)	14.0 to 20.0 g	12 to 17 g
Mass (8-cm disc)	-	5 to 8 g
Readout wavelength	405 nm (±5)	405 nm (±5)
Read power		0.35 (±0.1) mW (SL), 0.70 (±0.1) mW (DL)
Polarization	Circular	Circular
Numerical aperture (objective lens)	0.65 (±0.01)	0.85 (±0.01)
Wavefront aberration (ideal substrate)	≤0.033 λ rms	≤0.033 λ rms
Relative intensity noise (RIN) of laser	≤-125 dB/Hz	≤-125 dB/Hz
Beam diameter		0.58 μm
Optical spot diameter		0.11 μm
Reflectivity	45 to 85% (SL), 18 to 30% (DL)[a]	35 to 70% (SL), 12 to 28% (DL)
Refractive index	1.50 to 1.70	1.45 to 1.70
Birefringence	≤0.040 μm	<0.030 μm
Radial runout (disc)	≤0.3 mm, peak to peak	≤0.050 mm (SL), ≤0.075 mm (DL), peak to peak
Radial runout (track)	<50 μm, peak to peak	-
Axial runout (disc)	-	≤0.3 mm (≤0.2 mm for 80-cm disc)
System radial tilt margin (angular deviation, α)	±0.80°	±0.60°
System tangential tilt margin (angular deviation, α)	±0.30°	±0.30°
Disc radial tilt margin (α)	±0.70°	<1.60°, peak to peak
Disc tangential tilt margin (α)	±0.20°	<0.60°, peak to peak
Asymmetry	-0.10 to 0.10	-0.10 to 0.15

continues

Table 8.1 Physical Characteristics of HD DVD and BD (continued)

	HD DVD	BD
Track spiral (outer layer)	Clockwise (hub to edge)	Clockwise (hub to edge)
Track spiral (inner layer)	Counterclockwise (edge to hub)	Counterclockwise (edge to hub)
Track pitch	ROM: 0.4 μm R/RW: 0.34 μm	0.32 μm ($\pm$0.003)
Pit length	ROM:0.204 to 1.020 μm (2T to 13T) R/RW: 0.173 to 1.126 (zone 0) to 0.187 to 1.213 (zone 18)	0.149 to 0.695 μm (2T to 8T)
Data bit length (avg.)	ROM: 0.154 μm R/RW: 0.130 to 0.140 μm	0.11175 μm
Channel bit length (avg.)	ROM: - R/RW: 0.087 to 0.093 μm	0.745 μm
Jitter (of channel clock period)	<8%	<6.5%(DL or L0) or 8.5%(L1)[b]
Parity inner error (PIE)	$\leq$280	n/a
Correctable burst error	6 mm	7 mm
Symbol error rate	$<5\times10^{-5}$	$<2\times10^{-4}$
Maximum local defects	100 μm (air bubble), 300 μm (black spot), =20-mm scanning distance between max defects	100 μm (air bubble), 300 μm (black spot)
Rotation	Counterclockwise to readout surface	Counterclockwise to readout surface
Rotational velocity (CLV)[c]	R/RW: 944.4 (zone 18) to 2251.8 (zone 0) rpm	
Scanning velocity[c]	5.64 to 6.03 m/s	5.280 m/s (23.3 GB) 4.917 m/s (25.0 GB) 4.554 m/s (27.0 GB)
Storage temperature	-10 to 50°C (14 to 122°F), $\leq$ 10°C (50°F)/h change	-
Storage humidity	3 to 85% relative, 1 to 30 g/m^3 absolute, $\leq$ 10%/h change	-
Operating temperature	5 to 60°C (41 to 140°F), $\leq$ 10°C (50°F)/h change	-
Operating humidity	3 to 85% relative, 1 to 30 g/m^3 absolute, $\leq$ 10%/h change	

[a]SL = single layer; DL = dual layer.

[b]L0 = layer 0; L1 = layer 1

[c]Reference value for a single-speed drive.

HD DVD Physical Format

An HD DVD disc is composed of two bonded plastic substrates, similar to standard DVD (see Figure 7.1, Chapter 7). Each substrate can contain one or two layers of data. The first layer, called *layer 0*, is closest to the side of the disc from which the data is read. The second layer, called *layer 1*, is further from the readout surface.

A disc is usually read from the bottom, so layer 0 is below layer 1 when the disc is in the drive. The layers are spaced very close together, the distance between layers is 55 microns, which is less than one-tenth of the thickness of the substrate (less than one-twentieth of the thickness of the disc). The proposed three-layer HD DVD disc puts two layers on one substrate and a third on the second substrate, all readable from one side of the disc.

Although the substrates are the same depth, a smaller wavelength laser with a larger numerical aperture objective lens allows for higher recording density, resulting in larger capacity (see Table 8.2). Read-only (HD DVD-ROM) discs have lower capacity than write-once (HD DVD-R) and rewritable (HD DVD-RW). This decision was made early in the development process so that ROM discs could be made on existing DVD replication equipment with minimal changes. The production and replication process for HD DVD is quite similar to DVD (see Chapter 7).

Table 8.2 HD DVD Capacities

Type	Size	Sides and Layers	Billions of Bytes (10^9)	Gigabytes (2^{30})
HD DVD-ROM	12 cm	1 side, 1 layer	15	13.97
HD DVD-R/RW			20	18.62
HD DVD-ROM	12 cm	1 side, 2 layers	30	27.94
HD DVD-R/RW			50	37.25

HD DVD Recording

Recordable HD DVD discs are formed with wobbled grooves in a spiral. Data can be recorded in the groove or on the land between. The data area is divided into physical segments of 11,067 bytes each (132,804 channel bits). Physical segment address information is stored in the wobble modulation in *wobble address in periodic position* (WAP) information, which is 17 *wobble data units* (WDU). A wobble data unit is made up of 84 wobbles. Groups of seven physical segments make up one *data segment*. Each data segment holds 64K (65,536 bytes) of modulated user data. A *recording cluster* is one or more data segments followed by a 24-byte guard field that allows for overwriting and cluster shifting. The start point of a recording cluster is randomly shifted by 0 to 167 channel bits to avoid recording layer "burn in" when the same data is overwritten many times.

A *data frame* holds 2048 bytes of user data, 4 bytes of data ID, 2 bytes of ID error detection code (IED), 6 reserved bytes, and 4 bytes of error detection code (EDC) for the entire frame, making a total of 2,064 bytes. The data ID holds information such as sector type (CLV or zoned), tracking method (pit or groove), reflectivity (≥ 40 percent), recording type (general or real-time), area type (system lead-in area, data lead-in area, data area, or data lead-out area), data type (read-only or writable), layer number (0 or 1), and data frame number.

HD DVD Error Correction

Each data frame is scrambled to spread out the data. Then, 32 scrambled frames are combined for cross Reed-Solomon (RS) error correction coding. 172-byte chunks from the scrambled frame are grouped into two columns of 192 rows, then 16 bytes of outer parity (PO) codes are calculated and added to the bottom of each column, making them 208 rows tall, for an outer RS of (208, 192, 172). Then, 10 bytes of inner parity (PI) codes are calculated, including the PO rows, and added to the right of each column, making each one 182 bytes wide, for an inner RS of (182, 172, 11). The resulting array of 208 rows of 182×2 columns is then interleaved to spread the combination inner/outer parity codes between every 12 rows, and the resulting array of data is called a *recording frame*. Sync codes are added to the beginning of every 91 bytes of the recording frame, producing a *data field*. A *data segment* holds one data field with an additional 95 bytes of sync codes and buffer fields.

HD DVD Data Modulation

HD DVD uses *eight-to-twelve modulation* (ETM), an RLL[1](1, 10) code (2T to 11T marks) where each 8 bits of a data field are converted to 12 channel bits that are then converted to a *non return to zero inverted* (NRZI) channel bit stream that is recorded on the disc. A very complicated set of rules, states, and tables are used for modulation and demodulation. One physical sector of 29,016 channel bits is equivalent to 2,418 unmodulated bytes.

HD DVD-R and -RW Formats

HD DVD-R uses an organic dye recording layer similar to DVD-R.

HD DVD-RW uses a phase-change recording layer similar to DVD-RW. Sectors are written using a zoned constant linear velocity (ZCLV) arrangement similar to DVD-RAM (see Chapter 7). The data area is broken into 19 zones, numbered 0 through 18. Within each zone the rotation speed stays the same, maintaining a channel bit rate of 64.80 Mbps, and the number of physical segments per track stays the same. Zone 0, near the center of the disc, has 13 physical segments per track, read at 2,251.8 rpm. Zone 18, near the edge of the disc, has 31 physical segments per track, read at 944.4 rpm.

HD DVD-RW Defect Management

Defect management can be done by the file system or the drive. If the drive manages defects, it uses four *defect management areas* (DMAs) per disc side. Two DMA managers keep track of the state of the four DMAs. There are 10 DMA manager sets, each consisting of DMA manager 1 and DMA manager 2 fields. Each DMA manager field is one physical

---

[1]RLL (Run length limited). A data modulation technique that codes sequences of data as runs of zeros and ones. For example, the run lengths in 0001011 are 3, 1, 1, and 2, often expressed as RLL (d, k), where d is the minimum run length and k is the maximum run length.

segment. There are 100 DMA sets, each consisting of DMA1 and DMA2 fields near the inside of the disc and DMA3 and DMA4 fields near the outside of the disc. Each DMA field is two physical segments. DMA1 and DMA2 are followed by a physical segment block containing the *disc definition structure* (DDS) and *primary defect list* (PDL) and a physical segment block containing the *secondary defect list* (SDL). The DDS contains information such as number of zones, starting sector of each zone, location of PSA, etc. The *primary spare area* (PSA) is used during formatting. The *secondary spare area* (SSA) can be used during or after formatting. The disc must be formatted before use. Defective physical segment blocks are replaced by slipping replacement, linear replacement, or block skipping, which is designed for real-time recording. (See Chapter 7 for more on replacement techniques.)

Blu-ray Physical Format

Blu-ray discs (BD) are similar to DVDs and HD DVDs but with different substrate thicknesses (see Figure 8.1). The data substrate, often called the *cover layer*, is 0.1 mm thick, and the other substrate is 1.1 mm thick. The thin cover layer allows the laser to converge more sharply[2] without undue aberration from disc tilt.[3] This leads to smaller spot sizes and thus higher data density compared to DVD or HD DVD. Minor variations in pit length on the disc result in three slightly different capacities (see Table 8.3).

Figure 8.1 BD Disc Structure

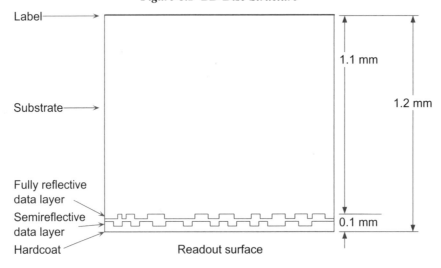

[2]Convergence performance of an objective lens is expressed as a unitless number called numerical aperture (NA), calculated as the sine of the half angle of convergence. The diameter of converging light is inversely proportional to the NA. That is, D = ? / NA, where ? = wavelength (405 nm, in this case). The larger the NA, the smaller the convergence.

[3]Think of looking at reeds in a pond. If you look straight down, the reeds look normal, but if you look at an angle, the reeds seem fractured at the point they penetrate the surface of the water because of light refraction. The deeper the pond, the more severe the distortion. Likewise, a laser beam passing through the plastic material of a disc is distorted if it doesn't enter at a perfectly perpendicular angle, and the thicker the plastic the more severe the aberration.

Table 8.3 Blu-ray Capacities

Name	Size	Sides and Layers	Billions of Bytes (10^9)	Gigabytes (2^{30})
BD-23	12 cm	1 side, 1 layer	23.305	21.704
BD-25			25.025	23.307
BD-27			27.020	25.164
BD-46	12 cm	1 side, 2 layers	46.610	43.409
BD-50			50.051	46.613
BD-54			54.040	50.329
BD-8	8 cm	1 side, 1 layer	7.791	7.256
BD-16	8 cm	1 side, 2 layers	15.582	14.512

Since the laser does not read through the second substrate it doesn't have to be clear, it can even be made from non-plastic material such as paper or corn starch. It's theoretically possible to make a two-sided, three-substrate BD with a 1.0 mm middle substrate sandwiched between two 0.1 mm cover layers, but this is not covered in the specification.

A standard feature of BD is a 2-nanometer hard coat at the readout surface to protect the disc from fingerprints and scratches. Think of it as transparent Teflon. Hard coatings can be applied to DVD and HD DVD as well.

Blu-ray discs can optionally use a cartridge to protect the disc (see Figure 8.2). There are two types of cartridge: sealed (from which the disc cannot be removed) and open (allowing the disc to be removed and replaced). The original BD-RE 1.0 format used a sealed cartridge, but since hard coat technology was adopted, the BDA recommends that cartridges should not be used for any of the BD formats except in special situations.

Figure 8.2 BD Disc Cartridges

Open cartridge   Sealed cartridge

BD-ROM Mastering

Different technologies are used for creating BD masters. One technology is *phase-transition metal* (PTM) mastering that uses a 405-nm laser to create a pattern in inorganic photoresist material similar to that used for phase-transition recordable DVD-RW and BD-RE. The PTM layers are placed on a silicon wafer instead of a glass plate. One advantage to PTM is that the pickup head can monitor the creation of pit patterns as they are created so that adjustments can be made in real time. Another approach is to modify a laser beam recorder

(LBR) to use *liquid immersion*, where a stream of liquid is injected in front of the recording head and retrieved behind it. The interface created by the liquid in contact with the recording lens and the master surface allows a deep-UV laser to create a sufficiently small spot size for BD-ROM. A third approach uses an electron beam recorder (EBR) to expose a photo-resist layer. Electron beam recording can take hours to cut a master, but researchers are making progress on enhanced photo-resists material to speed up the process to make it more commercially viable. Regardless of the approach, the resulting masters are used to create stampers for replication.

BD-ROM Composition and Production

BD-ROM discs are created, complete with tracks of pits holding data, by a stamping process. There are two basic approaches to producing Blu-ray discs: 1) transfer a data pattern to a 0.1 mm layer (typically by molding a PMMA base, injecting polycarbonate, then peeling off the PMMA) and bonding it to a 1.1 mm substrate, and 2) injection-molding a pattern directly on top of a 1.1 mm substrate, then adding a cover layer. The latter method, which is more popular, includes approaches to forming the cover layer such as 1) spin-coating the cover with an ultraviolet (UV) resin, 2) bonding a pre-made cover sheet with a UV resin, and 3) bonding a pre-made cover sheet with pressure sensitive adhesive (PSA) (see Figure 8.3). The resin coating process spins the disc to spread the resin from the hub to the edge, which makes it difficult to keep the thickness uniform. The two sheet coating processes require tight tolerances when manufacturing the sheets and performing the bonding.

Figure 8.3 BD Dual-layer Construction

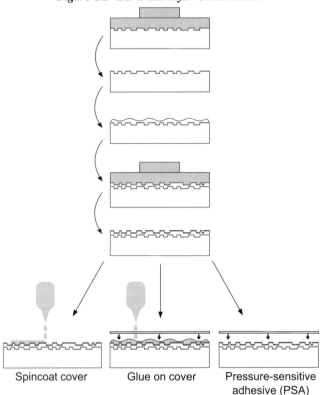

| Spincoat cover | Glue on cover | Pressure-sensitive adhesive (PSA) |

BD-RE Composition

Blu-ray rewritable discs have a structure similar to Blu-ray read-only (BD-ROM) discs. They are produced using one or two phase-change recording layers with a high-frequency modulated wobbled groove to provide addressing information and speed control for the recorder. Wobble period is approximately 5 μm, with embedded addresses in *address in pre-groove* (ADIP) units of 56 wobbles. ADIP information is stored using *minimum-shift-keying* (MSK) modulation and *saw-tooth wobble* (STW). Data is written in the groove rather than on the land between.

Table 8.4 BD-RE Characteristics

Recording material	Phase change
Transmission stack thickness	0.095 to 0.105 mm ($\pm$0.002) (TS0), 0.070 to 0.080 mm ($\pm$0.002) (TS1)
Recording method	In groove
Write power (1x)	<7 mW (SL), <12 mW (DL)
Address tracking	High-frequency modulated wobbled groove with addresses
Groove frequency (1x)	956.522
Channel bits per wobble	69
Storage temperature	-10 to 55°C (14 to 131°F), ≤ 15°C (59°F)/h change
Storage humidity	5 to 90% relative, 1 to 30 g/m^3 absolute, ≤ 10%/h change
Operating temperature	5 to 55°C (41 to 131°F)
Operating humidity	3 to 90% relative, 0.5 to 30 g/m^3 absolute

BD-R Composition

Blu-ray write-once discs are similar to Blu-ray rewritable (BD-RE) discs, although the recording layers can be formulated with organic dye, inorganic alloys such as Si/Cu, or write-once phase-change material. Data can be written in the groove or on the groove (on the land between).

Table 8.5 BD-R Characteristics

Recording material	Organic dye, inorganic alloy, or phase change
Transmission stack thickness	0.095 to 0.105 mm ($\pm$0.002) (TS0), 0.070 to 0.080 mm ($\pm$0.002) (TS1)
Recording method	On groove or in groove
Write power (1x)	<6 mW
Address tracking	High-frequency modulated wobbled groove with addresses
Groove frequency (1x)	956.522
Channel bits per wobble	69
Storage temperature	-10 to 55°C (14 to 131°F), ≤ 15°C (59°F)/h change
Storage humidity	5 to 90% relative, 1 to 30 g/m^3 absolute, ≤ 10%/h change
Operating temperature	5 to 55°C (41 to 131°F)
Operating humidity	3 to 90% relative, 0.5 to 30 g/m^3 absolute

BD Error Correction

Error correction is the same for all BD formats. Data is recorded in 64K partitions, called clusters, each containing 32 data frames with 2048 bytes of user data each. Clusters are protected by two error correction mechanisms. The first is a *long distance code* (LDC) using Reed-Solomon in a (248, 216, 33) structure. The second error correction mechanism multiplexes the data with a *burst indicator subcode* (BIS) using (62, 30, 33) Reed-Solomon (RS) codewords. BIS includes addressing information and application-dependent control data information (18 bytes per data frame). BIS pinpoints long burst errors that can then be removed to improve LDC error correction.

A data frame holds 2048 bytes of user data and 4 bytes of simple error detection code (EDC) for a total of 2052 bytes. Each data frame is scrambled to spread the bits around. Then, 32 data frames are combined into a data block with 216 rows of 304 columns. Each column is one byte. A data block is extended into an LDC block by appending the LDC codes for the data block as 32 rows of 304 columns. The LDC block is internally interleaved and shifted to improve burst error correction, resulting in an LDC cluster of 152 columns and 496 rows.

The 64KB physical cluster is divided into 16 *address units* (AU). The 4-byte address unit numbers are derived from the physical sector numbers and together with 1 byte of flags, 4 bytes of error correction, and user control data, they make up the data used for the BIS, which goes through a RS (62, 30, 33) coding and is arranged into a BIS cluster of 496 rows by 3 columns.

The LDC cluster is split into four groups of 38 columns, and each of the three columns from the BIS cluster is inserted between them, forming an ECC cluster. An additional column of frame sync bits is added at the beginning of the ECC cluster and DC control bits are inserted to form a *recording frame* of 496 rows by 155 columns, also called a physical cluster. Prohibit RMTR limits the number of consecutive minimum run lengths (2T) to six, which avoids low signal levels and improves readout performance.

BD Data Modulation

Data in a recording frame, except for the sync bits, are modulated using the 1-7PP technique, an RLL (1, 7) code (2T through 8T run lengths) where the first P stands for *parity preserve* and the second P stands for prohibit *repeated minimum transition run length* (RMTR). Parity preserve means that the parity of the data stream matches the parity of the modulated stream. That is, if the number of 1s in the selected chunk of data is even, then the number of 1s in the modulated bits is even, and the same if the number of 1s is odd. This is an efficient way to control the low-frequency content of the recorded signal. A modulation conversion table is used to map sequences of data bits to modulation bits, which are then converted to a *non return to zero inverted* (NRZI) channel bit stream that is recorded on the disc.

BD-R/RE Recording

The smallest unit for recording data is a *recording unit block* (RUB) consisting of 2,760 channel bits of *run-in*, followed by a 64KB physical cluster, followed by 1,104 channel bits of *run-out*. A continuously written sequence of one or more RUBs is terminated with a *guard_3* field of 504 channel bits. The run-in and run-out provide buffers so that clusters can be randomly written and rewritten, and to allow for *start position shift* (SPS), which randomly shifts the start position of each recording sequence by up to 128 channel bits before or 127 channel bits after the nominal start position to help the recording material last longer through multiple overwrites of the same data.

The lead-in at the beginning of the disc contains a pre-embossed section, called the *permanent information and control data* (PIC) zone, followed by a rewritable section. The PIC zone holds general information about the disc and includes a special section called the *emergency brake* (EB). Up to 62 emergency brake fields can be defined, which specify a drive manufacturer, drive model, and firmware version, with associated special handling procedures to avoid damage to the drive or disc. The rewritable section is used for optimum power control (OPC) tasks and to store information about the recorded data on the disc, including defect management information, physical access control, and drive-specific information. Discs must be initialized before use, which largely consists of creating a defect list, if any.

BD-R/RE Defect Management

Defect management during recording can be handled by the file system or by the drive. If the drive manages defects, it maps defective clusters using linear replacement and a single defect list into one or two optional spare areas per layer. The inner spare area of layer 0 is at the inner side of the data zone and has a fixed size of 4,096 physical clusters (256 megabytes). The outer spare area of layer 0 and the inner spare area of layer 1 each have a variable size of 0 to 16,384 clusters (1,024 megabytes). The outer spare area of layer 1 has a variable size of 0 to 8,192 clusters (512 megabytes). The total spare areas represent about 5 percent of the storage capacity of the disc. If the drive detects a defective physical cluster it may replace the cluster, mark it for future replacement, mark it as defective without replacement, or ignore the error.

Chapter 9
Application Details

The shiny disc evolution continues to usher in newer and more elegant solutions for data storage and content presentation. Blue laser and red laser disc progeny afford developers myriad opportunities for unique applications and enhanced uses. Building on the ground-breaking research and development of DVD, the latest optical discs provide exponential advances in capacity, speed, and flexibility.

The various disc specifications have been developed, not by regulation or legislation, but by groups of talented, imaginative, enlightened technologists, scientists, and engineers. Computer and electronics manufacturers, content developers, and intellectual property owners have maintained many of their associations and continue to build on the efforts that resulted in CDs and DVDs, while new alliances have been forged and former partners have moved on to separate pursuits. Other players, such as Microsoft and Sony, manuever to enhance their proprietary advancements.

It is worth noting that disc application specifications are not formal standards — that is, they have not undergone an open review process under the auspices of a standards organization. For example, the DVD-Video format specification is an application of DVD-ROM intended to provide high-quality audio and video in a format supported worldwide by consumer electronics groups, movie studios, video publishers, and computer makers. Likewise, the DVD-Audio specification is an application of DVD-ROM focused on high-fidelity multichannel audio.

Variations of DVD for professional use, video games, Web-connected content, and other areas also begat application specifications based on DVD-ROM or writable DVD. The DVD Forum defined some of these; other organizations have defined additional applications, such as SACD from Philips and Sony, and Sony's PlayStation® format.

The physical format specifications for DVD-ROM and writable DVDs were submitted to the ECMA and have become ISO/IEC standards, but the application formats are controlled by the DVD Forum — a consortium of companies with vested interests. The DVD Forum is the leading proponent of the emerging HD DVD technology, although Sony, Matsushita (Panasonic), and Pioneer have distanced themselves from the Forum with Blu-ray Disc.

UDF (Universal Disk Format) specifications establish criteria for the various optical disc formats. DVD subscribes to UDF 1.02, HD DVD is built around UDF 2.5, and BD is designed to meet the specifications in UDF 2.5 and 2.6. The DVD-VR (video recording) format adheres to UDF 2.0.

This chapter provides details of the disc application formats for DVD, HD DVD, and BD technologies and their derivatives.

DVD File Format

A DVD player must be capable of reading files and instructions that are embedded on a disc. DVD player sophistication is proportional to cost, but all players have a minimum level of onboard computing capability.

In order to limit the computing requirements for a home consumer DVD player, additional constraints were applied to the OSTA UDF file format. This constrained format is outlined in Appendix 6.9 of the UDF standard and is commonly referred to as *MicroUDF*. These constraints include the following:

- All DVD players shall support UDF and ISO 9660. Although ISO-9660-capable devices are no longer manufactured, many of those first-generation DVD players are still in operation.
- There should be one logical volume per side, one partition, and one file set.
- Individual files must be less than 1 gigabyte in length ($2^{30} - 1$).
- Each file must be a contiguous extent.
- File and directory names must use only 8 bits per character.
- Only short allocation descriptors are allowed.
- Fields such as OS and implementation use must be zero.
- No symbolic links (aliases) can be used.
- ICB strategy 4 is used.
- No multisession format is allowed.
- No boot descriptor is allowed.
- A specific directory (VIDEO_TS or AUDIO_TS) must be present and must contain a specific file (VIDEO_TS.IFO or AUDIO_TS.IFO).

These constraints apply only to the directory and to the files that the DVD player needs to access. Other files and directories may exist on the disc which are not intended for the DVD player and do not need to meet the preceding constraints. Such directories and files must be physically located on the disc following the audio zone and the video zone and are ignored by DVD players (unless the DVD player employs extensions to the DVD standard). This allows many kinds of data and applications to coexist on a disc.

The DVD-Video specification prescribes a directory and set of files within the UDF file format. Player-requisite files are stored in one directory (VIDEO_TS). In the case of a DVD-Audio disc, the DVD-Audio information is stored in a homologous directory (AUDIO_TS). An optional, root-level directory (JACKET_P) can contain identifying images for the disc in three sizes, including image thumbnails for graphic directories of disc collections.

For a DVD-Video volume, the required VIDEO_TS.IFO file contains the video manager title set information (VMGI). The contents of the Video Manager is analagous to the main menu for the disc. Additional .IFO files hold other title set information (VTSI), with backup copies in .BUP files. For each video title set on the disc, there can be up to 10 video object (.VOB) files that each hold one chunk from video object set blocks. Similarly, for DVD-

Audio volumes, the required AUDIO_TS.IFO file contains the audio manager (main menu) information (AMGI). Each audio object title set is broken into .AOB files (see Figures 9.1 and 9.2, as well as "Physical Data Structure," later in this chapter).

Real-Time Recording Mode

"If you're serious about doing real-time, why do you want an operating system?" — Anonymous[1]

The disc flavors capable of performing real-time recording rely on sophisticated software for data handling in combination with hardware flexibility that can keep up with the bandwidth and the rate of the data to be recorded. The UDF version 2.0 file system specifications for the various rewritable formats define a special real-time recording mode. Real-time stream recording and playback are critical applications, yet the drive units, especially those used in consumer players, perform poorly compared with hard-disk drives. In addition to slow seek times and lower data rates, dispersion of data across the medium may degrade performance.

A practical solution to the problem is to make sure that the streaming data is arranged in a continuous order on the disc so that no head seeks are required during recording or playback. This helps guarantee a minimum data rate. The real-time stream model of the version 2.0 file system specifies new methods to handle defective sectors on writable discs.

When a player or operating system specifies that a file containing real-time streaming content is being recorded, data is placed in the largest unallocated extents that provide a physically contiguous space. Since linear replacement causes discontinuities in the data, it is disabled for streaming write operations. (The linear replacement method marks a defective area and moves data to a non-contiguous, designated spare area.) Slipping, which is performed at formatting time, remains in effect, as it marks a defective area and sequentially moves data to the next good region. Even if the recorder encounters a defective block, it will not replace it, because the highest priority is given to continuity of data. Thus, using real-time stream operations may result in erroneous data. However, minor errors in an MPEG-2 video stream, for example, are often undetectable or not objectionable during playback and are preferable to loss of recording or playback continuity.

DVD-Video

DVD-Video provides one stream of MPEG-2 variable-bit-rate video with up to nine interleaved camera angles; up to eight audio streams of audio; and up to 32 streams of full-screen graphic overlay subpictures. That breadth of data is supplemented by navigation menus and still pictures, in addition to instructions for seamless branching, parental management, and rudimentary interactivity. The DVD-Video format is designed for playback on standard 525-line (NTSC) and 625-line (PAL) television displays via analog video connections. It also can be played in standard resolution, interlaced fields or progressive frame format, and on digital displays and high-definition televisions via analog or digital connections.

---

[1]Anonymous quote is located at Wikipedia.org, http://en.wikipedia.org/wiki/Real-time_operating_system.

Figure 9.1 DVD-Video and DVD-Audio File Structure

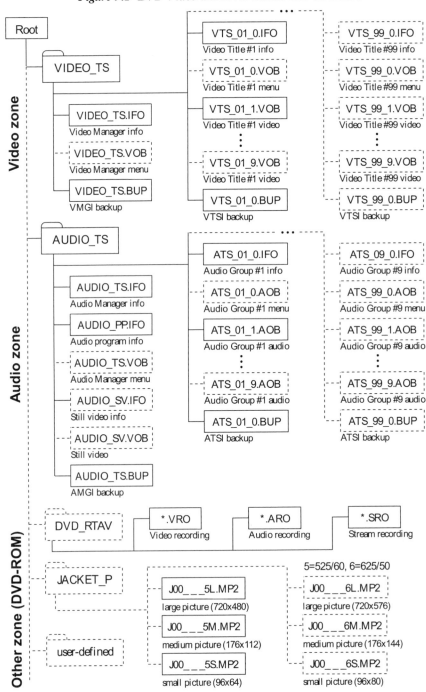

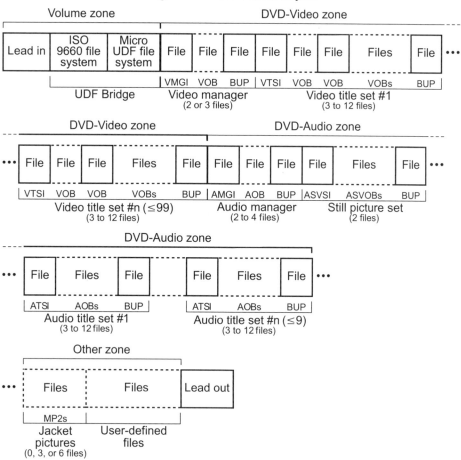

Figure 9.2 DVD Volume Layout

DVD-Video incorporates multichannel digital audio for reproduction on compatible digital audio systems and provides digital or analog audio support for standard stereo audio systems. DVD-Video also can be played on compatible computer systems (see Chapter 15 for details) and supports video and audio formats from various standards (see Appendix B).

Navigation and Presentation Overview

DVD players (including software players) are based on a presentation engine and a navigation manager. The *presentation engine* uses the information in the presentation data stream on a disc to control what is shown. The *navigation manager* uses information in the navigation data stream on a disc to provide a user interface, create menus, control branching, etc. In a general sense, a user selection on a player control determines the path of the navigation manager, which controls the presentation engine to create the display (see Figure 9.3).

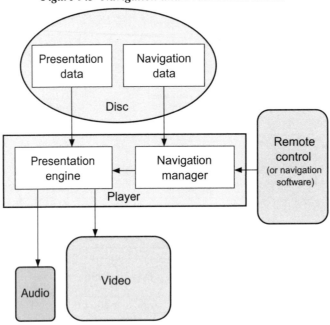

Figure 9.3 Navigation and Presentation Model

Navigation data includes information and a command set that provide rudimentary inter-activity. Menus are present on almost all discs to allow content selection and feature control. DVD-Video content is broken into titles (movies or other programs) and parts of titles (chapters). For example, a disc containing four television episodes could present each episode as a title. A disc with a movie, supplemental information, and a theatrical preview might be organized into three titles, with the movie title arrayed in chapters.

A disc can have up to 99 titles, but in many cases there will be only one. This would be straightforward were it not for the industry practice of calling discs *titles*. For example, *The Matrix* was the first title to ship a million copies. This particular title (disc) has 35 titles (video segments) on it, ranging from a title that holds the entire movie to titles containing short behind-the-scenes clips, logos, and the FBI warning.

Most discs have a main menu for the entire disc, from which titles are selected. Each title or group of titles can have its own menu. Depending on the complexity of the title, addition-al menus can be provided at any point. Each menu has a still picture or moving video back-ground and onscreen buttons. It is possible to use menus anywhere on the disc, such as to provide pop-up features over the top of any video.

Most remote-control units have four directional arrow keys for selecting onscreen buttons, plus numeric keys, a "Select" or "Enter" key, a "Title" or "Top Menu" key, a "Menu" key, and a "Return" key. Standard playback functions are play/pause, fast forward, fast reverse, next, and previous. Additional remote functions may include step, slow, fast, multispeed scan, audio select, subtitle select, camera-angle select, play mode select, search to title, search to part of title (chapter), and search to time. Any of these features can be disabled by the producer of the disc. Additional features of the navigation and command set are covered in the navigation and presentation sections that follow later in this chapter.

Material for camera angles, seamless branching, and parental control is interleaved together on the disc in chunks. The player jumps from chunk to chunk, skipping over unused angles, branches, or scenes to stitch together a seamless video presentation. These chunks are not multiplexed as individual streams of video and audio,[2] but are interleaved separately (see Figure 9.4). Each video angle has its own set of audio and subpicture tracks, although they are generally the same across all angles, and the angle streams must be of equal duration.

Figure 9.4 Multiplexing vs. Interleaving

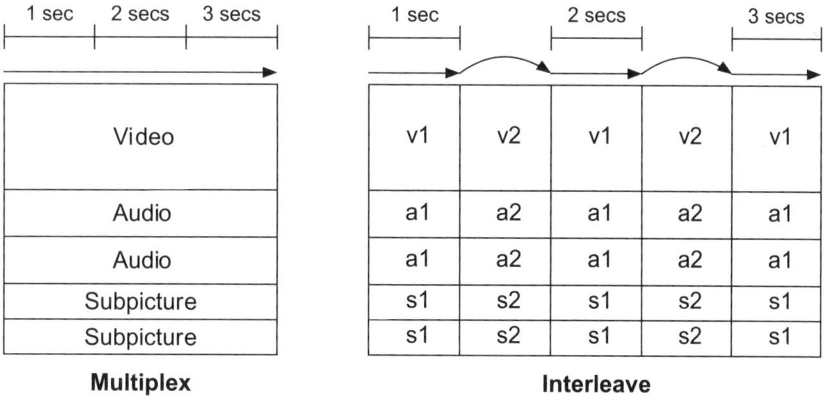

1 sec	2 secs	3 secs		1 sec		2 secs		3 secs
Video				v1	v2	v1	v2	v1
Audio				a1	a2	a1	a2	a1
Audio				a1	a2	a1	a2	a1
Subpicture				s1	s2	s1	s2	s1
Subpicture				s1	s2	s1	s2	s1
Multiplex				**Interleave**				

The interleaved chunks have no direct effect on the bit rate, but they do reduce the maximum allowable rate because the track buffer must be able to continue supplying data while intervening chunks are skipped over. Adding one camera angle for the duration of a title roughly doubles the amount of space it requires (and cuts the playing time in half).

DVD application data is organized into a complex structure representing the physical location of the data on the disc. Because data may be shared among different titles and programs, logical data structures are overlaid on the physical structure. The logical structures contain navigation information and determine the presentation order of information, which is independent of the physical order.

Physical Data Structure

The physical data structure determines the way data is organized and placed on the disc.[3] The standard specifies that data must be stored sequentially — physically contiguous — according to the DVD-Video physical structure. The top line of Figure 9.5 represents the

---

[2]Everything turns out to be interleaved if you dissect it far enough. At the MPEG packet level, the different program streams are interleaved, but the player sees them as individual streams coming out of the demultiplexer.

[3]Physical in this sense refers to the positional ordering of the data, not to the physical characteristics of the underlying storage medium.

order in which data is stored on the disc. The structure is hierarchical; each block can be broken into component blocks, which can be further subdivided, as illustrated in Figure 9.5 (also see Table 9.5).

Figure 9.5 DVD-Video Physical Data Structure

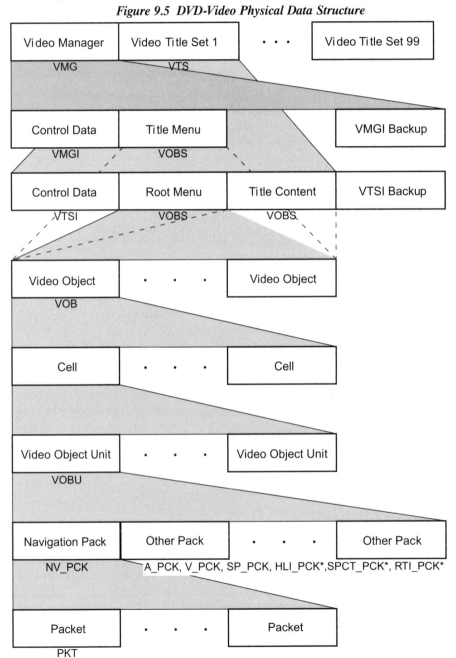

*DVD-Audio only

The primary block is a video title set (VTS), which carries internal information about the titles it contains (menu pointers, time maps, cell addresses, etc.), followed by video object sets (VOBSes), as shown in Figures 9.6 and 9.7. Since the VTS information applies to all the titles in the set, the titles must contain the same number, format, and order of audio tracks and subpicture tracks. MPEG-1 and MPEG-2 video cannot be mixed within a VOBS. The first video object set may be an optional menu (VTSM), called a *root menu*, followed by video object sets that contain the actual title content.

Figure 9.6 VTS Structure

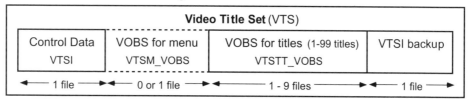

Figure 9.7 VOBS Structure

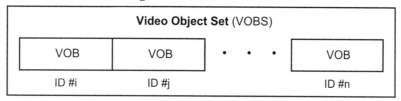

The video manager (VMG) is a special case of a video title set that optionally may contain a main menu for the disc, called the *title menu* or *top menu*. This is the table of contents for the disc. If this menu is present, it is usually the first thing the viewer sees after inserting the disc. Alternately, there can be a special autoplay piece that automatically begins playback when the disc is inserted.

Data at the VOBS level includes attributes for video, audio, and subpicture (see Tables 9.1 -9.3). The language of audio and subpictures can be identified with ISO 639 codes (see Table A.2) and also can be identified with an extension code as commentary, simplified audio, and so forth.

Table 9.1 VOBS Video Attributes

Compression mode	MPEG-1 or MPEG-2
TV system	525/60 (NTSC) or 625/50 (PAL/SECAM)
Aspect ratio	4:3 or 16:9
Display mode permission[a]	Letterbox, pan and scan, both
4:3 source is letterboxed	Yes or no

[a]When 16:9 aspect ratio is presented on a 4:3 display.

Table 9.2 VOBS Audio Attributes

Audio coding mode[a]	Dolby Digital, MPEG-1/MPEG-2, MPEG + extension, LPCM, DTS, or SDDS
DRC (dynamic range)	On, off
Quantization (PCM)	16, 20, or 24 bits
Sampling rate (PCM)	48 or 96 kHz
Number of channels[a]	1 to 8
Film/camera mode (625/50)	Film or camera
Application mode	Surround, karaoke, or unspecified
Code	ISO 639 language code or unspecified
Code extension	Unspecified, caption, for visually impaired, director comments 1, director comments 2

[a]A VOB containing a menu can use PCM or Dolby Digital audio. A menu cannot contain DTS or SDDS audio.

Table 9.3 VOBS Subpicture Attributes

Coding mode	2-bit RL (no other options)
Code	ISO 639 language code or unspecified
Code extension	Unspecified, normal caption, large caption, children's caption, normal closed caption, large closed caption, children's closed caption, forced caption, director caption, large director caption, director caption for children
Channel mixing info	Contents, mixing phase, mixed flag, mix mode

Identification codes are stored on the disc as binary representations of two-letter (lowercase) codes. For example, the code for English is *en*, and the code for Zulu is *zu*. Different authoring systems and production tools show these codes in different ways. The ASCII hexadecimal representation for *e* is 65 and for n is 6E, so the code might be displayed as 656E or as the decimal equivalent 25966. A few systems use an alternate code in which each letter is assigned a decimal value beginning with 1, so *en* would be 0114.

Each video object set (VOBS) is composed of one or more video objects (VOB(s)) (see Figure 9.8). A video object is part or all of an MPEG-2 program stream. The picture resolution and display rate are the same for all video objects in a video object set. Granularity at the VOB level is designed to group or interleave blocks for seamless branching and camera angles. Each VOB set contains one or more VOB blocks. A contiguous block contains one VOB in contiguous sectors on the disc. An interleaved block contains multiple VOBs broken into interleaved units (ILVUs) that are physically interleaved on the disc to enable seamless presentation. The size of the ILVUs is determined by the data rate of the streams, with the size minimized so that the pickup head has time to jump over the intervening units of other video objects to get the next unit in sequence before the track buffer is depleted.

Figure 9.8 VOB Structure

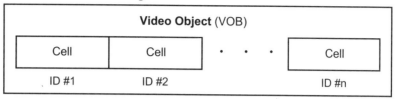

A VOB is made up of one or more cells (see Figure 9.9). A *cell* is a group of pictures or audio blocks and is the smallest addressable chunk. A cell can be as short as a second or as long as a movie. Some authoring systems call cells *scenes*. Cell IDs are relative to the video object in which they reside, so a cell can be uniquely identified with its cell ID and enclosing video object ID. Note that the player may need to get additional presentation information about the cell from the program chain (PGC, defined later). A cell cannot be spread across both layers of a disc. This rule is also supposed to apply to a pair of cells intended to play seamlessly, but so-called seamless layer changes can be created by violating this rule.

Figure 9.9 Cell Structure

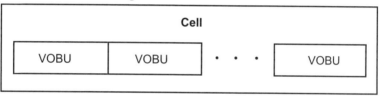

Figure 9.10 VOBU Structure

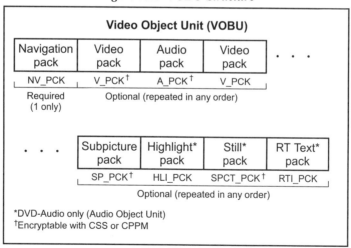

Each cell is further divided into video object units (VOBU(s)) as shown in Figure 9.10. A VOBU is the smallest unit of playback. Despite its name, a VOBU does not always contain video. A VOBU is an integer number of video fields and is from 0.4 to 1 second long, unless it is the last VOBU of a cell, in which case it can be up to 1.2 seconds long. A VOBU contains zero or more GOPs, usually one. If this is the case, it must begin with an MPEG sequence header, followed by a GOP header, followed by an I frame. If the last GOP in the VOBU does not align with the presentation end time of the VOBU, or if the VOBU is followed by a VOBU with no video data, then there must be an MPEG sequence_end code. Only one sequence_end code is allowed per VOBU. Analog protection system (APS) data is stored at the video object unit level and specifies whether analog protection (Macrovision) is off or is one of three types (see Table 9.4). Also, see the copy protection data in Chapter 5.

Table 9.4 VOBU APS (Macrovision) Attributes

0	Off
1	Pseudosync and AGC
2	Pseudosync and AGC, plus Colorstripe type 1 (inverted split burst on 2 lines of every 17)
3	Pseudosync and AGC, plus Colorstripe type 2 (inverted split burst on 4 lines of every 21)

Finally, at the bottom of the heap, VOBUs are broken into packs of packets (see Figures 9.11 and 9.12). The format of packs and packets is compliant with the MPEG program stream standard. Packs include system clock reference (SCR) information for timing and synchronization. Each packet identifies which stream it belongs to and carries a chunk of data for that stream. Packs are stored in recording order, interleaved according to the different streams that were multiplexed together. Different packs contain data for navigation, video, audio, and subpicture. Additional packs are used by the DVD-Audio and DVD recording formats, as detailed later in this chapter. DVD-Video physical units are outlined in Table 9.5.

Figure 9.11 Pack Structure

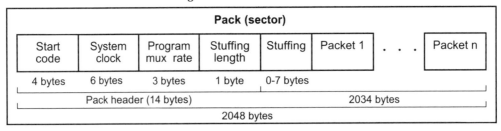

Figure 9.12 Packet Structure

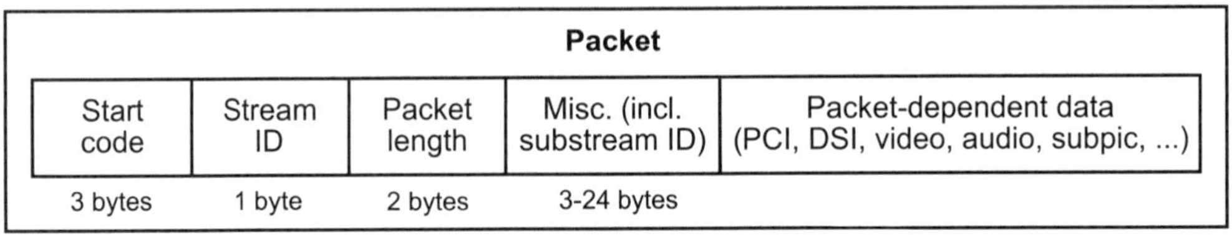

Table 9.5 DVD-Video Physical Units

Unit	Maximum
Video title set (VTS)	99 per disc
Video object set (VOBS)	99 per VTS
Video object (VOB)	32767 per VOBS
Cell	255 per VOB
Video object unit (VOBU)	n/a
Pack (PCK)	2048 bytes
Packet (PKT)	n/a

Domains and Spaces

The concepts of *domains* and *spaces* relate to DVD navigation. Domains and spaces are not physical structures, nor are they logical structures defined by information on the disc. Rather, they are abstract groupings of data structures. Technically, a domain is a set of similar *program chains* (PGCs), although it may be better to think of a domain as a state that the DVD player monitors in order to keep track of the type of content being displayed (see Table 9.6). The stop state is not actually a domain, since it is an absence of PGCs, but it is useful to classify it with the other four domains noted in the table.

Table 9.6 Domains

Domain	Abbreviation	State	Valid commands	Number
First play domain	FP_DOM	Preparing to play	Non-playback commands	None or one
Video manager menu domain	VMGM_DOM	Displaying the title menu or its submenus	Menu-related commands	None or one for each title menu language
Video title set menu domain	VTSM_DOM	Displaying a root menu or one of its submenus (subpicture, language, audio, or angle)	Menu-related commands	None or one for each menu and each language
Title domain	TT_DOM	Playing video content in a title	Playback commands	One for each title
Stop state		Stopped. The head is retracted from the disc	Play	n/a

Navigation commands are constrained by domains to be appropriate for the current state. For example, a "select button" command is only meaningful if a menu is being displayed, and a fast-forward command makes no sense if the player is stopped or in a menu with a still image. Some authoring systems further define subdomains for each menu language in the VMGM and VTSM domains.

Spaces are a way of grouping domains into functionally similar units, as shown in Table 9.7 (also see Table 9.12, which shows how the spaces overlap).

Table 9.7 Spaces

Space	Scope	Number
System space	Everything other than individual titles FP_DOM + VMGM_DOMs + VTSM_DOMs	None or one
Menu space	All menus VMGM_DOM + VTSM_DOMs	None or one
VMG space	Intro and title menu FP_DOM and VMGM_DOMs	None or one
VTS space	Menus and titles of one VTS VTSM_DOMs + TT_DOMs	One for each VTS

Presentation Data Structure

The presentation data structure is a logical hierarchy that overlays the physical data structure (see Table 9.8).

Table 9.8 DVD-Video Logical Units

Unit	Maximum
Title	99 per disc
Parental block (PB)	n/a
Program chain (PGC)	999 per title, 16 per parental block, 32767 per VTS
Part of title (PTT)	999 per title, 99 per one-sequential-PGC title
Program (PG)	99 per PGC
Angle block (AB)	n/a
Interleave block (ILVB)	n/a
Interleave unit (ILVU)	n/a
Cell pointer	255 per PGC

The presentation data structure determines the grouping of video sequences and the playback order of each block of video in a sequence (see Figures 9.13 and 9.14). The top level comprises titles. Each title contains up to 999 program chains (PGCs). The first PGC in a title is called the *entry* PGC. Linking between PGCs is not seamless. The title menu and each root menu are entry PGCs. A program chain contains 0 to 99 *programs* (PGs), which are groupings of cells. A PGC with no programs (no VOBs), called a *dummy* PGC, contains only navigation commands.

Figure 9.13 Presentation Data Structure

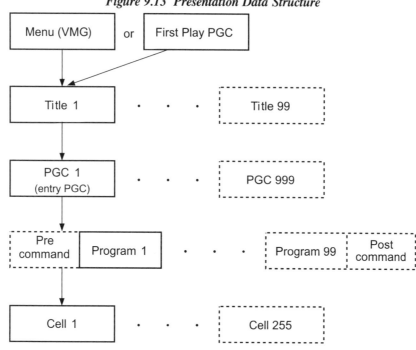

Figure 9.14 Relationship of Presentation Data to Physical Data

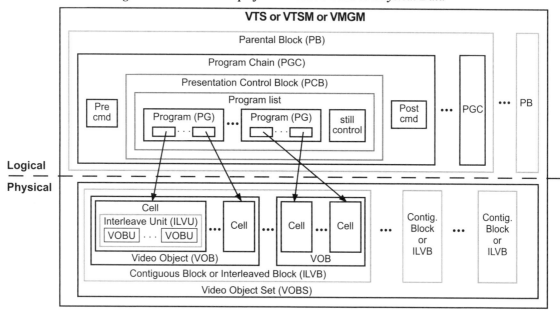

The physical data and the logical presentation data structure converge at the cell level. Each PGC contains one *presentation control block* (PCB), which is an ordered list of pointers to cells that indicates in what order the programs and cells are to be played. Programs within a PCB can be flagged for sequential play, random play (programs are selected randomly and may repeat), or shuffle play (programs are played in random order without repeats). Individual cells may be used by more than one PGC, which is how parental management and seamless branching are accomplished — different program chains define different sequences, primarily through the same material.

The presentation data structure includes additional groupings at various levels to provide additional organization (see Figure 9.14). Groupings include parental blocks for parental management, angle blocks for multiple camera angles, and language units for menus in multiple languages.

There is also the part-of-titles (PTT) construct, commonly called a *chapter*. A *part of title* is a marker or branch point, not a container. A PTT marker can only go at the beginning of a program.[4] For a multi-PGC title, the user may take different paths from PGC to PGC, so chapter 4 via one path may be made up of different PGCs than Chapter 4 via another path (see Figure 9.15).

Figure 9.15 Example of Part-of-title Markers

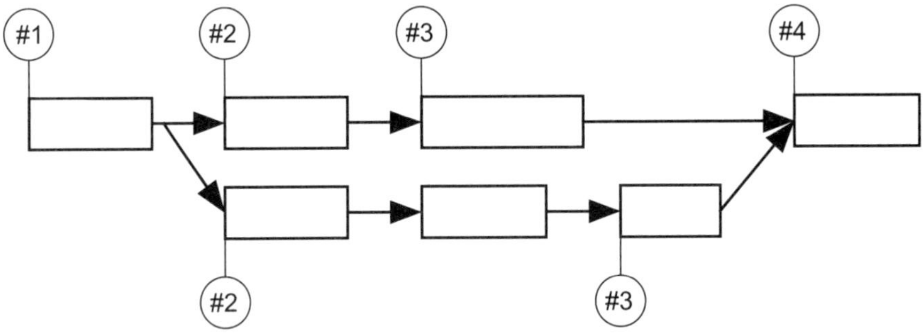

Three types of titles exist: a monolithic title meant to be played straight through (one_sequential_PGC title), a title with multiple PGCs for varying program flow (multi_PGC title), and a title with multiple PGCs that are automatically selected according to the parental restriction setting of the player (parental_block title) (see Figures 9.16 and 9.17). One_sequential_PGC titles are the only kind that have time maps for timecode display and searching.

---

[4] In general, because chapters must start on a program boundary, a chapter is equivalent to a program. Specifically, every chapter is a program, but not every program is a chapter. In practice, the "Next" and "Previous" keys on remote controls jump between programs, not chapters.

Figure 9.16 Example Title Structures

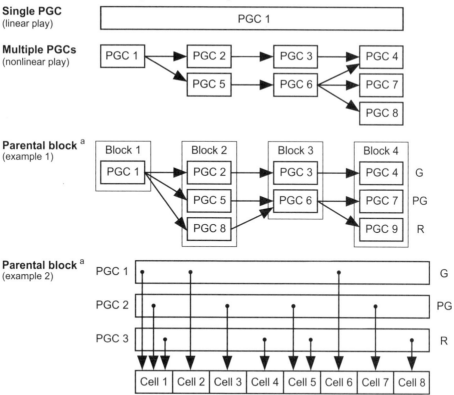

Single PGC
(linear play)

Multiple PGCs
(nonlinear play)

Parental block [a]
(example 1)

Parental block [a]
(example 2)

[a]Playing time of cells and PGCs may not be the same

Figure 9.17 Example Presentation Structures

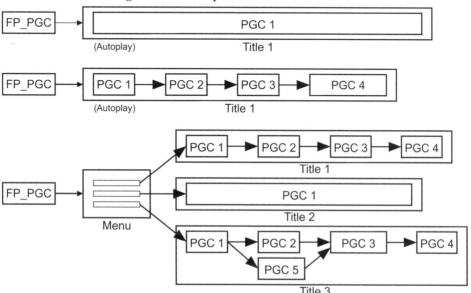

Note: First play(FP) PGC automatically begins playback on disc insertion

Navigation Data

Navigation data is a collection of information that determines how the physical data is accessed. In a sense, navigation data is built on top of presentation data. Navigation data controls access and interactive playback. It is grouped into three categories: control, search, and navigation commands. The menus in the search category compose the user interface (see Figure 9.18).

Figure 9.18 Navigation Data Structure

Control
- Format (NTSC/PAL)
- Language
- Audio selection
- Subpicture selection
- Parental management
- Karaoke
- Display mode and aspect

Navigation
- General parameters
- System parameters
- Navigation timer
- Buttons

Search
- PGCI search (jump to menu)
  - Title (disc menu; VMG)
  - Root (local menu; VTS)
  - Audio (audio submenu)
  - Subpicture (subpicture submenu)
  - Angle (angle submenu)
  - Part of Title (chapter submenu)
- Presentation data search
  - Title
  - Part of Title (chapter)
  - Program (next/previous)
  - Time
  - Angle
  - VOBU (trick play)

There are five levels at which navigation data exists: *video manager information* (VMGI), stored in the VMG, which controls the video title sets and the title menu; *video title-set information* (VTSI), stored in each VTS, which controls the titles and menus in a VTS; *program chain information* (PGCI), stored in the PGC, which controls access to components of a PGC such as audio and subpicture streams; *presentation control information* (PCI), stored in packets spread throughout the data stream (one per VOBU), which controls menu display and program presentation in real time; and *data search information* (DSI), also stored in packets spread throughout the data stream (one per VOBU), which controls forward/reverse scanning and seamless branching (see Table 9.9).

Table 9.9 Navigation Data

Name	Abbreviation	Important Content
Video manager information	VMGI	Number and attributes of title sets (VTS_ATRT); pointers to titles (VTS_PTT_SRPT); parental management table (PTL_MAIT); text data (TXTDT_MG); attributes of title menu (VMGM) video stream, audio streams, and subpictures; title menu (VMGM) cell pointers and VOBU maps

continues

Table 9.9 Navigation Data (continued)

Name	Abbreviation	Important Content
Video title set information	VTSI	Pointers to chapters (TT_SRPT); pointers to program chains; time maps; attributes of root menu (VTSM) video stream, audio streams, and subpictures; root menu (VTSM) cell pointers and VOBU maps; video title set (VTS) cell pointers and VOBU maps
Program chain information	PGCI	Number and length of programs; permitted user operations (UOPs); links between program chains (previous, next, return/go up); playback mode (sequential, random, shuffle); PGC still time; commands (pre, cell, and post); program maps and cell pointers; cell still times
Presentation control information	PCI	For each VOBU: APS, user operations (UOPs), nonseamless angle jump pointers, button information (rectangle, color, highlight, directional links, associated aspect ratio, command, etc.), presentation times, ISRC for video/audio/subpicture streams
Data search information	DSI	For each VOBU: Reference picture pointers (I or P frames), link to next interleaved unit (ILVU), seamless angle jump pointers, presentation times, audio gap lengths, VOBU pointers for forward/reverse scanning, video synchronization pointers to audio and subpicture packs

Previous, next, and return (go up) PGC information is stored in the PGCI. The return link corresponds to the "Return" button on the remote control. The "Previous" (I◄◄) and "Next" (►►I) buttons on a remote control jump between programs until there is no previous or next program, in which case they follow the previous and next links (if present) in the PGCI.

Control Information Control information describes the data. This includes characteristics such as format (525/60 or 625/50), aspect ratio (4:3 or 16:9), language, audio and subpicture selection, and moral codes for parental management.

Search Data Search data defines a navigational structure that can be navigated with the remote control or by program control. There are seven search types, described in Table 9.10, which are associated with a key or combination of keys on the remote control. These searches are associated with a PGC; the search map is stored in the PGCI. Additional search types are provided for parts of titles (chapters), time, angles, and VOBUs (for trick play modes such as slow and fast).

If no search data is present for a particular search type, the player is unable to provide that function. It is possible to create a disc that plays from beginning to end and cannot be interrupted by the user (other than by ejecting the disc).

Table 9.10 Search Types

Search Type	Example Keypresses (Remote Control)
PGCI search	
Title menu (top menu; VMG)	Top (or Title)
Root (menu; VTS)	Menu
Audio (audio menu)	Audio+Menu
Subpicture (subpicture menu)	Subtitle+Menu
Angle (angle menu)	Angle+Menu
Part of title (chapter menu)	Chapter+Menu
Data search	
Title	Search+Title+2+Enter
Part of title (chapter search)	Search+Chapter+8+Enter
Time	Search+Time+1+2+0+0+Enter
Angle	Angle

Navigation Commands and Parameters Commands provide the interactive features of DVD. Each PGC optionally can begin with a set of precommands, followed by cells that can each have one optional command, followed by an optional set of postcommands. In total, a PGC cannot have more than 128 commands, but since commands are stored in a table at the beginning of the PGC and referenced by number, they can be reused many times within the PGC. Cell commands are executed after the cell is presented. PGC commands are stored in the PGCI. Each menu button also may contain a single command, stored in the PCI (see "Buttons" later in this chapter).

The commands are similar to computer CPU instruction words and are, in fact, instructions to the processing unit of the player. Each command is up to 8 bytes long and may consist of one, two, or three instructions. Instructions include (see Table 9.11 for a complete list of commands):

- Math operations: add, subtract, multiply, divide, modulo, random
- Logical (bitwise) operations: AND, OR, XOR
- Comparisons: equal, not equal, greater than, greater than or equal to, less than, less than or equal to
- Register operations: load, move, swap
- Program flow control: goto, break
- Video presentation control: link, jump, call, resume, exit

Table 9.11 Navigation Commands

GoTo group	Commands for controlling program flow within a pre or post sequence
GoTo	Jump to a specified command number
Break	Stop executing commands in pre or post section
NOP	No operation (do nothing)
SetTmpPML	Ask user to confirm temporary parental level change. If ok, change level and go to specified command
Link group	Commands for moving within the current domain
LinkPGCN	Start at program chain number
LinkPTTN	Start at part-of-title (chapter) number
LinkPGN	Start at program number
LinkCN	Start at cell number
LinkTopPGC	Restart current PGC
LinkTailPGC	Go to end of current PGC (execute postcommands)
LinkPrevPGC	Start at previous PGC
LinkNextPGC	Start at subsequent PGC
LinkGoUpPGC	Start at higher PGC
LinkPGCN	Start at specific PGC number
LinkPTTN	Start at specific chapter number
LinkTopPG	Restart current program
LinkPrevPG	Start at previous program
LinkNextPG	Start at next program
LinkPGN	Start at specific program number
RSM	Resume at location where playback was suspended by CallSS or MenuCall user operation (Destination not limited to current domain.)
Jump group	**Commands for moving out of the current domain**
JumpTT	Start at title number (from VMG)
JumpVTS_TT	Start at title number (in same VTS)
CallSS	Start at a menu PGC number in system space,[a] saving resume state
JumpSS	Start at PGC number in system space[a] (from system space)
JumpVTS_PTT	Start at part-of-title (chapter) number in title number (in same VTS)
Exit	Stop (enter stop state)
Compare group	**Commands for comparing values and parameters**
BC	Compare bitwise (logical and)

continues

Table 9.11 Navigation Commands (continued)

EQ	Test if equal
NE	Test if not equal
GE	Test if greater than or equal
GT	Test if greater than
LE	Test if less than or equal
LT	Test if less than
SetSystem group	**Commands for setting system parameters and general parameters**
SetSTN	Set audio, subpicture, or angle number (SPRM 1, 2, and 3)
SetNVTMR	Set navigation countdown timer (SPRM 9 and 10)
SetHL_BTNN	Set selected button (SPRM 8)
SetAMXMD	Set karaoke audio mixing mode (SPRM 11)
SetGPRMMD	Set general parameter value and mode (register or counter) (GPRM 0 to 15)
Set group	**Commands for setting and manipulating values in general parameters**
Mov	Set GPRM value (from a constant, GPRM, or SPRM)
Swp	Exchange the values in two GPRMs
Add	Add a value (constant or GPRM) to a GPRM
Sub	Subtract a value (constant or GPRM) from a GPRM
Mul	Multiply a GPRM by a value (constant or GPRM)
Div	Divide a GPRM by a value (constant or GPRM)
Mod	Take the remainder of dividing a GPRM by a value (constant or GPRM)
Rnd	Set GPRM to a random number between 1 and a value (constant or GPRM)
And	Take the bitwise product of a GPRM and a value (constant, GPRM, or SPRM)
Or	Take the bitwise sum of a GPRM and a value (constant, GPRM, or SPRM)
Xor	Exclusive or a GPRM with a value (constant, GPRM, or SPRM)

[a]The command can jump to the first play PGC, the entry PGC (or entry PGC block) of a menu (in VMGM or VTSM domain) or a title (in VTS domain), or any PGC in the VMGM.

Video presentation control commands (the link group and the jump group) are limited by the current domain and the destination domain. Table 9.12 shows the commands necessary to move from one domain to another.

Table 9.12 Domain Transitions

From	To			
	System Space			
	Menu Space			
	VMG Space		VTS Space	
	FP_DOM	VMGM_DOM	VTSM_DOM	TT_DOM
FP_DOM	n/a	JumpSS	JumpSS	JumpTT
VMGM_DOM	JumpSS	n/a	JumpSS	JumpTT, Link (GoUpPGC or RSM)
VTSM_DOM	JumpSS	JumpSS	n/a	JumpVTS_TT, JumpVTS_PTT, Link (GoUpPGC or RSM)
TT_DOM	CallSS	CallSS	CallSS	JumpVTS_TT, JumpVTS_PTT

Of course, there are myriad restrictions and gotchas. For example, a PGC cannot link directly to a PGC in a different domain,[5] and one VTS cannot link directly to another VTS. Many discs are authored with dummy PGCs in the VMGM that serve as switching points between PGCs and VTSs. Presentation control with cell commands is not seamless. Branching within a title can be done with link commands. The JumpVTS_PTT command is different from LinkPTTN in that it always executes the precommands. LinkPG... commands are limited to the current PGC, except that LinkPrevPG can go to the beginning of the previous PGC.

There are 24 system-use 16-bit registers (called *system parameters*, or SPRMs; affectionately known as "sperms") that hold information such as language code, audio and subpicture settings, and parental level (see Table 9.13). Some SPRMs can only be used to determine the state of the player (read-only), whereas others can be set by commands (read/write), allowing programs to control presentation of audio, video, subpicture, camera angles, and so on. SPRMs cannot be set directly. They can only be set by special commands such as SetNVTMR that sets both SPRM 9 (countdown timer) and SPRM 10 (timer jump destination) or by commands that move to a different place on the disc, potentially causing SPRM 4 (title number in volume), SPRM 5 (title number in VTS), SPRM 6 (PCG), and SPRM 7 (chapter) to change. SPRMs 11 through 20 are called *player parameters* because they reflect the settings of the player, although SPRM 0 ought to be in the same group.

There are 16 general-use 16-bit registers (called *general parameters*, or GPRMs; also known as "germs") that can be used by on-disc programs for such things as keeping score, storing viewer responses, or tracking what sections of the disc have been seen. Each parameter holds an unsigned integer, with a value from 0 to 65,535, but, with clever programming, any number of smaller values may be combined in a single register; up to 16 one-bit flags. GPRMs also may be used in counter mode, where the value increases by one each second. A limitation is that all GPRMs are cleared when a title search or title play command is executed, when the stop (or eject) command is executed, and when the player is turned off.

[5]This is mostly true, except for the JumpSS/CallSS "PGC in System Space" command, which allows a PGC to jump to any specific PGC in the VMGM_DOM.

Table 9.13 Player System Parameters (SPRMs)

No.	Description	Access	Values	Default Value
0	Preferred menu language	Read-only	Two lowercase ASCII letters (ISO 639)	Player-specific
1	Audio stream number	Read/write	0 to 7 or 15 (none)	15 (Fh)
2	Subpicture stream number and on/off state	Read/write	b_0 to b_5: 0 to 31 or 62 (none) or 63 (forced subpicture) b_6: display flag (0 = do not display)	62 (3Eh)
3	Angle number	Read/write	1 to 9	1
4	Title number in volume	Read/write	1 to 99	1
5	Title number in VTS	Read/write	1 to 99	1
6	PGC number	Read/write	1 to 32,767	Undefined
7	Part of title number	Read/write	1 to 99	1
8	Highlighted button number	Read/write	1 to 36	1
9	Navigation timer	Read-only[a]	0 to 65,535 (seconds)	0
10	PGC jump for navigation timer	Read-only[a]	1 to 32,767 (PGC in current title)	Undefined
11	Karaoke audio mixing mode	Read/write	(0 = do not mix) b_2: mix ch2 to ch1 b_3: mix ch3 to ch1 b_4: mix ch4 to ch1 b_{10}: mix ch2 to ch0 b_{11}: mix ch3 to ch0 b_{12}: mix ch4 to ch0	0
12	Parental management country code	Read-only	Two uppercase ASCII letters (ISO 3166) or 65,535 (none)	Player-specific
13	Parental level	Read/write	1 to 8 or 15 (none)	Player-specific
14	Video preference and current mode	Read-only	$b8$—$b9$: current video output mode 0 (00b)= normal (4:3) or wide (16:9) 1 (01b)= pan-scan (4:3) 2 (10b)= letterbox (4:3) 3 (11b)= reserved b_{10}—b_{11}: preferred display aspect ratio 0 (00b)= 4:3 1 (01b)= not specified 2 (10b)= reserved 3 (11b)= 16:9	Player-specific

continues

Table 9.13 Player System Parameters (SPRMs) (continued)

No.	Description	Access	Values	Default Value
15	Player audio capabilities	Read-only	(0 = cannot play) b_2: SDDS karaoke b_3: DTS karaoke b_4: MPEG karaoke b_6: Dolby Digital karaoke b_7: PCM karaoke b_{10}: SDDS b_{11}: DTS b_{12}: MPEG b_{14}: Dolby Digital	Player-specific
16	Preferred audio language	Read-only	Two lowercase ASCII letters (ISO 639) or 65,535 (none)	65,535 (FFFFh)
17	Preferred audio language extension	Read-only	0 = not specified 1 = normal audio 2 = audio for visually impaired 3 = director comments 4 = alternate director comments	0
18	Preferred subpicture language	Read-only	Two lowercase ASCII letters (ISO 639) or 65,535 (none)	65,535 (FFFFh)
19	Preferred subpicture language extension		0 = not specified 1 = normal subtitles 2 = large subtitles 3 = subtitles for children 5 = normal captions 6 = large captions 7 = captions for children 9 = forced subtitles 13 = director comments 14 = large director comments 15 = director comments for children	0
20	Player region code (mask)	Read-only	One bit set for corresponding region (00000001 = region 1, 00000010 = region 2, etc.)	Player-specific
21	Reserved			
22	Reserved			
23	Reserved for extended			

[a]Bits within the word are referred to as b_0 (low order bit) through b_{15} (high order bit).

Summary of Data Structures

To recap, building up in order from the lowest level: video, audio, subpicture, and presentation/control information is divided and interleaved into packets (see Figure 9.15, pg 9-16). Each chunk of content is organized as an MPEG *group of pictures* (GOP), and is usually 0.5 seconds in length.[6] GOPs are arranged in a *video object unit* (VOBU), with usually one GOP per VOBU. VOBUs are collected into cells. A sequence of cells and cell commands forms a program, whose audio/video content is stored in a *video object* (VOB). A program usually corresponds to one scene. Simple programs are usually held in a single cell. Chapter (part-of-title, PTT) markers are added to create access points, usually at program and cell boundaries. Programs are linked in order into a *presentation control block* (PCB). The PCB and additional command and *control information* (PGCI) make up a *program chain* (PGC). The audio/video content of a PGC is stored in a *video object set* (VOBS). PGCs contain video for the first play sequence, menus, and titles. PGCs may be grouped into logical parent blocks. PGCs are grouped conceptually into domains with related navigational features. Domains are grouped conceptually into spaces. Similar titles are grouped logically along with their menus into a *video title set* (VTS).

Menus

Each disc may have one main menu, called the *title* (or *top*) *menu*. Please note, though, that a title menu is optional. Additional menus, called *root menus*. A disc may have no root menus, or it may have up to 99. Menus may either be in a simple structure or organized into a hierarchy or linked willy-nilly into any tortuous structure the author desires. There is nothing in the DVD specification to support hierarchies of menus — they are created only by linking buttons to menus and assigning destinations for MenuCall and GoUp commands.

There are four other kinds of menus that are essentially submenus of a root menu: part-of-title (chapter) menu, audio menu, angle menu, and subpicture menu. These menus reside in a VTS with their associated root menu. Most players do not provide a way to access the submenus directly, so many discs include buttons, positioned on the root menu, that take the user to the submenus, if present on the disc. For example, if a title has multiple audio language tracks and includes an audio menu, the root menu usually includes a button to get to the audio menu where the user may select a language track. Also, a disc can be authored to automatically show a menu for available angles when it enters an angle block.

The buttons and selection highlights used to create menus are actually subpictures, which may be put anywhere in a video program. In-play menus offer countless possibilities for interactive viewing because they can appear, disappear, and change during playback. The reason for a designated title menu (VMGM) and for root menus (VTSMs) is to provide menu access with the press of a single remote control key.

DVD menu nomenclature is confusing, to say the least. The title menu really should be called the disc menu or top menu because it is the top-level menu for the disc, not the menu for a title. Root menus (which are not at the root) are the menus for various titles and title sets. Pressing the "Top" ("Title") key on a remote control accesses the main menu, and press-

---

[6]GOP headers are optional in MPEG-2 MP@ML. Entry points occur at MPEG-2 sequence headers.

ing the "Menu" key accesses the root menu of the current title.[7] When the "Top" ("Title") or "Menu" key is pressed a second time (after going to the menu), playback resumes at the point where it was interrupted. There is also a "Return" or "Go Up" button that — sometimes — returns from a submenu to its parent menu or to the current title if the menu is a top-level menu.

Some remotes or software players use other labels such as "Guide" or "Setup" for the "Top" ("Title") key and "Root" or "Digest" for the "Menu" key. To make things worse, many Hollywood movies do not have a title menu, so the "Top" ("Title") key has no effect, and the "Menu" key acts like the "Top" ("Title") key. Confused? So are the owners of most DVD players! Unfortunately this condition has not ameliorated over the years.

The DVD Forum issued a recommendation for standard remote control key names in an effort to clarify the matter.[8] See Table 9.14 for a summary. More consistent use of menus by disc producers also will help a great deal. Chapter 16 discusses consistency of menu navigation design.

Table 9.14 Remote Control Labeling

Internal Name	Technical Name	Recommended Key Label	Other Labels in Use
Title menu	VMGM (video manager menu)	Top Menu or Top	Title, Title Menu, Guide, Info, Setup
Root menu	VTSM (video title set menu)	Menu	Menu, Root, Root Menu, Digest
GoUp	MenuCall (GoUp)	Return	Back, Previous

Buttons

Information to create onscreen buttons is included in the navigation data. Up to 36 highlightable, rectangular buttons may be positioned on the screen and displayed concurrently. In the case of widescreen content (with anamorphic, automatic letterbox, or automatic pan and scan modes), only 18 buttons are allowed per screen when any two modes are used, and only 12 buttons are allowed per screen when all three modes are used. In this case, a separate set of buttons for each display mode is required so that the highlights are drawn in the correct area of the picture. Button locations are relative to the final displayed picture, after possible formatting with pan and scan or letterbox, not to the anamorphic picture stored in

---

[7]Some Pioneer remote controls require two keypresses to get to the title menu — the "Title" key and then the "Menu" key — unless the disc is stopped, in which case only the "Title" key is needed. The "Menu" key works with a single press, except that when the disc is stopped it goes to a player-generated menu that is neither the title menu nor the root menu.

[8]Unfortunately, in typical DVD specification style, the "clarification" is almost impenetrable. Here is an example, complete with typos, of a definition given for button behavior; which is one of 18 paragraphs defining four rules for the "Top Menu" and "Menu" buttons: "When [MEMU] is pressed, Menu_Call() (the transition to Root menu) is executed and then the function of [TOP MENU] is changed to Menu_Call() and the function of [MENU] is changed to RSM. In case that no entry of Root menu exists, the transition to Root menu is not actually executed although Menu_Call() is executed."

the video stream. For example, many movie discs use a single, widescreen picture for the menu, but also enable pan and scan mode to crop to the center of the menu for 1.33 televisions. Because the picture is cropped, there must be a second set of buttons with different screen positions.

The display and the highlighting of menu buttons are done with subpictures. The subpicture is composed of a foreground and a background. The subpicture foreground pixel types are used to draw the buttons on the video background (which may be a still image or motion video). The subpicture background pixel type is invisible so that the video will show through.

Invisible buttons can be created by setting the subpicture foreground and background pixel contrast to 0. In this case, high-quality buttons usually are rendered with the menu video instead of low-quality subpicture graphics, and subpictures are used only to create the highlight art that indicates which button is selected.

The arrow keys on a remote control or a software player interface are used to highlight buttons by jumping from one to another. Each button includes four directional links that determine which button on the screen is selected when the corresponding arrow keys are pressed. This creates a complex web of links between buttons that may or may not correspond to their physical arrangement. When a button is selected (highlighted), its color and contrast values (four each) are changed to those defined for the selected state. The selected button is activated by pressing the "Enter" or "Select" key on the remote control. Alternatively, a button may be activated by pressing the corresponding number keys on the remote control. Some remotes activate buttons 1 through 9 with a single keypress; others require multiple keypresses. When the button is activated, its pixels are momentarily displayed in a new set of color and contrast values defined for the *action* state. See the section on subpictures for details about subpicture pixels, colors, and contrasts.

Each button has one command associated with it. This is generally a flow-control command that links to a title or a PGC. In many cases, a button is linked to a dummy PGC that contains no programs (physically does not contain any VOBs) but is just a set of precommands and postcommands, usually culminating in a jump to a title or PGC with video in it. PGCs can be linked together, allowing a button to trigger an arbitrarily large sequence of commands.

A button can be set for *auto action*, which means that selecting (highlighting) it will activate it immediately. This is useful when creating menus where things change when buttons are selected or for creating menu navigation implementations where merely pressing a directional arrow will move to a new place, without requiring that the "Enter" key be pressed.

Each menu can have one button designated for *forced selection* and one button designated for *forced activation*. The latter can be set to occur after a specific amount of time. This feature is commonly called *idle out*, where something occurs if the user does nothing for a certain period of time. The time is specified in increments of 1/90,000 of a second, up to about 795 minutes ($2^{32}/90,000$).

Stills

Still frames (encoded as MPEG-2 I frames) are supported and can be displayed indefinitely. Still frames can be accompanied by audio for a slideshow type of presentation. The DVD-Video presentation format allows automatic freeze frames at the end of any video segment (a PGC, a cell, or a VOBU).

Most still images on DVDs are authored as menus and, in fact, this is a good way to put still images on a disc. Each still video image or graphic is encoded as a single MPEG video frame. Thousands of still images can be linked together with small or invisible menu buttons.

There are three types of stills: a PGC still, a cell still, and a VOBU still. Each has similar functionality, with the primary difference being the way the still is intended to be used and the location where the still information is stored. In all cases, the still occurs at the last PTM of a VOBU. A PGC still causes a still at the last VOBU of the PGC, a cell still causes a still at the last VOBU of the cell, and a VOBU still causes a still at the end of a specific VOBU. VOBU stills and cell stills occur before the cell command is executed. PGC stills, which can be used in random or shuffle PGCs but not sequential PGCs, occur after cell commands and PGC looping but before the postcommands. During a still, the navigation countdown timer and any GPRMs in counter mode continue to count. PGC and cell stills can be held indefinitely or from 1 to 254 seconds. VOBU stills are always indefinite. Stills are a handy way to show nonmotion video (such as a logo) for a period of time without wasting space encoding hundreds of frames of the same image.

The still feature is not the same as the pause or freeze-frame feature. Stills are implemented automatically by the navigation system, whereas pauses are the result of the viewer pressing the "Pause" key. Depending on player design, the user may be able to continue past a still by pressing "Play," "Pause," or "Next."

User Operations

User operations are the low-level functions defined for a player. A player can implement any sort of remote control design or user interface, but it must then translate all user input into one of the defined user operations. Some operations are mandatory for all players, others are optional (see Table 9.15). Because user operations are defined in Annex J of the DVD specification, they are sometimes referred to as *annex J operations*.

User operation controls (UOPs) are flags that the DVD author sets to restrict a viewer's navigation options at any particular place on the disc. For example, many discs do not allow the viewer to fast forward or jump to a menu at the beginning of the disc (such as the FBI warning). Each disc can enable or disable a user operation at any point, even if the operation would otherwise be valid within the domain. For instance a disc may be authored to disallow fast forwarding in certain places or to prevent jumping to a menu once a title begins playing.

User operation controls are implemented as 1-bit flags, each identified with a bit number from 0 to 25 (see Table 9.16). These are identified as UOP0, UOP1, and so on. If the bit is set, the operation is prohibited. Some UOP flags control more than one user operation. User operation controls can be placed at the title level, PGC level, and VOBU (PCI) level. This creates nested scopes where lower-level prohibitions take precedence. In other words, if a user operation is not prohibited, it can be prohibited at lower levels, but once a user operation is prohibited, it cannot be "unprohibited" by UOPs at a lower level.

Table 9.15 User Operations

User Operation	Equivalent Navigation Command	Control	Mandatory in Player
Title play	JumpTT or JumpVTS_TT	UOP2	Mandatory
Time play	None	UOP0	Optional
Time search	None	UOP0, UOP5	Optional
PTT play	JumpVTS_PTT	UOP1	Optional
PTT search	LinkPTTN	UOP1, UOP5	Optional
Stop	Exit	UOP3	Mandatory
GoUp	LinkGoUpPGC	UOP4	Mandatory
PrevPG search	LinkPrevPG	UOP6	Mandatory
TopPG search	LinkTopPG	UOP6	Optional
NextPG search	LinkNextPG	UOP7	Mandatory
Forward scan	None	UOP8	Optional
Backward scan	None	UOP9	Optional
Menu call (Title)	JumpSS or CallSS	UOP10	Mandatory
Menu call (Root)	JumpSS or CallSS	UOP11	Mandatory
Menu call (Subpicture)	JumpSS or CallSS	UOP12	Optional
Menu call (Audio)	JumpSS or CallSS	UOP13	Optional
Menu call (Angle)	JumpSS or CallSS	UOP14	Optional
Menu call (PTT)	JumpSS or CallSS	UOP15	Optional
Resume	RSM	UOP16	Mandatory
Upper button select	SetHL_BTNN	UOP17	Mandatory
Lower button select	SetHL_BTNN	UOP17	Mandatory
Left button select	SetHL_BTNN	UOP17	Mandatory
Right button select	SetHL_BTNN	UOP17	Mandatory
Button activate	None	UOP17	Mandatory
Button select and activate	None	UOP17	Optional
Still off	None	UOP18	Mandatory
Pause on	None	UOP19	Optional
Pause off	None	None	Optional
Menu language select	None	None	Mandatory
Audio stream change	SetSTN	UOP20	Mandatory
SP stream change	SetSTN	UOP21	Mandatory

continues

Table 9.15 User Operations (continued)

User Operation	Equivalent Navigation Command	Control	Mandatory in Player
Angle change	SetSTN	UOP22	Mandatory
Parental level select	None	None	Optional
Parental country select	None	None	Optional
Karaoke audio presentation mode change	SetAMXMD	UOP23	Optional
Video presentation mode change	None	UOP24	Mandatory

Table 9.16 User Operation Control

User Operation	Bit	Controlled in		
		Title	PGC	VOBU
Time play or seach	UOP0	√	√	–
PTT play or search	UOP1	√	√	–
Title play	UOP2	–	√	–
Stop	UOP3	–	√	√
GoUp	UOP4	–	–	√
Time or PTT search	UOP5	–	√	√
TopPG or PrevPG search	UOP6	–	√	√
NextPG search	UOP7	–	√	√
Forward scan	UOP8	–	√	√
Backward scan	UOP9	–	√	√
Menu call (Title)	UOP10	–	√	√
Menu call (Root)	UOP11	–	√	√
Menu call (Subpicture)	UOP12	–	√	√
Menu call (Audio)	UOP13	–	√	√
Menu call (Angle)	UOP14	–	√	√
Menu call (PTT)	UOP15	–	√	√
Resume	UOP16	–	√	√
Button select or activate	UOP17	–	√	–
Still off	UOP18	–	√	√
Pause on	UOP19	–	√	√
Audio stream change	UOP20	–	√	√
SP stream change	UOP21	–	√	√
Angle change	UOP22	–	√	√
Karaoke audio presentation mode change	UOP23	–	√	√
Video presentation mode change	UOP24	–	√	√

Video

See Table 9.17 for a summary of DVD-Video format details.

Table 9.17 DVD-Video Format

Multiplexed data rate	Up to 10.08 Mbps
Video data rate	Up to 9.8 Mbps
TV system	525/60 (NTSC) or 625/50 (PAL)
Video coding	MPEG-2 MP@ML/SP@ML VBR/CBR or MPEG-1 VBR/CBR
Internal picture rate	24 or 23.976 fps(film)[a], 29.97 fps[b] (525/60), 25 fps[b] (625/50)
Display frame rate	29.97 fps[b] (525/60), 25 fps[b] (625/50)
MPEG-2 resolution	720×480, 704×480, 352×480 (525/60); 720×576, 704×576, 352×576 (625/50)
MPEG-1 resolution	352×240 (525/60); 352×288 (625/50)
MPEG-2 GOP max.	36 fields (525/60), 30 fields (625/50)
MPEG-1 GOP max.	18 frames (525/60), 15 frames (625/50)
Aspect ratio	4:3 or 16:9 anamorphic[c]
Pixel aspect ratio	Refer to Table 9.20

[a]Progressive (decoder performs 2-3 pulldown from 24/23.976 fps for interlaced 29.97 fps display or 2-2 pulldown from 25 fps for interlaced 25 fps display).

[b]Interlaced (59.94 fields per second or 50 fields per second).

[c]Anamorphic only allowed for 720 and 704 resolutions.

Video Stream DVD-Video is based on a subset of MPEG-2 (ISO/IEC 13818) Main Profile at Main Level (MP@ML) or Simple Profile at Main Level (SP@ML). Constant and variable bit rates (CBR and VBR) are supported.

DVD-Video also supports MPEG-1 video at constant and variable bit rates (see Table 9.18). DVD adds additional restrictions, which are also detailed in Table 9.18.

Before MPEG-2 compression occurs, the video is subsampled from ITU-R BT.601 format at 4:2:0 sampling with 8 bits of precision, which allocates an average of 12 bits per pixel. The actual color depth of the samples is 24 bits (1 byte for Y', 1 byte for C_b, and 1 byte for C_r), but the C samples are shared by 4 pixels.

The uncompressed source data rate is 124.416 Mbps ($720 \times 480 \times 12 \times 30$ or $720 \times 576 \times 12 \times 25$). For 24-fps film, the source is typically at video frame rates of 30 fps (a telecine pulldown process has added duplicate fields). The MPEG encoder performs an inverse telecine process to remove the duplicate fields. Therefore, it is appropriate to consider the uncompressed film source data rate to be 99.533 Mbps ($720 \times 480 \times 12 \times 24$) or 119.439 Mbps ($720 \times 576 \times 12 \times 24$).

Table 9.18 Differences between DVD and MP@ML

MPEG Parameter	DVD	MP@ML
Display frame rate (525/60)	29.97	23.976, 29.97, 30
Display frame rate (625/50)	25	24, 25
Coded frame rate (525/60)	23.976, 29.97	23.976, 24, 29.97, 30
Coded frame rate (625/50)	24, 25	24, 25
Data rate	9.8 Mbps	15 Mbps
Frame size (horizontal size × vertical size) (525/60)	720 × 480, 704 × 480, 352 × 480, 352 × 240	From 16 × 16 to 720 × 480
Frame size (horizontal size × vertical size) (625/50)	720 × 576, 704 × 576, 352 × 576, 352 × 288	From 16 × 16 to 720 × 576
Aspect ratio	4:3, 16:9	4:3, 16:9, 2.21:1
Display horizontal size	540 (4:3), 720 (16:9)	Variable
GOP maximum (525/60)	36 fields/18 frames (30/15 recommended for video source, 24/12 recommended for film source)	No restriction
GOP maximum (625/50)	30 fields/15 frames (24/12 recommended)	No restriction
GOP header	Required (first GOP in VOBU)	Optional
Audio sample rate	48 kHz	32, 44.1, 48 kHz
Packet size	2048 bytes (one logical block)	Variable
Color primaries and transfer characteristics (525/60)	4 (ITU-R BT.624 M), 6 (SMPTE 170 M)	1, 2, 4, 5, 6, 7, (8)
Color primaries and transfer characteristics (625/50)	5 (ITU-R BT.624 B or G)	1, 2, 4, 5, 6, 7, (8)
Matrix coefficients (RGB to $Y'C_bC_r$)	5 (ITU-R BT.624 B or G), 6 (SMPTE 170 M)	1, 2, 4, 5, 6, 7
Low delay	Not permitted	Permitted

The maximum video bit rate is 9.8 Mbps (but must be less to allow for audio). The "typical" bit rate varies from 3.5 to 6 Mbps, but the rate depends on the length of the original video, the quality and complexity of the video, the amount of audio, etc. The canonical 3.5 Mbps average video data rate is a 36:1 reduction of an uncompressed video source (124 Mbps) or a 28:1 reduction of a film source (100 Mbps). Video running near 9 Mbps is compressed at less than a 14:1 ratio (refer to Table 3.1).

Variable bit rate (VBR) encoding allows more data to be allocated for complex scenes and less data to be used during simple scenes. By lowering the average data rate, longer amounts of video may be accommodated, and the extra headroom allows the encoder to maintain video quality when a higher data rate is needed. See Figure 9.19 for an illustration.

Figure 9.19 Example of Variable Bit Rate Video

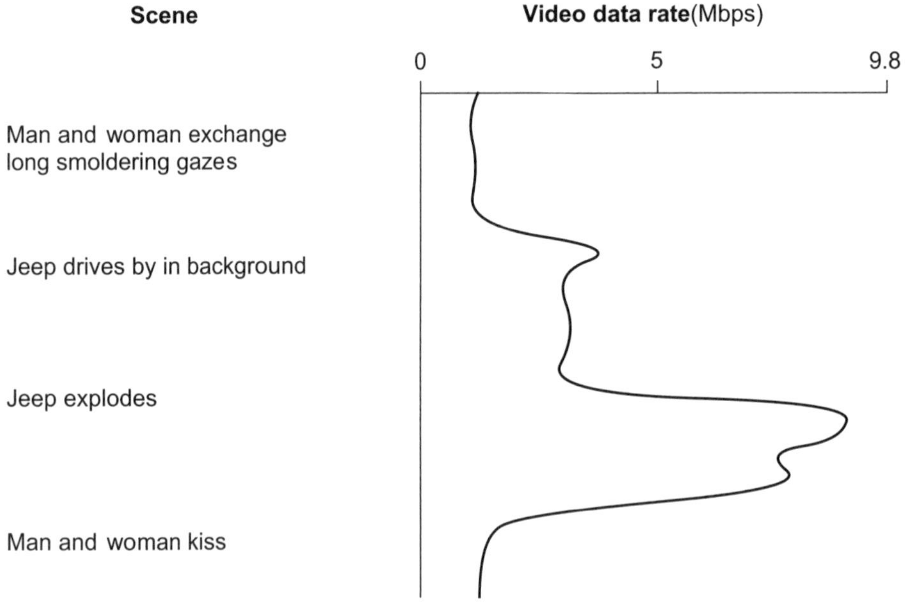

Scene   **Video data rate**(Mbps)

```
                                    0              5              9.8
Man and woman exchange
long smoldering gazes


Jeep drives by in background


Jeep explodes


Man and woman kiss
```

VBR is like riding a bicycle up and down hills. If you pedal at the same rate going up and down hills, you will run out of energy sooner than if you coast downhill and pedal easier when it's flat. VBR is in contrast to the statistical multiplexing scheme used by most digital video transmission systems (digital satellite, digital cable), which allocates bits across multiple channels into a fixed transmission rate. When one channel demands more quality, bits are stolen from other channels.

Scanning and Frame Rates DVD-Video supports two display television systems, 525/60 (NTSC, 29.97 interlaced fps) and 625/50 (PAL/SECAM, 25 interlaced fps). Internal coded frame rates are typically at 29.97 fps interlaced scan from NTSC video, 25 fps interlaced scan from PAL video, and 24 fps progressive scan from film. In the case of 24 fps, the MPEG-2 encoder adds repeat_first_field flags to the data to make the decoder perform 2-3 pulldown for 60 (59.94) Hz displays. For 50 Hz (PAL) display, the change from 24 to 25 fps causes a 4 percent speedup. Audio must be adjusted to match, resulting in a pitch shift (one semitone sharp) if it is not digitally readjusted.

A total of 480 lines of active video from a 525/60 video source are encoded and regenerated. Video encoding starts at line 23. For a 625/50 source, 576 lines of active video are encoded.

Very few DVD players convert from PAL video or film rates to NTSC display format or from NTSC video or film rates to PAL display format. Almost all PAL DVD players are able to produce video in NTSC scanning format transcoded to PAL color format for display on televisions supporting 4.43 NTSC signals (also called 60 Hz PAL). Some PAL players convert NTSC discs to standard PAL output, and a few NTSC players convert PAL discs to NTSC output. Computers are not tied to TV display rates, so most DVD computer software and hardware can play both NTSC and PAL. Some DVD computers can only display the converted video on the computer monitor, but others can output it as a video signal for a TV.

The actual coded picture rate in the MPEG-2 stream does not have to be exactly 24 or 30. Other coded picture rates or even varying rates will work, as long as the MPEG-2 repeat_first_field and top_field_first flags are set properly to produce either 25 or 29.97 fps display rates.

Resolution DVD-Video supports numerous resolutions designed for the NTSC and PAL television display systems (see Table 9.19). The lower resolutions are intended primarily for compatibility with MPEG-1 video formats. Most material uses a raster of 720 × 480 for 525/60 display and a raster of 720 × 576 for 625/50 display.

Table 9.19 DVD-Video Resolutions (Rasters)

	(16:9 aspect ratio allowed)		(16:9 aspect ratio not allowed)	
525/60 (NTSC)	720 × 480	704 × 480	352 × 480	352 × 240
625/50 (PAL/SECAM)	720 × 576	704 × 576	352 × 576	352 × 288

DVD pixels are not square — 525/60 pixels are tall, whereas 625/50 pixels are short. There are eight pixel aspect ratios (see Table 9.20) depending on the raster size and the picture aspect ratio. These ratios vary from the tallest of 0.909 to the widest of 2.909 (almost three times as wide as it is tall). Obviously, the pixels are wider for 16:9 anamorphic form than for normal 4:3 form. The 720- and 704-pixel rasters produce identical pixel aspect ratios because the 720-pixel version includes more of the horizontal overscan area (with a scanning line period of 53.33 microseconds), but the 704-pixel version is a tight scan (with a line period of 52.15 microseconds).

There is an alternate way of calculating pixel aspect ratios that divides the horizontal count by the vertical count and then divides the result by the picture aspect ratio. Confusingly, this gives a height-width aspect ratio rather than a consistent width-height aspect ratio. Table 9.20 includes values from the alternate method in parentheses as reciprocals and also adjusts them to match television scanning rates. The table also shows the integral conversion ratios used by most video digitizing hardware that converts between square and nonsquare pixels. Of course, there are no horizontal pixels in analog signals, although scan lines correspond to vertical pixels.

Table 9.20 DVD-Video Pixel Aspect Ratios

Resolution	4:3 Display	Standard 4:3 Integral Ratio	16:9 Display
720 × 480,[a] 704 × 480,[b] 352 × 240[b]	0.909 (1/1.095)	10/11	1.212 (1/0.821)
720 × 576,[a] 704 × 576,[b] 352 × 288[b]	1.091 (1/0.9157)	59/54	1.455 (1/0.687)
352 × 480[b]	1.818 (1/2.19)	20/11	2.424 (1/0.411)
352 × 576[b]	2.182 (1/1.831)	118/54	2.909 (1/0.343)

[a]Overscan

[b]Exact scan

Using the traditional (and rather subjective) television measurement of *lines of horizontal resolution per picture height* (TV lines, or TVL), DVD has a theoretical maximum of 540 lines on a standard TV [720/(4/3)] and 405 on a widescreen TV [720/(16/9)]. Lines of horizontal resolution also can be approximated at 80 per MHz. DVD's MPEG-2 luma component is sampled at 13.5 MHz, which results in 540 lines. The actual observable lines of horizontal resolution may be closer to 500 on a standard TV due to low-pass filtering in the player. Typical luma frequency response maintains full amplitude to between 5.0 and 5.5 MHz. This is below the 13.5 MHz native frequency of the MPEG-2 digital signal. In other words, most players fall short of reproducing the full quality of DVD. Chroma frequency response is half that of luma.

For comparison, video from a laserdisc player has about 425 lines of horizontal resolution (5.3 MHz), SuperVHS and Hi8 have about 400 (5 MHz), broadcast television has about 335 (4.2 MHz),[9] and video from a VHS VCR has about 240 (3 MHz).

DVD resolution in pixels may be roughly compared with the resolution of analog video formats by considering pixels to be formed by the intersections of active scan lines and lines of horizontal resolution adjusted for aspect ratio. Table 9.21 shows how the resolution of DVD compares with other formats, including the two high-definition formats of the U.S. ATSC proposal (labeled DTV3 and DTV4), which also correspond to the H0 and H1 levels of the Microsoft/Intel/Compaq "Digital TV Team" proposal. Also see Table A-17 for additional video resolution figures.

Table 9.21 Resolution Comparison of Different Video Formats

Format	VCD (16:9)	VCD (4:3)	VHS (16:9)	VHS (4:3)	LD (16:9)	LD (4:3)	DVD (16:9/4:3)	DTV3 (16:9)	DTV4 (16:9)
Horizontal pixels	352	352	333	333	567	567	720	1280	1,920
Vertical pixels	180	240	360	480	360	480	480	720	1,080
Total pixels	63,360	84,480	119,880	159,840	204,120	272,160	345,600	921,600	2,073,600
x VCD (16:9)		*1.33*	1.89	*2.52*	3.22	*4.30*	5.45	14.55	32.73
x VCD (4:3)			*1.42*	1.89	*2.42*	3.22	4.09	*10.91*	*24.55*
x VHS (16:9)				*1.33*	1.70	*2.27*	2.88	7.69	17.30
x VHS (4:3)					*1.28*	1.70	2.16	*5.77*	*12.97*
x LD (16:9)						*1.33*	1.69	4.51	10.16
x LD (4:3)							1.27	*3.39*	*7.62*
x DVD (16:9/4:3)								2.67	6.00
x DTV3 (16:9)									2.25

Note: 16:9 aspect ratios for VHS, LD, and VCD are letterboxed in a 4:3 picture. Comparisons between different aspect ratios are not as meaningful. These are shown in italics. Comparisons at 1.85 or 2.35 letterbox aspect ratios are essentially the same as at 1.78 (16:9).

---

[9] Measurements for broadcast are usually tighter than those for recorded media, so broadcast quality is closer to SuperVHS and Hi8 than the numbers would indicate.

Widescreen Format Video may be stored on a DVD in aspect ratios of 4:3 or 16:9. The 16:9 format is anamorphic, meaning the picture is squeezed horizontally to fit a 4:3 rectangle and then unsqueezed during playback. DVD players can produce video in four different ways:

- 4:3 normal (for 4:3 or 16:9 displays)
- 16:9 letterbox (for 4:3 displays)
- 16:9 pan and scan (for 4:3 displays)
- 16:9 widescreen (anamorphic, for 16:9 displays)

DVD-Video segments may be marked for the following display modes:

- 4:3 full frame
- 4:3 letterboxed (for automatically setting display mode on widescreen TVs)
- 16:9 letterbox only (player not allowed to pan and scan)
- 16:9 pan and scan allowed (viewer can select pan and scan or letterbox on 4:3 TV)

Some players send a signal to the television indicating that the picture is in anamorphic widescreen form so that widescreen televisions can adjust automatically. In some cases, the player also can inform the television that the 4:3 picture was transferred to video in letterbox format so that the television can expand the picture to remove the mattes. In Europe, the widescreen signaling system (WSS) may be used to convey this type of information (Tables 9.22 and 9.23). This standard is recommended for use by NTSC systems as well.

Table 9.22 Widescreen Signaling Information

Aspect Ratio	Range	Format	Position	Active Lines
4:3 (1.33)	≤ 1.46	Full	Center	576 (480)
14:9 (1.57)	>1.46, ≤ 1.66	Letterbox	Top	504 (420)
14:9 (1.57)	>1.46, ≤ 1.66	Letterbox	Center	504 (420)
16:9 (1.78)	>1.66, ≤ 1.90	Letterbox	Top	430 (360)
16:9 (1.78)	>1.66, ≤ 1.90	Letterbox	Center	430 (360)
>16:9 (>1.78)	>1.90	Letterbox	Center	Undefined
14:9 (1.57)	>1.46, ≤ 1.66	Full[a]	Center[a]	576 (480)
16:9 (1.78)	>1.66, ≤ 1.90	Full (anamorphic)	n/a	576 (480)

[a]Shoot and protect 14:9. Soft matte format intended to be displayed with top and bottom cropped on a 16:9 display.

Table 9.23 Widescreen Signaling on SCART Connectors

SCART Pin 8 Voltage	Meaning
0 V	Normal
+6 V	16:9
+12 V	4:3 letterbox

There is also a convention for signaling widescreen format by adding a 5V direct current (DC) component to the chroma (C) line of the Y/C (s-video) output. This tells the widescreen equipment to expect video in anamorphic form. Most DVD players and newer video displays support this technique. Unfortunately, there remains a great deal of existing equipment that the video signal might pass through on its way to the display that is designed to filter out such DC "noise." Newer A/V receivers and video switchers are being designed to recognize the widescreen signal and either pass it through or recreate it at their output.

In anamorphic mode, the pixels are fatter (see Table 9.20), but due to the high horizontal resolution of DVD, they are not objectionably noticeable.

Video in anamorphic form causes no problems with line doublers because they simply double the lines on their way to the widescreen display that then stretches out the lines.

Letterbox Conversion For automatic letterbox mode, the player uses a letterbox filter that creates mattes at the top and bottom of the picture (60 lines for each matte in NTSC, 72 lines for each matte in PAL). This leaves three-quarters of the height remaining, creating a shorter but wider rectangle for the image. In order to fit the shape of this shorter rectangle, the player squeezes the picture vertically by combining every four lines into three. The vertical downsampling compensates for the anamorphic horizontal distortion and results in the movie being shown in its full width but with a 25 percent loss of vertical resolution. Some players simply throw away every fourth line. Better players use weighted averaging to give smoother results, albeit with a softer picture (see Figures 9.20 and 9.21). Some DVD players do a better job of letterbox filtering than others, perhaps by compensating for interlace jitter effects.

Figure 9.20 Weighted Letterbox Conversion

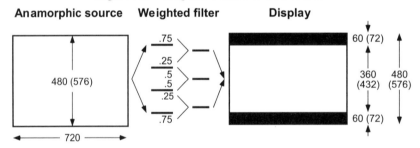

Figure 9.21 Letterbox Math

$$\frac{\frac{4}{3}}{\frac{16}{9}} = \frac{1.33}{1.78} = 0.75 = 75\% = 25\% \text{ reduction}$$

0.75 x 480 = 360

0.75 x 576 = 432

Pan and Scan Conversion For automatic pan and scan mode, a portion of the film image is shown at full height on a 4:3 screen by following a center-of-interest offset that is encoded in the video stream according to the preferences of those who transferred the film to video. The pan and scan image window is 75 percent of the full image width, which reduces the horizontal resolution from 720 to 540 (see Figures 9.22 and 9.23), causing a 25 percent loss of horizontal resolution. Expanding the window by 33 percent compensates for the anamorphic distortion and achieves the proper 4:3 aspect ratio. The 540 extracted pixels on each line are interpolated by creating four pixels from every three to scale the line back to the full width of 720. Weighted averaging gives smoother results.

Figure 9.22 Weighted Pan and Scan Conversion

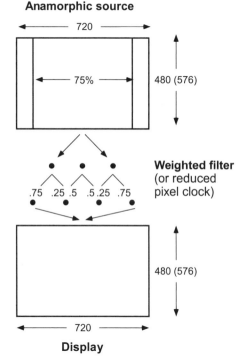

Figure 9.23 Pan and Scan Math

$$\frac{\frac{4}{3}}{\frac{16}{9}} = \frac{1.33}{1.78} = 0.75 = 75\% = 25\% \text{ reduction}$$

0.75 x 720 = 540

$$\frac{\frac{16}{9}}{\frac{4}{3}} = \frac{1.78}{1.33} = 1.33 = 133\% = 33\% \text{ increase}$$

1.33 x 540 = 720

Unlike letterbox conversion, which is new to DVD, pan and scan conversion is part of the MPEG-2 standard. Most MPEG-2 decoder chips include a pan and scan conversion feature. The offset is specified in increments of one-sixteenth of a pixel. DVD allows only horizontal adjustments; that is, the MPEG-2 frame_center_vertical_offset must be 0, while frame_center_horizontal_offset can vary from -1440 to $+1440$ for 720-pixel frames and from -1312 to $+1312$ for 704-pixel frames. It is also possible to stretch out the pixels by increasing the pixel clock in the TV encoder chip.

Video Interface The MPEG video stream is decoded into 4:2:0 digital component format. This format can be sent directly from the player (with accompanying content protection information) to a digital connection such as IEEE 1394/FireWire or SDI. However, in most cases the player is connected to an analog video display or recording system and requires that the signal be converted to analog form through a digital-to-analog converter. Although the original BT.601 video values are sampled with 8 bits of precision, better players use 10 bits or more in the digital-to-analog converter to provide headroom for more accurate calculations, which can help produce a smoother picture.

For analog component output, the digital signals are scaled and offset according to the desired output format(s) of $Y'P_bP_r$ or RGB, blanking and sync information is added, and the digital values are converted to analog voltage levels. Some players label the analog component output as $Y'C_bB_r$, which is incorrect.

For analog s-video and composite baseband video, the digital signal is sent through a TV encoder to produce an NTSC signal from 525/60 format data or a PAL or SECAM signal from 625/50 format data. Almost all TV encoders support both NTSC and PAL, but most NTSC players do not enable PAL video signal output (see Figure 9.24). See Chapter 10 for more information about video connections.

Figure 9.24 Video Block Diagram

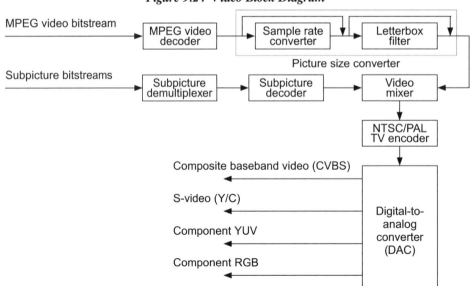

Audio

DVD-Video supports three primary audio standards: Dolby Digital, MPEG-2, and linear PCM (LPCM). Two optional audio formats are included — DTS (Digital Theater Sound) and SDDS (Sony Dynamic Digital Sound) — but players are not required to support either one (see Table 9.24).

Table 9.24 DVD-Video Format, Audio Details

Audio	0 to 8 streams
Audio coding	Dolby Digital, MPEG-1, MPEG-2[a], LPCM, DTS, SDDS[a]
Audio bit rate	32 kbps to 6.144 Mbps, 384 kbps typical

[a]not used

Dolby Digital, MPEG-2, LPCM, DTS, and even SDDS can provide discrete multichannel audio.[10] This gives clean sound separation with full dynamic range from each speaker. The result is a more realistic soundfield in which sounds can travel left to right and front to back.

Discs containing 525/60 (NTSC) video are required to include at least one audio track in Dolby Digital or PCM format. After that, any combination of formats is allowed. Discs containing 625/50 (PAL/SECAM) video are required to have at least one track of Dolby Digital, MPEG, or PCM audio. Because Dolby Digital decoders greatly outnumber MPEG-2 audio decoders, most disc producers also use Dolby Digital audio tracks on PAL discs instead of MPEG audio tracks. Audio streams are encoded at various data rates, depending on the number of channels.

Dolby Digital Audio Details Dolby Digital (AC-3) is the format used for audio tracks on almost all DVDs. It is a multichannel digital audio format, lossily compressed using perceptual coding technology from original PCM with a sample rate of 48 kHz at up to 24 bits of precision. The Dolby Digital standard provides for other sampling rates of 32 and 44.1 kHz, but these are not allowed with DVD. Frequency response is 3 Hz to 20 kHz for the main five channels and 3 to 120 Hz for the low-frequency effects (LFE) channel (Table 9.25).

Table 9.25 Dolby Digital Audio Details

Sample frequency	48 kHz
Sample size	Up to 24 bits
Data rate	64 to 448 kbps
Typical 5.1 data rate	384 or 448 kbps[a]
Typical 2.0 data rate	192 or 256 kbps(for music)
Typical 1.0 data rate	64 or 96 kbps (for music)
Channels (front/rear)[a]	1/0, 2/0, 3/0, 2/1, 2/2, 3/1, 3/2, 1 + 1/0 (dual mono)
Karaoke modes	L/R, M, V1, V2

[a]Dolby Laboratories recommends 448 kbps

[b]LFE channel can be added to all variations.

---

[10]The only format that is not discrete multichannel is Dolby Prologic (aka Dolby Surround) which uses analog matrixing. DVD players do not natively support Dolby Prologic, though it may be embedded into stereo analog audio for use on a DVD and interpretation by external receivers.

The Dolby Digital bit rate is 64 to 448 kbps, with 384 or 448 kbps being the typical rate for 5.1 channels. The 448 kbps rate results in a compression ratio of 10:1 (90 percent) from 5.1 channels of 48/16 PCM. The typical bit rate for stereo (with or without Dolby Surround encoding) is 192 kbps. Monophonic audio is usually at 96 kbps for music or 64 kbps for voice.

There can be 1, 2, 3, 4, or 5 channels. The LFE (.1) channel can be added optionally to any combination. Two-channel mode can either be stereo or dual mono (where each channel is a separate track). An extra rear center channel is possible with the Dolby Digital Surround EX format, which is compatible with all DVD discs and players and with existing Dolby Digital decoders. The added channel is not an additional full-bandwidth discrete channel; rather, it is phase matrix encoded into the two rear channels in the same way Dolby Surround is matrixed into standard stereo channels. A new (or additional) decoder is needed to extract the rear center channel.

All Dolby Digital decoders are required to perform a downmixing process to adapt 5.1 channels to 2 channels for stereo PCM and analog output. The downmixing process matrixes the center and surround channels onto the main stereo channels in Dolby Surround format for use by Dolby Pro Logic decoders. The LFE channel is omitted from the downmix because most audio systems without six speakers cannot reproduce the low bass. This can help keep sound from becoming "muddy" on typical home audio systems.

When the audio is encoded, the downmixed output is auditioned by using a reference decoder. If the quality is not adequate, the encoding process can be tweaked, the 5.1-channel mix can be adjusted, or a separate Dolby Surround track can be added (in either Dolby Digital or PCM format). Most modern action movies require minor adjustments to the 5.1-channel mix to ensure that the dialogue is audible. Some disc producers prefer to create a separate two-channel surround mix rather than letting the decoder do the downmixing. Two-channel Dolby Surround streams are also used when only the Dolby Surround mix is available or where the disc producer does not want to remix from the multitrack masters.

Dolby Digital is not synonymous with 5.1 channels. A Dolby Digital soundtrack can be mono, dual mono, stereo, Dolby Surround stereo, etc. For example, older movies have only monophonic soundtracks encoded as Dolby Digital 1.0.

Dolby Digital also provides dynamic range compensation. DVDs have soundtracks with much wider dynamic range than most recorded media. This improves performance for a quality home theater setup but may make dialogue and other soft passages too low to hear clearly on less-than-optimal audio systems. Information is added to the encoded data to indicate what parts of the sound should be boosted or cut when the player's dynamic range compression setting is turned on. Downmixing automatically turns on dynamic range compression to help maintain a high average loudness without inducing peak overloads.

Additional information can be carried in the Dolby Digital stream, such as a copyright flag and a flag that identifies when Surround EX encoding has been used. See Chapter 3 for more details on Dolby Digital encoding.

MPEG Audio Details MPEG audio is multichannel digital audio, lossily compressed using perceptual coding from original PCM format with a sample rate of 48 kHz at 16 bits. MPEG-1 Layer II and MPEG-2 backward-compatible (BC) are supported. The variable bit rate is 64 to 912 kbps, with 384 kbps being the normal average rate. An MPEG-1 stream is limited to 384 kbps (see Table 9.26).

Table 9.26 MPEG Audio Details

Sample frequency	48 kHz only
Sample size	Up to 20 bits
MPEG-1	Layer II only
MPEG-1 data rate	64 to 192 kbps (mono), 64 to 384 kbps (stereo)
MPEG-2	BC (matrix) mode only
MPEG-2 data rate[a]	64 to 912 kbps
Extension streams[b]	5.1-channel, 7.1-channel
Channels (front/rear)[c]	1/0, 2/0, 2/1, 2/2, 3/0, 3/1, 3/2, 5/2 (no dual channel or multilingual)
Karaoke channels	L, R, A1, A2, G
Emphasis	None
Prediction	Not allowed

[a]MPEG-1 Layer II stream + extension stream(s).
[b]AAC (unmatrix, NBC) not allowed.
[c]LFE channel can be added to all variations.

There can be 1, 2, 3, 4, 5, or 7 channels. The LFE (.1) channel is optional with any combination. The 7.1-channel format adds left-center and right-center channels but is not intended for home use. Stereo channels are provided in an MPEG-1 Layer III stream. Surround channels are matrixed onto the MPEG-1 stream and duplicated in an extension stream to provide discrete channel separation. The LFE channel is also added to the extension stream. An additional extension layer can be added for 7.1 channels. The extension streams are carried by MPEG packets. This layering and packetizing process makes MPEG-2 audio backward-compatible with MPEG-1 hardware; an MPEG-1 decoder will see only packets containing the two main channels.

Stereo output includes surround channel matrixing for Dolby Pro Logic processors. The MPEG signal already has the center and surround channels matrixed onto the main stereo channels, so no special downmixing is required.

Since MPEG-2 decoders were unavailable at the introduction of DVD, first-generation PAL players contain only MPEG-1 decoders.

The MPEG-2 standard includes an AAC (advanced audio coding) mode. In order for DVD-Video discs to be playable in players with only MPEG-1 decoders, the AAC mode is not allowed for the primary MPEG audio track, which always must be compatible with MPEG-1. The AAC mode originally was known as NBC (non-backward-compatible) and also is referred to as unmatrix mode. MPEG Layer 3 audio mode, also known as MP3, is also prohibited. Of course, AAC and MP3 tracks can be recorded on a disc outside the DVD-Video or DVD-Audio zones. See Chapter 3 for more details of MPEG audio encoding.

PCM Audio Details Linear PCM (pulse-code modulation) is lossless, uncompressed digital audio. The same format is used on CDs. DVD-V supports sampling rates of 48 or 96 kHz with 16, 20, or 24 bits per sample. (Audio CD is limited to 44.1 kHz at 16 bits.) There can be from 1 to 8 channels in each track (see Table 9.27).

Table 9.27 PCM Audio Details

Sample frequency	48 or 96 kHz
Sample size	16, 20, or 24 bits
Channels	1, 2, 3, 4, 5, 6, 7, or 8
Karaoke channels	L, R, V1, V2, G

The maximum PCM bit rate is 6.144 Mbps, which limits sample rates and bit sizes with 5 or more channels (see Table 9.28). It is generally believed that the 96 dB dynamic range of 16 bits, or even the 120 dB range of 20 bits combined with a frequency response of up to 22,000 Hz from 48 kHz sampling, is adequate for high-fidelity sound reproduction. However, additional bits and higher sampling rates are useful in studio work, noise shaping, advanced digital processing, and three-dimensional sound field reproduction.

Table 9.28 Allowable PCM Data Rates and Channels

kHz	Bits	Number of Channels, Data Rate (kbps)							
		1	2	3	4	5	6	7	8
48	16	768	1536	2304	3072	3840	4608	5376	6144
48	20	960	1920	2880	3840	4800	5760		
48	24	1152	2304	3456	4608	5760			
96	16	1536	3072	4608	6144				
96	20	1920	3840	5760		These combinations exceed the			
96	24	2304	4608			maximum allowed data rates			

DVD players are required to support all the variations of PCM, but most models subsample the 96 kHz rate down to 48 kHz, and some may truncate extra bits above 16 or 20. High-end players pass 96 kHz audio to the digital audio outputs, but only from tracks that are not CSS encrypted, because the CSS license limits protected output to 48 kHz.

DTS Audio Details Digital Theater Systems Digital Surround is an optional multichannel (5.1 or 6.1) digital audio format, lossily compressed from PCM at 48 kHz and up to 24 bits. The data rate is from 64 to 1536 kbps, with 768 or 1536 kbps being the typical rates for 5.1-channel audio. The 768 kbps rates results in a compression ratio of 6:1 (83 percent) from 5.1 channels of 48/16 PCM. The 1536 kbps rate results in a compression ratio of 3:1 (67 percent) from 5.1 channels of 48/16 PCM. DTS supports up to 4096 kbps variable data rate for lossless compression, but DVD does not allow this. DVD also does not allow sampling rates other than 48 kHz (see Table 9.29).

Table 9.29 DTS Audio Details

Sample frequency	48 kHz
Sample size	Up to 24 bits
Bit rate	64 to 1536 kbps; 768 or 1536 kbps typical
Channels (front/rear)[a]	1/0, 2/0, 3/0, 2/1, 2/2, 3/2, 3/3 (no multilingual)
Karaoke modes	L/R, M, V1, V2

[a]LFE channel can be added to all variations.

Channel combinations are (front/surround): 1/0, 2/0, 3/0, 2/1, 2/2, 3/2, and 3/3. The LFE channel is optional with all seven combinations. A rear center channel is supported in two ways: (1) a Dolby Surround EX matrixed channel that is compatible with existing decoders (*DTS-ES Matrix 6.1*) or (2) a discrete seventh channel (*DTS-ES Discrete 6.1*). DTS also has a 7.1-channel mode (*DTS-ES Discrete 7.1*, with 8 discrete channels), but it is not used commonly. The 7-channel and 8-channel modes work with existing DTS decoders but require a new decoder to take advantage of all channels.

The DVD standard includes an audio stream format reserved for DTS, but many older players ignore it and are thus unable to recognize and output the DTS stream. These players play only the Dolby Digital or PCM stream that is mandatory on all discs, including DTS discs.

The encoding system used on DVDs (DTS Coherent Acoustics, a psychoacoustic transform coder) is different from the one used in theaters (Audio Processing Technology's apt-X, an ADPCM coder). All DVD players can play DTS audio CDs because the standard PCM stream holds the DTS code. See Chapter 3 for more details on DTS encoding.

SDDS Audio Details SDDS is an optional multichannel (5.1 or 7.1) digital audio format, lossily compressed from PCM at 48 kHz. The data rate can go up to 1280 kbps. SDDS is a theatrical film soundtrack format based on the ATRAC transform coder compression format that is also used by Sony's MiniDisc.

Karaoke Modes All five DVD-Video audio formats support karaoke mode, which has two channels for stereo (L and R) plus an optional melody channel (M) or a guide channel (G), and two optional vocal channels (V1 and V2, aka A1 and A2). The karaoke modes generally are implemented only by DVD player models with karaoke features for mixing audio and microphone input. Karaoke mixing is implemented as a multichannel track that is selectively mixed to two output channels. Source channels 1 and 2 are usually the background instrumentals, channels 3 and 4 are typically vocal channels (usually a harmony channel and a melody channel), and channel 5 is the guide or melody helper channel (usually a single instrument playing the melody).

Codes are included to identify audio segments such as master of ceremonies intro, solo, duet, male vocal, female vocal, climax, interlude, ending, and so on.

Audio Interface The digital audio signal from the currently selected audio track is sent to the audio subsection. Multichannel (Dolby Digital or MPEG-2) audio signals are directed to the digital audio output jacks for decoding by external equipment. The multichannel signals are also sent to the appropriate decoder where they are downmixed to two channels and converted to PCM signals. PCM signals from the decoder or directly from the disc are also routed to the digital audio output jacks for processing by external equipment with digital-to-analog converters. (See Figures 9.25 and 9.26.)

Figure 9.25 Dolby Digital Audio Block Diagram

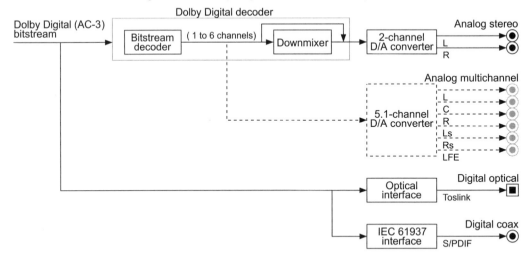

Figure 9.26 MPEG-2 Audio Block Diagram

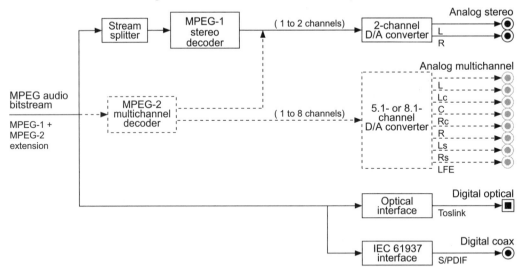

Some players also include built-in digital-to-analog audio converters and external jacks for discrete output from the decoder. Most players do not include built-in DTS decoders, but instead pass it directly out the digital audio output jacks for processing by external equipment. Built-in DTS decoders operate the same as Dolby Digital decoders.

Some players provide PCM audio and multichannel audio signals on separate connectors; others provide dual-purpose connectors that must be switched between PCM and multi-channel output. A standard extension to IEC 958, designed by Dolby and described in ATSC A/52, is used for Dolby Digital audio signals. This extension, with additions from Philips and others, has been formalized as IEC 61937.

PCM audio signals are also routed to a digital-to-analog converter for standard analog stereo output. See "How to Hook Up a DVD Player" in Chapter 10 for more information about audio connections. See "Digital Connections" in Chapter 10 for information about other interface options.

Subpictures

DVD-Video supports up to 32 subpicture streams that overlay the video for subtitles, captions, karaoke lyrics, menus, simple animation, and so on. These are part- to full-screen, run-length-encoded bitmaps limited to four pixel values (see Table 9.30).

Table 9.30 Subpicture Details

Data	0 to 32 streams
Data rate	Up to 3.36 Mbps
Subpicture Unit (SPU) size	53,220 bytes (up to 32,000 bytes of control data)
Coding	RLE (max. 1440 bits/line)
Resolution[a]	Up to 720×478 (525/60) or 720×573 (625/50)
Bits per pixel	2 (defining one of four pixel types)
Pixel types	Background, foreground (pattern), emphasis-1, emphasis-2
Colors[a]	4 of 16 (from 4-bit palette[b], one per pixel type)
Contrasts[a]	4 of 16 transparency levels (one per pixel type), 0=invisible, 15=opaque

[a]Area, content, color, and contrast can be changed for each field.

[b]Color palette can be changed every PGC.

The maximum data rate for a single subpicture stream is 3.36 Mbps, with a maximum size per frame of 53,220 bytes. Each run-length-encoded line can be up to 1440 bits (720×2). Associated display control sequences (DCSQs) can be up to 32,000 bytes (see Figure 9.27).

As noted in Table 9.30, each pixel is represented by 2 bits, allowing four types. The four pixel types are officially designated as *foreground* (also called *pattern*), *background*, *emphasis-1*, and *emphasis-2*, but are not strictly tied to these functions. The highlight functions are part of the menu button feature (described earlier). Each pixel type is associated with one color from a palette of 16 and one contrast or transparency level. The 24-bit $Y'C_bC_r$ color palette entries provide selections from more than 11 million colors. The transparency is set directly, from invisible (0), through 14 levels of transparency (1-14), to opaque (15).

Figure 9.27 Subpicture Unit Structure

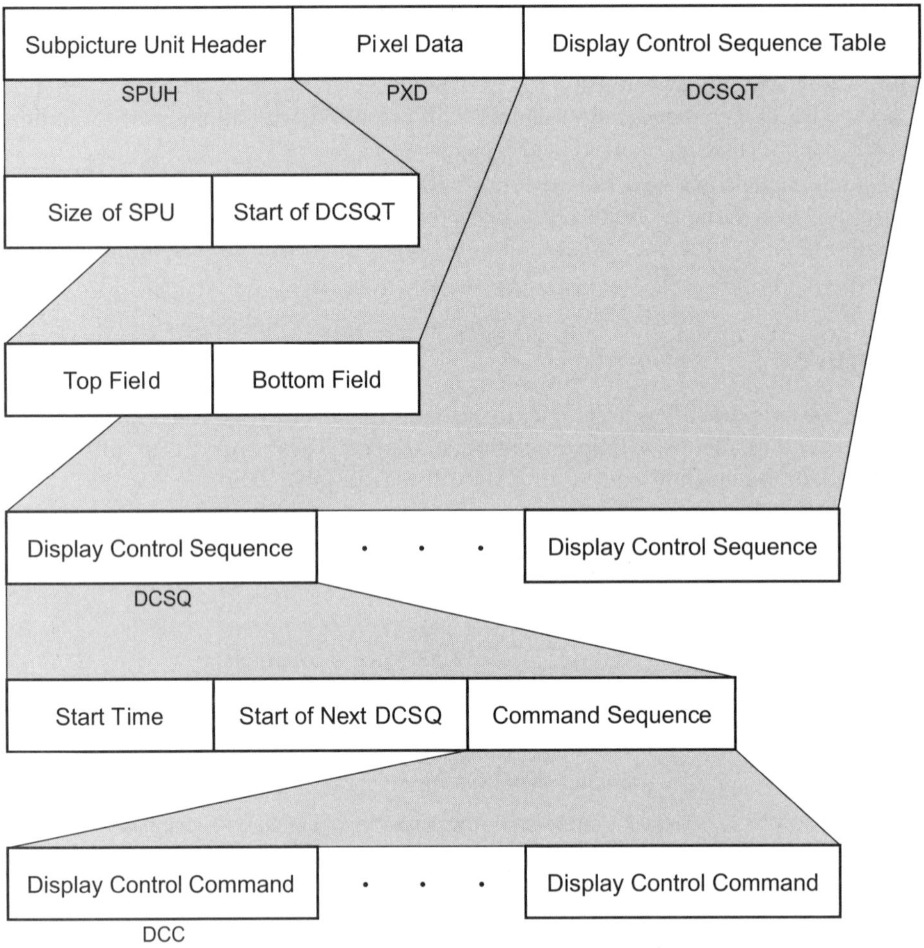

Subpicture Unit Header	Pixel Data	Display Control Sequence Table
SPUH	PXD	DCSQT

Size of SPU	Start of DCSQT

Top Field	Bottom Field

Display Control Sequence	· · ·	Display Control Sequence

DCSQ

Start Time	Start of Next DCSQ	Command Sequence

Display Control Command	· · ·	Display Control Command

DCC

The display area (starting coordinates, width, and height) can be specified up to almost a full-screen rectangle (0,0,720,478 or 0,0,720,573). The display area and content (bitmap) may be changed for each frame or field. The color and contrast of the four pixel types may be changed for each frame or field; and, the palettes can be changed for each PGC.

Using a subpicture stream, a typical set of subtitles are presented at about 14 per minute, yielding approximately 1500 subtitle frames per movie.

The subpicture display commands (DCCs) are used to change the location, scroll position, transparency, etc. A sequence of commands can be used to create effects such as color changes, moving highlights, fades, crawls, and so on.

Closed Captions

DVD includes support for NTSC Closed Captions, although some DVD players may not output NTSC Closed Captions. Closed Captions (CC) are a standardized method of encoding text into an NTSC television signal. The text can be displayed by a TV with a built-in decoder or by a separate decoder. All TVs larger than 13 inches sold in the United States since 1993 must have Closed Caption decoders.

Even though the terms *caption* and *subtitle* have similar definitions, *captions* commonly refer to onscreen text specifically designed for hearing-impaired viewers, whereas *subtitles* are straight transcriptions or translations of the dialogue. Captions are usually positioned below the person who is speaking, and include descriptions of sounds and music. Closed captions are not visible until the viewer activates them, whereas open captions are always visible.

Closed Captions on DVDs are carried in the MPEG-2 video stream and are sent automatically to the TV. They cannot be turned on or off from the DVD player. They are stored as individual character codes (one character per field). Subtitles, on the other hand, are DVD subpictures. Subpictures also can be used to create captions. To differentiate NTSC Closed Captions from subtitles, captions created as subpictures are usually called "captions for the hearing impaired."

DVD does not support PAL Teletext, the much-improved European equivalent of Closed Captions. The Advanced Television Enhancement Forum (ATVEF) recommends the use of EIA-768 TV links, which are URLs stored in the line 21 T2 service, to add Web links to video. See Chapter 15 for more information.

Camera Angles

A DVD video program can have up to nine camera angles, which are essentially interleaved video tracks. The viewer may switch among the angles by pressing the "Angle" key on the remote control. The angle also can be changed with commands embedded on the disc. When the angle feature is employed on a disc, the alternate chunks of video must all be the same length because the viewer cannot switch to a different camera angle and have the action be out of sync (unless the video itself was not produced in sync). Each angle is technically required to have identical audio tracks, but it is possible to circumvent this requirement and assign a different audio track to each angle. However, care should be exercised when attempting to change audio streams at random points, as it may cause sudden "pops" in the audio and could lead to speaker damage. This is particularly true for LPCM audio.

Video objects, one for each angle, are interleaved into an interleaved block (ILVB). ILVBs are required for camera angles and optionally can be used for parental branching and seamless branching (see Figure 9.28).

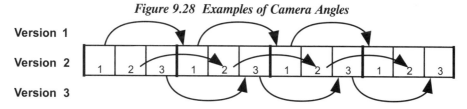

Figure 9.28 Examples of Camera Angles

A major advantage of camera angles is that they are supported directly by the buffer management feature of DVD, which automatically leapfrogs through the interleaved video objects, reading only the ones assigned to the selected angle. This means that adding an angle does not reduce the data rate limit, although it does reduce playing time. For example, a single-angle program that runs for five minutes at 7 Mbps would be about 262 megabytes long. If it were changed to a two-angle program, each angle could still run at 7 Mbps, but the program would take up 524 megabytes. Limitations are imposed on the maximum data rate depending on the number of angles (see Table 9.31). The data-rate limit applies to the angle block and for 2.5 seconds preceding the angle block.

Table 9.31 Data-Rate Limitations for Camera Angles

Number of Angles	Maximum Data Rate for Each Angle
2	8.00 Mbps
3	7.83 Mbps
4	7.66 Mbps
5	7.49 Mbps
6	7.31 Mbps
7	7.14 Mbps
8	6.97 Mbps
9	6.80 Mbps

Parental Management

Playback restrictions can be placed on a player by choosing a parental level from the settings menu. There are eight management levels, each less restrictive of content than the level below. Some players may be set to any of the eight levels, others limit the selection to the five Motion Picture Association of America (MPAA) levels, and still others offer three levels: all titles, no adult titles, only children's titles. The MPAA standard movie ratings are mapped to specific parental levels (see Table 9.32).

Table 9.32 MPAA Ratings and Corresponding Parental Levels

MPAA Rating	Parental Level	General Description
	8	Unrated (most restricted audience)
NC-17	7	Adult audience
R	6	Mature audience
	5	Mature teenage audience
PG-13	4	Teenage audience
PG	3	Mature young audience
	2	Most audiences
G	1	General (unrestricted audience)

The DVD Forum has not established an official correspondence to classification systems of other countries, such as the United Kingdom's BBFC, Germany's FSK, or France's FSS. In any case, most discs released outside the United States show a classification code on the package but do not set a corresponding parental level on the disc.

Parental level support is optional for players, but most support it. Parents can set a password or code in the player to prevent the setting from being changed, but resetting the player generally clears the password. A disc with no parental management information on it is treated by some players as being at the highest level of restriction.

Parental levels restrict the playback either of an entire disc or of certain scenes. Parental codes are placed on the disc in each parental block (a group of PGCs) so that the player can automatically select the proper path from scene to scene. This allows multiple ratings versions of a movie to be put on a single disc. For this to work, the video must be broken carefully into scenes. Objectionable scenes must be coded so that they can be skipped, or alternate versions of the scenes must be provided and coded appropriately. Video, audio, and subpictures need to be orchestrated so that there is no discontinuity across scene splices. Obviously, this is a significant undertaking. As a result, few discs are produced with multiple versions for different parental levels.

Because chapters can exist within a parental block, there may be two versions of the same chapter in a title, each assigned a different parental management level and in a different parental block. For example, a child who logs in and plays the disc would see one version of chapter 3, and an adult who logs in would see a different version, assuming that the authoring application supports parental management levels.

A command can be inserted in the middle of a video sequence to temporarily change the parental management level. If the current level is too low for the new level, the player will pause and ask the user to change the level. If the user is unable to raise the level, playback will stop. Temporary parental levels generally are authored as angle blocks, so a scene in a film may have two versions, one rated for younger viewers and one for adults.

Seamless Playback

Seamless presentation, also called *multistory*, allows for the disc playback to follow multiple paths through largely the same material by jumping from place to place without a break or pause in the video (see Figure 9.29). This is useful for movies with alternate endings, directors' cuts, and so on.

Figure 9.29 Example of Seamless Playback

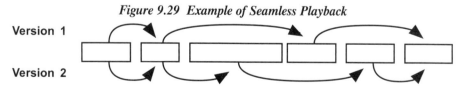

This seamless branching presentation style is accomplished in a manner similar to parental management. In fact, since commands exist to set the parental level, it is possible to use the parental features to control seamless branching. Camera angles may also be included in the seamless branching process. For example, a dual-language video that shows a scene of a book can be authored with multiangle sections wherever the book is shown. The angle can be set to match the language, and as playback moves from a single-angle section to a multiangle section, it will seamlessly select the correct angle. Alternatively, the video may be authored in *partial interleave* mode, where one PGC is used for one language and a second PGC is used for the other language. The different book scenes are interleaved together so that the PGC can jump over one set of cells while showing the other.

Seamless branching can be accomplished within a PGC. Cells may be contiguous or non-contiguous, but the distance between them is restricted depending on the data rate (see Table 9.33). At higher data rates, the distance must be smaller so that the pickup head can jump to the new position and begin reading data before the track buffer runs out. Seamless branching (and camera angles) generally impose a combined stream data-rate limit of 8 Mbps or lower. Also, video objects for seamless playback must be within a single PGC and on the same layer.

Table 9.33 Allowable Distance Between Cells for Seamless Playback

	Bit Rate (Mbps)			
	8.5	8	7.5	7
Maximum jump sectors	5,000	10,000	15,000	20,000
Minimum buffer sectors	201	221	220	216

Seamless branching generally cannot be done on the fly in response to user interaction. Each path is encapsulated by a PGC which contains cell pointers that link together the desired chunks of video. Each PGC must be created during the authoring process and stored on the disc. Cell commands cannot be used for seamless branching. Jumping between PGCs is also not guaranteed to be seamless.

Text

A disc can contain a wide variety of text data (TXTDT) that is identified for its purpose and the part of the content to which it applies. This includes information to be displayed for the user, such as title names, song names, and production credits; identifying information from the production process, such as version numbers and product codes; and extra information for enhanced use of the disc, such as Web links. DVD text is organized in a way that mirrors the logical hierarchy of the DVD volume (see Figure 9.30).

Text is stored in language blocks so that the contents can be described in different languages. There is no requirement that the information be the same in each language. For example, detailed information could be presented about every title and chapter in English but only title information in other languages.

Figure 9.30 DVD Text Data Example

Text	Identifiers	Structure
	<Volume>	Volume
DVD Demystified 3	Name	
Omnibus	General category	
	<Title>	Title 1
About DVD	Name	
	<Title>	Title 2
Recipe 4 DVD	Name	
	<Chapter>	Chapter 1
Introduction	Name	
	<Chapter>	Chapter 2
Plan	Name	
	<Chapter>	Chapter 3
Encode	Name	
	<Chapter>	Chapter 4
Author	Name	
	<Chapter>	Chapter 5
Multiplex	Name	
	<Chapter>	Chapter 6
Deliver	Name	
	<Title>	Title 3
In A New York Minute	Song	
Eagles: Hell Freezes Over	Album	
Music	General category	
Pop Music	Music category	
	<Audio stream>	Audio 1
DTS 5.1	Name	
	<Audio stream>	Audio 2
Dolby Digital 2.0	Name	

<> = structure identifier

DVD has two types of text items: *structure identifiers* and *application content items*. Each text item includes a numerical code that identifies it. Text items with a code of 1 through 32 are structure identifiers. They have no associated text. The numerical code identifies the logical structure to which any following content items belong (see Table 9.34). A hierarchical structure is used that corresponds closely to the logical structure of a DVD disc contents: volume, title, chapter, etc. Codes above 32 identify content items that hold text information. The exact way in which content strings are used is not rigidly defined, so DVD authors may use them in various ways.

Correspondence of text items to logical objects on the disc is established by the number of times the structure identifier is repeated. For example, the first occurrence of a title structure identifier indicates that all following content items apply to title 1. The second occurrence of a title identifier indicates that all following content items apply to title 2. The first occurrence of a chapter identifier after a title identifier indicates that all following content items apply to chapter 1 of that title, and so on.

DVD text is optional. It does not have to be included on the disc, and players are not required to do anything with it. Historically, DVD text has been used almost exclusively on karaoke discs, and these discs usually use structure identifiers 1 and 2 and content type 48. However, as players and jukeboxes become more sophisticated, and as computer-based DVD players proliferate, DVD text is becoming more important.

Table 9.34 DVD Text Data Codes

Item	Code (hex)	Code (decimal)	Description
General Structure Identifiers			
Volume	01h	1	Indicates that following items pertain to the entire disc side
Title	02h	2	Indicates that following items pertain to a title
Parental ID	03h	3	Indicates that following items pertain to a particular part of a parental block
Chapter	04h	4	Indicates that following items pertain to a chapter
Cell	05h	5	Indicates that following items pertain to a cell (usually one scene from a movie)
Stream Structure Identifiers			
Audio	10h	16	Indicates that following items pertain to an audio stream
Subpicture	11h	17	Indicates that following items pertain to a subpicture stream
Angle	12h	18	Indicates that following items pertain to an angle block
Audio Channel Structure Identifiers			
Channel	20h	32	Indicates that following items pertain to one channel in an audio stream
General Content			
Name	30h	48	The most common name for the volume, title names, chapter names, song names, and so on
Comments	31h	49	General comments about the title, chapter, song, and so on
Title Content			
Series	38h	56	Title of the series that the volume, title, or chapter is part of
Movie	39h	57	Title of the content if it is a movie
Video	3Ah	58	Title of the content if it is a video
Album	3Bh	59	Title of the content if it is an album
Song	3Ch	60	Title of the content if it is a song
Other	3Fh	63	Other title of the volume, title, or chapter if it belongs to some other genre or category
Secondary Title Content			
Series	40h	64	Secondary or alternate title of the series that the volume, title, or chapter is part of
Movie	41h	65	Secondary or alternate title of the content if it is a movie

continues

Table 9.34 DVD Text Data Codes (continued)

Item	Code (hex)	Code (decimal)	Description
Video	42h	66	Secondary or alternate title of the content if it is a video
Album	43h	67	Secondary or alternate title of the content if it is an album
Song	44h	68	Secondary or alternate title of the content if it is a song
Other	47h	71	Secondary or alternate title of the volume, title, or chapter if it belongs to some other genre or category
Original Title Content			
Series	48h	72	Original name of the series that the volume, title, or chapter is part of
Movie	49h	73	Original name of the content if it is a movie
Video	4Ah	74	Original name of the content if it is a video
Album	4Bh	75	Original name of the content if it is an album
Song	4Ch	76	Original name of the content if it is a song
Other	4Fh	79	Original name of the volume, title, or chapter if it belongs to some other genre or category
Other Info Content			
Other scene	50h	80	Additional information about a scene in a title or chapter
Other cut	51h	81	Additional information about a cut in a title or chapter
Other take	52h	82	Additional information about an alternate take in a title or chapter
Other label	53h	83	Other label for a volume, title, or chapter
Language	58h	88	Language of the volume, title, or chapter
Original Language	59h	89	Original language of the volume, title, or chapter
Work	5Ch to 67h	92 to 103	Color, shooting location, production company, and so on
Character	6Ch to 8Dh	108 to 141	Leading actor, leading actress, supporting actor, supporting actress, producer, conductor, orchestra, animator, and so on
Data	90h to 92h	144 to 146	Production, award, historical background
Karaoke	94h to 99h	148 to 153	Male melody, female harmony, and so on
Category	9Ch to 9Fh	156 to 159	Category
Lyrics	A0h to A2h	160 to 162	Lyrics
Document	A4h	164	Liner notes

continues

Table 9.34 DVD Text Data Codes (continued)

Item	Code (hex)	Code (decimal)	Description
Other	A8h	168	Other
Administration	B0h to C9h	176 to 201	Product number, SKU, ISRC, copyright, release date, encoding information, master tape information, and so forth
Vendor unique	E0h to Efh	224 to 239	Available for private use by disc author
Sorting	F0h	240	Special extension code for adding a second entry that provides an alphabetically or numerically sortable variation of the previous entry

DVD-Audio

DVD-Audio is a separate application format from DVD-Video. The DVD-Audio specification includes different data types and features (see Table 9.35), with content stored in a separate audio zone on the disc (the AUDIO_TS directory) (see Figure 9.1). The DVD-Audio format includes a subset of DVD-Video features to provide onscreen menus and motion video. DVD-Audio discs that include video are sometimes called *DVD-AudioV* discs.

Table 9.35 Comparison of Audio Features

Feature	DVD-Video	DVD-Audio
Number of streams	0 to 8	1 to 2
Channels/stream	1 to 8	1 to 6
PCM sample size	16, 20, 24 bits	16, 20, 24 bits
PCM sample frequency	48, 96 Hz	44.1, 48, 88.2, 96, 176.4, 192 Hz
Lossless compression (MLP)	No	Yes

Understanding the DVD-Audio specification first requires understanding the basics of the DVD-Video specification, because DVD-Audio inherits much of the architecture and navigation features of DVD-Video. Any use of video with the audio also depends on the DVD-Video format. This section of the book generally describes the parts of DVD-Audio that are new or different. Other features, such as karaoke and text data, are essentially the same for both formats, so they are only described in the preceding DVD-Video section.

DVD-Audio players come in two flavors: *audio-only players* (AOPs), such as portable players, and *video-capable audio players* (VCAPs) that also can show onscreen information and play a subset of the DVD-Video format, as defined in the DVD-Audio specification. In addition, there are *universal* players, or *DVD-Audio/Video* players, that play both DVD-Video and DVD-Audio discs. Audio-only players may have a simple display, such as a one-line liquid-crystal display (LCD), that can display text information (song name, artist name, etc.) and playback information (group number, track number, etc.). Audio-only players are not required to display still pictures or motion video from DVD-Audio discs. Since visual menus are not always available, a simplified interface allows direct access to tracks and groups of tracks.

It is also possible to make a "universal disc" that plays in all DVD-Audio and DVD-Video players and computers by including a Dolby Digital version of the audio in the video zone. Anyone producing a DVD-Audio disc should strongly consider investing the small amount of extra work necessary to make a disc that plays in all DVD players.

DVD-Audio players work with existing receivers. They output PCM and Dolby Digital, and some support the optional DTS and DSD (direct stream digital) formats. However, most current receivers are not designed with high-definition multichannel PCM audio. DVD-Audio players with high-end digital-to-analog converters (DACs) may be hooked up to receivers with two- or six-channel analog inputs, but some quality is lost if the receiver converts back to digital for processing. Improved receivers with advanced digital connections such as IEEE 1394 (FireWire) or HDMI are needed to use the full digital resolution of DVD-Audio.

Data Structures

The DVD-Audio specification defines physical and logical data structures similar to those in DVD-Video (see Figure 9.31). The audio manager (AMG) is the top-level table of contents for a disc. It contains the navigation data for accessing smaller audio units. The audio manager optionally may contain a DVD-Video-style video object that provides a visual menu for the disc. One or more audio title sets (ATSs) hold audio objects (AOBs) and also can contain visual menus. Each audio object contains one or more tracks of audio, a set of still images, and text.

Figure 9.31 DVD-Audio Logical Structure

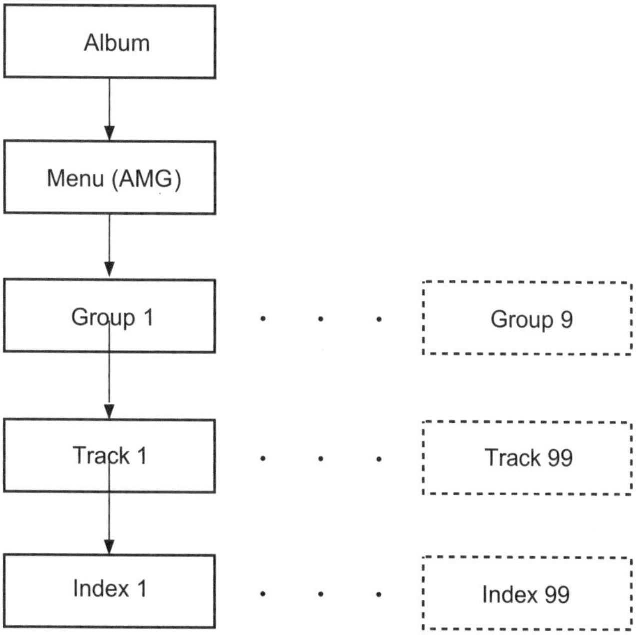

In general, one AOB is used for each track on the disc. An AOB is limited to two audio streams, usually a multichannel version and a stereo version. Still images related to an audio title set are stored in an audio still video set (ASVS). Still pictures are stored in the multiplexed stream in SPCT packs. Real-time text, such as synchronized lyrics, is stored in real-time information (RTI) packs in the multiplexed stream (refer to Figure 9.11).

The structure of AOB streams is simplified from video object streams, with less interleaving of data in order to support continuous high-resolution audio playback. Links can be defined in the DVD-Audio structure that tie it to the DVD-Video content in the video zone.

DVD-Audio discs are broken into groups of tracks (songs). Tracks on a DVD-Audio disc are similar to tracks on a CD; each one is usually a single song or musical performance. If there are many tracks or tracks of similar nature (such as a set of stereo songs and a set of the same songs in multichannel versions), they may be allocated into audio title groups (ATTGs). A disc (album) can have up to nine groups, although many discs have only one group. Each group can have up to 99 tracks. As with CDs, optional indexes allow direct access to points within a track. Each track can have up to 99 indexes. In DVD-Video terms, a group is a title, a track is a program (or a PGC), and an index is a cell (see Table 9.36).

Table 9.36 DVD-Audio Terminology Recommendations

Friendly Term	Technical Term from DVD-Audio Specification	Example	Equivalent in DVD-Video
Album	Volume	Album, disc	Volume
Group	Title group	Stereo play list, multi-channel play list	Title
Bonus group (enter four-digit key number to access)	Hidden group	Bonus tracks, hidden tracks	None
Track	Track	Song, cut	Program
Index	Index	Music marker	Cell
Spotlight	Spotlight	Highlighted lyrics	None
Album text	ATXTDT	Artist info, liner notes	TXTDT
Track text	RTXTDT	Lyrics, song credits	Note
Slideshow	Slideshow	On-screen lyrics	Slideshow
Browsable pictures	Browsable pictures	Gallery	None

A track may contain an audio selection block, which allows different channel configurations of the same music. The user can select stereo or multichannel, and the appropriate version of the track will be played. This is an alternative to including coefficients for downmixing a multichannel track to a stereo track.

DVD-Audio defines a *silent cell*, at index 0, that is the first cell in a program. Silent cells contain no audio data and are specially recognized by the player to provide a break between tracks. When tracks are played sequentially, the player "plays" the silent cell. When tracks are accessed randomly or skipped over, the silent cell is not played. If a still picture is associated with the silent cell, it will not be shown.

Navigation

Because DVD-Audio-only players have no video screen, and because users may wish to play DVD-Audio discs without turning on the television, a simplified navigation method is specified, called *simplified audio program play* (SAPP). The basic functions are play/pause, group access (jump to the first track of a group), track access (skip to the next or previous track), index access (skip to the next or previous index), and audio selection (choose between audio streams). Some of these functions, such as group access and index access, are optional. A set of still picture access commands also is defined: next, previous, and home. Next and previous are equivalent to "next page" and "previous page," whereas home is equivalent to "return." The disc author specifies which is the home image.

Bonus Tracks

The last group on a DVD-Audio disc may be designated for conditional access. The group is hidden and is not normally accessible to the user unless they know a four-digit access code. Once the user enters the access code, the tracks in the last group become playable.

The idea behind this feature is to provide bonus tracks that may be used for prizes or incentives. There could be a series of puzzles in the disc packaging (or even in the disc menus) that have to be solved to reveal the access code, or customers may need to visit a band or record label Web site to obtain it. Of course, any secret such as this will not remain a secret for long, especially in the age of the Internet. It is inevitable that Internet sites will post lists of all known access codes for DVD-Audio discs.

Audio Formats

Linear PCM (LPCM) is mandatory on DVD-Audio discs, with up to 6 channels at sample rates of 44.1, 48, 88.2, 96, 176.4, and 192 kHz, with sample sizes of 16, 20, or 24 bits. This allows a theoretical frequency response of up to 96 kHz and a dynamic range of up to 144 decibels. Multichannel PCM tracks are downmixable by the player, although at 192 and 176.4 kHz, only two channels are available. The maximum data rate for a PCM stream is 9.6 Mbps.

To provide for longer playing times without compromising audio quality, *Meridian Lossless Packing* (MLP) can be used to compress the data. MLP removes redundancy from the signal to achieve a compression ratio of about 2:1 while allowing the PCM signal to be completely recreated by the MLP decoder. DVD-Audio players are required to include an MLP decoder.

The other audio formats of DVD-Video (Dolby Digital, MPEG audio, and DTS) are optional on DVD-Audio discs, although Dolby Digital is required for audio content that has associated video. A subset of DVD-Video features is allowed on the disc, while those that are not are listed in Table 9.37. See Table 9.38 for compression ratios and Table 9.39 for playing times.

Table 9.37 DVD-Video Features Not Allowed in DVD-Audio

Branching (multistory)

Camera angles

Parental management

Region control

User operation control

Table 9.38 MLP Compression

PCM Source	Minimum	Typical
44.1 kHz, 16 bits	25%	50%
48 kHz, 16 bits	10%	50%
96 kHz, 20 bits	40%	55%
96 kHz, 24 bits	38%	52%
192 kHz, 24 bits	43%	50%

Table 9.39 DVD-Audio Playing Times

Audio Format	Playing Time (PCM)	Typical Playing Time (MLP)
6 channels, 96-kHz, 24-bit	45 minutes	85 minutes
5.1 channels, 96-kHz, 24-bit	45 minutes	105 minutes
5.1 channels + 2 channels, 96-kHz, 24-bit	33 minutes	75 minutes
2 channels, 192-kHz, 24-bit	1 hour	2 hours
2 channels, 96-kHz, 24-bit	2 hours	4 hours
2 channels, 44.1-kHz, 16-bit	7 hours	12 hours
1 channel, 44.1-kHz, 16-bit	14 hours	25 hours

Sampling rates and sizes can vary for different channels by using a predefined set of two groups, as shown in Table 9.40. For mono and stereo configurations, there is only one group. For other configurations, there are two . Group 1 always has sampling rates and sizes that are higher than or the same as group 2. Note that channel configurations 8 to 12 are the same as 13 to 17, other than the channel groupings. Groups apply to PCM streams and MLP streams.

TABLE 9.40 DVD-Audio Channel Assignments

Channel Assignment	Ch 0	Ch 1	Ch 2	Ch 3	Ch 4	Ch 5
0	C					
1	L	R				
2	L	R	S			
3	L	R	Ls	Rs		
4	L	R	LFE			
5	L	R	LFE	S		
6	L	R	LFE	Ls	Rs	
7	L	R	C			
8	L	R	C	S		
9	L	R	C	Ls	Rs	
10	L	R	C	LFE		
11	L	R	C	LFE	S	
12	L	R	C	LFE	Ls	Rs
13	L	R	C	S		
14	L	R	C	Ls	Rs	
15	L	R	C	LFE		
16	L	R	C	LFE	S	
17	L	R	C	LFE	Ls	Rs
18	L	R	Ls	Rs	LFE	
19	L	R	Ls	Rs	C	
20	L	R	Ls	Rs	C	LFE
	Group 1				Group 2	

Note: L = left; R = right; C = center; LFE = low-frequency effects; S = surround; Ls = left surround; Rs = right surround.

High-Frequency Audio Concerns

Most of the existing audio equipment was designed with a top-end frequency of 20 kHz in mind, based on performance limitations of vinyl records, cassette tapes, CDs, and so on. Now, with sampling frequencies of 176.4 and 192 kHz, DVD-Audio theoretically can reproduce frequencies of 88.2 or 96 kHz, with attendant supersonic signals. Existing equipment generally has not been tested for power-handling capacity when processing sustained supersonic signals. System components such as tweeters and supertweeters, passive speaker crossover networks, and power amplifier output stages are particularly vulnerable to overheating and damage from high-frequency signals.

The DVD Forum, on advice from audio experts, has recommended that warnings be placed on test discs containing test signals with very high frequencies or very large amplitudes. The DVD Forum also recommends that DVD-Audio players include a protection device, such as a low-pass filter. The protection device may be turned off when the player is used with equipment that has been adequately tested. Audio engineers and developers of audio mastering equipment also should take care not to introduce dangerous high-energy signal levels that might come from sources such as excessive noise shaping in signal processing or conversion, aliasing from upsampling without adequate filtering, equipment or hookup faults producing electronic oscillations in the supersonic region, and supersonic output from electronic instruments.

Downmixing

DVD-Audio includes specialized downmixing features for PCM channels. PCM downmixing was included in DVD-Video but was never well defined or well supported. DVD-Audio includes coefficient tables to control mixdown and avoid volume buildup from channel aggregation. Up to 16 tables may be defined by each audio title set (album), and each track can be identified with a table. Coefficients range from 0 to 60 decibels. This feature goes by the horribly contrived name of SMART (system-managed audio resource technique).

MLP includes a separate downmixing feature. MLP downmix coefficients can range from $+6$ to -24 decibels or 0 (infinite attenuation). In place of downmixing, the coefficients can be used to control positive or negative phase.

Stills and Slideshows

DVD-Audio defines the oddly named term *audio still video* (ASV). An ASV is a still image (an MPEG I frame) that is displayed on a video screen while the audio plays. ASVs may include buttons, in which case they include subpicture and highlight information similar to that of DVD-Video. ASVs can be presented synchronously in slideshow mode or asynchronously in browse mode, depending on how the disc was authored. Slideshow mode uses predefined display times for each image, synchronized to the music. Browse mode allows the viewer to step forward and back through the images, independent of the music playback. In either mode, the images may be displayed either in sequential order or in random or shuffle order.

There can be up to 99 still images per audio track. The still pictures are loaded into a 2MB buffer in the player before the track begins playing, so the number of stills depends on how heavily compressed they are. At typical compression levels, there is room for about 20 stills per track. A predefined set of transitions can be used between stills: cut in, cut out, fade in, fade out, dissolve, and six wipe patterns.

Text

DVD-Audio provides two types of text: general text (*album text*) and real-time synchronized text (*track text*). General text data is stored in a single place on the disc and uses the same format as DVD-Video, covered earlier in this chapter. It is intended to hold general information about the content of the disc, such as album and track names, artists, producers, and ISRC or UPC_EAN codes. General text data also can hold URLs of related Web sites.

DVD-Audio real-time text is stored in packets along with the audio and video. It is intended to be displayed onscreen (or on the small display of an audio-only player) while the music plays, showing lyrics, credits, song notes, and other kinds of information directly related to the music. In a way, real-time text is like a character-based version of the DVD-Video subtitle feature.

Characters are encoded in ISO 8859-1 (Latin) for alphanumeric European languages or in Music Shift JIS for the Japanese language. Audio-only players can display European (single-byte) text using 4 lines of 30 characters each. Japanese (double-byte) text is displayed on 2 lines of 15 characters each.

Super Audio CD (SACD)

Sony and Philips (both members of the DVD Forum) developed an independent high-fidelity audio format for DVD media called *Super Audio CD* (SACD). SACD uses direct stream digital (DSD) encoding with sampling rates of up to 100 kHz. DSD is based on the pulse-density modulation (PDM) technique that uses single bits to represent the incremental rise or fall of the audio waveform. This supposedly improves quality by removing the "brick wall" filters required for PCM encoding. It also makes downsampling more accurate and efficient. Many modern analog-to-digital conversion systems use delta sigma oversampling, which is the technology used by DSD. In conventional digital audio systems, the bit stream is decimated to a PCM representation. Since DSD retains the 1-bit signal, it avoids decimation and interpolation steps, which can result in a more accurate and natural representation of the original waveform. DSD provides frequency response from DC to more than 100 kHz, with a dynamic range exceeding 120 decibels. DSD includes a lossless encoding technique that produces approximately 2:1 data reduction by predicting each sample and then run-length encoding the error signal. Maximum data rate is 2.825 Mbps.

SACD is intended to provide "legacy" discs that have two layers: a CD-style layer that can be read by existing CD players and a DVD-style layer for SACD players. A 0.6 mm substrate containing CD data on the outside surface is bonded to a 0.6 mm DVD substrate. This places the CD content at the standard 1.2 mm distance from the read-out surface. The DVD layer is designed to be reflective at 650 nanometers (DVD laser wavelength) but transparent at 780 nanometers (CD laser wavelength). SACD players must be designed to recognize the DVD layer rather than first checking for CD data and assuming that the disc is a CD. Technical difficulties have limited the release of dual-format discs, especially considering that the initial price of these dual-layer discs was higher than for a standard CD plus a standard DVD.

Both stereo and multichannel versions of each track can be provided. SACD includes text and still graphics, but no video. SACD includes a physical watermarking feature called *pit signal processing* (PSP), not to be confused with PlayStation Portable, discussed later in this chapter. PSP modulates the width of pits on the disc to store a digital watermark. Since normal user data are stored using the pit length, modifying the pit width has no effect on the data. The optical pickup must contain additional circuitry to read the PSP watermark, which is then compared with information on the disc to make sure that it is legitimate. Because of the requirement for PSP detection circuitry, protected SACD discs cannot be played in standard DVD-ROM drives.

SACD technology is available to existing Sony/Philips CD licensees at no additional cost. Pioneer, which released the first DVD-Audio players in Japan at the end of 1999, included SACD support in its DVD-Audio players. If other manufacturers were to follow suit, the entire SACD versus DVD-Audio format debate would become moot because DVD-Audio players will play both types of discs. Once again, the marketplace appears to be proving itself as the final arbiter, as neither format has garnered appreciable market share.

DVD Recording

The DVD recording application formats are designed to work on all the various writable DVD versions. Because the DVD Forum does not endorse DVD+RW, no effort has been made to ensure compatibility, but there is no reason the recording formats cannot be used with DVD+RW media.

Of course, compatibility issues do exist, both at the physical level and at the application level. More than 15 million DVD-Video players (more than 50 million, counting computers) were in the marketplace before the first wave of DVD recorders hit in late 2000 and early 2001. None of those 50 million existing players could read the DVD recording formats. The recording formats have been largely supplanted by unofficial use of the DVD-Video format for recording. Clever formatting tricks allow recorders to use the DVD-Video format for real-time recording with only a quick fixup pass at the end to update content pointers. Some of the fancy features of the DVD recording formats, such as custom playlists, are not available when using the DVD-Video format, but this pales against compatibility with existing players and computers. The various compatibility issues are covered in more detail in the compatibility sections of Chapter 6.

DVD-VR

The DVD Video Recording application format is designed to record video from analog and digital sources in real time. The original DVD-Video format requires that some information, such as time maps, pointers to video objects, and so on, be determined after the size and running time of the video are known. When recording real-time streaming data, this information is not available, so DVD-VR includes alternative ways to include navigation and search information. Instead of VTSI time maps and multiplexed navigation packs that DVD-Video and DVD-Audio use, DVD-VR defines a new VOBU map that stores timestamps along with the size of the video recorded since the last timestamp. The player can then calculate access points by adding up sizes. VOBU time maps are stored in video object information (VOBI) structures in a new object management layer that sits between the PGCI and the VOB levels. The object management layer also includes video object group information (VOGI) structures to keep track of still pictures.

New picture coding resolutions of 544×480 and 480×480 are defined for NTSC signals, and new still picture VOBs allow easy recording of MPEG I-frames for use with digital cameras. Audio recording in PCM is limited to one or two channels.

DVD-VR includes a data structure to record vertical blanking interval (VBI) signals that contain NTSC Closed Captions, PAL Teletext, aspect ratio signals, APS, CGMS, Web links, and so on. VBI data are recorded in real-time data information (RDI) packs in the multiplexed stream.

DVD-VR also supports user editing of recorded video. Users can delete recorded programs and create custom playlists. A *playlist* is a collection of time pointers — a user-defined program chain — that allows video scenes to be presented in any order without modifying the original recorded data. After programs are recorded, DVD-VR recorders can create menus to provide instant access to each program. Using standard DVD-Video menu features, they can even create menus with thumbnail stills representing each video program.

DVD-AR

The DVD Audio Recording application format has still not been officially approved. It is a modified version of the DVD-Video and DVD-Audio specs designed for real-time audio recording from analog and digital sources. The navigation structure is taken from DVD-Video, and the presentation structure is taken from DVD-Audio. Of course, it would be too much to hope that the discs would be compatible with DVD-Audio or DVD-Video players.

Incoming audio signals can be sampled and recorded at the same resolutions as DVD-Audio (44.1, 44, 88.2, 96, 176.4, or 192 kHz, and 16, 20, or 24 bits). Only one audio stream can be recorded at a time. The audio can be stored as uncompressed PCM or in compressed formats such as Dolby Digital and MPEG audio.

Still pictures can be recorded along with the audio as MPEG-1 or MPEG-2 I-frames. A memory buffer, as in DVD-Audio players, allows a group of still pictures to be displayed without causing interruption of the audio. Text information (album name, track name, lyrics, and so forth) also can be recorded, along with the audio as album text or as real-time track text, similar to DVD-Audio.

As with DVD-VR, user-defined play lists can be created to customize playback of the recorded audio. Segments in a play list can be added, deleted, reordered, combined, and divided, all without changing the original recorded audio.

DVD-SR

The DVD Stream Recording application format was introduced at the end of 2000 but has yet to be finalized. The general idea of DVD-SR is to provide a standardized way to record streaming data from any source, such as a digital satellite receiver, digital cable tuner, digital video camera, or even streaming content from the Internet. The recorder does not "understand" the contents of what it is recording, whether it is audio, video, stock tickers, or something else, thus it is incapable of directly playing back what it has recorded. It must pass the stream on to some other device that knows how to interpret, decode, and present the recorded content. DVD-SR players use IEEE 1394/FireWire to connect to other devices that supply or play back streams of digital content, and they also will implement content protection using CPRM and DTCP.

DVD-SR uses the same general structure as DVD-Video but is much simplified. Data is stored in stream recording objects (SROs). Instead of recording video packs, audio packs, subpicture packs, and so on, DVD-SR records only application data packs. Incoming data, such as MPEG transport streams, MPEG program streams, Microsoft ASF streams, etc., is chopped into 1998-byte chunks and wrapped in packs. If the recorder is designed to recognize high-level information in the incoming stream, such as timestamps, it can generate additional navigation data to support searching and trick play (fast forward, reverse, slow, and so on).

DVD-SR content may be mixed with DVD-VR content. They can share play lists and can have mixed program chains that include both VR cells and SR cells. Draft versions of the DVD-SR format defined streaming objects, or SOBs, causing much hilarity among those more familiar with the English language.

Universal Media Disc (UMD™)

First released in December 2004 in Japan, followed in March 2005 by its introduction in the United States, the *Universal Media Disc* (UMD) format is experiencing a significant and surprising expansion in the marketplace. UMD is a proprietary disc format created by Sony Computer Entertainment to support their new PlayStation Portable (PSP) game console. Measuring only 6 cm across and stored in an isometrically shaped caddy, UMD discs contain two data layers which are capable of holding up to 1.8 gigabytes of data — PSP games, video or audio. Within the first nine months of its release, Sony has sold more than five million PSP consoles in the US and Japan, and the European market is due to open in September 2005. By March 2006, the installed base is expected to reach 16 million units, with a market size projected to be in excess of 100 million units over the next five to ten years.[11]

Compared to other portable gaming systems currently available, the PlayStation Portable is uniquely designed to play movies and other video content, in addition to games. One of the most surprising points about the PSP market is that UMD Video products (movies authored for exclusive use by the platform) have begun to outsell PSP games. Sporting an impressive

---

[11]PSP sales are expected to follow the same pattern as the PlayStation 2, which sold 90 million units in the five years after its launch.

widescreen format LCD display, light in weight and with a long battery life,[12] the PSP has become a wildly successful portable video platform. Several major Hollywood studios have begun releasing titles on UMD Video, with some individual titles quickly selling more than 100,000 discs and counting.

UMD Video

So what is UMD Video? Table 9.41 shows the general specifications for the format.[13]

Table 9.41 General Characteristics of UMD Video Presentation Data

Disc capacity	1.8 Gbytes
PSP display size	480x272 widescreen
Video codec	MPEG-4 AVC Main Profile Level 3
Picture size	720×480, 352×480, 352×240[a]
Display aspect ratio	4:3 or 16:9 anamorphic
Frame rate	23.976 fps (progressive), 29.97 fps (interlaced)
Audio codec	ATRAC3plus
Audio channel configuration	2-ch. (6-ch., 7.1-ch.)[b]
Audio sampling frequency	48 kHz
Maximum audio bitrate	128 kbps (768 kbps)[c]
Subtitle data compression	4-bit PNG[d]
Stream structure	MPEG-2 Systems Program Stream
Max. data rate	10.0 Mbps
Max. number of video streams	9
Max. number of audio streams	8
Max. number of subtitle streams	32
Estimated playing time	135 minutes[e]

[a]Progressive only.

[b]Although the codec supports up to 7.1-channel encoding, UMD only supports stereo.

[c]Although the codec supports up to 768 kbps encoding, 128 kbps is the maximum per stream for stereo.

[d]Although 16-color subtitles can be used, 4-color subtitles are still common.

[e]Estimated playing time based on 1 video, 1 audio and 5 subtitle streams.

To the unsuspecting eye, UMD Video looks quite similar to DVD-Video. This was intended to help ease the transition from DVD to UMD for content producers. However, UMD Video differs greatly in many important ways. For example, in order to squeeze 135 minutes of video onto a 1.8 GB disc, UMD employs MPEG-4 AVC (Advanced Video Coding) tech-

[12]The PSP display is a 4.3-inch, 16:9 widescreen TFT LCD, 480×272 pixels, producing 16.77 million colors. The approximate weight of the PSP is 260 grams, including the battery.

[13]Despite Sony's use of separate branding for UMD Video and UMD Audio discs, both fall under the UMD Video format.

nology. Although AVC is considered to be twice as efficient as MPEG-2, the encoding process requires pushing the video data rate down to approximately 1.5 Mbps (roughly equivalent to 3 Mbps MPEG-2). Oddly enough, the video is typically encoded from a 720×480 anamorphic source (squeezed for 16:9 display), even though the PSP screen size is only 480×272 pixels. The PSP scales the decoded video to less than 40% of its original resolution during playback. Fortunately, the device includes a powerful graphics engine that is able to perform the scaling operation while maintaining excellent quality.

While AVC is able to provide dramatic improvements in efficiency for compressing video, UMD's ATRAC3plus[14] audio codec is not much more efficient than Dolby Digital for DVD. Whereas a Dolby Digital 2.0 stream would typically compress using 192 kbps as the data rate, the corresponding ATRAC3plus track would require using 128 kbps as the data rate. As a result, additional audio streams will have a dramatic impact on UMD disc space. For example, adding a second 135-minute stereo audio stream to a UMD Video project would consume more than 120 MB of additional space (approximately 7% of the overall disc capacity).

Subtitle streams operate in a similar fashion on the PSP as they do for DVD-Video. Given the smaller screen size and 4-bit image compression, the data rate of a typical subtitle stream is about 10 kbps, approximately one-fourth of an equivalent DVD-Video subtitle stream. For this reason, UMD Video titles will often use additional subtitle streams to provide localization of movie content for different regions rather than providing alternate language audio streams.

Organizational Structure

Similar to DVD, UMD Video separates the physical layout of multiplexed A/V streams from the logical arrangement that defines how those streams are played back. This separation results in three primary layers (see Figure 9.32).

Figure 9.32 Structural Organization of UMD Content

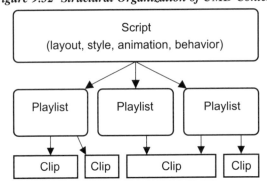

---

[14]ATRAC3plus is a successor to the ATRAC and ATRAC3 coding formats used on Sony MiniDisc (MD) and Memory Stick Audio products.

Script. The script layer defines all of the user interface, playback and navigation controls for the UMD disc. This includes the graphic and audio elements, markup files, and programmatic scripting that determines how the content appears, how it behaves, and how the user may interact with the material on the disc. In this sense, the script layer loosely corresponds to the definition of menus, button highlights, and command sequences in DVD-Video, but with some major differences. For example, the script layer provides full playback control and can manipulate the user interface, such as displaying new graphics and programmatically animating them.

Playlist. A playlist in UMD Video is roughly equivalent to a program chain in DVD-Video. A playlist defines the playback order of clips and may be composed of one or more clips or portions of clips called *play items*.[15] These playlists are stored on the disc and called by the interactive scripts when it is time to play the video.

Clip. This layer corresponds to the *Clip AV Stream* files, which contain the multiplexed audio, video and subtitle stream data, similar to the VOB files on a DVD-Video disc. The clip layer also includes the *Clip Information* files, which provide an index of valid entry points into the streams, mapping time values to specific byte offsets within the program stream.

Interactivity on UMD

When it comes to menus, UMD Video diverges significantly from its DVD counterpart. Because the PSP has a powerful graphics and sound engine for games, the format is able to leverage this in its presentation of menus. Instead of limiting the unit to one full screen graphic or video at a time, as in DVD, the PSP allows you to dynamically assemble graphic, audio and video elements in real-time during playback. For example, instead of a single video background with a subpicture overlay that defines button highlights as you have in DVD, you can layer multiple graphic elements with varying amounts of transparency over a video background and mix in sound effects, such as button clicks. Each graphic element may be independently positioned, scaled, rotated, and animated over time. Element color and opacity can be manipulated by script, and even pseudo-3D graphic transforms (sometimes referred to as "2.5D") may be applied to create depth and perspective.

The layout, appearance and behavior of interactive menus in UMD Video is defined by a combination of HTML-like markup and program scripting language. An XML derivative of XHTML is used to define the layout of elements on screen, using a page metaphor similar to web pages. However, this XML format also includes tags (keywords) for defining style (appearance) and timing (animation). An ECMAScript-based scripting engine is used to define all of the dynamic behavior, such as navigation among buttons, video and audio playback control, sound effects, and complex animations.[16]

Each element on a UMD menu is referred to as a "widget." There can be up to 272 widgets in existence at any one time, including every page element and graphic (visible or not) that is defined in the current XML file. As the number of widgets increases, more process-

---

[15]Initial versions of the UMD Video player only support one play item per playlist.

[16]ECMAScript is an open standard language that is derived from Netscape's JavaScript language. See Myths: "Javascript is Java" and "Javascript is the same as ECMAScript" in Chapter 11.

ing power is required to maintain them. Eventually, this begins to starve the audio and video decoders of critical processing power needed to maintain smooth playback of the audio and video elements. In addition, the loss of processing power impacts the speed with which animations can be presented, so there are definite practical limitations to what may be accomplished.

All of the XML code, ECMAScript programming, graphic, and audio elements for a menu set are packaged together in a *resource file*. The multiplexed video, audio and subtitles remain separate. Each resource file is limited to a size of 512 KB, but multiple resource files may be stored on a disc. Only one resource file can be loaded and active at a time, and it can take several seconds to switch to another resource file (during which all audio and video will be paused). When a resource file is loaded, the graphic elements it contains are decoded to a portion of memory called the *surface pool*. The surface pool has an upper limit of 3 MB, so all of the uncompressed graphic images within a resource file need to fit within this amount of memory, while a small amount of the memory space is needed for system overhead. The producer of a UMD Video disc must constantly balance these limitations on widget count, file size and memory capacity while trying to achieve the intended look and behavior of the disc.

What UMD Is Not

If you are a content producer, your mouth might be watering at the thought of using all the features of a game console for supplementing your video program! Well, keep in mind that the PSP platform is a restricted space. Just because the console has certain capabilities does not necessarily mean that you get to use all of them. For instance, the PSP includes wireless networking and persistent storage. However, as an author of UMD Video content, you have no access to the networking capabilities, and extremely limited access to persistent storage. (The PSP offers a set of general player registers that may sometimes be preserved across playback sessions.) In addition, although the PSP includes an impressive 3D engine capable of real-time playout of complex scenes, UMD Video content is still limited to pseudo-3D operations with no real capability of constructing meaningful 3D objects or environments. Likewise, although the PSP has a dynamic text rendering engine, no mechanism is provided that allows UMD Video authors to render text within the application.

UMD does not, and probably never will, have a recordable version. In order to play UMD Video content, it must be molded onto replicated discs in an encrypted form. The custom security mechanisms built into PSP players require that UMD Video content only come from the UMD drive (i.e., not a Memory Stick or network port), and that it must be properly encrypted in order to play. This has been done to help protect against any illegal piracy of UMD content, or the ripping of DVDs to UMD format. If the security on UMD Video discs is breached, the security mechanism can be updated via a firmware upgrade of the player. To ensure that consumers update their PSP firmware, a *forced update* program is in place, which requires that the PSP be updated before it can play newly released UMD titles.

Where Is UMD Going?

As part of the UMD security plan, Sony has stated that the UMD Video format is intended to be a portable video format only, and that it will never be playable on a normal television. Some people argue that a proprietary format entirely controlled by one company and limited in scope is destined to fail. Yet, others argue that this limited scope of UMD Video helps guarantee that UMD titles do not poach the sales of corresponding DVDs. As a result, UMD sales are considered by many to be an incremental increase in the existing home video market, thus ensuring its future existence. The jury is still out, however, and all generally agree that the sales numbers for UMD Video titles at the end of 2005 will demonstrate whether the format can be expected to have any real longevity.

HD DVD

HD DVD represents the next major expansion of the DVD family of formats, starting with the publication of the first HD DVD-ROM specification in November 2003, which was quickly followed by a recordable version specification. HD DVD is based on much of the fundamental technology of standard DVD. Using the familiar 12 cm disc with data sandwiched between two 0.6 mm substrates, HD DVD uses a shorter wavelength blue laser to read finer pits and lands on a disc, allowing for a significant increase in disc capacity. With the physical disc format specifications completed in June 2004, work began in earnest to complete the application layer specification for this new high-capacity disc.

HD DVD-Video

Work on the HD DVD-Video specification started as a simple extension of the DVD-Video specification that would add support for high definition video and fix some of the more glaring problems with the original specification, although it soon became clear that the new format would need to go far beyond what DVD-Video had become. With the DVD industry exceeding $10 billion in revenue and beginning to overcome the theatrical market, there was a concern that DVD's successor might not be compelling enough to drive new business and match the surprisingly successful growth curve that DVD enjoyed.

In March 2003, the DVD Forum convened a group representing major motion picture studios and both hardware and software player manufacturers to compile a list of functional and performance requirements that a new format should meet in order to increase the chances for success (see Table 6.9). In time, the DVD Forum mandated that the new format would comply with the requirements, which included features like advanced programmability, support for networking and downloadable content, enhanced graphics capabilities, and secondary audio and video decoding. Further, the DVD Forum concluded that the HD DVD-Video format would need to address both video and audio applications, thus becoming the joint successor to both the DVD-Video and DVD-Audio formats. Table 9.42 shows the general specifications for the resulting format.

Table 9.42 General Characteristics of HD DVD-Video Presentation Data

Disc capacity	4.7 Gbytes / 8.54 Gbytes for 3X DVD-ROM (red laser) 15 Gbytes / 30 Gbytes for HD DVD-ROM (blue laser)
Video codecs	MPEG-4 AVC: HP@4.1 (HD) and HP@3.2 (SD) SMPTE VC-1: AP@L3 and AP@L2 MPEG-2: SP@ML, MP@ML and MP@HL
Picture size	1920 × 1080 down to 352 × 240[a]
Display aspect ratio	4:3 or 16:9 anamorphic
Frame rate	29.97 / 59.94 fps (NTSC regions) 25 / 50 fps (PAL regions)
Audio codecs	Linear PCM Dolby Digital Plus DTS-HD[b] (CBR mandatory, VBR lossless optional) Dolby TrueHD (aka MLP) MPEG-1 and MPEG-2 (mandatory for PAL regions) aacPlus v2, mp3 and WMA Pro (optional for streaming)
Audio channel configuration	1.0 up to 7.1 channels[c]
Audio sampling frequency	48 kHz, 96 kHz and 192 kHz
Subtitle	2-bit and 8-bit subpictures (run-length encoded) Text-based Advanced Subtitles (OpenType font rendering)
Stream structure	MPEG-2 Systems Program Stream
Max. data rate	30 Mbps
Max. number of video streams	9
Max. number of audio streams	8
Max. number of subtitle streams	32

[a]HD DVD-Video supports approximately 40 different picture size and frame rate combinations, including 720 × 480 (NTSC), 720 × 576 (PAL), 1280 × 720 (60/50p), and 1920 × 1080 (60/50i).

[b]DTS-HD LBR (low bit rate) is another version of the codec that has been approved as an option for streaming audio content.

[c]A wide range of possible channel configurations exist, including mono, stereo, 5.1 and 7.1, among others.

Video Formats

Unlike DVD-Video, which supports a single video codec (MPEG-2), the DVD Forum has mandated that HD DVD-Video support Four separate video codecs — MPEG-4 AVC (aka H.264), SMPTE VC-1 (based on Microsoft's Windows Media Video codec), MPEG-1 and MPEG-2. This means that every HD DVD-Video player will need to include decoders for all three codecs, although it is up to the content producer to decide which one(s) to use for their project.

The MPEG-2 codec was selected primarily because it is a mature, well-understood technology with a wide range of supporting tools, including encoders, decoders, verifiers and editors. In addition, the quality of MPEG-2 compressed video has continued to improve over the years, as it has been widely used in DVD and broadcast applications. Despite the improvements, however, the MPEG-2 coding structure and algorithms are limited in how efficiently they can compress video.

The SMPTE VC-1 standard, based on Microsoft's Windows Media Video codec, was developed well after MPEG-2, and has benefited from the lessons learned during the development of that technology. Implementations of Windows Media Video have been on the market for quite some time, though not specific implementations of SMPTE VC-1, which was just recently standardized. The VC-1 codec employs many of the same advanced coding techniques as those used in MPEG-4 AVC (described below), and is thus able to achieve similar coding efficiencies.

As video coding technologies have advanced, the MPEG-4 Advanced Video Coding (AVC) standard (ISO/IEC 14496-10) has risen to the top as one of the most promising next-generation video codecs. Considered to be twice as efficient as MPEG-2, AVC has greater flexibility and more coding tools and algorithms available to it, allowing it to achieve levels of quality that MPEG-2 and even VC-1 are unable to reach. MPEG-4 AVC requires dramatically more processing power than MPEG-2 for both encoding and decoding, and is a new codec with relatively few established toolsets available. On the other hand, AVC offers standard post-processing features such as a deblocking filter, which can dramatically reduce the appearance of coding artifacts. (Blocking was always one of the most noticeable artifacts of MPEG-2 video compression, particularly at lower data rates.)

Another feature of AVC that has been adopted by HD DVD-Video (but not its Blu-ray counterpart) is a post-process called *Film Grain Technology* (FGT), which provides special coding tools specifically for the efficient handling of film grain. It was determined that for high definition video, film grain is an important component to the movie watching experience. Many directors and producers specifically choose film stocks in part because of the quality of the film grain. This, in turn, becomes part of the visual narrative that communicates the director's vision to the viewer. For this reason, the MPEG-4 AVC specifications include a standardized mechanism for transmitting film grain characteristics in the form of a *supplemental enhancement information* (SEI) message. HD DVD-Video implements a post-process that decodes this information and uses it to simulate the original look of the film grain and apply it to the decoded picture. This makes it possible for encoders to characterize and then filter out the film grain prior to encoding, and then restore the film grain characteristics via the very efficient film grain SEI message. As such, the encoder can concentrate on the actual picture and avoid trying to also compress the film grain in the picture. The end result is a higher quality picture at lower data rates which still conveys the experience that the director intended.

Audio Formats

As with video, the DVD Forum has mandated a plethora of audio codecs for HD DVD. The Dolby Digital codec of DVD-Video has been updated to Dolby Digital Plus, which is a superset of the original Dolby Digital codec (meaning you can still use Dolby Digital compressed audio on HD DVD-Video without modification). The DTS codec, which was optional for DVD-Video, has now been mandated as DTS-HD Constant Bit Rate (CBR), which is another superset of its predecessor. Meridian Lossless Packing (MLP), a required codec in the DVD-Audio format, has been renamed Dolby TrueHD and has also been adopted as a mandatory codec in HD DVD-Video. In addition, a whole suite of optional codecs has been approved, including the DTS-HD Variable Bit Rate (VBR) statistically lossless extensions and Low Bit Rate (LBR) extensions for streaming content. Also, aacPlus v2, mp3, and Windows Media Audio Pro (WMA Pro) have all been approved as optional codecs for streaming content.

So what does all this mean? It means that players will be required to support several audio codecs, and have the option to support numerous others. In the end, the market will decide which codecs get used and which do not. As for the mandatory codecs, they each address a specific market segment. For example, Dolby Digital Plus is still a solid, relatively low data rate codec that will probably remain a mainstay for easily adding lots of audio streams to a disc. DTS-HD, on the other hand, provides for that next level up in data rate for high quality surround sound mixes, and will provide a key marketing component for next-generation players and discs. Dolby TrueHD (aka MLP) had been selected as the codec of choice by the recording industry for DVD-Audio, and that choice has now carried over to HD DVD-Video, providing the highest quality audio reproduction available.

For streaming content, there will be a lot of competition. On the one hand, you have well established codecs like mp3, which have been around for years and will probably be provided in a player simply because the player will most likely support mp3 CDs. On the other hand, you have newer codecs like aacPlus v2 and WMA Pro, which are able to offer better coding efficiency at lower data rates. Finally, you have relative unknowns like Dolby Digital Plus and DTS-HD LBR. Dolby Digital Plus has already been mandated in the player, which could make it an industry standard, but only if the necessary tools are available to the industry. DTS-HD LBR may also end up in the player merely due to favorable licensing conditions and the fact that DTS decoder chips will likely implement all of the DTS extensions "flavors" in order to have broader appeal. We will have to wait and see what the industry becomes comfortable with and which codecs become part of the *de facto* standard.

Content Types on HD DVD

The HD DVD-Video specification defines two types of content, *standard* and *advanced*. Standard content is essentially an extension of the original DVD-Video specifications. It provides for the additional video and audio codecs noted above, as well as 256-color subtitles (an improvement over the old 4-color subtitles), but the navigation structure and capabilities are relatively unchanged.

Advanced content defines a new paradigm for interaction and content production, with features like network connectivity, access to persistent storage, interactive graphics over video, audio mixing capabilities and secondary audio and video playback. An HD DVD-Video disc may include either standard or advanced content, or both (in which case the advanced content loads first). In fact, it is possible for advanced content applications to directly access standard content presentation data, or for an advanced content application to jump to standard content and later return. This combination allows for a wide range of application scenarios.

Standard Content

In general, the organizational structure for standard content is left unchanged from DVD-Video. There is still a *first play program chain* (FP_PGC), a *video manager* (VMGM) and multiple *video title sets* (VTS) containing titles, program chains, programs and cells. However, the number of allowed video title sets and titles has been increased from 99 to 511; the number of available *general parameters* (GPRMs) has increased from 16 to 64; and, additional operations for accessing the GPRMs have been added to allow for more sophisticated manipulation of program variables. Additional *system parameters* (SPRMs) have been added to properly accommodate many of the other changes that have been made.

The first play program chain has been modified to allow it to contain presentation like a language selection page for menus. In DVD-Video, the FP_PGC could only contain command sequences. The DVD-Video format's mechanism for automatically selecting menu language based on player preferences turned out to be less than adequate in practice. With some players not providing access to this setup information and others providing only a limited set of choices, most international DVDs opted to simply put all of the different language content together in one area and use an on-screen menu to choose, bypassing the built-in mechanism. For HD DVD-Video, it was decided that the FP_PGC could house just such a menu, and then be permitted to change the preferred menu language system parameter, which allows the built-in menu selection mechanism to be meaningful again.

The sometimes confusing restrictions on content in the menu domain have also been relaxed. In DVD-Video, the menu domain (which includes the video manager and VTS menu domains) was limited to only allow access to a single audio, video and subpicture stream, unlike the title domain, which allowed access to the full eight audio streams, nine video streams and 32 subpicture streams. In HD DVD-Video, this arbitrary restriction has been removed. All domains now have access to the full range of audio, video and subpicture streams, including both seamless and non-seamless multiangle functionality. Additional system parameters have been added to track these values in each domain so that menus can operate more or less independently of the title domain.

Some restrictions on seamless multiangle video have also been relaxed. Seamless multiangle allows a disc to have multiple parallel video streams whose playback may be switched on-the-fly between navigation command or user operation. This "edit your own video" capability has become a popular feature for concert DVDs as well as other content where it is logical to let the user control the point of view. The restriction in standard DVD, however, was that when the video angle changed, the audio had to remain identical, thus allowing the video to play seamlessly (without pause or interruption). With the new HD DVD-Video specifications, this restriction has been relaxed to allow Dolby Digital Plus and DTS audio content to vary among the different angles. So, for example, if you had a multiangle concert video with close-ups of each musician, you can now have the audio stream in each angle emphasize the musician that you are watching, yet still switch among angles seamlessly. This is possible because Dolby Digital Plus and DTS both incorporate filters and protection circuitry that can prevent unwanted pops or spikes in the audio which could cause speaker damage. However, the restriction remains in place for Linear PCM and Dolby TrueHD audio, which do not incorporate such protections.

Many DVD authors will be glad to see that DVD-Video's resume functionality has been revamped. Resume had been defined as an automated behavior, controlled almost entirely by the player, which could lead to inconsistencies in how complex DVD titles were authored. As a result, producers would have to jump through extraordinary hoops to detect when a resume operation had been triggered and then redirect the player to an appropriate path. With HD DVD-Video, resume behavior still exists, but more options are offered to the content author for controlling how it functions. For example, a new *resume command sequence* has been added to the definition of program chains, which allows up to 1,023 commands to be executed prior to resuming playback of a program chain. This command sequence could, for example, be used to reset audio and subtitle settings or redirect playback to another location, if needed. Each PGC in the title domain also includes a new resume permission flag, which indicates whether this PGC updates the resume information that is tracked by the player. When a PGC has this flag disabled, the PGC will have no impact on the resume information, so the next time the resume function is triggered, playback will resume only to the last PGC for which the resume permission flag was enabled.

Another convenient change to the specifications has come in the form of navigation command support for jumping to different parts of the disc. In DVD-Video, it was not possible to jump directly from one video title set to another without first going through the video manager. On more complicated discs, this resulted in a great deal of overhead as the player would have to frequently jump to a PGC in the video manager only to be immediately redirected to another PGC elsewhere. In HD DVD-Video, jump commands can link directly to the intended destination. Also, several of the navigation commands have been updated to allow GPRM values to define the destination of a link operation. For example, instead of having to use an immediate value to identify the specific PGC to play, the value may be read out of a general parameter by the link command itself. This allows for simpler command scripts and more sophisticated programming on the disc.

Finally, and perhaps most notably, the HD DVD-Video specification allows up to eight commands per button and cell command sequence, as opposed to DVD-Video's one. This allows producers to perform sophisticated actions within a button or cell command sequence that were previously not possible. Cell commands now execute seamlessly, which means that they will no longer interrupt the playback of video content, unless they specifically jump to another location. This modification to the DVD-Video specifications will allow producers to create more dynamic content while maintaining the seamless user experience that is so important.

Advanced Content

The most dramatic change between DVD-Video and HD DVD-Video is the addition of advanced content, referred to by the term iHD. Advanced content provides a totally new way to author interactive graphics and applications through XML markup and ECMAScript working in concert with a specialized video title set. *Extensible markup language* (XML) provides a standardized syntax for defining the declarative layout, appearance and behavior of content, while ECMAScript provides the programmability necessary for truly dynamic applications. Advanced content is composed of four primary elements (see Figure 9.33):

Playlist. Advanced content on an HD DVD-Video disc can have one or more *playlists*, which define the playback structure of the content. A playlist is an XML file that specifies the timelines for one or more titles (similar to the standard content concept of titles) and defines which audio, video and subpicture streams are mapped to each title from the advanced VTS as well as, which advanced applications and advanced subtitles are mapped to the timeline. In addition, the playlist defines where chapter points exist in the title timeline and where to begin playback.

Advanced Applications. An *advanced application* is a combination of XML markup, graphic and sound elements, and ECMAScript code that defines the layout, appearance, timing and behavior of an interactive application that runs within the time period of a title.

Advanced Subtitles. Essentially a restricted version of an advanced application, *advanced subtitles* provide a rendered text-based approach to presenting subtitles to the viewer. Being text-based rather than picture-based, advanced subtitles can be much smaller and more ideally suited for network delivery than, say, a subpicture stream.

Advanced VTS. The *advanced VTS* is a variation of the video title set used for standard content that provides the multiplexed presentation data (audio, video and subpicture streams) for playback. Playlists map the audio, video and subpicture streams defined in the advanced VTS to each title's timeline. In some cases, the advanced VTS may have multiple streams, but the playlist may only map a subset of those streams to a given title. An advanced VTS also has the unique ability to map to content within a standard VTS. For example, it is possible to have a movie that is recorded only once on the disc but is used by both standard content and advanced content.

Figure 9.33 Structural Organization of Advanced Content

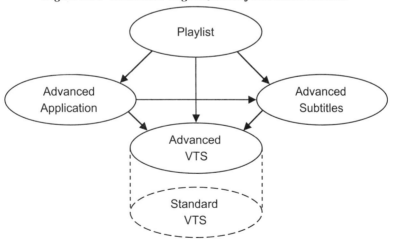

Components of an Advanced Application

Advanced applications are themselves composed of four primary elements (see Figure 9.34):

Manifest. The *manifest* is an XML file that provides the initialization information for the application. The player loads the application based on the values specified in the manifest, which define the script files and initial markup file to load and execute. The manifest also defines the initial screen region that the application will occupy, relative to which all of its graphics will be drawn.

Script. One or more ECMAScript files may be defined per application which are loaded into that application's script context. This programming code handles user operation events, playback control of video and audio content, and access to the network interface and persistent storage. In addition, script code can be used to perform animations, manipulate on-screen graphics, and perform general computational functions.

Markup. Advanced applications follow a page model similar to that of the World Wide Web. For each page, there exists a *markup* file, an XML file whose syntax is derived from XHTML (*Extensible Hypertext Markup Language*), XSL (*Extensible Stylesheet Language*), and SMIL (*Synchronized Multimedia Integration Language*). The markup file describes the initial layout of elements within the application (XHTML), the appearance of those elements such as size and position (XSL), and the timing relationship and animations of the elements (SMIL). In addition, the markup supports XPath (*Extensible Path Language*) expressions, a power mechanism for providing declarative behavioral control without requiring script code.[17]

Resources. The final component of an advanced application, *resources*, is the conglomeration of media elements that the application requires. This includes graphics elements (PNG and JPEG images), animations (MNG files), sound effects (WAV files), and OpenType fonts. It can also include any type of application-specific data files, such as additional XML files or raw text. The markup and script refer to these resources, which represent the interactive content elements beyond the stream content defined in the advanced VTS.

Figure 9.34 Components of an Advanced Application

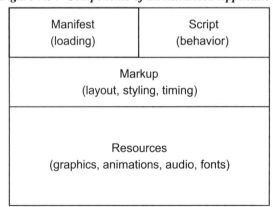

Manifest (loading)	Script (behavior)
Markup (layout, styling, timing)	
Resources (graphics, animations, audio, fonts)	

[17]XHTML, XSL, SMIL, and XPath are all standards of the World Wide Web Consortium (W3C). However, the HD DVD-Video specification only uses subsets of each standard and invents quite a few new tags, attributes, relationships, and behaviors that that go beyond these specifications.

Interactivity on HD DVD-Video

In many ways, advanced content on HD DVD-Video is similar to the UMD Video format. In both cases, you have an XML and ECMAScript environment working in conjunction with a set of multiplexed stream data. Both use the script context to handle events and respond to user operations, as well as to control playback. Likewise, both use the markup environment to establish the layout and appearance of the application using a page metaphor. UMD Video offers greater expressiveness in terms of advanced graphics operations and animation, while HD DVD-Video offers greater application. In other words, UMD Video gives you all the tools to make a menu really interesting, but it will still just be a menu. In HD DVD-Video, the menu may not look as pretty but it can be far more than just a menu!

From a graphics processing standpoint, there is a tremendous difference between pushing pixels to fill a tiny 480×272 PSP screen and trying to fill a 1920×1080 high definition screen. That's almost *sixteen times* more pixels that need to be drawn in the same fraction of a second, so it shouldn't come as a surprise that an HD DVD-Video player will not be able to offer quite the same level of graphics processing as a PSP. Although, the HD DVD-Video specification does offer much of the same processing when it comes to layering images, utilizing a technique known as alpha blending over video. In fact, the HD DVD format goes beyond UMD in that it offers full text rendering capabilities. One weakness will be in the implementation by players, where memory size and bandwidth constraints will limit what can be done with graphics.[18] HD DVD players do not have a powerful 3D graphics pipeline installed, so UMD wins hands down.

Where HD DVD-Video supersedes UMD is in application functionality. While both PSP and HD DVD players have network capability and persistent storage, the UMD Video format does not provide signficant access to these features; HD DVD-Video certainly does. With HD DVD, you can stream audio and video content from network servers and integrate it with material playing from a disc. You can also store content or data, such as movie trailers or game scores, into the persistent storage of the player so that it is available the next time you play the disc.

Another area where HD DVD clearly surpasses UMD is in its enhanced video engine. Although both formats have the ability to scale and position the primary video anywhere on the screen, only HD DVD includes the ability to decode and play multiple A/V streams simultaneously. For example, it is possible to have a secondary video at standard definition resolution (or optionally HD, depending on the player) playing back over the primary HD video. This secondary video stream can be multiplexed into the Advanced VTS along with the primary video, or it can be streamed from another location (either persistent storage or from the network). The secondary video can even have a *luma key* applied, in which any pixel below a certain black level becomes transparent. Imagine a Director's Commentary, for instance, where the director appears over the primary video whenever you want, and they can literally point to areas of the screen concurrent with their commentary. The DVD Forum has defined this and 80 other applications as typical usage scenarios for the HD DVD-Video format, so hold onto your socks when you finally get to see it!

---

[18]The HD DVD specification provides a method for defining player profiles, so that as technology advances and prices drop on memory and processing power, players will be able to do more while remaining backward compatible with earlier profiles.

HD DVD-VR

A key obstacle to the success of HD television is the inability to record high definition content. Even with HD personal video recorders (PVRs), the public wants a way to record content for long-term access, similar to the role that video cassette recorders still play for standard television. The HD DVD-VR (video recording) format essentially combines the capabilities of DVD-VR and DVD-SR (stream recording), while addressing the need to handle high definition video. As this book went to press, the specifications for HD DVD-VR neared completion, with publication anticipated in late 2005. This recording format is designed to work with HD DVD-R and HD DVD-RW media, providing real-time recording structured on a UDF 2.6 file system.

Recording Modes

The HD DVD-VR format has been designed to support three types of content to address today's diverse range of applications:

Self-Encoded Content. This mode is designed to address the traditional application in which the video signal is delivered in analog or digital baseband form and is then compressed by a built-in MPEG-2 encoder for writing to disc. This mode is analogous to the traditional VCR approach and allows the greatest flexibility, such as trick play modes (fast forward, rewind, etc.).

Digital Broadcast Content (with stream analysis). This second mode is adapted for today's digital broadcast systems in which the content arrives in digital form, such as an MPEG transport stream. Rather than decode the video only to immediately encode (causing a generational loss in quality), this recording mode allows the digital bitstream to be recorded in its delivered form (as defined by the broadcast system). The recorder analyzes the stream as it comes in, cataloging time maps and other information necessary to provide a similar level of capability as the self-encoded mode. For this mode, the recorder will generally have all of the decoding hardware necessary to play back the stream, without having to rely on an external or peripheral device.

Bitstream Content (without analysis). The third, and perhaps simplest, mode involves recording an incoming digital stream without the recorder necessarily having any knowledge of the contents of the stream. The recorder simply packages the incoming digital bits so that it will be able to play them back to an external device at a later time. The recorder performs no particular analysis of the incoming stream and does not necessarily have any way of decoding the data in the stream. It can only recreate it, bit-for-bit, and deliver it to the external device, which will be able to decode and play the bitstream. This is also the most restrictive mode because without knowledge of the bitstream contents, it is essentially impossible to support stream editing, trick play modes, or other functionality that would be available in the other recording modes.

Compatibility with HD DVD-Video

Unlike its DVD predecessors (DVD-VR/SR/AR), which were developed long after the DVD-Video format was established, the HD DVD-VR format was developed at the same time as the HD DVD-Video format. As a result, there has been the opportunity to ensure a greater level of compatibility between HD DVD recorders and standalone players. Central to achieving this compatibilty is the *interoperable VTS*. Both formats recognize the interoperable VTS as an exact subset of the advanced VTS defined for HD DVD-Video. When an HD DVD-VR recording session in the self-encoding mode is complete, the recorder has a relatively simple method for translating the VR format's navigation information into a subset of the advanced VTS navigation information, which can be read by an HD DVD-Video player. Because of the designed similarities between the HD DVD-VR and HD DVD-Video multiplexed stream data, the translation process only requires creation of a few additional information files, which is a relatively quick and smooth process. Of course, in order for the resulting disc to be playable on an HD DVD-Video player, it will be necessary that the player be able to read recordable media (HD DVD-R/RW), not just HD DVD-ROM discs. (This alone would be a stumbling block in the DVD arena, as many DVD-Video players tend to have a difficult time reading DVD recordable media.)

Blu-ray Disc (BD)

Blu-ray Disc (BD) is a new format family created by the *Blu-ray Disc Founders* (BDF), spearheaded by Sony, Philips and Panasonic. A vast majority of BD member companies, including the founders, were (and still are) DVD Forum members, however the BDF chose to pursue this new format beyond the walls of the DVD Forum and have set their format against HD DVD and its variations as a result. The soap opera of events that has transpired between these two camps over the past several years could fill a book, but we won't speak to that now. Suffice to say that many of the companies that brought you the very successful DVD format have decided to try something entirely new again for high definition, but they are playing by different rules this time. And this time, they put "BD" in front of all the terms. See Table 9.43 for a listing of the BD acronyms.

Table 9.43 Guide to BD Alphabet Soup

Acronym	Full Name, Meaning
BD	Blu-ray Disc, the overall format family
BDF	Blu-ray Disc Founders group, original BD member companies
BDA	Blu-ray Disc Association, the larger association open to new members
BD-RE	Blu-ray Disc Rewritable, initial physical disc format created by the BDF
BD-R	Blu-ray Disc Recordable, record-once version of Blu-ray discs
BD-ROM	Blu-ray Disc Read Only Memory, read-only version of Blu-ray discs
BD-FS	Blu-ray Disc File System, original file system for BD-RE (not PC compatible)
BD-AV	Blu-ray Disc Audio/Visual, application format for BD-RE and BD-R discs
BD-MV	Blu-ray Disc Movie, application format for BD-ROM (uses BD-AV streams)
HDMV	HD Movie mode, declarative navigation environment for the BD-MV format
BD-J	Blu-ray Java, procedural software environment for the BD-MV format
BoD	Blu-ray Board of Directors, top-level voting group in the BDA

The BD formats break down into a pattern similar to the HD DVD formats. There are three physical disc formats, BD-RE (rewritable), BD-R (record once) and BD-ROM (pre-recorded, read-only), which are similar to HD DVD-RW, HD DVD-R, and HD DVD-ROM, respectively. There are three variations of the file system format, one for each physical disc format, each of which is based on UDF 2.5, as are the HD DVD file systems.[19] There is also a video recording specification, BD-AV, and a pre-recorded content specification, BD-MV, which is similar to HD DVD-VR and HD DVD-Video. To the BDF's credit, most of their disc format specifications were started prior to the HD DVD formats and lead down fundamentally different paths (though on the surface, they still address the same consumer needs). At the time of this writing, only the BD-RE 1.0 specifications for the physical, the file system and the application layer have been published, and products have been launched. The other BD format specifications are nearing completion, with most anticipated in late 2005.

Unlike DVD, which began with the pre-recorded DVD-ROM disc format long before recordable discs were considered, BD started with a focus on recording digital broadcast content. The combination of BD-RE, BD-FS and BD-AV (physical, file system and application formats) was developed first and has become the basis for the other BD formats. For example, BD-R is a write-once variation of BD-RE, designed to use the same file system and application format for recording. BD-ROM is a pre-recorded media format that maintains many of the same physical characteristics as BD-RE and uses an application layer (BD-MV) that is built upon the same transport stream foundation as BD-AV. The intention was to facilitate compatibility among the formats so that standalone players would be more likely to play back BD-RE and BD-R discs, while recorders would more easily be able to support BD-ROM playback capabilities.

BD-AV

The *Blu-ray Disc Audio/Visual* (BD-AV) recording format is primarily designed to address the recording of digital broadcast content at high definition resolution to BD-RE (and later BD-R) discs. This focus led to several key technological decisions that have gone on to propogate other formats in the BD family. Unlike DVD, which is based on MPEG-2 Program Streams (PS), BD-AV uses an *MPEG-2 Transport Stream* (TS) multiplexing format. Transport streams use a much smaller 188-byte fixed packet size and are capable of multiplexing multiple channels and *electronic program guide* (EPG) information, making them far more suitable for broadcast applications. The BD-AV format defines a method for extracting, recording, and playing a partial transport stream composed of a specific content channel from a full transport stream, which may contain several channels of content.

Although the BD-AV format was primarily intended for recording digital broadcast, it also supports recording of self-encoded streams (i.e., from analog standard definition video inputs) and, optionally, from DV video streams, such as a camcorder. In the case of self-encoded streams, the incoming analog video is fed to a built-in MPEG-2 video encoder and mulitplexed with the incoming audio to form a transport stream similar to a digital broadcast. In the case of DV, the video content already exists in an MPEG-2-based format and can be directly recorded without any picture or quality degradation.

---

[19]The BDFS 1.0 file system for BD-RE was designed for real-time digital broadcast recording, however it was not compatible with any computer systems and was later supplanted by UDF 2.5.

Organizational Structure

Like other recordable application formats, BD-AV has been designed to allow seamless, non-destructive editing of recorded stream data by the user. To do this, the format adopts a general structure that is quite similar to that of UMD Video. Figure 9.35 depicts the structural organization of BD-AV content, in which the player provides a user interface for editing the recorded content.

Figure 9.35 Structural Organization of BD-AV Content

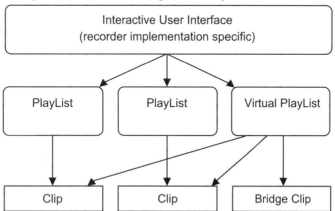

Interactive User Interface. Unlike UMD's script environment, in which all of the navigation and behavior for the disc is recorded on the disc itself, the BD-AV format puts this responsibility onto the individual player implementations.

Playlist. BD-AV supports two types of playlists: "real" and "virtual." *Real playlists* have a one-to-one correspondence to recorded clips on the disc. *Virtual playlists* are created as a result of user editing of the content, and refer to portions of pre-existing clips on the disc. As with UMD, each playlist is composed of one or more *playitems*, clip intervals to be played. For a virtual playlist, multiple playitems that reference portions of existing clips may be tied together into a single, seamless presentation.

Clip. Each clip is composed of *Clip AV Stream* data and a corresponding *Clip Information* file. For each recorded clip, there exists a playlist with a single playitem. For each edited sequence, however, the clips are not changed, but a new virtual playlist is created that links to the desired portions of each clip for seamless playback. In some cases, a new *bridge clip* may be created in which small portions of recorded clips are re-recorded to facilitate a seamless transition from one playitem to another playitem in the virtual playlist.

This organizational structure not only applies to BD-AV, but has become the basis for the organization of the BD-MV format for pre-recorded media as well. The principal difference for pre-recorded content is that the navigation and user experience is defined by the content producer, not the player implementation. Therefore, the user interface defined by the player in BD-AV gets replaced by sophisticated execution environments for authored content already on the disc, just as was done in UMD.

BD-MV

Built on the same key technologies as the BD recording formats, the *Blu-ray Disc Movie* (BD-MV) format is designed to ensure compatibility between recordable and pre-recorded BD formats, and is based on the same MPEG-2 transport stream multiplexing format. In addition, it borrows the playlist and clip concepts as well. Where BD-MV most differs is in how the user interface is defined. Whereas a BD recording format relies heavily on the recording device to provide the user interface, pre-recorded content requires far more of the user interface to be defined by the content on the disc. For example, an ideal recorder would have essentially the same user interface no matter what content is being recorded or played. This is essential for the usability of the recorder. However, when delivering movies on pre-recorded media, there is a strong need to allow thematically driven user interfaces, as has been prevalent with DVD-Video. Table 9.44 describes the general characteristics of the BD-MV format.

Table 9.44 General Characteristics of BD-MV Presentation Data

Disc capacity	25 Gbytes / 50 Gbytes
Video codecs	MPEG-4 AVC: HP@4.1/4.0 and MP@4.1/4.0/3.2/3.1/3.0 SMPTE VC-1: AP@L3 and AP@L2 MPEG-2: MP@ML and MP@HL profiles
Picture size	1920×1080, 1440×1080, 1280×1080, 720×576, 720×480
Display aspect ratio	4:3 or 16:9 anamorphic
Frame rate	24 / 23.976 fps (film) 29.97 / 59.94 fps (NTSC regions) 25 / 50 fps (PAL regions)
Audio codecs	Linear PCM Dolby Digital DTS DTS-HD (core + extension[a]) Dolby Digital Plus[a] Dolby TrueHD[a]
Audio channel configuration	1.0 up to 7.1 channels[b]
Audio sampling frequency	48 kHz, 96 kHz and 192 kHz[a]
Subtitles / Graphics	8-bit Interactive Graphics (IG) stream for menus 8-bit Presentation Graphics (PG) stream for subtitles HDMV Text Subtitle stream[a]
Stream structure	MPEG-2 Systems Program Stream
Max. data rate	40 Mbps
Max. number of video streams	9
Max. number of audio streams	32
Max. number of subtitle streams	32

[a]Support for these features is optional in the player.

[b]A wide range of possible channel configurations exists, including mono, stereo, 5.1 and 7.1, among others.

Although there is general consistency in available user operations and navigation, the unique look and feel of each movie title becomes a strong selling point for the product. For this reason, BD-MV adds a great deal on top of the BD-AV presentation layer in order to create a flexible, programmable environment that can facilitate a wide range of creative user interfaces for the content on each disc. In addition, the BD-MV specification defines a wide range of content types, such as additional audio and video codecs, graphics, and subtitles, among others.

Video Formats

Like HD DVD-Video, BD-MV supports three video codecs — MPEG-4 AVC, SMPTE VC-1 and MPEG-2 HD. There are, however, a number of subtle differences between the two format families. For instance, the peak data rate for BD-MV approaches 40 Mbps; HD DVD limits the peak video data rate to 29.4 Mbps (more than 25% lower than BD). BD-MV allows for 24 fps (or 23.976 fps) material to be encoded and included on the disc at that frame rate, so that the player will add 2-3 pulldown when it needs to convert the video to 59.94i output. The HD DVD-Video specifications require that material originating at 24 fps (or 23.976 fps) must be marked as a 29.97 fps (or 59.94 fps, depending on resolution) stream. Given the higher data rates and disc capacities available on BD-ROM discs, for special coding technologies like *Film Grain Technology* (FGT) are not needed.

Audio Formats

Similar to the HD DVD formats, many different audio codecs were considered for inclusion, but the specific arrangement of mandatory versus optional codecs is still under discussion. What has been decided so far is that DTS and Dolby Digital audio would be supported as mandatory in BD players. Dolby Digital Plus, Dolby TrueHD, DTS-HD, and DTS-HD LBR would be considered optional for the player.

Content Types on BD-ROM

Although many of the specific features and profiles for BD players are still under discussion, the capabilities of BD-MV closely match those of HD DVD-Video. Just as HD DVD-Video has its standard content and advanced content modes, BD also has a declarative, DVD-like mode (HDMV or *HD Movie* mode) and an enhanced, programmatic mode (BD-J, or *BD-Java* mode).

While HD Movie mode is described as a *declarative* execution environment, it means that there is relatively little support for dynamic control of the presentation. A content author must essentially anticipate all of the possible scenarios for a given application and prepare all of the graphics and elements ahead of time for each scenario. This is in contrast to a *procedural* environment, in which software can be written to handle the different scenarios dynamically during playback. The benefit of a declarative environment like HDMV is that it is relatively straightforward to create content, to verify that it complies with the specifications, and to ensure consistent performance across a wide range of device implementations. However, the victim of this predictability is flexibility.

In order to support both predictable, easy-to-author content and complex, dynamic content, BD-MV adds BD-J, a Java-based procedural environment that is derived from DVB's

Multimedia Home Platform (MHP).[20] BD-J provides a rich collection of *application pro-gramming interfaces* (APIs) for multimedia content on top of a Java virtual machine. In short, BD-J defines a complete software execution environment in which custom applications can generate graphics, interact with the user, and control media playback from within the context of the BD-MV content framework. Unlike HD DVD-Video's iHD advanced content environment, which is primarily declarative with ECMAScript providing procedural extensions, BD-J is a fully procedural environment with no assumed declarative portion. This produces the benefit of being extremely flexible, as it sheds the basic assumptions that any declarative environment must make. It also means, however, that any content created in this environment must be built from the smallest of building blocks, which any software developer will tell you is a "non-trivial process."

Organizational Structure

BD-MV allows playback to jump back and forth between modes, defines a sophisticated framework in which content can interact between the different modes, and allows for transitioning from one mode to another. The organizational structure of BD-MV is similar to HD-AV, though the upper layers contain a much greater level of sophistication and functionality. Figure 9.36 depicts a conceptual view of the key layers of the BD-MV environment, which are described below.

Figure 9.36 Structural Organization of BD-MV Content

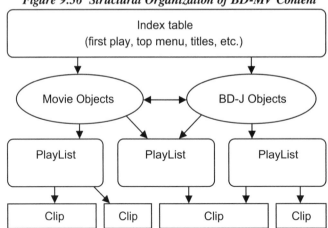

Index Table. The *index table* contains the entry points for the top menu and all of the titles of a BD-ROM disc. In addition, the index table defines where the player should go upon initial playback of the disc. The player references the index table at start-up, whenever a title or menu is to be executed, and during title search and menu call operations.

Movie and BD-J Objects. *Movie objects* refer to HDMV movie objects, and *BD-J objects* refer to Java applications. Navigation commands in a movie object can launch playlists or other movie objects. Likewise, BD-J objects can launch playlists or other BD-J or movie objects via the programming interface. The index table binds movie objects and BD-J objects to specific titles.

---

[20]The Digital Video Broadcasting Project (DVB) is a European industry consortium originally formed in 1993 to study how to establish a concerted pan-European platform for digital terrestrial television. In June 2000, the DVB released the first DVB-MHP specifications, upon which BD-J is based.

PlayList. A *playlist* is a collection of playing intervals (called *playitems*) within the clips. A playlist can be composed of multiple playitems, whose in and out points do not necessarily need to correspond to the specific start and end of a given clip. Likewise, multiple playlists may contain overlapping playitems that refer to the same clips. By default, playlists refer to a *main path* of playitems. However, *subpaths* of out-of-mux data such as text subtitles can be synchronized with the main path.

Clip. A *clip* is composed of a *Clip AV Stream* file and its associated *Clip Information* file. The former is a multiplexed MPEG-2 Transport Stream that complies with the BD-ROM AV specifications. The latter is the time map and other information associated with the transport stream that defines valid entry points into the stream.

HDMV (HD Movie mode)

HD Movie (HDMV) mode takes the functionality of DVD-Video and recasts it as the presentation framework for BD-AV. It supports several DVD-like features, including seamless multiangle video, multistory video branching, multiple parallel audio and subtitle streams, interactive menus, and user operational controls during playback. It does so under a new set of terms and definitions which, although very similar to the old DVD-Video terminology, will require a fair amount of re-schooling for the consummate DVD author.

HDMV includes a series of features that go well beyond DVD-Video's capabilities. The format supports two separate high definition graphics planes, one for frame synchronized *presentation graphics* (PG) and one for *interactive graphics* (IG). Both are loosely related to DVD-Video's subpicture concept in that presentation graphics are used for delivering subtitle content overlaid on the video and interactive graphics are used for defining menu button highlights. The primary difference is that where as DVD-Video only offered a single 4-color graphics plane to serve both purposes, HDMV is able to offer two independently controlled 256-color graphics planes so that subtitles and interactive menu graphics do not have to be mutually exclusive, as they are in DVD-Video.

In addition, HDMV offers advanced constructs such as multi-page menus, in which interactive menu graphics (i.e., buttons and background images) may change from page to page without interruption of the background A/V playback. Using this feature, an author could design a menu set in which a single background video and audio loop plays continuously while the user navigates among pages of different menu options, resulting in a far more seamless user experience.

Pop-up menus and click sounds are another feature added with HDMV. Pop-up menus allow the producer to define menu constructs that may appear and dissappear over the video without interrupting playback. If the user were watching a movie and pressed the Menu key, instead of interrupting playback and jumping out to a separate full screen menu, the pop-up menus feature would allow the producer to define a menu that appears over the film, such as in the lower one-third of the screen. This would allow the user to quickly navigate to and select the desired option without abruptly interfering with their movie watching experience. Another feature called *click sounds* could provide clear, immediate feedback to the user when menu buttons are being actioned, which can greatly improve the usability of the menu as well as the satisfaction of the experience.

Another interesting feature of HDMV seems to be taken from the DVD-Audio feature of the browsable slideshow. In essence, this features allows the producer to create a slideshow of high definition images that the viewer may interactively navigate while music plays uninterrupted. In DVD-Video, the content producer was left with the choice of either having undisturbed audio and a fixed presentation of slides, or user-navigable slides with no background audio (or audio that is interrupted at each slide change).

One more feature that HDMV adds over traditional DVD-Video style authoring is support for separately delivered content such as text-based subtitles, which could be provided as a separate file on the disc or downloaded from a network source and then played back in sync with the video. In this way, HDMV provides a method for relaxing the release window so that additional features can be added to a disc after it has already been distributed.

BD-Java (BD-J)

Despite all of the new features offered by HDMV, there was a general concern that they did not go far enough to establish the kind of truly new experience necessary to insure a truly successful format. New interaction paradigms have been developed by the major studios that involve network-based content, online "freshening" of disc-based content, and enhanced applications and games, among others. Many of these paradigms were derived from experiences with PC-based DVD playback using software like InterActual Technology's PCFriendly and Microsoft's WebDVD. These PC-bound applications made it possible to unite the powerful general processing, networking and local storage of the PC with the high quality A/V media of the DVD-Video system, resulting in an extremely rich multimedia experience. The studios sought a means to get this type of experience into the living room in a manner even more cleanly integrated than the earlier DVD/PC systems could offer. Figure 9.37 shows a conceptual overview of how the BD-J system integrates with HDMV and other BD players.

Figure 9.37 BD-J System Overview

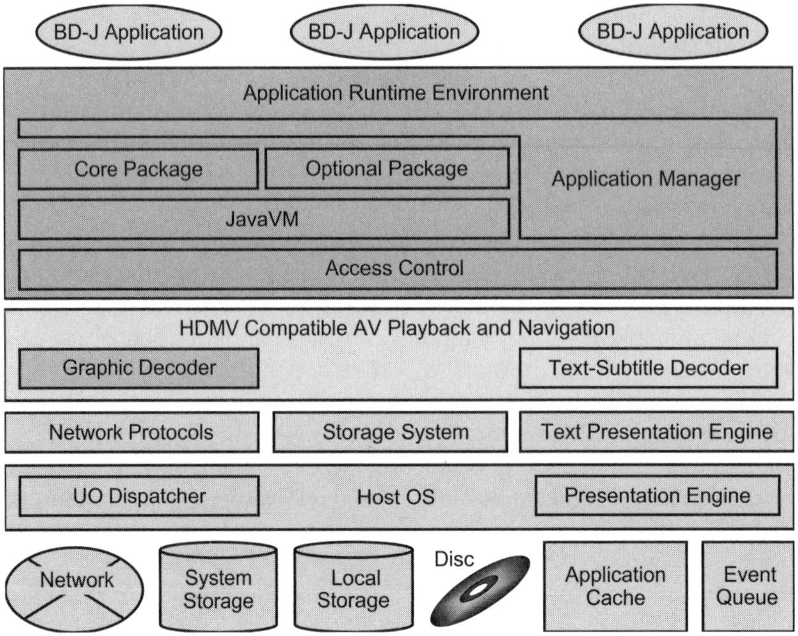

Derived from the DVB's Multimedia Home Platform (MHP), BD-J offers a fully pro-grammable environment in which custom software applications can be created to execute within the realm of a constrained device, such as a settop BD-ROM player.

With Java now appearing in hundreds of millions of constrained devices (cell phones, PDAs, cable and satellite boxes, etc.), this seems to be the best solution for Blu-ray. Based on the Java 2 Micro Edition (J2ME) and Personal Basis Profile (PBP), BD-J offers support for network connectivity, local storage access (either hard disk or flash memory), advanced graphics and animation, media control (via the Java Media Framework), and all of the general processing capabilities that one would normally expect.

The challenge is specifying an environment in which applications can operate with relative consistency across a wide range of implementations. Unlike cable and satellite boxes, in which a service provider may only support a handful of devices, or cell phones where application builds may be individually targeted for each proprietary implementation, BD-J must anticipate that hundreds or even thousands of different makes and models of BD players will be created over time. Each player will have its own chipsets and performance characteristics, which will improve dramatically over the years as technologies continue to advance. And yet, the BD-J application written today will be expected to operate with relative consistency across all of these players. With a declarative system like HDMV, it is a far simpler task to define performance constraints and compliance tests. In a full software development environment, however, the complexity goes up exponentially with flexibility, making it extremely difficult to test for functional compliance, much less performance consistency. This is one of the greatest challenges that BD-J will need to meet in order to become a successful application format.

Other Formats

While the DVD Forum and Blu-ray Disc Association slug it out to become the next dominant optical disc format in the Western world, an entirely different format war waging in the Eastern world. China represents a potential market with more than one billion consumers. As the government slowly opens its markets to outside suppliers, companies within mainland China, Hong Kong and Taiwan are battling it out to establish their own HD optical disc formats specifically geared for the Chinese market.

With an emphasis on low cost and complexity, these systems are generally based on the DVD red-laser physical disc format and MPEG-2 high definition video, with only limited support for interactive menus, multiple audio languages and subtitles. Going by the acronyms EVD, FVD, HVD and others, these formats are shooting for the heavily commodotized, low-cost, high-consumption market that China may be able to offer, but they have to do so quickly in order to establish themselves as the *de facto* standard before the market is fully opened to the rest of the world.

While Japan, the US, and Europe are distracted with a repeat of the "VHS vs. Betamax" debate, it will be interesting to see what unfolds in this giant economic superpower-on-the-rise, where they've gone back to basics and built a simple player that gives a good experience, and where the content is still king.

Chapter 10
Players

Player capabilities have expanded considerably since the introduction of DVDs. Player price range now starts below $50US and rises to over $3,000US for "reference" players.

The more inexpensive players are simple DVD-Video players that provide analog connections, with the limited cross-media capability of CD and, if you're lucky, MP3 disc compatibility.

A high-end reference player, however, does it all! Check out this litany of playable formats from one player manufacturer: DVD-Video, DVD-R, DVD-RW, DVD+R, DVD+RW, Video CD, audio CD, CD-R, CD-RW, MP3, WMA CD-R & CD-RW, DVD-Audio, SACD, photo CDs, Kodak Picture CDs, and Fujicolor CDs. Wait, there's more. This reference player performs upscaling of the digital video output, has every connector known to man, and only weighs 40 pounds!

This just goes to show that you get what you pay for. Well, almost. What it also reveals are the specifications that separate the "-philes" from the philistines.

We don't desire to imply that you will only get the best output from the most expensive players on the market. Manufacturers are sprinkling their player options across a wide spectrum of prices and features. Several players with prices of a few hundred dollars provide digital upscaling and the latest digital connections, the newest and most important features that should be considered when deciding what DVD player to buy, though, they are not required. Excellent players are available without either feature.

What we hope to impart in this chapter is a solid understanding of the technology, tools, and techniques that a customer should have in order to make an informed decision about what player is right for them and if it will remain so for an appreciable period of time.

Getting Connected

Every technological advance is aggressively evaluated by content providers before they agree to participate in the latest and greatest. The primary concern of movie studios centers on their all-consuming need to retain control of the display and duplication of their proprietary content. The digital revolution has created digital dilemmas that could open a portal to pristine replication of copyrighted content.

So, how do we keep everyone on board the technological express? Sophisticated connections! That's right, it's all in how we hook everything together. The latest generation of digital connections raises the bar in player sophistication, the most recent relevant acronym being HDMI, which supplants DVI and can be augmented with HDCP, thus creating the buzz of HDMI/HDCP. Walk in to your neighborhood electronics boutique, drop that term, and see how they clamor for your attention.

As players proceeded along their evolutionary trail, the connections provided with them progressed commensurately. Early players offered various analog outputs for audio and video. Even the earliest DVD-V players had optical or coaxial digital audio outputs. What came later were players that supported multichannel analog outputs, and eventually multi-channel PCM via high bandwidth connections, often based on IEEE 1394 or some form of proprietary digital link.

Digital video connections between players and display devices are the Achilles heel for content providers. Beginning with *DVI (Digital Visual Interface)*, the breadth of video displays — HDTVs, LCDs, plasma displays, projectors — provide stunning visuals, with crisp colors and expansive latitude in brightness and contrast, when connected to digital sources. However, tapping into this digital conduit with a high-speed recording device, such as a digital video recorder (DVR) or a large capacity hard drive, could allow for unlimited copying of the digital information. Thus, the dilemma becomes how to protect the content while providing the user with the best possible image for their viewing pleasure. After all, they paid for it, they should get it.

The content protection schemes described in Chapter 5 provide a degree of security for the copyright holders, but some of those protections may be defeated by hackers and software cracking applications.

Moving on from DVI, the electronics manufacturers, in consultation with the movie studios, have developed an ingenious connection scheme called *HDMI, High-Definition Multimedia Interface*. Whereas DVI only accommodates video, HDMI is an audio-video interface with enough bandwidth to accommodate uncompressed high-definition digital video and multi-channel audio in a single cable. The ingenious element of HDMI is that in addition to the audio and video data, communication is established that certifies the connection as legitimate between source and display. Buried within the devices is circuitry that locks and unlocks the data passed from source to display. If either end of the path is not legit, then either no video or video of a less than optimal quality is allowed to pass between devices. A similar scheme is also present with DVI.

When either DVI or HDMI is coupled with *HDCP (High-bandwidth Digital Content Protection*, see Chapter 5 for details), a home theater enthusiast can enjoy the best possible sights and sounds of the digital experience and Hollywood is assured that their property is not compromised. HDMI can automatically establish the highest data rate and display characteristics possible for the source and display combination. For example, if a DVD player is capable of upscaling the video output to provide a widescreen match for a high-definition display, that connection will automatically be established.

Finalized as a specification near the end of 2003, more than six million HDMI-enabled consumer electronics devices shipped in 2004, and market analysts project 125 million HDMI-enabled devices in the marketplace by 2007. From settop boxes to DVD players to digital televisions to monitors and projectors, the HDMI standard offers a solution to all segments of the market. The movie industry gets its copyright protection, manufacturers get a widely adopted interface standard, and users get a compact, elegant solution that provides them with the best visual and aural entertainment possible.

Additionally, the FCC has ruled that all digital cable-ready televisions sold after July 1, 2005, are required to have either an HDMI/HDCP or a DVI/HDCP input.

Digital Upscaling

Frankly, when we first saw this term, another phrase came to mind: silk purse, only to realize that some of the biggest players in the player arena are deadly serious about providing this feature. As this book goes to press in the fall of 2005, what everyone must recognize is NO DVD-VIDEO PLAYER CURRENTLY ON THE MARKET PLAYS NATIVE HIGH DEFINITION FORMAT VIDEO BECAUSE DVD-VIDEO IS ONLY STANDARD DEFINITION![1] (Okay, we feel better now.)

Digital upscaling? What can we say? Well, at least this much.

Electronics manufacturers have hit on a way to provide purchasers of high definition televisions with a display size matching video output from a standard DVD-Video player. Through the use of specially designed integrated circuit chipsets, the DVD video output may be scaled to match the desired display setting. These chipsets utilize various techniques to achieve the higher display rates, such as line-doubling, pixel interpolation or pixel repetition, and some chipsets provide better pictures than other chipsets.

The DVD contains MPEG-2 video encoded as either NTSC video at 720 pixels by 480 lines or PAL format video of 720 pixels by 576 lines, and the digital upscaling recomposes the video to match the 720p, 1080i, or 1080p presentation settings of the display.

The advantage of having a DVD player perform the upscaling function is that the data remains in the digital domain. Previously, the DVD video would be converted to analog and sent to the display as composite, s-video, or component video. The display would then perform an analog-to-digital conversion prior to presenting the rescaled imagery on the screen.

An improved viewing capability may be most noticeable when watching a digitally upscaled output of a progressive scan DVD player using the HDMI or DVI link. In fact, the digitally upscaled output is not viewable on any of the player's analog outputs.

With the introduction of the next generation optical discs, HD DVD and BD, new players will be capable of presenting higher bandwidth video content.

Choosing a DVD Player

It is not imperative that the selection of a DVD player rest exclusively on whether the player has the latest digital connections and digital options. Personal preferences, your budget, and your existing home viewing environment play the most important roles when deciding what player would be best. Even in a high-end home theater setup, analog connections have proven quite satisfactory for even the most discerning of viewers.

Make a list of player features that are important to you, a list of players in your price range, and then eliminate or downgrade the ones that do not match your list. If possible, visit an electronics store and try out some of the players, focusing on ease of use (remote control design, user interface, and front-panel controls).

Because there is not much variation in picture quality and sound quality within a given price range, convenience features play a big part. Pay special attention to the remote control.

---

[1]Actually, a few players are capable of playing Windows Media High Definition Video (WMV HD) files, from DVD-5 or DVD-9 discs..

Are the buttons easy to find, with clear labels? Is it comfortable to hold, and can you reach the important buttons with your thumb? Are the controls illuminated? Do you prefer a jog-shuttle knob? You will use the remote control all the time, and it will drive you crazy if it does not suit your style. See Table 10.1 for items that should be considered in a player decision.

Table 10.1 Checklist of Player Features

Do I want progressive-scan video?	Gives the best-quality picture, but only if you have a progressive-scan TV (DTV, HDTV).
Do I care about black-level adjustment?	Check for a 0/0.75 IRE setup option.
Do I want to zoom in to check details of the picture?	Look for players with picture zoom.
Does my receiver have only optical or only coaxial digital audio inputs?	Make sure the player has outputs to match.
Do I want DTS audio output?	Look for a player with the "DTS Digital Out" logo.
Do I want virtual surround?	Useful if you only have two speakers.
Do I want 96-kHz, 24-bit audio decoding?	For high-end audio systems (analog connection).
Do I want 96-kHz, 24-bit PCM digital out?	For high-end audio systems (digital connection).
Do I need an internal 6-channel Dolby Digital or DTS decoder?	Important if you have a multichannel ("Dolby Digital ready") amplifier.
Do I want to play video CDs?	Check the player specs for video CD compatibility. Or, look for the "Video CD" logo. Approximately two-thirds of DVD players can play video CDs.
Do I want to play SVCDs?	Check for the "SVCD" logo.
Do I want to play SACDs or DVD-Audio discs?	Not all DVD players can play these types of discs. Look for the appropriate logos.
Do I want to play homemade CD-R audio discs?	Look for the "dual laser" feature.
Do I want a player that holds more than one disc?	Three- or five-disc players are useful for music. Players that hold 100 or more discs allow you to make your entire collection quickly available.
Am I bothered by bright front panel displays when watching movies?	Check for an option to dim or turn off the lighted display.
Do I need a headphone jack?	Handy for late-night viewing.
Do I want on-screen player setup menus and displays in languages other than English?	Look for multilanguage setup feature. (Note: All players support multilanguage menus when provided on the disc.)
Do I want to control all my entertainment devices with one remote control?	Look for a player with a programmable universal remote, or make sure your existing universal remote will run the new DVD player.

Realistically, player selection should also be driven by the connection options incorporated in your current or planned display. Future-proofing your selections takes on a whole new meaning when factoring the impending HDCP influence on what you'll be allowed to see due to the interconnection methodology you adopt.

Just as most CD players met the needs of the average CD buyer, most DVD players have the features and quality that most buyers desire. Video and audio performance in modern DVD players are excellent, and a higher price may buy an improvement in video and audio quality, in addition to increased reliability and sturdiness. However, just because a player is expensive does not mean it is good. Informal studies have shown that expensive players were often just as bad (or worse) as some of the cheapest players when it came to simply playing a disc. We suggest that you check the return policy of the vendor prior to finalizing your purchase.

Connection Spaghetti - How to Hook Up a DVD Player

The DVD format is designed around some of the latest advances in digital audio and video, yet the players are also designed to work with TVs, video systems of all varieties, as well as audio systems from monaural to surround. As a result, the back of a DVD player can have a confusing diversity of connectors producing a potpourri of signals.

Most DVD players produce the following output signals:

- *Analog stereo audio*. This standard two-channel audio signal can include Dolby Surround encoding.

- *Digital audio*. This raw digital signal can connect to an external digital-to-analog converter or to a digital audio decoder. There are two signal interface formats for digital audio: S/PDIF and Toslink. Both formats can carry *pulse-code modulated* (PCM) audio, multichannel Dolby Digital (AC-3) encoded audio, multichannel DTS encoded audio, and multichannel MPEG-2 encoded audio.

- *Composite baseband video*. This is the standard video signal for connecting to a TV with direct video inputs or an audio-visual (A/V) receiver.

- *S-video (Y/C)*. This is a higher-quality video signal in which the luminance and chrominance portions of the signal travel on separate wires within a bundle or wrapper.

With a slightly higher level of sophistication, some players may produce these additional signals:

- *Six-channel analog surround*. These six audio signals from the internal audio decoder connect to a multichannel amplifier or a "Dolby-Digital-ready" (AC-3-ready) receiver.

- *AC-3 radio-frequency (RF) audio*. The Dolby Digital [QPSK] audio signal from a laserdisc player connects to an audio processor or receiver with an AC-3 [RF] demodulator and decoder.

- *Radio-frequency (RF) audio/video*. These older-style combined audio and video signals are modulated onto a VHF RF-carrier for connecting to the antenna leads of a TV tuner or to a TV cable system connection. Connectors of this type are provided as a legacy holdover for use with older televisions, as the image via an RF connector tends to be of a poorer quality.

■ *Component analog video (interlaced scan)*. These three video signals (RGB or Y′P$_b$P$_r$) can connect to a high-end TV monitor or video projector.

■ *Component analog video (progressive scan)*. These three video signals (RGB or Y′P$_b$P$_r$) can connect to a progressive scan monitor or projector.

At the upper range of features and sophistication, but not necessarily higher priced, players may provide one or more of the following digital connections:

■ *IEEE 1394*, aka *FireWire*. This signal type, although in use on a multitude of consumer electronics devices, is generally used to interface devices with computers.

■ *DVI, Digital Visual Interface*. Present on newer-model DVD players and display devices, DVI provides a secure HDCP-compliant data path but is video only.

■ *HDMI, High-definition Multimedia Interface*. Also a secure HDCP-compliant data interface, HDMI carries both full-bandwidth uncompressed digital video and multichannel digital audio.

■ *Ethernet, LAN*. When available, an ethernet signal can be used to interface a DVD player to a computer or a local area network.

Connector Soup

DVD player connections come in a variety of types. The different audio and video signals may be present on the following types of connectors:

■ *RCA phono* (Figure 10.1). This is the most common connector, used for analog audio, digital audio, composite video, and component video. The term *cinch* is also used for this connector.

Figure 10.1 RCA Phono Connector

■ *BNC* (Figure 10.2). This connector carries the same signals as RCA connectors but is more popular on high-end equipment.

Figure 10.2 BNC Connector

■ *Phono* or *miniphono* (Figure 10.3). This connector carries stereo analog audio signals and may be used by portable DVD players. It also may appear on the front of a DVD player for use with headphones.

Figure 10.3 Phono/Miniphono Connector

■ *S-video DIN-4* (Figure 10.4). This connector, also called Y/C, carries separate chroma and luma video signals on a special four-conductor cable.

Figure 10.4 DIN-4 (S-video) Connector

■ *Toslink fiberoptic* (Figure 10.5). This connector, developed by Toshiba, uses a fiber optic cable to carry digital audio. One advantage of the fiber optic interface is that it is not affected by external interference and magnetic fields. The cable should not be more than 30 to 50 feet (10 to 15 meters) in length.

Figure 10.5 Toslink Connector

■ *IEEE 1394* (Figure 10.6). Also known as FireWire or i.Link, these connectors are actually an external bus carrying all types of digital signals using only one cable for all the signals, including audio and video. The larger 6-pin cable also carries power; the smaller 4-pin cable does not.

Figure 10.6 IEEE 1394 Connectors

- *DB-25* (Figure 10.7). This 25-pin connector, adapted from the computer industry, is used by some audio systems for multichannel audio input.

Figure 10.7 DB-25 Connector

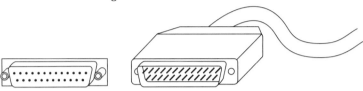

- *SCART* (Figure 10.8). This 21-pin connector, used primarily in Europe, carries many audio and video signals on a single cable: analog audio, composite RGB video, component video, and RF. Also called a Euro or Peritel connector.

Figure 10.8 SCART Connector

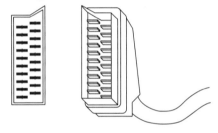

- *Type F* (Figure 10.9). This connector typically carries a combined audio and video RF signal over a 75-ohm cable. A 75- to 300-ohm converter may be required.

Figure 10.9 Type F Connector and Adapters

- *RJ-45* (Figure 10.10). This connector type contains eight very fine leads and was initially developed for the telecommunications industry. It is now ubiquitous in ethernet and network installations.

Figure 10.10 RJ-45 Connector

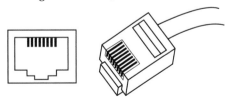

- *DVI* (Figure 10.11). This connector accommodates uncompressed high-definition video connections between DVI-capable devices. The initial application for DVI was to connect a computer with a fixed pixel monitor, such as an LCD display or a projector.

Figure 10.11 DVI Connector

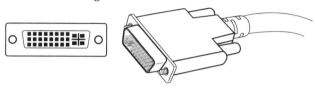

- *HDMI* (Figure 10.12). This connector supports the transmission of standard, enhanced and high-definition video combined with multichannel digital audio on a single cable.

Figure 10.12 HDMI Connector

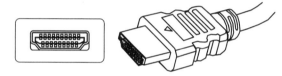

Audio Hookup

In the interests of simplicity, the following sections occasionally will use the term *multichannel audio* to refer to Dolby Digital (AC-3) audio, DTS audio, and MPEG-2 audio. The term is not used to refer to Dolby Surround audio, which has only a two-channel signal.

To hear the audio from a DVD player, the player must be connected to an audio system: a receiver, a control amp or preamp, a digital-to-analog converter, an audio processor, an audio decoder, an all-in-one stereo, a TV, a boombox, or other equipment designed to process or reproduce audio. If your audio system provides both Dolby Digital and Dolby Pro Logic, connect the DVD player to the Dolby Digital (S/PDIF or Toslink) inputs for the best result.

Discs for NTSC players are required to provide at least one audio track using either Dolby Digital or PCM audio. Discs for PAL players are required to provide at least one Dolby Digital, MPEG audio, or PCM audio track. Not all NTSC players are able to play MPEG audio tracks, although all PAL players are able to play both MPEG and Dolby Digital audio tracks. Dolby Digital has become the dominant multichannel audio standard for DVD and other digital video formats, and it is not likely that MPEG-2 will ever see much use. DVD also supports optional multichannel formats, including DTS and SDDS. See Chapter 3 for details on different audio formats, including the difference between Dolby Surround/Pro Logic and Dolby Digital.

Some players also provide the Dolby Headphone feature, which processes the multichannel audio to create a more three-dimensional sound field for headphones. The Dolby Headphone feature works with all standard headphones.

Some audio subwoofers accommodate connections to other speaker inputs. If the audio amplifier in your setup has a subwoofer output, and you have a subwoofer, and your subwoofer has a direct (coaxial) audio input, use that connection rather than connecting any speaker outputs from/to the subwoofer. Unless the subwoofer is much more expensive than the amplifier, the amplifier will do a better job of bass management (see the following A Bit About Bass section). Most DVD players also provide two or three audio hookup options. These options are detailed in the following sections.

Digital Audio

The digital audio outputs provide the highest-quality audio signal. This is the preferred connection for audio systems that have them. Almost all DVD players have digital audio outputs for PCM audio and multichannel audio. These outputs carry either the raw digital audio signal directly from the digital audio track or the two-channel downmixed PCM signal from the internal multichannel decoder.

For multichannel audio output, the encoded digital signal bypasses the player's internal decoder, and the appropriate decoder is required in the audio receiver or separate audio processor. For PCM audio output, the PCM signal is sent directly to the digital audio output. Alternatively, the multichannel decoder in the player may produce the PCM signal, but it is restricted to a maximum of two output channels when using S/PDIF or Toslink. In either case, a digital-to-analog converter (DAC) is required, which may be built in to the receiver or may be a peripheral device connected to the receiver. Some players provide separate outputs for multichannel audio and PCM audio. Other players have either a switch on the back or a section in their onscreen setup menu where you can choose between PCM output and multichannel output (undecoded Dolby Digital, DTS, or MPEG audio). The multichannel output menu option is usually labeled "AC-3" or "Dolby Digital."

All NTSC DVD players include a two-channel Dolby Digital decoder, so they can produce PCM audio from Dolby Digital audio. Most PAL DVD players include both two-channel Dolby Digital and two-channel MPEG audio decoders, so they can produce PCM output from either format. Many players (NTSC and PAL) also provide DTS audio output, but only for connection to an external DTS decoder. Some players have built-in DTS decoders.

The digital audio output is also used for PCM audio from a CD. Players that can play Video CDs also may produce PCM audio output converted from the MPEG-1 audio signal. Combination laserdisc/DVD players, if you still have one of these please let us know, also use this output for the laserdisc's PCM audio track, but not the AC-3 track, as will be explained.

The direct output from PCM tracks on a DVD is at a 48 kHz or 96 kHz sampling rate with 16, 20, or 24 bits. The converted PCM output from multichannel audio tracks is at 48 kHz and may be up to 24 bits. The PCM output from a CD is at 44.1 kHz and 16 bits. The PCM output from a laserdisc player is also at 44.1 kHz and 16 bits. The connected audio component does not need to be able to handle all these variations, but the more the better. A system capable of 16 and 20 bits at sampling rates of 48 and 96 kHz is recommended. Some DVD players are incapable of properly formatting a PCM signal for output at the higher sampling rates or bit sizes. If you have an audio system capable of 24 bits or 96 kHz, make sure the player can correctly produce the corresponding digital audio signal. Be aware that players are required by the CSS/CPPM license to restrict digital output of 96 kHz audio when the disc is encrypted. In this case, most players downsample to 48 kHz.

The digital audio output must be connected to a system designed to accept either PCM digital audio, Dolby Digital (AC-3), or both. Most modern digital receivers can automatically sense the type of incoming signal.

Some players include a dynamic range control setting (also called *midnight mode*) that boosts soft audio and reduces loud sound effects. It is recommended that this setting should be turned off to achieve the best effects with a home theater system, but you may wish to activate it for situations where the dialogue cannot be clearly heard, such as when everyone else in the house has gone to bed and the volume is down low.

Additionally, Dolby Digital has a feature called *dialog normalization*, which is designed to match the volume level from various Dolby Digital programs or sources. Each encoded audio source includes information about the relative volume level. Dialog normalization automatically adjusts the playback volume so that the overall level of dialog remains constant. It makes no other changes to the audio, including dynamic range; it is equivalent to manually turning the volume control up or down when a new program is too soft or too loud. Usually, only one setting exists in a program, which means the volume control does not change in the middle. Dialog normalization is especially useful with a digital television source to handle variations in volume when changing channels. It is less useful for DVDs, unless they have many separate programs on them.

Although the use of DTS is growing steadily it is still not used on most DVD discs, and SDDS is not used on any discs; they are optional multichannel surround formats that are allowed in the DVD format but are not directly supported by most players. Each requires an appropriate decoder in the receiver or a separate audio processor.

Connecting Digital Audio

Two different standards govern the digital audio connection interface: coaxial and optical. The arguments are many and varied as to which is superior, but since they are both digital signal transports, high-quality cables and connectors will deliver the exact same data. Some players have only one type of connector, although many players have both.

Coaxial digital audio connections use the IEC-958 II for PCM, also known as *S/PDIF* (Sony/Philips Digital Interface Format). Most players use RCA phono connectors, but some use BNC connectors. Use a 75-ohm rated cable to connect the player to the audio system. Multichannel connectors usually are labeled "Dolby Digital" or "AC-3." PCM audio connectors usually are labeled "PCM," "digital audio," "digital coax," "optical digital," etc. Dual-purpose connectors may be labeled "PCM/AC-3," "PCM/Dolby Digital," or something similar. Be sure to use a quality cable; a cheap RCA patch cable may degrade the digital signal to the point that it will not allow the signal to pass.

Optical digital audio connections use the EIAJ CP-340 standard, known as *Toslink*. Connect an optical cable between the player and the audio receiver or audio processor. The connectors are labeled "Toslink," "PCM/AC-3," "optical," "digital," "digital audio," or the like.

If the connection (either coaxial or optical) is made to a multichannel audio system, select Dolby Digital/AC-3 (or DTS or MPEG multichannel) audio output from the player's setup menu or via a switch on the back of the player. If the connection is to a standard digital audio system (including one with a Dolby Pro Logic processor), select the PCM audio output instead. In cases where a player has an optical (Toslink) connection but the audio system has a coaxial (S/PDIF) connection, or vice versa, a converter may be purchased for a nominal cost.

When you have used a digital connection, there is no need to make an analog audio connection. The exception is when certain players are set to output PCM audio at a 96 kHz sampling rate. When an encrypted disc prohibits 96 kHz output, some players will only produce analog audio output. In this case, a separate analog connection will be needed.

Multichannel Analog Audio

A component multichannel audio connection can be as good as a digital audio connection. However, such outputs use the digital-to-analog converters that are built into the player, and they may not always be of the highest quality, especially on a low-cost player. The analog signal must be converted back to digital when connected to a digital receiver. If you have an amplifier with multichannel inputs or a Dolby-Digital-ready receiver, six-channel audio connections are an appropriate choice, because a Dolby Digital decoder will not be required.

All DVD players include a built-in two-channel Dolby Digital audio decoder. Only some players include a full six-channel decoder along with the multichannel digital-to-analog converters and external connectors necessary to make the decoded audio available. Some players support the Dolby EX or DTS ES formats, which add a rear center channel. In this case, a seventh audio connection is necessary to use the added channel.

DVD-Audio, with more emphasis on multichannel PCM audio, makes a multichannel amplifier more important. Six analog connections or three pairs of two-channel digital audio connections are needed to carry all six PCM audio signals, otherwise the player must downmix to two-channel audio, which in most players can be either a standard stereo or a Dolby Surround compatible downmix.

Connecting Multichannel Analog Audio

A multichannel capable DVD player typically has six RCA or BNC jacks (seven for EX/ES formats), one for each channel. Hence, a receiver/amplifier with six audio inputs — or more than one amplifier — is needed. Hook six audio cables to the connectors on the player and to the matching connectors on the audio system. The connectors typically are labeled for each speaker position: L, LT, or Left; R, RT, or Right; C or Center; LR, Left Rear, LS, or Left Surround; RR, Right Rear, RS, or Right Surround; Subwoofer or LFE; and sometimes, CR, Center Rear, CS, or Center Surround. Some receivers use a single DB-25 connector instead of separate connectors. An adapter cable would be required to convert from DB-25 on one end to multiple RCA connectors on the other.

Stereo/Surround Analog Audio

A two-channel audio connection is the most widely used option, but it does not have the quality and discrete channel separation of a digital or multichannel audio connection. All DVD players include at least one pair of RCA (or sometimes BNC) connectors for stereo output. Any disc with multichannel audio will be downmixed automatically by the player to Dolby Surround output for connection to a regular stereo system or a Dolby Surround/Pro Logic system.

When making connections for a stereo/surround analog audio environment, connect two audio cables with RCA or BNC connectors to the player. Connect the other ends to a receiver, an amplifier, a TV, or other audio amplification system. Connectors may be labeled "audio," "left," or "right." The connector for the left channel is usually white, and the connector for the right channel is usually red.

In some cases, the audio input on the stereo system (such as a boom box, if it can be rightly called a stereo system) will be a phono or miniphono jack instead of two RCA jacks, thus an adapter cable becomes necessary. If the player is a portable player with a miniphono connector, a phono-to-RCA adapter cable is usually required to connect the player to the audio system. If the player includes a phono or miniphono connector for headphones, it generally is not recommended that the headphone output be used to connect the player to a stereo system because the line levels are not appropriate.

AC-3 RF Digital Audio

An AC-3 RF digital audio output is provided only by combination laserdisc/DVD players. The audio signal from the FM audio track of a laserdisc is presented at this output. Laserdisc AC-3 audio does not appear at the standard PCM/Dolby Digital output (although the stereo PCM audio track does). Audio from a DVD does not come out the AC-3 RF output, it is a special output designed solely for the AC-3 signal from a laserdisc, which is in a different format from DVD's Dolby Digital signal.

Hook a coaxial cable from the AC-3 RF output of the player to the AC-3 RF input of the receiver or Dolby Digital processor. Make sure the receiving end is set to RF mode or can automatically adapt to an AC-3 RF signal.

In order to receive all of the audio signals from a combination laserdisc/DVD player, three separate audio hookups are required: a PCM/Dolby Digital connection (for DVD digital audio and laserdisc PCM digital audio), an AC-3 RF connection (for laserdisc AC-3 audio), and an analog stereo connection (for the laserdisc analog channels, which often contain supplemental audio).

A Bit About Bass

The heart-thumping, seat-shaking excitement of action movies relies heavily on deep, powerful low-frequency audio effects. DVD provides audio quality that is actually superior to what comes on film for theaters. It is up to the home theater owner (subject to a spousal approval and a neighbor tolerance) how close he or she wants to get to a theater sound system.

All the ".1" sound encoding formats on DVD provide special channels for low-frequency effects (LFE). Despite becoming a standard feature, the LFE channel is frequently misunderstood by DVD producers and listeners. Part of the problem is the LFE channel's overuse by audio engineers. It is possible, and quite normal, for all of the bass in a movie to be mixed in the five main channels, because all are full frequency channels. The LFE channel should be reserved for extraordinary bass effects, the type that only work well in a full discrete surround system with at least one subwoofer. Again, though, the same bass effects could be mixed in a 5.0 configuration with no loss or compromise, because modern receivers, particularly those with Dolby Digital and DTS decoders and a separate subwoofer output, have integrated bass management. Depending on the speaker configuration, the receiver automatically filters and routes bass below a certain frequency to the speakers that can reproduce it. For example, if an audio system has five small bookshelf speakers and a subwoofer, the receiver should send all the bass below 80 Hz or so to the subwoofer. In an audio system with a few large speakers, some smaller speakers, and a subwoofer or two, the receiver will route low-frequency audio from all channels to the large speakers and the subwoofers. It does not matter what channel the bass comes from; all low frequencies from the main channels and the LFE channel will be sent to every speaker that can handle them.

This is the key to understanding why certain complaints about bass and LFE are groundless. A 5.0 mix does not compromise the audio or cheat owners of high-end audio systems, because all the necessary bass is still in the mix. Omitting the LFE channel when downmixing to two channels is not a terrible thing, because nothing vital rests in the LFE channel — only, extra "oomph" effects that few two-channel systems can do justice to. This does assume that the engineer creating the audio mix understands the purpose of the LFE channel and does not blindly move all low frequencies into it. See Chapter 16 for more about bass allocation and LFE mixing.

Video Hookup

To see the video from a DVD player, the player must be connected to a video system: a television, a video projector, a flat-panel display, a video processor, an audio/visual (A/V) receiver or video switcher, a VCR, a video capture card, or other equipment capable of displaying or processing a video signal. For those with a widescreen TV, connection details can become more confusing. (See Chapter 3 for information about aspect ratios and wide-screen display modes.)

With the exceptions of RF video or HDMI-capable players, audio cables are also required when connecting a DVD player to your system, in addition to a video connection because the video connections do not carry audio. And, communication becomes a bit trickier when there are multiple devices fighting over a single TV.

For example, a DVD player, VCR, cable box, and video game console may all need to be connected to a TV that has only one video input. In this case, the best option is to use an A/V receiver, which will switch the video along with the audio. When buying a new A/V receiver, get as many video inputs as you can afford. You will almost always end up with more video sources than you initially plan for. If an A/V receiver is out of your price range, then you may get either a new TV with more video inputs or a manual video switching box. Alternatively, if you have only a DVD player and a VCR or cable box, you can connect the VCR or cable box to the antenna input of the TV and connect the DVD player to the video input. Using the TV remote control, you can then switch between channel 3 (or 4) and the auxiliary video input.

Do not connect the DVD player through the VCR. Most movies on DVD use Macrovision protection (see Chapter 5) which affects VCRs and may cause problems such as a repeated darkening and lightening of the picture. You also may have problems with a TV/VCR combo, because many of them route the video input through the VCR circuitry. In this case, the only solution would be to obtain a device that removes Macrovision from the signal, but such a device may prove difficult to locate.

Most DVD players provide two or three video connection options, which are detailed in the following sections.

Component Video

Prior to the introduction of digital connections, component video was the preferred method for connecting a DVD player to a video system. A component video output provides three separate video signals in RGB or $Y'P_bP_r$ (or Y', B-Y', R-Y') format. These are two different formats that are not directly interchangeable.

Unlike composite or s-video connections, component signals are discrete and do not interfere with each other. Thus, a component connection is not subject to the picture degradation that might be caused by video crosstalk.

Two versions of component video exist: interlaced scan and progressive scan. Interlaced scan is standard television. Progressive scan component video produces a picture with significantly more detail than interlaced scan component video and requires a progressive scan display.

Not all DVD players provide component video output. Further, few televisions have component video connections, and few receivers or A/V controllers can switch component video inputs.

Connecting Component Video

Some DVD players, notably US and Japanese models, have $Y'P_bP_r$ component video output in the form of three RCA phono or three BNC connectors. The connectors may be labeled "Y," "U," and "V," or "Y," "Pb," and "Pr," or "Y," "B-Y," and "R-Y,"[2] they may be colored green, blue, and red, respectively.

Some DVD players, notably European models, have RGB component video output via a SCART connector or via three RCA phono or three BNC connectors. The RGB connectors are generally labeled "R," "G," and "B" and may be colored to match. Hook a SCART cable from the player to the video system, or hook three video cables from the three video outputs of the player to the three video inputs of the video system.

S-Video

All DVD players have an s-video (Y/C) output, which generally provides a better picture than composite video output, unless the s-video cable is very long. In fact the picture from an s-video connection is only slightly inferior to a component connection. S-video provides more detail, better color, and less color bleeding than composite video. Another advantage of s-video is that the luma (Y) and chroma (C) signals are carried separately. The best results are achieved when they are not combined by the player and then reseparated by a comb (or similar) filter in the TV. S-video is sometimes erroneously referred to as S-VHS because it was popularized by S-VHS VCRs. Please note that most low-end A/V receivers are unable to switch s-video signals, or may only allow for it on a few inputs.

Connecting S-Video

When making an s-video connection, use an s-video cable from the player to the video system. The round, 4-pin connectors may be labeled Y/C, s-video, or S-VHS.

---

[2]Many DVD players label the YPbPr connectors as YCbCr. This is incorrect because YCbCr refers only to digital component video signals, not analog component video signals.

Composite Video

This is the most common but lowest-quality connection. All DVD players have standard baseband video connectors, the same type of video output provided by most VCRs, low-cost camcorders, and video game consoles. This signal is also called *composite video baseband signal* (CVBS).

Connecting Composite Video

For composite connections, use a standard video cable from the player to the video system. If the connector is an RCA phono type, it is usually yellow. Connectors may also be BNC type. Either connector type may be labeled "video," "CVBS," "composite," or "baseband," etc.

RF Audio/Video

This is the worst way to connect a DVD player to a television and is only provided by a few players for compatibility with older televisions that only have an antenna or cable connection. The RF signal carries both audio and video modulated onto a VHF carrier frequency, the type of output frequently provided by VCRs and cable boxes.

Connecting RF Audio/Video

For an RF connection, use a coaxial cable with type F connectors from the player to the antenna input of the TV. The connectors may be labeled "RF," "TV," "VHF," "antenna," "Ch. 3/4," or something similar. If the TV antenna connection has two screws rather than a screw-on terminal, a 75- to 300-ohm adapter is needed. Set the switch near the connector on the back of the player to either channel 3 or channel 4, whichever is not used to broadcast a local television station in your area. Tune the TV to the same channel. If you have a TV with only RF antenna inputs, you will need to either get a DVD player with RF output or an RF modulator ($20 to $30). We must warn you that going the DVD to RF route may result in exceptionally poor video quality.

Digital Connections

When DVD players were first introduced, none included digital video connections or digital bitstream connections, in part because the copyright protection and encryption systems for digital video had not yet been established and partly because almost no other consumer equipment provided digital connections.

With the emergence of a variety of digital display technologies, and with the adoption of DVI and HDMI connection standards, DVD players may now be connected to a digitally equipped display, with no loss of video quality. The digitally encoded data on the disc stays in a digital realm for presentation on the display.

HDMI is fully backward compatible with DVI, which allows a DVI capable DVD player to be connected to a display with HDMI, using an adapter cable for DVI to HDMI. The reverse condition is also possible, where the DVD player is HDMI capable and the display is DVI capable.

The primary difference between DVI and HDMI is that DVI is video only, but both standards are capable of handling uncompressed digital video. The HDMI advantage is that it can contain multichannel digital audio, too.

Please revisit the earlier sections in this chapter — Getting Connected, Digital Upscaling, and Choosing A DVD Player — for additional information.

Understanding Your DVD Player

Remote Control and Navigation

The "Title" key, which is frequently referred to as the "Top Menu" key, should take you to the main menu for the disc. The "Menu" key is intended to take you to the menu that is most appropriate for your current location on the disc. Discs with only one program on them often disable the "Title" key and use the "Menu" key to get to the main menu.

Menus can have submenus for audio, subtitle, and chapter selection. Some remote controls let you directly access these submenus by pressing a combination of keys. If you cannot execute direct access from the remote control, you must use the onscreen buttons to get to the submenus.

The "Return" key should perform the valuable function of taking you up a level, somewhat like the "Back" button on a Web browser, but unfortunately it is still not understood by some disc authors and is not well supported.

Player Setup

Field/Frame Still

Some players let you choose between showing fields or frames when paused. Frame stills look better because they show all lines from both fields rather than half the lines from one field, but with an interlaced video source, a frame still can produce odd twitter effects. Some players have an auto setting that enables the player to determine which type of still is best for the current video.

Digital Audio Output

Most players give you a choice of PCM or Dolby Digital output on the digital audio connectors. The Dolby Digital setting also selects DTS, when available. Some players enable you to specify DTS as the primary choice when available. Some players can automatically choose the DTS track; others require you to choose it manually.

In addition, some players can automatically choose the first 5.1-channel soundtrack, given that many discs come with both 2-channel Dolby Digital and 5.1-channel Dolby Digital. This solves the annoying problem of being partway into a movie with a nagging thought in the back of your mind that it does not sound quite right, only to realize that the 2-channel Dolby Digital track was selected instead of the 5.1-channel track.

Future Connections

When HD DVD and BD players hit store shelves, there will be a whole new set of connection challenges. One of the greatest difficulties for player manufacturers will be determining how best to connect legacy equipment. For example, if you are one of the million or so video enthusiasts who purchased an "HD Ready" television back when that meant the TV did not contain an HD tuner and only supported analog component ($Y'P_bP_r$) high definition inputs, you should know that there is a battle being fought in your name. Unfortunately, many of these displays were sold before there was any sort of analog content protection for high definition video signals. The content industry wants to protect its content. The consumer device manufacturers want to protect their customers. As a result, many in the content industry want to prevent the upcoming high definition players from having (unprotected) high definition analog video outputs. The consumer display and device manufacturers, on the other hand, are fighting for your right to use the television you bought in the manner expected.

Unfortunately, that's not the only challenge that must be overcome. At last count, there were more than 39 million A/V receivers on the market that had built-in decoders for Dolby Digital and/or DTS audio. Regrettably, none of these receivers directly support Dolby Digital Plus, DTS-HD, or Dolby TrueHD decoding, and very few provide a mechanism for delivering up to eight channels of uncompressed audio. In fact, most continue to sport S/PDIF or Toslink connectors, which are only able to carry up to two channels of uncompressed audio data. This could prove to be particularly problematic for future players because the new HD disc formats are expected to include audio mixing capabilities that would allow commentaries and sound effects to be mixed in with the feature audio content during playback. This could pose a problem, because in order to mix audio from different encoded sources, it must all be decoded in the player and mixed in the uncompressed audio domain. If any of the audio sources has more than two channels, the mixed result will exceed the capacity of a standard digital audio connection.

Several solutions are being considered for this audio bandwidth problem, among them an audio encoder embedded in the player that would recompress the high bandwidth audio data into, for example, a Dolby Digital 5.1 stream for delivery to the legacy decoder. Other solutions involve implementing a bypass option, in which the user can choose to bypass audio mixing and simply send the primary Dolby Digital or DTS-encoded audio stream to his or her external A/V receiver. Still other options require new A/V receivers to support HDMI or other high bandwidth methods for delivering up to eight channels of uncompressed digital audio.

A new addition to next-generation players will be an increased level of support for interaction devices, such as mice, joysticks, game controllers and keyboards. Many of the applications envisioned for next-generation players require a far more sophisticated level of control than what the traditional infrared remote control offers. At the very least, one can expect to see diagonal arrow keys on the remote in addition to the standard Up, Down, Left and Right keys. It is speculated that many players will offer standardized Universal Serial Bus (USB) ports to allow connections from a wide range of standardized devices. Both HD DVD and BD also support generic controller programming interfaces that make it possible for advanced applications to support new types of controllers yet to be conceived.

The Player Inside

Look inside a DVD player and you'll see a complex collection of microchips and circuitry. (Well, what did you expect to find?) Although you may not be able to tell just by looking, all DVD players have components that serve essentially the same functions (see Figure 10.13).

Figure 10.13 DVD-Video Player Block Diagram

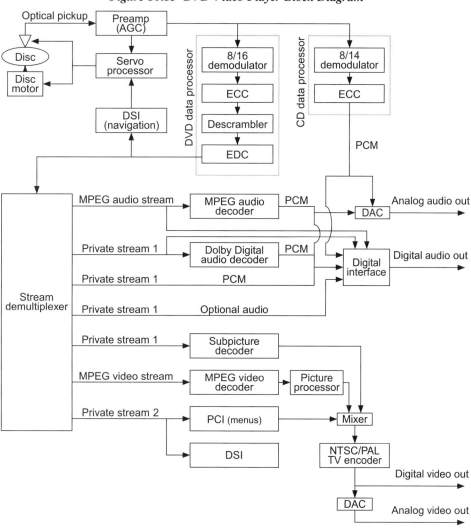

Most DVD player functions have been pulled together into a handful of chips, and they can be represented by the following major components:

Laser Pick-up and Servos. The laser pick-up and servos control the physical reading of pits and lands on the disc. Coherent light of an appropriate wavelength is cast onto the disc and the signal generated by its reflection is processed through various algorithms which collect the raw data that is physically molded on the disc.

Demodulator & Error Control Blocks. In order to make sense of the raw signal being picked up from the disc, the data must be passed through a demodulator and error control blocks. Here, the stored data is unwrapped and checked for integrity. If errors are detected, various procedures are available for calculating the correct values. Failing that, the player can make another attempt to read what may have been a damaged or obscured area on the disc.

Data Decryption Circuitry. For a disc that uses security mechanisms like content scrambling system (CSS) in DVD-Video, data is encrypted on the disc and must be decrypted before it can be of use to the rest of the system. This process usually takes place as the data is read from the disc, before the rest of the system has access to it.

Track Buffer. Decrypted data coming from the disc is placed into a track buffer, the first of a series of queues that help isolate and contain the challenges of the physical world. When errors are encountered while reading the physical disc, the rest of the player can merrily go about its business taking data from the track buffer while the laser pick-up and error control blocks busily attempt to fix the problem. If the problem is resolved before the track buffer empties, then the rest of the player need never know there was ever any trouble. The only time the player circuitry needs to bother the laser pick-up is when it is time to jump to another location on the disc, such as starting to read another movie or navigation file.

Demultiplexer. When multiplexed video data is read from the track buffer, the first stop it makes is the demultiplexer. This component identifies each pack of data, whether it is video, audio, subpicture or navigation data, and directs it to the appropriate location for further processing.

Stream Buffers. Presentation data (audio, video and subpictures) coming out of the demultiplexer moves to individual stream buffers, one for each type of data. This is a second set of queues that helps isolate the different sections of player circuitry. These queues also assist by helping to maintain consistent timing. For instance, the player will often receive presentation data packets before they are needed in which case they wait in their corresponding stream buffers until called upon. One often hears the "leaky bucket" analogy in which data is filling one end of the stream buffer just as it is being drained from the other. The challenge is to ensure that the bucket never empties prematurely.

Decoders. Reading the presentation data coming out of the stream buffers are the decoders. Whether video, audio or subpicture decoders, their basic function is essentially the same, to read coded data from the stream buffer and translate it according to an often complex set of rules into the baseband presentation data (video frames, audio waveforms, subpicture overlays, etc.).

Presentation and Timing Circuitry. Of course, it's not enough to simply decode the individual elementary streams. The resulting baseband data still needs to be delivered to the user in a meaningful way — displaying the pictures on screen, playing the audio out to speakers or headphones, overlaying the subpictures on the video, etc., and it all has to be done at the right time. If the video frames are displayed before the corresponding audio is heard, the audio and video have lost synchronization. Likewise, if the subpictures display at the wrong time during the feature, a discontinuity is created between what is happening on screen and what the viewer is reading. Therefore, a key component to the presentation circuitry is the system clock, the master timekeeper that ensures that the individual decoded streams are presented with continuity.

Navigation Control Circuitry. Orchestrating this series of events is the navigation control circuitry, which receives the user operation commands (e.g., Play) and translates them into a series of low-level actions, such as instructing the laser pick-up to jump to a particular location on the disc and begin reading, or telling the audio subsystem to decode the second audio stream. In addition to responding to user actions, the control circuitry also interprets the navigation structures and commands authored on the disc, as well as executing predefined player logic detailed by the DVD specifications. In the end, it is the navigation control circuitry that turns a simple, linear player into a random access, highly interactive device.

Granted, this description is oversimplified, but it gives an overview of the inner workings of a DVD player. A DVD recorder, by comparison, essentially reverses the process, moving data received via the audio and video inputs to stream encoders, to multiplexers, to track buffers, and eventually burning the data via laser onto a disc.

Next-Generation Players

So what does a next-generation player look like? Well, there are some general differences depending on whether you are looking at an HD DVD player or a BD player, but there are also quite a few similarities. One aspect that both formats have in common is far greater complexity!

One of the most notable additions to next-generation players is the ability to access data from locations other than the disc itself. Specifications for both HD DVD and BD include persistent storage and network interfaces. This makes it possible to deliver applications and presentation data (video, audio and subtitles) via home networks and the Internet, as well as to store and later retrieve that data from persistent storage. Imagine, for example, being able to download the latest theatrical previews to an internal hard drive or flash memory while you watch a movie, and then have those trailers appear automatically at the end of the show.

Another key difference between the internal operation of a standard DVD player and a next-generation player is the content protection system. Instead of a relatively simple decryption system for the disc content, the new players will feature far more sophisticated systems that will allow each player to be uniquely identified and granted access to the disc's contents. In addition, there will be more complex processing of the security information, intended to better combat both casual and commercial piracy. On the surface, this may not seem to yield any significant benefit for the viewer, but think of it this way: Would you want to distribute your feature film, which took two years and several million dollars to create, in a form that has high enough resolution and quality suitable for a full theatrical presentation, without content protection? Once put in those terms, it becomes clear that the enhanced security of the new formats is a cornerstone to enabling the studios to release their movies in high definition.

Another interesting feature that you'll find in a next-generation player is a much more complex demultiplexer and presentation engine, that includes both primary and secondary decoders, and complex data pathways that quickly adapt to different content scenarios. For instance, both formats support some form of secondary video that can play simultaneously with the feature video in a sort of picture-in-picture (PIP) configuration. Likewise, a secondary audio decoder can be used to stream network-delivered audio commentaries that are decoded and mixed in real-time with the movie as it plays back. Speaking of which, the new players will also support real-time audio mixers that can combine both primary and secondary audio content along with additional sound effects.

Perhaps the most obvious addition to next-generation players differentiating them from the standard model is an interactive graphics engine. Although the specific implementation differs for each format, both next-generation disc types support strong general processing capabilities for creating advanced, media-rich user experiences. This includes a script processor or virtual machine, large amounts of system memory, complex input controllers, and a whole suite of graphics scalers and compositors that can blend high definition video, sub video, interactive graphics, subtitles, and even cursors for real-time presentation.

When compared with today's standard DVD player and a computer system, you'll find that a next-generation player is looking more and more like a computer system wrapped up in a convenient box that fits in your living room. After all these years, the overused concept of media convergence will finally be manifesting itself in a very real and tangible way.

Licensing

No single company "owns" DVD. The official format specification was developed initially by a consortium of 10 companies: Hitachi, JVC, Matsushita, Mitsubishi, Philips, Pioneer, Sony, Thomson, Time Warner, and Toshiba. Various working groups within the DVD Forum are responsible for different parts of the DVD specification, and representatives from many other companies have contributed in working groups since the original DVD-ROM and DVD-Video specifications were produced. Although people around the world have contributed to the DVD format, it is not an international standard. Some of the DVD physical format specifications were submitted to ECMA and ISO for international standardization, but they are only a small part of the complete, multivolume DVD specification.

The DVD format specification books are only available from the DVD Format and Logo Licensing Corporation (FLLC)[3] after signing a nondisclosure agreement and paying a $5000 fee for the first book, plus $500 for each additional book. Manufacture of DVD products and use of the DVD logo for nonpromotional purposes requires an additional format and logo license, with a $10,000 fee for each format. For example, a combination DVD-Video and DVD-Audio player requires a license for the base DVD-ROM format, plus a DVD-Video format license and a DVD-Audio format license, for a total of $30,000. The format books do not include information about content protection systems, which are licensed and documented separately.

The term DVD is too common to be trademarked or owned. Time Warner originally trademarked the DVD logo and has since assigned it to the DVD Format and Logo Licensing Corporation. Hardware manufacturers must license the DVD logo and certify compatibility of their products. Authoring tool developers may license the logo and have their software certified. Logo licensing is not required for certain promotional uses. Hardware distributors and retailers may use the DVD logo without a license if their product is manufactured by a licensee. System integrators may use the DVD logo without a license if the DVD components are manufactured by licensees and no additional logo is added (for example, an integrator cannot add a DVD-Video logo to a system that includes a DVD-

---

[3]Before April 14, 2000, logo and format licensing was administered by Toshiba in an interim capacity.

ROM drive unless they sign a DVD-Video license). Content producers, title distributors, and retailers are allowed to use the DVD logo without a license if the disc is replicated by a licensee.

The format and logo license do not convey any patent royalties, which are claimed by dozens of companies, some of whom banded together and pooled their patents to make licensing easier, but there is still a bewildering array of companies holding out their hands for their slice of royalties (see Table 10.2).

Table 10.2 DVD Patent Licensing

Licensing Entity	License	Cost	Who Pays
3C: LG, Philips, Pioneer, Sony	DVD and optical disc technology patents	3.5% per player, minimum $3.50; additional $0.75 for Video CD compatibility	Player manufacturers, software player developers
		$0.0375 per disc	Disc replicators
6C: Hitachi, IBM, Matsushita, Mitsubishi, Sanyo, Sharp, Time Warner, Toshiba, Victor	DVD technology patents	4% per player or drive, minimum $3, maximum $8; 4% per DVD decoder, minimum $1	Player manufacturers, software player developers
		$0.045 per ROM/Video/Audio disc; 4% per DVD-R/RW/RAM disc, minimum $0.065	Disc replicators
Thomson	DVD technology patents	~$1 per player/drive	Player manufacturers
Discovision	Optical disc technology patents	~$0.20 per disc	Disc replicators
DVD CCA	CSS, CPRM, CPPM	$15,000 annually per license category	Player manufacturers, software player developers, disc replicators, large content developers
Macrovision	Macrovision APS	$30,000 initial charge; $15,000 yearly renewal $0.04-$0.10 per disc	Hardware manufacturers (players, graphics cards) Content developers
Dolby	Dolby Digital decoding patents MLP technology	$0.66 per 2-channel decoder, $0.71 per 2-channel decoder + 2-channel encoder; ~$1.50 per multichannel decoder	Player manufacturers, software player developers
MPEG LA	MPEG-2 patents	$0.004 per disc $2.50 per player	Disc replicators, Player manufacturers, software player developers

continues

Table 10.2 DVD Patent Licensing (continued)

Licensing Entity	License	Cost	Who Pays
Verance	DVD-Audio watermarking patents	$0.04 per disc or per program	Disc replicators or content developers
		$25,000 per year (for source code) or $10,000 per year (for object code)	Player manufacturers, software player developers
		$50 per watermarked track	Audio production houses
		25% of revenue on watermarking equipment	Manufacturers of authoring or mastering systems
Nissim	Parental management patents	$0.25 per player	Player manufacturers, software player developers
Various companies and licensing entities	Packaging patents		Disc replicators and fulfillment houses

Essential DVD technology patents must be licensed from an LG, Philips, Pioneer, Sony pool (known as "3C" since there were originally three companies), a Hitachi, IBM, Matsushita, Mitsubishi, Sanyo, Sharp, Time Warner, Toshiba, Victor pool (known as 6C for the original 6 companies), and from Thomson (jokingly referred to as the "1C pool"). Patent royalties also may be owed to Discovision Associates, which owns about 1300 optical disc patents. The licensor of CSS, CPPM, and CPRM encryption technology is DVD CCA (Copy Control Association).[4] Macrovision licenses its analog antirecording technology to hardware makers but does not charge a royalty per unit. Macrovision charges a per-disc royalty to content publishers. Dolby licenses Dolby Digital decoders and encoders on a per-channel basis, and licenses MLP on a per-product basis. Philips, on behalf of CCETT and IRT, also charges per player and per disc for patents underlying Dolby Digital. DTS licenses optional DTS decoders. MPEG-LA (MPEG Licensing Administrator) represents most MPEG-2 patent holders, with licenses per player and per disc, although there seems to be disagreement on whether content producers owe royalties for discs. Nissim claims per-player and per-disc royalties for patents on the optional parental management features and for other DVD-related patents. Many DVD players are also Video CD (VCD) players. Philips licenses the Video CD format and patents on behalf of itself, Sony, JVC, Matsushita, CNETT, and IRT. Philips and Sony charge per-disc royalties for DVD+R. Hewlett Packard and Philips state that there are no licensing costs for manufacturers who include DVD+RW read capability in their units. Implementation of the DVD-RAM specification incurs no royalties so long as no patented technologies are used.

The various essential licensing fees add up to more than $14 in royalties and about 20 cents per disc. Disc royalties are paid by the replicator.

---

[4]Before December 15, 1999, CSS licensing was administered on an interim basis by Matsushita.

New Format Licensing

Many of the patents that apply to DVD also apply to subsequent optical disc formats, although some of the original CD-related optical disc patents are beginning to expire. As per usual, patent pools and licensing details won't be settled until after the formats are launched. The primary new licenses are for additional audio and video codecs: DTS is now mandatory, Coding Technologies licenses patents for aacPlus, Via Licensing (a subsidiary of Dolby) charges for AVC/H.264 and DVB-MHP (which might apply to BD-J), MPEG LA also covers AVC/H.264 and has started the process to pool patents for VC-1, Microsoft holds VC-1 patents and charges royalties for implementations outside of Windows. The good news is that many of the newer technology license programs include caps, where payments stop after a few million dollars. The interactive features of new formats are also grounds for patent royalties from companies such as Microsoft, Panasonic, Toshiba, Sonic Solutions, Sony, Sun (for Java), and others.

Chapter 11
Myths

Well, where should we start? Do we start with setting bad information straight, or launch right into debunking those "gee, it sounds right" sorts of nonsense? The myth chasm is wide and deep when it comes to DVD, and now that HD DVD and BD are at your door the "oh, good golly, where did you hear that, for cryin' out loud" stories are already comin' at us thick 'n fast.

Apparently, some people had nothing better to do while waiting for DVD to appear than to sit around and misconstrue its characteristics. And, we can rest assured that some of those same folk will conjure fantastically on the new shiny disc flavors, too. Some myths quickly met their deserved deaths once DVD proved itself, but many others continue to circulate, like urban legends of microwaved cats and kidney thefts.

Presented in no special order, this chapter tackles myths that won't die, myths that sound true but just think about 'em for a minute, and myths that even some really respectable people bought into whose names we're keeping secret for a whole bunch o' money.

Myth: "DVD is not HDTV Compatible"

This is one of those "think about it for a minute, and it'll come to you" kind of myths. Do you have a VCR? Do you have an HD-capable television? Are you watching VHS tapes playing from your VCR while sitting on your couch watching your really too big for the house television display? Well, current DVD players are compatible with HDTVs in the same way that VCRs and camcorders are compatible. Manufacturers have incorporated the signal processing technology that allows us to connect standard definition video devices to their latest high definition displays.

In order to take advantage of the digital nature and progressive scan of HDTV, new DVD players are now on the market that make current discs look even better on HDTVs. The disc is standard resolution, but the combined presentation of a progressive scan disc on a progressive scan display markedly raises the image quality level. Eventually, HD DVD and BD players and discs will become available, with the result that those who have HDTVs will start to build a new library of their favorite titles in the new disc formats. Thus making the content producers and copyright holders just as happy as the television manufacturers.

Myth: "DVD Is Revolutionary"

DVD is evolutionary, not revolutionary. The printing press was revolutionary. Television was revolutionary. Even CDs can be considered revolutionary because they were a completely new way of storing digital audio and computer data on a compact optical disc. But DVD is not fundamentally more than the evolution of CD and the refinement of Video CD.

Other than digital video and some clever features, nothing is radically different between DVD and VHS, between DVD and laserdisc, or between DVD-ROM and CD-ROM.

Myth: "DVD Will Fail"

Yes, it seems that there are still some people predicting the demise of DVD. It is now way more than past the time to put this myth in the dumper. As DVD was being developed, many pundits predicted that it would be a flop, joining the ranks of other neglected consumer electronic innovations such as quadraphonic sound, the 8-track tape, the Tandy/Microsoft VIS, and the digital compact cassette. In less than three years, however, DVD became the most successful consumer electronics product ever. Hundreds of companies supply DVD products and services: all major consumer electronics manufacturers (and many minor ones), all major Hollywood movie studios (and scores of independent filmmakers), many major music labels (as well as indie labels), all major computer hardware manufacturers, countless audio/video production houses, and the rapidly growing ranks of corporate A/V departments. On the consumer entertainment side, DVD continues to fulfill its destiny of replacing VHS tape in a decade or two. On the computer side, DVD-ROM drives and recordable DVD drives are inexorably replacing CD technology, to the point where it is now almost impossible to buy a PC without a DVD drive.

The possibility did exist that DVD-Video would never capture more than a niche market, similar to laserdisc, but the success of DVD-ROM was virtually secured from the beginning. The ever-expanding needs of computer data and multimedia require a capacious medium for storage and distribution. CD-ROM has rapidly been supplanted by DVD-ROM, and the data hoarders are lusting for even greater capacity discs. We can only hope they will be sated when BD and HD DVD are fashioned into -ROM flavors.

The window of opportunity for new technology grows smaller all the time, as evidenced by such not-quite-failures as S-VHS, DAT, and MiniDisc. But none of these can be compared with DVD, which had a mainstream computer counterpart holding open the door to acceptance. In fact, DAT is arguably the most successful of these other products because it also can be used for computer data backup.

DVD-Video has more industry support than any new entertainment product in the history of consumer electronics. The annual sales income of the 10 founding DVD companies alone is over $350 billion, more than the gross domestic product of many countries. Staggering amounts of money were spent to develop DVD, and even more has been spent to produce and market it.

Myth: "DVD Is a Worldwide Standard"

If only this were so. DVD is still closely tied to the NTSC and PAL television formats. All PAL DVD players can play NTSC discs, but very few NTSC DVD players can play PAL discs. Even worse, DVD includes regional codes that can prevent a disc from being played on players sold in other countries. See Chapter 5 for further explanation.

Technically, DVD is not a "standard" at all in the formal sense. Just like CD, it is a proprietary but open standard created by a group of companies motivated by mutual interests and anticipated profits. Existing standards such as MPEG video and the UDF file format were adopted for DVD. Some of the fundamental parts of the DVD specification, such as the physical formats for read-only and writable discs, have been submitted and approved by official standards bodies such as ECMA and ISO. However, the important parts of the standard, such as the application formats for video and audio, are proprietary to the DVD Forum. Both the official standards and the proprietary specifications are subject to patent royalties.

Myth: "Region Codes Do Not Apply to Computers"

Regional codes apply to DVD-Video discs played in DVD-ROM drives. Every DVD-ROM drive is either set to a region by the manufacturer or must be set by the user before a region-coded disc may be played. Newer (RPC2) drives allow up to five region changes before the region code is set permanently. Of course, there are ways around regional restrictions, just as with standalone DVD-Video players. Regional codes do not apply to PC software or DVD-Audio discs, only to DVD-Video.

Myth: "A DVD-ROM Drive Makes Any PC a Movie Player"

Most DVD computers can play DVD-Video discs, but some, especially those which had a DVD-ROM drive installed post-purchase of the computer, do not have everything that is needed. A computer can play DVD-Video movies only if it has all the right stuff. A computer, even a clunker such as a 350-MHz Pentium II with a modicum of video hardware or a Mac G4, only needs a DVD-ROM drive and DVD playback software. Slower computers require additional DVD playback hardware and, even then, must run faster than 100 MHz to handle the load. See Chapter 15 for more about DVD and PCs.

Myth: "Competing DVD-Video Formats are Available"

This statement is a myth in the sense that inadequately educated friends and "advisors" sometimes cautioned others not to buy DVD players because supposed competing formats existed. As far as DVD-ROM and DVD-Video go, one and only one format exists. The confusion seems to be partly a carryover from the earlier competition between DVD's progenitor prototypes, SD and MMCD. Alas, in the case of competing recordable formats and DVD-Audio discs that will not play in DVD-Video players, this is anything but a myth, as discussed in Chapters 4 and 6.

Myth: "DVD Players Cannot Play CDs"

Even though the DVD specification makes no mention of CD compatibility, all DVD players can play audio CDs — as long as they are the commercially stamped kind. Many DVD-Video players cannot read CD-R discs, although most can read CD-RW discs. Compatibility with other CD formats varies. Only about 50 percent of DVD players can read Video CDs (assuming they are not on CD-R media), and only a very few players can play Super Video CDs or MP3 CDs. And, of course, DVD-Video players cannot play CD-ROMs.

Many people pushed the idea of putting DVD-Video content onto a CD-R. Aside from physical compatibility problems, these so-called MiniDVDs only played in DVD computers and in a few odd player models designed around DVD-ROM drives.

Some early DVD-ROM drives could not read CD-Rs, but now all modern DVD drives — apart from the very cheapest models — can read CD-Rs. Most DVD computers can play Video CDs and MP3 CDs, but many lack the software needed to play Super Video CDs.

Myth: "DVD Is Better Because It Is Digital"

Nothing is inherent to digital formats that magically makes them better than analog formats. The celluloid film used in movie theaters is analog, yet few people would say that DVD-Video is better than film. Japan's HiVision television had much higher video quality than DVD, but it was analog. Conversely, the quality of digital video from CD-ROMs is certainly nothing to write home about.

The way DVD stores audio and video in digital form has advantages, not the least of which is the ability to use compression to extend playing times by lowering data rates. The quality and flexibility of DVD stand out when compared with similar analog products. It is a mistake, however, to make the generalization that anything digital must be superior to anything analog.

Myth: "DVD Video Is Poor Because It Is Compressed"

During the roll-out of DVD, much ado was made of the "digital artifacts" that supposedly plague DVD-Video. While it is true that digital video can appear blocky or fuzzy, a well-compressed DVD exhibits few discernible artifacts on a properly calibrated display. Many early discs, especially demonstration discs, were created with hardware or software that was partially finished or not fully tested. Compression techniques have improved exponentially; hardware video encoders have improved, producing better pictures within the same compression constraints. Nowadays, the cutting edge MPEG-2 encoder is software only. The improvements have benefited all existing players, and remaining minor glitches and quality problems are disappearing as compression engineers refine their craft.

The term *artifact* refers to anything that was not in the original picture. Artifacts can come from film damage, film-to-video conversion, analog-to-digital conversion, noise reduction, digital enhancement, digital encoding, digital decoding, digital-to-analog conversion, NTSC or PAL video encoding, Macrovision, composite signal crosstalk, connector problems, impedance mismatch, electrical interference, waveform aliasing, signal filters, television picture controls, tube misconvergence, projector misalignment, and much more. Many people blame all kinds of visual deficiencies on the MPEG-2 encoding process. Occasionally, this blame is placed accurately, but usually it is not. Only those with training or experience can tell for certain the origin of a particular artifact. If an artifact cannot be duplicated in repeated playings of the same sequence from more than one copy of a disc, then it is clearly not a result of MPEG encoding. Here are a few of the most common artifacts:

- *Blocks* are small squares in the video. These may be especially noticeable in fast-moving, highly detailed sequences or video with high contrast between light and dark. This artifact appears when not enough bits are allocated during MPEG compression for storing block detail.

- *Halos* or *ringing* are small areas of distortion or dots around moving objects or high-contrast edges. This is called the *Gibbs effect* and is also known as *mosquitoes*, or *mosquito wings*. This is an artifact of MPEG encoding, but it is easy to confuse with edge enhancement.

- *Edge enhancement* is a digital picture-sharpening process that is frequently over-done, causing a "chiseled" look or a ringing effect like halos around streetlights at night. This happens before MPEG encoding.

- *Posterization* or *banding* which results in bands of colors or shading in what should be a smooth gradation. This can come from the MPEG encoding process or the digital-to-analog conversion process in the player. It also can happen on a computer when the number of video colors set for the display is too low.

- *Aliasing* occurs when angled lines have "stair steps" in them. This artifact is usually caused by lines that are too sharp to be properly represented in video, especially when interlaced.

- *Noise* and *snow* refer to the gray or white spots scattered randomly throughout the picture, or graininess. This may result from film grain or low-quality video.

- *Blurriness* refers to low detail and fuzziness of video. This results from low-quality video or too much filtering of the video before encoding.

- *Worms* or *crawlies* are squirming lines and crawling dots. Usually this results from low-quality video or bad digitizing. This also may be the result of chroma crawl in composite video (either in the original source or from the connection from the DVD player).

However, the number one cause of bad video is a poorly adjusted TV! The high fidelity of DVD video demands much more from the display. Lower the sharpness value and turn the brightness down, and you'll be surprised at the improvement these simple adjustments have wrought. See "How to Get the Best Picture and Sound" in Chapter 14 for more information.

Myth: "Compression Does Not Work for Animation"

It is often claimed that animation, especially hand-drawn cell animation such as cartoons and Japanese anime, does not compress well with MPEG-2. Others claim that animation is so simple that it compresses better. Neither is generally true.

Supposedly, jitter between frames that may be caused by differences in the drawings or in their alignment causes problems. Modern animation techniques produce very exact alignment, so usually no variation occurs between object positions from frame to frame unless it is an intentional effect. Even when objects change position between frames, the motion estimation feature of MPEG-2 can easily compensate. That said, we should point out that many MPEG-2 encoders were optimized for live action and may have trouble with animated content.

Because of the way MPEG-2 compresses video, it may have difficulty with the sharp edges common in animation. This loss of high-frequency information can show up as ringing or blurry spots along high-contrast edges. However, at the data rates commonly used for DVD, this problem normally does not occur. The complexity of sharp edges tends to be balanced out by the simplicity of broad areas of single colors.

Myth: "Discs Are Too Fragile to Be Rented"

The Blockbuster Video chain allegedly took a stance early on that it would not rent DVDs unless the format included a protective caddy. Just look at what they are doing now. Sheesh, try to rent a VHS from 'em. Designers of DVD, having learned from the bad experience of CD-ROM caddies and not wishing to more than double the cost of discs by requiring a caddy or protective shell, politely ignored such requests. Within two years after DVD was released nearly every video rental chain and outlet carried the discs.

DVDs are, of course, liable to scratches, cracks, accumulation of dirt, and fingerprints. But these occur at the surface of the disc where they are out of focus to the laser. Damage and imperfections may cause minor channel data errors that are easily corrected. A common misperception is that a scratch will be worse on a DVD than on a CD because of higher area density and because the audio and video are compressed. DVD data density is about seven times that of CD-ROM, so it is true that a scratch will affect more data. But DVD error correction is more than 10 times more effective than CD error correction. This improved reliability more than makes up for the density increase. Major scratches on a disc may cause uncorrectable errors that will cause an *input-output* (I/O) error on a computer or show up as a momentary glitch in the DVD-Video picture, but many schemes exist for concealing errors in MPEG video.

Laserdiscs, music CDs, and CD-ROMs are likewise subject to scratches, but many video stores and libraries offer them. DVD manufacturers are fond of taking a disc, rubbing it vigorously with sandpaper, and then placing it in a player, where it plays perfectly. Disc cleaning/polishing products can repair minor damage. Commercial polishing machines can restore a disc to pristine condition after an amazing amount of abuse. DVD does not perform any worse in a rental environment than tapes.

Myth: "Dolby Digital Means 5.1 Channels"

Do not assume that the "Dolby Digital" label is a guarantee of 5.1 channels of digital surround style audio. Dolby Digital is an encoding format that can carry anywhere from 1 to 6 discrete channels. A Dolby Digital soundtrack may be mono, stereo, Dolby Surround stereo, Dolby Surround EX, and so on.

Most movies produced before 1980 had a monophonic soundtrack only. When these movies are put on DVDs, unless a new soundtrack is mixed, the original soundtrack is encoded into a single channel of Dolby Digital.

In some cases, more than one Dolby Digital version of a soundtrack is available: a 5.1 channel track and a track specially remixed for two-channel Dolby Surround. Please note, it is normal for a DVD player to indicate playback of a Dolby Digital audio track while the concurrently employed receiver indicates Dolby Surround. This means that the disc contains a two-channel Dolby Surround signal encoded in Dolby Digital format. The same applies to DTS, although very few DTS tracks are encoded with fewer than 5.1 channels.

Myth: "The Audio Level from DVD Players Is Too Low"

People complain that the audio level from DVD players is too low. In truth, the audio level on everything else is too high! Movie soundtracks are extremely dynamic, ranging from near silence to intense explosions. In order to support an increased dynamic range and hit peaks (near the 2V RMS limit) without distortion, average sound volume must be lower. This is why the line level from DVD players is lower than from almost all other sources. The volume level among DVDs varies, but it is more consistent than on CDs and laserdiscs. If the change in volume when switching between DVD and other audio sources is annoying, check the equipment to see if you are able to adjust the output signal level on the player or the input signal level on the receiver.

Myth: "Downmixed Audio Is Not Good Because the LFE Channel Is Omitted"

The LFE channel is omitted for a good reason when Dolby Digital 5.1-channel soundtracks are mixed down to two channels in the player. The LFE channel is intended only for extra bass boost, because the other 5 channels carry full-range bass. Audio systems without Dolby Digital capabilities generally do not have speakers that can properly reproduce very low frequencies, so the designers of Dolby Digital chose to have the decoders throw out the LFE track to avoid muddying the sound on an average home system. Anyone who truly cares about the LFE channel should invest in a receiver with Dolby Digital, bass management, and a separate subwoofer output.

Myth: "DVD Lets You Watch Movies as They Were Meant to Be Seen"

This refers to DVD's 1.78 anamorphic widescreen feature, which is close to the most common movie aspect ratio (1.85). However, many movies have a wider shape than widescreen TVs. Thus, even though they look much better on a widescreen TV, they still have to be formatted to fit the less oblong shape, usually with black bars at the top and bottom. See "Aspect Ratios" in Chapter 3 for more information.

Myth: "DVD Crops Widescreen Movies"

As mentioned in the preceding paragraph, some movies are wider than DVD's widescreen format. Some people assume that the only way to make them fit is to crop the sides of the picture. However, in almost all cases, Cinemascope and similarly wide movies are letterboxed to fit the entire original width within DVD's widescreen picture shape. Although, it is true that for standard TV display, widescreen movies are often cropped. Again, please see "Aspect Ratios" in Chapter 3 for more information.

Myth: "DVD Will Replace Your VCR"

When DVD was first released, this was a misleading statement because DVD players could not record. DVD video recorders were introduced in 2000, and as they dropped in price over the first few years of the new millennium, they have become viable replacements for VCRs. However, it will take years for recordable DVD to make an appreciable dent in the installed base of VCRs. Plus, other technologies — personal video recorders, computers, digital videotape — are also vying to be the VCR of the present/future.

The incompatibilities between the various recordable DVD formats initially delayed their acceptance. In the case of videotape, once the battle between VHS and Betamax was over, you could expect that any VHS tape would work in any other VCR. DVD's version of the battle of the formats, which at first blush seemed to be bigger and more brutal, rapidly deflated with the introduction of multiformat recorders.

Myth: "People Will Not Collect DVDs Like They Do CDs"

A common argument against the success of video is that it is not as collectible as music. Music may be listened to over and over, whereas most movies only bear watching a few times. Music can play in the background without disrupting everyday tasks, but movies require devoted watching.

It is true that the average household will own more CDs than DVDs. However, music combined with video is more collectible than music alone — and more playable than movies. Research shows that televisions are often left tuned to music channels such as MTV with no one in the room.

Beyond music, the extra features of DVDs make them much more collectible than videotapes. "Special edition" DVDs packed with audio commentaries, outtakes, interviews, featurettes, and other goodies are often too much to be digested in a rental period, making them more likely to be purchased.

Myth: "DVD Holds 4.7 to 18 Gigabytes"

As mentioned in the Introduction, the abbreviation GB, when referring to storage capacity, sometimes stands for gigabytes, which are measured in powers of 2, and sometimes stands for billions of bytes, in powers of 10. A DVD holds 4.4 to 15.9 gigabytes, which is the same as 4.7 to 17 billion bytes. Advertisers favor the bigger numbers, and in some cases the marketing maniacs pushed the boundaries of creative mathematics by rounding 17.01 up to 18, when describing the dual-layer/double-side variety of DVD. Just for grins, keep on the lookout for how the latest and greatest capacities in HD DVD and BD are going to be, and are already being, described. We humbly suggest that you refer to the tables and figures in other chapters of this book for a reality check on disc capacities.

Myth: "DVD Holds 133 Minutes of Video"

The oft-quoted length of 133 minutes for DVDs is apocryphal. It is simply a rough estimate based on a less-than-optimal average of 3.5 Mbps for a video track and three 384 kbps audio tracks. If there is only one audio track, the average playing time goes up to 159 minutes. The video rate is highly variable — a single-layer DVD-5 actually can hold over nine hours of VHS-quality video. There are two constants: disc capacity and maximum data rate (which is 9.8 Mbps for video and 10.08 Mbps combined with audio and subpictures). All the rest is variable. Even the capacity varies depending on the number of sides and layers. A dual-layer DVD-9 holds over four hours of high-quality video, and a double-sided, dual-layer DVD-18 holds over eight hours. Using MPEG-1, a DVD-18 can contain a mind-numbing 33 hours of video, which also would be butt-numbing if you tried to watch all of it in one sitting.

It is said that the figure of 133 minutes was originated by the trade press. The original SD proposal achieved approximately 142 minutes of playing time, but by adopting 8/16 modulation from the MMCD format, manufacturers sacrificed 6.3 percent of disc capacity. Supposedly, a clever but clueless journalist applied 6.3 percent to 142, and the meaninglessly exact figure of 133 minutes has stuck ever since.

Anyone talking or writing about DVD-Video should make things easier for themselves and their audience by simply stating that a single layer holds over two hours of video. If more precision is required, the nice round figure of two hours and 15 minutes (135 minutes) is just as accurate.

It should be noted that this applies only to DVD-Video. A DVD-ROM may hold any sort of digitized video to be played back on an endless variety of computer hardware or software. If someone developed a revolutionary new holographic wavelet compression algorithm, a DVD-ROM might hold three hours of film-quality video. This is unlikely, but it is important to differentiate between the deliberate restrictions of DVD-Video and the limited capacity of DVD-Video players compared with the wide-open digital expanse of DVD-ROM and computers.

Myth: "DVD-Video Runs at 4.692 Mbps"

This figure is about as meaningless as 133 minutes. The figure of 4.692 Mbps is presented as the average data rate for DVD-Video. But that figure requires assumptions of content and further would require that all DVDs be produced alike. Table 11.1 reflects one approach that would result in the figure of 4.692 Mbps.

Table 11.1 How to Create a Meaninglessly Exact Number

Bit Rate	Count	Total
3.5 Mbps average video	1	3.500 Mbps
384 kbps audio	3	1.152 Mbps
10 kbps average subpicture	4	0.040 Mbps
		4.692 Mbps

But what if there was no subpicture track? Then the pristine sum is off by an egregious 0.04 Mbps. And if only one audio track and one subpicture track were used, the so-called average data rate would only go up to 3.924, an error of almost two-tenths, when juxtaposed with 4.692!

The probable genesis for this number was that it was calculated from the required 133 minutes of length (which was calculated from the original 135) by figuring out what video data rate was left over after accounting for the audio. Then, the subpicture track was thrown in to even things up.

Sarcasm aside, what usually happens is that the content is compressed to fit the capacity of the disc. If the movie is 110 minutes long and has two audio tracks, the video bit budget can be set at a much higher 4.9 Mbps to achieve better quality. Or, a 2 1/2-hour movie might be compressed slightly more than usual if the disc producer determines that the video quality would remain acceptable.

The maximum video data rate of DVD is limited to 9.8 Mbps by the DVD-Video specification. The maximum combined rate of video, audio, and subtitles is limited to 10.08 Mbps. The average data rate is almost always lower, usually between 4 and 6 Mbps. Some people assume that DVD is therefore unable to sustain a continuous rate of 9.8 Mbps or higher. This is not the case. All DVD players and drives can maintain an internal data rate of at least 11.08 Mbps. DVD-Video players have a 1 Mbps overhead for navigation data. A movie compressed to a constant bit rate of 10.08 Mbps would play for 62 minutes.

Single-speed DVD-ROM drives can sustain a transfer rate of 11.08 Mbps, with burst rates as high as 100 Mbps or more, depending on the data buffer and the speed of the drive connection. DVD-ROM drives with higher spin rates are accordingly faster, although most multispeed DVD-ROM drives cannot maintain the maximum quoted speed across the entire surface of the disc — they only achieve the maximum data rate when reading from the outer edge.

Myth: "Some Units Cannot Play Dual-Layer or Double-Sided Discs"

Dual-layer compatibility is required by the DVD specification. Almost every DVD-Video player and DVD-ROM drive, even the first ones sold, can read dual-layer discs. Occasional problems with dual-layer discs are caused by faulty disc production, flawed players (which often can be fixed with a firmware upgrade), or bugs in DVD-ROM driver software (which can be upgraded to fix the problem).

All players and drives can read double-sided discs — as long as you flip the disc over. So far, only DVD jukeboxes can switch automatically to the other side of a disc. This capability eventually may appear on a standard player, but since a single-sided/dual-layer disc can hold four hours or more of continuous video, demand for it is not very high. Most people appreciate the bathroom break. Some combination laserdisc/DVD players can play both sides of a laserdisc but not both sides of a DVD.

Myths: "JavaScript is Java" and "JavaScript is the same as ECMAScript"

The first myth is pretty easy to deal with. Java is an island in Indonesia south of Borneo, as well as the name of an object-oriented programming language primarily used for network applications and applets. Programs written in Java do not rely on a specific operating system, so they are platform independent. Javascript is an unfortunately confusing name given to a very dynamic scripting programming language used to enable scripting access to objects embedded in other applications.

JavaScript initially was used to expand the flexibility, look and style of HTML pages on websites and in browser windows. Now, JavaScript is also being utilized to augment optical disc programming capabilities, providing enhanced features for content presentation, notably with web and kiosk interactivity.

Java programming language was developed by Sun Microsystems, Inc., while JavaScript was parented by Netscape Communications Corporation. Reportedly, Netscape was allowed to co-opt Java and create the word JavaScript, as part of a deal with Sun in exchange for providing support of Java applets. Sun has regretted the confusion and misbranding ever since.

Okay, what about that second myth, the ECMAScript/JavaScript duality? This one is akin to "people in Great Britain speak English and people in the United States speak English" but sometimes the language is not the same. The reality behind the myth is that all ECMAScript is JavaScript but not all Javascript is ECMAScript.

The flexibility of the Javascript programming structure accommodates a myriad of instructions. Programmers and content developers continue to push the boundaries of JavaScript instruction sets. Whereas, ECMAScript is based on the core JavaScript language that was standardized in ECMA-262, as adopted by the ECMA General Assembly in June 1997. There have been two editions of ECMA-262 published since then, and ECMA-357, defining an extension to ECMAScript, known as E4X, was published in June 2004. In short, ECMAScript is standardized by an approval process, while JavaScript is an amorphous non-standardized language structure. By the way, ECMA is the acronym for European Computer Manufacturer's Association.

Chapter 12
What's Wrong with DVD

"The best-laid plans o' mice an' men, gang aft a-gley..."

— Robert Burns[1]

Having been bloodied by the videotape pirates, the movie studios, in league with the technology companies, were determined to birth a magic bullet that would save their assets whilst vanquishing their foes. VHS had beaten Betamax, but it was analog videotape, ubiquitous and easily duplicated.

As noted in an earlier chapter, DVD was developed with the expectation that the new digital media would provide a family of applications that would satisfy the needs of a diverse consortium of companies and copyright holders. DVD has sucessfully met a great number of the objectives, yet there are more than a few inherent drawbacks, insurmountable travails, and downright failures with the media.

DVD-ROM is a well-designed update of CD-ROM. Storage space and speed are much improved, and many of the shortcomings of CD-ROM were rectified. DVD-Video, on the other hand, is a collection of compromises in which commercially and politically motivated restrictions produced technical limitations. In an overzealous attempt to protect their intellectual property, Hollywood studios unfortunately crippled the format that could have given them the best venue of expression outside a theater.

Amazingly, the most prominent shortcoming of DVD-Video is ease of piracy, or, more accurately, the perception of potential losses from piracy. Ironically, the superior digital quality of DVD has proven to be its Achilles heel. The thought of discs, legal or not, available for sale anywhere in the world was a huge concern. The ease with which unscrupulous or unthinking people might be able to make a perfect digital copy and spread it across the Internet caused many sleepless nights for studio executives. As a result their prior fears and overreactions spawned restrictive regional codes, an analog content protection system that can degrade the picture, and limits on digital audio output quality. These restraints, along with encryption and watermarking, also complicated the otherwise simple use of DVDs in computers.

DVD's shortcomings are the primary motivations behind the urgency to develop and market the next generation of optical discs, HD DVD and BD. See Chapter 6 for details on the next generation disc technologies that, it is hoped, will alleviate the failings of DVD.

This chapter presents some of the arguments that have spurred the latest retooling of optical disc technology.

---

[1]From the poem, "To a Mouse", by Scottish poet Robert Burns (1759-1796).

Regional Management

Movies are not released all over the world at the same time. A major motion picture may come out on video in the United States — months after its theatrical release — at the same time it premieres in Europe or Japan. Movie distributors worried that if copies of a movie were available in areas before their theatrical release, it would reduce the success of the film in the theaters. In addition, different distribution rights and licenses are established in different countries. For example, Disney may own distribution rights of a movie in the United States, but Studio Ghibli may own them in Japan.

Because of this, studios and distributors wanted a way to control distribution, and the DVD Forum proceeded to divide the world into six regions.[2] As a result, every DVD player has a code that identifies the geographic region in which it was sold. Discs intended for use only in certain countries have codes stored on them that identify permitted and banned regions. Thus, a DVD player will not play a disc that is not permitted to be played in the region in which the player was sold.

This regional management system wreaks havoc with import markets. An otherwise legally imported disc often may not work because of its regional codes. A more grievous problem occurs when discs and players purchased in one country do not work when taken to another. If someone in Japan moves to the United States and takes their locally purchased DVD player along, that person will not be able to play most movies purchased in the United States. Likewise, if that person buys a new DVD player in the United States, they would not be able to play many of the discs bought in Japan.

Regional management was implemented to preserve a parochial system of distribution. As the Internet breaks down national boundaries of commerce, and as digital cinema allows movies to debut in theaters worldwide at the same time, region codes will become mostly irrelevant. Nevertheless, DVD movies undoubtedly will continue to be circumscribed geographically.

DVD disc regional locks are optional; any disc can be designed to play in all players. However, most movie releases — even those which have been available for years — include regional locks. As DVD was being developed and then released, some studios stated that older movies that had already made their theatrical run would not be restricted by region. However, none of the studios have lived up to this promise.

Outside the United States, because of the inconvenience of regional codes, more than 50 percent of players sold in most countries are modified (or modifiable) to disable region coding. DVD-ROM drives also are regionally coded, and computer DVD-Video player applications check for regional codes before playing movies. Not too surprisingly, though, computer software on DVD-ROM does not use regional codes.

Copy Protection

The movie industry claims that more than $700 million is lost worldwide each year to casual illegal copying of videotapes. Surveys of video rental outlets show that more than half

---

[2]Technically, there are eight regions defined for DVD distribution — Six are geographical, the seventh is reserved, and the eighth is special nontheatrical venues (airplanes, cruise ships, hotels).

believe that their business is hurt by consumers who make copies of rental tapes. The impact of DVD recorders has yet to be assessed, as well as the impact of bootleg DVDs.

The industry's own data shows that 30 percent of consumers who were thwarted by video-tape content protection subsequently rented or bought a legitimate copy, and video retailers believe that copy-protected tapes increased their revenue by 18 percent. Regardless of the effectiveness of content protection schemes, DVD includes a number of them.

Every DVD-Video player must include Macrovision or similar analog content protection technology to deter copying of DVD onto videotape. Macrovision adds pulses to the video signal that confuse the recording circuitry of VCRs and renders a tape copy unwatchable, supposedly without affecting the picture when the DVD is played on a television. The truth is some pieces of equipment, especially line doublers and high-end televisions, are unable to cleanly display video from a DVD that has been altered by the Macrovision process.

Analog and digital copying of DVD is controlled by information specifying whether the video may be copied and, if so, once or unlimited times. Digital recording equipment must respect this information. The "copy once" setting is intended to allow consumers to make copies for personal use but not allow copies of the copies (the copy is marked for "no copies").

Digital copying is also prevented by the encryption of critical sectors on the disc. Video data that has been copied from a disc using a DVD-ROM drive will not play without the decryption keys hidden on the original disc. During normal playback, the video data stays scrambled until just before its displayed, so it cannot be intercepted and copied easily within the computer.

Even the people who designed the content protection techniques (the industry's Copy Protection Technical Working Group) freely admit that they do not expect content protection to hinder professional thieves. The MPAA claims that more than $3 billion a year is lost to professional piracy worldwide. Anyone equipped with the proper equipment to defeat Macrovision or to make a bit-by-bit digital copy of a disc can circumvent content protection barriers. In fact, the targets of these measures are people who would make a copy or two for friends. The stated goal is "to keep the honest people honest."

Movie studios and consumer electronics companies promoted legislation to make it illegal to defeat DVD content protection. The result is the *World Intellectual Property Organization* (WIPO) Copyright Treaty, the WIPO Performances and Phonograms Treaty (December 1996), and the compliant U.S. *Digital Millennium Copyright Act* (DMCA), passed into law in October 1998. Software or devices intended specifically and primarily to circumvent content protection are now illegal in the United States and many other countries. A co-chair of the legal group of the content protection committee stated, "In the video context, the contemplated legislation should also provide some specific assurances that certain reasonable and customary home recording practices will be permitted, in addition to providing penalties for circumvention." Given that most DVD movies are designed to prevent any sort of copying, even single copies for personal use, it is not at all clear how this might be "permitted" by a player.

All content protection systems are optional for DVD publishers. Decryption in players is also optional, so it is possible that a small number of DVD-ROM computers and possibly even DVD-Video players will not be able to play protected movies.

Hollywood Baggage on Computers

Given the ability of computers to play movies from DVD, the regional management and content protection requirements apply to computers as well as home players. Manufacturers of hardware or software involved in the playback of scrambled movies are required to obtain a license from the DVD Copy Control Authority. Their products must ensure that decrypted files cannot be copied, that digital outputs contain proper content protection information, and that analog outputs are protected either by Macrovision or a similar process.

The upside is that these safeguards assure Hollywood that its property will not be plundered and spread illegally from computer to computer. The downside is that most law-abiding computer owners are inconvenienced and may pay slightly more because of these protection measures. This is especially irritating to those who have no interest in watching commercial movies on their computer screens. In addition, copying and editing of legitimate video, as from digital camcorders, is made difficult or impossible because of copyright safeguards. See Chapter 15 for more details on content protection and regional management issues with computers.

NTSC versus PAL

Because DVD-Video is based on standard MPEG-2 digital video, it could have become a worldwide standard capable of working with both 525/60 (NTSC) and 625/50 (PAL/SECAM) television systems, thus allowing discs produced anywhere in the world to be played anywhere else in the world. Sadly, this is not the case. Partly in an attempt to limit the widespread distribution of discs, but primarily because it was easier, the two video formats are used for DVD discs and players. Two different surround audio standards are also used: Dolby Digital and MPEG-2 audio. Movies intended for distribution in the United States, Japan, and other countries using the NTSC system are encoded for display at 30 frames per second along with Dolby Digital audio. Whereas movies intended for release in Europe and in countries using the PAL standard are encoded for display at 25 frames per second. PAL discs can use the MPEG audio format, which many NTSC players do not recognize. On top of the format differences, playback speed is slightly faster for movies in PAL format. Because of these differences, countries with NTSC televisions require NTSC DVD playback hardware and countries with PAL televisions require PAL DVD playback hardware.

Because NTSC is the dominant standard, almost all DVD players released in PAL countries can play both types of discs as long as the right kind of television is connected. When playing NTSC discs, PAL players output a 4.43 NTSC signal (called *60-Hz PAL*) that combines the NTSC scanning rate with the PAL color-encoding system. Most PAL televisions sold since mid 1990 can handle this type of signal. Some players also provide the option to produce a pure 525/60 signal from an NTSC disc, but this requires a fully NTSC-compatible TV.

There are a few standards-converting PAL players that completely convert 525/60 NTSC video to standard 625/50 PAL output, a process that is quite complex. Achieving high quality requires expensive hardware to handle scaling, temporal conversion, and object motion analysis. The audio speed also must be adjusted to conform to the different display rate, which may prevent the use of a digital audio connection. Because most DVD players use shortcuts and inexpensive circuitry to convert video, using 60-Hz PAL output with a PAL-compatible TV provides a better picture than converted PAL output.

Such conversion is much less compatible the other way around. Most NTSC players cannot play PAL discs. A very small number of NTSC players can convert 625/50 PAL to 525/60 NTSC, but again, video quality suffers from low-cost conversion solutions.

The restrictions of television systems do not apply to computer monitors. Properly equipped computers can play movies from both NTSC and PAL discs and can decode both Dolby Digital and MPEG audio.

Tardy DVD-Audio

The DVD-Audio standard was not formalized until four years after the introduction of DVD-Video. The delay in bringing DVD-Audio to market, coupled with the ability of the already existing DVD-Video standard to deliver very high quality audio, dramatically affected the adoption of DVD-Audio by consumers. What could have been a heralded achievement of a new audio standard has been left in the dust, given the lackluster participation by music producers and performers.

Incompatible Recordable Formats

Despite many obstacles, the companies backing DVD managed to come out with a single standard for DVD-ROM and DVD-Video. The incredible success of DVD clearly proves that customers will readily adopt a common format that has widespread support. Unfortunately, this lesson seems to have been lost on the very companies that should have learned it. There are no fewer than five recordable variations of DVD. Each recordable format has different incompatibilities with the other formats as well as with existing players and DVD-ROM drives.

Although ameliorated by the introduction of multiformat DVD recording drives, there remains a less than satisfactory reliability rate when trying to play a recordable disc. DVD-capable computers, even those recently purchased, that did not perform the disc recording may require software upgrades before they are able to play recordable format DVDs. Further, standard DVD-Video players may play recordable discs as long as the content is in the DVD-Video format, but will not play a disc created in the DVD-VR (video recording) format.

On top of the physical incompatibilities, are three separate recording application formats: one for video, one for audio, and one for "streaming data" such as from a camcorder or a digital video receiver. None of these recording file formats are readable by standard DVD video or audio players. Further details of these incompatible formats are covered in Chapter 6.

Late-Blooming Video Recording

Recordable DVD arrived about a year after DVD-ROM and DVD-Video, but at that time it was only capable of recording data on computers. It took another three years for DVD home video recorders to arrive, accompanied by very high price tags and, as mentioned in the preceding section, plagued by compatibility problems.

Some people believed that DVD home recorders would never be much of a success because digital tape is more cost-effective, yet it lacks many of the DVD's advantages, such as seamless branching, instant rewind/fast forward, instant search, and durability, not to mention the appeal of shiny discs. Consequently, as the encoding technology and recording hardware became fast and cheap and the blank discs decreased in price, DVD recorders were positioned to become the VCRs of the new millennium. However, due to the stupidity of not defining a recording format before millions of players and drives were sold, and as the competition between manufacturers introduced a jumble of proprietary recorders, it is taking recordable DVD much longer to succeed than it should have.

The competition from digital tape has not materialized, either. DV and DVCPro tapes are used for acquisition, footage is ingested to a computer, and enthusiasts generate a home-produced DVD.

Playback Incompatibilities

The DVD-Video specification was put together like a patchwork quilt, with different parts coming from dozens of engineers speaking Japanese, Dutch, and English. The books detailing the format specifications were assembled and produced in Japanese and then translated into English. Engineers from a large number of companies all over the world implemented the specs in player designs. It is no surprise, therefore, that the DVD-Video feature set is incompletely or improperly implemented in many players. Even so, many incompatibility problems have resulted simply from laziness or carelessness. Many early DVD title developers had to scale back their plans after discovering that their ingeniously designed discs worked differently or not at all in different players.

For example, the DVD specification allows 999 chapters per title, but some players cannot handle more than 511. Likewise, players are supposed to support 999 program chains, but some early players balk at 244. Some cannot handle full video data rates of 9.8 Mbps, whereas others cannot deal with extra files in the root directory.

The release of *The Matrix* in 1999 brought widespread publicity to the general malady. Because *The Matrix* was the first million-seller DVD, and an unusually complex title at that, which also contained PC enhancements, more people than ever before had problems playing the disc. Some problems were caused by authoring and formatting errors on the disc, but most problems were caused by flaws in a surprisingly high number of player models.

Taking the long view, *The Matrix* and other "problem discs" did the DVD world a favor by exposing flaws in players that failed to properly play discs authored according to the DVD specification. Although the backlash to the studios, production houses, and InterActual (the company providing the PCFriendly computer enhancement software) was painful, pushing the envelope early ensured that DVD manufacturers would be more responsible about making players that work correctly.

The number of players that continue to have design flaws, even after many iterations, is inexcusable because they fail on discs that have been available for years. Some problems were caused by bugs in DVD authoring software, but as authoring software programs matured, most of them incorporated workarounds to avoid known errors and deficiencies in players. The situation continues to slowly improve with each new release of players, but producers who want their discs to work in their customers' players must still accommodate older models.

Synchronization Problems

When some discs are played in certain players, there is a "lip sync" breakdown where audio lags slightly behind the video or, in some cases, slightly precedes the video. Perception of the sync problem is highly subjective — some people are bothered by it, others cannot discern it at all. The cause is a complex interaction of as many as four factors:

1. Poor sync techniques during film production or editing (especially postdubbing or looping)

2. Improper matching of audio and video tracks in the DVD encoding/authoring process

3. Loose sync tolerances in the DVD player

4. Delay in the external decoder/receiver

Factor 1 or 2 usually must be present in order for factor 3 or 4 to become apparent. Some discs with severe sync problems have been reissued after being re-encoded to fix the problem.

In some cases the sync problem in the player (factor 3) can be fixed by pausing or stopping playback and then restarting or by turning the player off, waiting a few seconds, and then turning it back on. Unfortunately, there is no simple answer and no single, easy fix for this continuing problem. Player manufacturers and disc producers need to continue their cooperation identifying and eliminating the various causes.

Feeble Support of Parental Choice Features

Hollywood was quick to request a parental management feature for DVD but very slow to actually use it. DVDs can be designed to play a different version of a movie depending on the parental level that has been set in the player. By taking advantage of the seamless branching feature of DVD, objectionable scenes can be skipped automatically or replaced during playback. This requires that the disc be carefully authored with alternate scenes and branch points that do not cause interruptions or discontinuities in the soundtrack. Unfortunately, very few multirating discs have been produced.

As the studios do not support the built-in parental management feature, one alternative is to use a software player on a computer that can read a "play list" that tells it where to skip scenes or mute the audio. Play lists can be created for the thousands of DVD movies that have been produced without parental control information. Devices such as TV Guardian that connect between the player and the TV and read Closed Captions in order to filter out profanity and vulgar language also work with DVD, as long as the disc contains Closed Captions.

A "cottage" industry has developed where independent companies re-edit and remaster "cleaned" DVDs. Hollywood challenged these practices in court, but recent rulings have supported these post-release applications. Perhaps if the studios had taken advantage of the DVD parental choice feature in the beginning, the purveyors of these sanitizing applications and "cleansed" releases would never have evolved.

Hollywood remains unconvinced that there is a large enough demand to justify the extra work involved (shooting extra footage, recording extra audio, editing new sequences, creating branch points, synchronizing the soundtrack across jumps, submitting new versions for MPAA rating, dealing with players that do not properly implement parental branching, having video store chains refuse to carry discs with unrated content, etc.). The result is that a distinctive feature of DVD is largely wasted.

Not Better Enough

Part of the reason for the success of CD was that the technology was not overly advanced. Technical difficulties in production and playback were minimized, tolerances could be met easily, and costs could be kept low. DVD took the same approach, which helped it become cheap and widespread, but that also meant that the bar was not raised terribly high. High-definition television (HDTV) is now available and DVD could have been the first format to fully support it, rather than DVD becoming passe as newer technologies are introduced that support the data rates and capacities required for HD presentation. In some respects, it is a shame that designers did not try harder to achieve more with the DVD generation of optical discs.

Because HDTV sets include analog video connectors that work with all DVD players and other existing video equipment such as VCRs, existing DVD players and discs work perfectly with HDTV sets and provide a much better picture than any other prerecorded consumer video format. Progressive scan DVD players provide an even better picture when connected to a progressive scan display. Unfortunately, some DVDs contain video from interlaced sources, which is difficult to present well on a progressive display. Even for progressive scan source video, certain steps in the DVD production process are oriented toward optimizing video quality on interlaced displays, which means that detail may be reduced from what it could be in a pure progressive format. Details of progressive scan are covered in Chapter 3.

No Reverse Gear

Because of the way MPEG-2 compressed video builds frames by using the differences from previous frames, it is impossible to smoothly play in reverse. To accomplish that feat, either a large memory buffer in which to store a set of previous frames or a very complex high-speed process of jumping back and forth on the disc to build a frame sequence based on a key frame, is required. Although RAM for video buffers has gotten less expensive, only DVD computers and some very high-end DVD players have the capacity to play backwards at normal speed.

Some players can move backwards through a disc by skipping between key frames. Attempting to display these frames at the proper time intervals results in jerky playback with delays of about 1/2 second between each frame. Smooth scan can only be achieved by showing the frames at 12 to 15 times normal speed, thus speeding up the action.

Only Two Aspect Ratios

DVD is limited to 1.33 (4:3) and 1.78 (16:9) aspect ratios, even though MPEG-2 allows a third aspect ratio of 2.21. Better yet would be the ability to support any aspect ratio. Because of the need for a standard physical shape, televisions essentially come in two shapes: 4:3 and 16:9. However, if the player or TV were able to unsqueeze an anamorphic source of any ratio, it would provide better resolution because pixels would not be wasted on letterbox mattes. Letterbox mattes would still need to be generated by the player or the display, but high-resolution displays would be able to make the most of every pixel of the anamorphic signal. The obvious disadvantage to this feature is that variable-geometry picture scaling circuitry is more expensive than the fixed-geometry scaling of DVD players and existing widescreen TVs.

A related problem is that the DVD specification includes a provision for tagging preletterboxed video so that widescreen displays can adjust automatically, but some discs are authored without this flag. That is, the television should automatically enlarge the picture to fill the screen and get rid of the letterbox mattes that were encoded with the video, but because the flag was not set properly during disc production, the player does not know to send the signal to the television. See Chapter 3 for details on aspect ratios.

Deficient Pan and Scan

DVD uses a feature of MPEG-2 to store picture offset information that allows anamorphic widescreen video to be converted automatically to full-frame pan and scan. However, the horizontal-only limitation of this feature has caused it to be widely ignored. The full-frame conversion process in a studio not only pans from side to side but also moves up and down and zooms in and out and usually includes extra picture from above and below the widescreen area.

Disc producers are not happy with DVD's limited automatic pan and scan feature, so they either create a widescreen-only disc or they include a second full-frame pan and scan version of the movie on the disc. The designers of DVD could have included the option for full-frame video storage at high resolution to allow extraction of 4:3 pan and scan, 4:3 letterbox, and 16:9 anamorphic with no loss of resolution. This would require more pixels and thus more storage space (39 percent more for 720×666 anamorphic and 71 percent more for 888×666

full frame) but would be more efficient than encoding two (or three) separate versions on the disc. The added video processing complexity would have only slightly increased player cost in early generations.

Inefficient Multitrack Audio

MPEG-2 audio is inefficient in MPEG-1 compatibility mode. Parts of the audio signal must be duplicated to achieve full channel separation, thus creating an overhead of about 10 percent (see Chapter 3 for details). MPEG-2 provides a non-backward–compatible format (AAC), but because this system was not finalized when DVD was introduced, and since MPEG-2 decoders were not available, DVD-Video discs intended for PAL players are saddled with a less-than-optimal algorithm on MPEG-2 multichannel tracks in order to support MPEG-1 decoders.

Other ideas for improving the efficiency of multiple audio tracks were considered. Alternate language sound tracks contain the same music and sound effects, differing only in the dialogue. It could be possible to store all dialogue tracks in mono or stereo form and have the player use a two-track decoder to mix them together. The problem is that dialogue is not isolated to one or two tracks. Using a mono or nondiscrete dialogue track could detract from the audio presentation. Another possibility is differential encoding, which allows additional speech tracks to be encoded relative to the main sound track containing music and special effects, rather than recoding an entire track for each language. This results in a lower data rate for additional language tracks. These ideas have the benefit of reducing data rate (and thus increasing quality or playing time) but the drawback of adding complexity to the audio decoding component of players.

On a related note, there is no provision in DVD for automatically dealing with the difference in playback speed when a 24 frame/sec (fps) film is displayed at 25 fps (PAL). The 4 percent speedup must be compensated before the audio is encoded. If Dolby Digital and DTS were able to automatically conform the audio at playback time for different speeds, production would be greatly simplified, as would the manufacture of players that convert from PAL to NTSC or vice versa.

Inadequate Interactivity

The ability of DVD-Video players to process and react to user control is limited. There is little beyond the bare basics of menus and branching. Features such as navigation, user input, score display, indexes, searching, and the like are possible with DVD-Video, but each permutation must be anticipated and installed when the disc is produced. All video displayed by the player has to be created ahead of time, unlike other systems such as CD-i or video game consoles that can generate graphics and text on the fly.

Consider, for example, a DVD-Video quiz game designed for a standard DVD player. Commands on the disc can program the player to keep track of the player's score, but there is no way to directly display the score. Instead, a set of screens must be created ahead of time to cover all the possible results. To give a score to within 5 percent, 21 screens are needed (0,

5, 10, ..., 90, 95, and 100 percent). Given the amount of work required to create the screens and to design the code for displaying them, it is more likely that a compromise of three or four screens would be used, giving a coarse result such as "bad," "fair," "good," and "great." This type of compromise is likely to pervade all attempts at interactivity in DVD-Video.

Searches or lookups can only be done with DVD-Video in a hierarchical or sequential manner. For example, to look up a word in a small 1000-word glossary, the first screen would list the 26 letters of the alphabet (DVD allows up to 36 buttons on a screen). After the first letter was selected, additional screens of words would be displayed. Ten or more screens might be required to list all the words beginning with a common letter. The "Next" and "Previous" buttons on the remote control would be used to page through the screens one at a time. Finally, after a word was selected, a screen with the definition would be shown. This would require the laborious preparation of 1000 screens for the definitions and more than 100 menu screens to access them. A more powerful system than DVD simply could read the text of a definition from a database file and display it on a generic definition screen.

The use of DVD-Video for education, productivity, games, etc. is disappointingly limited. The feature set was clearly designed for the limited goals of on-screen menus, simple branching, and karaoke. Clever use of the rudimentary interactive features of DVD can accomplish a truly surprising amount, but most producers have been hard pressed to put in the extra time and effort required, resulting in reduced usability. A DVD-Video cookbook, for instance, may provide a simple index search of ingredients or dishes that would include chickpeas and stew but would frustrate someone looking for garbanzo beans or goulash and may not even include the carrots that are in the stew. All this has been fixed by BD and HD DVD.

Limited Graphics

The subpicture feature of DVD could be extended far beyond captions, menus, and crude animation if it were not limited to four colors and four transparency levels at a time. Richer visual interaction could be achieved with more colors, and advanced features such as sprites and rudimentary three-dimensional (3D) capabilities also would expand DVD's repertoire.

The designers obviously chose to limit the subpicture format to save cost and bandwidth, but even a small improvement of 16 simultaneous colors would have made for significantly better-looking subtitles, more sophisticated highlighting, and a superior graphic overlay environment.

Small Discs

Part of the appeal of DVD is its small, convenient size. Ironically, however, this is also a drawback. People are psychologically averse to paying the same amount for something smaller. Even though the smaller item may be better, the larger item somehow seems to be worth more.

The small size of DVDs also limits the cover art. Gone are the days of innovative art on LP jackets or sizable movie art and descriptive text on laserdisc covers. Liner notes are also limited, but this shortcoming can be compensated for by including still pictures, short clips, and other material in video form on the disc itself.

No Bar Code Standard

One of the most powerful features of laserdisc players used in training and education is bar codes. Printed bar codes can be scanned using a wand that sends commands to the player via the infrared remote interface, telling it to search to a specific picture or to play a certain segment. A simple player becomes a powerful interactive presentation tool when combined with a bar code reader. Bar codes may be added to textbooks, charts, posters, lesson outlines, storybooks, workbooks, and much more, enhancing them with quick access to pictures and movies.

Some industrial/educational DVD players from companies such as Pioneer and Philips support bar code readers, but the lack of standardized support for bar code readers, even as add-ons, ultimately denies the advantages to average player owners.

No External Control Standard

Most consumer laserdisc players included an external control connector, and all industrial laserdisc players included a serial port for connection to a computer. An entire genre of multimedia evolved during the 1980s using laserdisc players to add sound and video to computer software. Reaching back nearly 20 years may be a real stretch and, admittedly, this is less important today as the multimedia features of computers improve, but many applications of DVD such as video editing, kiosks, and custom installations are limited by the lack of an external control standard. As with bar codes, some industrial/educational DVD player models include RS-232 or similar external control ports, but each player manufacturer uses a different proprietary command protocol.

Poor Computer Compatibility

The multimedia CD-ROM industry was long plagued by incompatibility problems. In 1995, return rates of CD-ROMs were as high as 40 percent, primarily because customers were unable to get them to work on their computers. Compatibility problems were caused by incorrect hardware or software setups, defects in video and audio hardware, bugs in video and audio driver software, and the basic problem that hardware such as CD-ROM drives and microprocessors often were not powerful enough for the tasks demanded of them.

The potential for problems with DVD-ROM has been even worse. In addition to all the compatibility problems of CD-ROMs, DVD-ROMs have to deal with defects in video and audio decoder hardware or software, incompatibilities of proprietary playback implementations, decoder software that cannot keep up with full-rate movies, DVD-Video navigation software that does not correctly emulate a DVD-Video player, and so on. Although less so nowadays, not all computers with a DVD-ROM drive will play movies from a DVD-Video disc, especially computers that were upgraded from a CD-ROM to a DVD-ROM drive. Someone buying or upgrading a computer may not understand this.

No WebDVD Standard

One of the most interesting uses of DVD is in combination with HTML and the Internet. Hollywood studios, corporations, educators, and thousands of other DVD makers are excited about the potential of combining the best of DVD with the best of the Internet. Unfortunately, the DVD-Video specification has no provisions to make this easier or standardized. The DVD-Audio specification does have a URL link feature, but it hardly scratches the surface. Other groups are working on standards for Web-connected DVDs, but most enhanced discs only work on certain platforms. See Chapters 13 and 15 for more on WebDVD.

Escalated Obsolescence

DVD is attached at the hip to the computer marketplace, which has a nasty habit of evolving faster than the average customer would prefer. The greatest advantage of digital information — its flexibility — is, in a way, its greatest shortcoming. It is too easy to tinker with the formula. TV, radio, and other systems have lasted for years with only minor improvements because it was too difficult to make major improvements without starting over.

The digital clay of the DVD format, however, is so malleable that in too short a time the temptation to revamp it becomes irresistible. This does not mean that existing discs will not play in future players, but it does mean that new features will be added that will require new or updated equipment, to the delight of those who sell new equipment, but to the dismay of their customers.

The next generation of blue laser optical discs was under development even before the first generation of DVD was released. Discs in the new HD DVD format will not play on older DVD players unless they are made as hybrids with an "old" side and a "new" side. Average American consumers replace their TVs every 7 to 8 years. Let us hope that the phases of DVD last at least that long.

Summary of the Old DVD's Shortcomings

The creators of DVD are acutely aware of its limitations. Some limitations were deliberate compromises to keep costs low and to reduce complexity. Others were simply beyond what the originators wanted to tackle. And, some of the flaws were the result of shortsightedness coupled with a lack of fervor for pushing the technology envelope and exploiting seemingly inherent capabilities as the development of DVD progressed.

Considering that DVD was a compromise solution beaten into a form that was acceptable to hundreds of people from dozens of companies in the consumer electronics, movie, and computer industries, all with different priorities and different motivations — other than profit — it is amazing that DVD turned out as well as it did.

DVD was never intended as the be-all and end-all medium for entertainment or computer data storage. Technology had reached the point where it was time to introduce a new format, free of the drawbacks and dead ends of compact disc, videotape, and laserdisc.

Yet, detractors complained that DVD should have waited longer and taken advantage of new technology such as shorter-wavelength lasers in order to store even more data, or that DVD-Video was not sufficiently improved over laserdisc. These arguments leave one forever poised on the brink of a leap that is never taken because something better is always around the corner. Once blue lasers are available commercially, ultraviolet lasers will be in the laboratories, promising even greater storage density. And how much better is "enough?" Improving the quality would have made DVD more expensive and less reliable for little or no visible gain.

All in all, DVD was a major step forward in many areas. As noted in the preceding examples, there is certainly room for improvement, but even the Brave Little Tailor killed only seven in one blow.[3]

What's Wrong with the New DVD

With the advent of next-generation optical disc formats like HD DVD and Blu-ray Disc (BD), all of the earlier problems go away, right? Nope. In fact, most of the old problems remained, and more than a few new ones have been added. For instance, remember that nagging video format problem, NTSC versus PAL? Yep, still there. But, as if that weren't enough, the industry has added half a dozen new HD video formats as well.

If you are a content producer you still have to worry about NTSC versus PAL for standard definition video and, now, will also have to worry about 60 Hz versus 50 Hz for high definition content. In addition, you will have to chose whether to produce your content at 1280×720 resolution (progressive) and risk that some displays will automatically downscale to 720×480 because that's the only resolution they have that supports your frame rate. Or, you'll consider going with 1920×1080 resolution (interlaced) and run the risk that the display either has to deinterlace the video or scale to a lower resolution, or both because the display isn't capable of showing 1920×1080 interlaced video.

Two Too Many

The problems that come with the new disc formats can be summarized with the phrase "too many." For starters, there are too many disc formats! With HD DVD and BD contending for the same market, while sharing many of the same features and technologies, there is no clear reason why a producer, retailer, or consumer would want to take the risk of producing, distributing, or buying either one.

DVD achieved its success largely because it had consensus in the industry. The fact that both HD DVD and BD are being developed concurrently proves that a similar consensus does not yet exist for a new format. (With any luck, by the time you read this the format wars will be over and there will be just one successful format. "Two formats enter, one format leaves!" Although, as we write this, that outcome is very unlikely.)

---

[3]As related in the fairy tale, "Das tapfere Schneiderlein"(The Brave Little Tailor), by Jacob and Wilhelm Grimm, Household Tales, trans. Margaret Hunt (London: George Bell, 1884), 1:85-93. Reference weblink at http://www.ucs.mun.ca/~wbarker/fairies/grimm/ .

Too Many Video Formats

Independent of the fact that there are too many disc formats, there are also too many video formats. As mentioned above, not only were the developers not able to resolve the problem of NTSC versus PAL, but the industry has created several video formats that are divided between 60 Hz implementations (for NTSC regions) and 50 Hz implementations (for PAL/SECAM regions). In addition, there are two HD video flavors:, 1280×720 and 1920×1080, both of which come in two flavors, progressive and interlaced. Further, certain televisions tell you that they can accept all of the above, but won't tell you what they have to do to the picture before you actually get to see it.

Too Many Encoding Formats

Beyond the plethora of video formats that have to be supported looms a variety of video codecs: take your pick of MPEG-2, VC-1, or AVC for high-definition content, which are in addition to MPEG-1 or MPEG-2 for standard-definition content. It's not too bad for those producing discs because they can pick one and stick with it, but those making players — either hardware consumer devices or software for PCs — have to implement, test, and pay royalties for three different video decoders.

New audio codecs have been added as well, including a new version of Dolby Digital, a new version of DTS, a renamed version of MLP, a lossless variation of DTS, plus a set of optional audio codecs for secondary audio streams: aacPlus2, mp3, and WMA Pro. While it's good to have choices, and the inclusion of lossless audio is mucho preferable to an incompatible second format such as DVD-Audio, the end result is that players and authoring tools are more complicated and more expensive.

Too Many Inputs

Even before serious work began on the new disc formats, the seeds were sown for new problems. In the period of time that has transpired since the introduction of HD television sets, a multitude of new video input types have also been created. In addition to the 75-ohm VHF RF-carrier (RF), composite, S-video (Y/C), and component (Y'PbPr) connectors defined for standard defintion video, HD television sets now may incorporate HD analog component (also Y'PbPr), DVI (including DVI-A and DVI-D), and HDMI, among others. Due to content protection issues, it is possible that HD televisions previously sold as "HD Ready" (meaning they had no HD tuner and only HD component analog video or DVI inputs) either will not be able to connect to next generation players, or may only display a degraded picture. Consumers will have to educate themselves about these issues in order to get what they have paid for from their televisions.

Too Many Channels

What about audio? The new specifications now support high-bandwidth multichannel lossless audio compression, giving them the ability to deliver up to eight channels of pristine audio ecstasy. However, it is still not clear how all these channels will get to your speakers. Most current A/V receivers support S/PDIF or Toslink digital connections, which have enough bandwidth to deliver two channels of uncompressed PCM audio data or 5.1 channels of Dolby Digital or DTS-encoded audio data. In order to carry eight channels of uncompressed PCM audio, you would need four S/PDIF or Toslink connections all working together.

Some suggest that HDMI could be adapted to deliver eight channels of uncompressed PCM audio to an A/V receiver. Another approach could be to fall back to using eight analog RCA phono plugs, or perhaps FireWire. Alternatively, the audio data could be delivered encoded, though that would still require more bandwidth than most receivers currently support, as well as built-in support for DTS-HD and/or Dolby TrueHD decoding.

Once a viable solution (or two) is found for how to get the audio data to your receiver, there is still the question of where to send it from there. In other words, where are those eight speakers? At this point in time, there is no specific standard for 7.1 speaker placement. In some configurations, the typical 5.1 configuration of Left, Center, Right, Left Surround, Right Surround may be augmented with either additional side or center speakers. In other configurations, the extra channels are used to provide altitude. Frankly, even for theatrical presentations, the 7.1 channel configurations are relatively rare. A great number of decisions have yet to be made on how eight-channel audio will impact the home environment.

Not Enough Interactivity

The new formats are leaps and bounds beyond old DVD when it comes to interactivity. BD and HD DVD use powerful programming languages (Java and ECMAScript), multiple planes of video, and sophisticated layout control to provide animated menus that can pop-up over the top of running video, game-like control of onscreen graphics, Web-like hyperlinks, Internet connectivity, and much more (see Chapters 4 and 13 for more information). However, many limitations have been set and a huge numbers of compromises have been made during the process of defining these interactive features, most in order to meet the restrictions of insuring that low-cost consumer devices will be produced. Regarding DVD, many of the earlier choices of this nature inevitably seem short-sighted just a few years after launch, as new generations of silicon can do much more for much less. Next-generation DVD players will do very cool things, but they'll still be a far cry from game consoles and PCs.

Too Much Interactivity

Notwithstanding the previous paragraph, there's a danger in power. The more complex and flexible a system is, the higher the probability that something will go wrong. The old DVD format was simple, yet there were still many cases where discs — sometimes million-sellers from major studios — would not play properly in some players. Virtual machines, complex programs, mixed video planes, and other interactive features multiply the potential for playback problems by a few orders of magnitude. Unless player makers and content creators are a lot more careful when designing and testing their wares, a serious backlash could result from consumers perceiving the newfangled format to be unreliable and finicky.

In some ways, the original DVD format suffered from many of the same problems. It, too, had too many video formats (NTSC and PAL). Likewise, it supported too many audio codecs. After all, was it really necessary to require PCM and Dolby Digital and MPEG (for PAL players) and also have optional support for DTS and SDDS audio? But, in the end, the market itself decided which of these capabilities and features were most important, and which could be ignored. Perhaps we can expect new and creative solutions to the "too many" problems of the new formats as well.

Chapter 13
New Interaction Paradigms

When DVD first appeared on the scene, it offered enhanced video and audio quality, random access to content, and interactive features, all in a slim package that took up less shelf space than videotapes. These were all valid advantages over VHS, but for end-users to fully benefit, other equipment was needed. Old televisions that only had antenna inputs, for example, could not handle the composite video or s-video output of DVD players. Converters were available, but they had a negative effect on picture quality, and running the signal through a VCR usually resulted in display problems as the Macrovision analog content protection confused the VCR's signal processing circuitry. Most home stereos were exactly that — stereo — so users could not benefit from the new advances in discrete, multi-channel audio. All told, one had to spend hundreds (if not thousands) of dollars to acquire the equipment necessary to experience the benefits of DVD. In addition, there were all of those movies that the public already owned on tape, but would have to purchase again in the new DVD format. What could possibly drive people to jump onto this new format with such enthusiasm, making it the fastest growing consumer electronic device ever?

Most people look back at the first DVD titles and think how simple they were. Consisting of a movie, a few trailers and basic menus, the only new feature (that quickly became overused to death) was the director commentary. Of course, none of this was really that new. Laserdisc, for example, had sported high quality audio and video along with secondary audio tracks, like commentaries, for years before DVD. So what made DVD so successful?

More speculation has been made on that subject than worth putting in print, but one point is worth making: DVD offered a new look and feel for content. As simple as the menus were, and as basic the extra features, DVD offered a quantum leap beyond the slower, linear videotape machine, and a level of interactivity that jumped far beyond laserdisc. The fact that DVDs almost immediately display a set of menus that ask a viewer what would they like to do, coupled with the convenience of no longer needing to rewind the movie — these changes represent a complete transformation in how people interact with the content. While this new *interaction paradigm* may not have been the only driving force behind DVD's success, it was certainly important when distinguishing DVD from the other mainstream video products that had come before.

Recreating DVD's Success

DVD is content driven. When the content runs out, so does DVD's growth and success. At current production volumes, the day is rapidly approaching when the movie and television vaults will be drained and all the re-releases will have been re-released. The inevitability of that day makes the major motion picture studios more than a little nervous. After all, DVD has grown into a $24 billion industry, out-shining VHS and even theatrical releases. As a judge of the 2005 DVD Awards put it, "Hollywood is no longer in the business of making movies but is now in the business of making DVDs." So what happens when it all dries up — when the only content coming out on disc each week is last week's theatrical release?

Not yet ready to see the DVD cash cow head for greener pastures, the studios have been a major backer in the development of next-generation discs. All agree that the new format needs to *feel* different to the end user. It must seem fundamentally different and *better* to the buying public. It also needs to be new and exciting and worth every penny, even though the new discs will serve essentially the same purpose — to play movies.

The New Paradigm

A basic characteristic of DVD interactivity is that everything can be done by menus (selecting audio and subtitle streams, playing the movie, accessing bonus features, etc.). This interaction, however, is generally a *modal* experience: you are either on a menu or playing a feature, but not both. One must leave the movie to get to the menus to do something else with the disc content. This somewhat jarring (and sometimes annoying) experience interferes with the continuity of the film. Even with highly stylized, thematic menus, the experience can be disruptive. Some attempts were made to circumvent the traditional DVD menu experience. The InfiniFilm™ feature on some New Line Home Video titles, for example, uses the subtitle track to provide pop-up menus over the film. The *Follow the White Rabbit* feature on the original Warner Brothers DVD *The Matrix* used a similar technique to provide interactive jump points from within the movie. Other movies followed suit, but in each case the content conflicted with the format. Having buttons appear over the video required a subpicture stream. More importantly, it meant that no other subpicture stream (i.e., subtitles) could play while the feature was active. As a result, if subtitles were needed in addition to the interactive buttons, then much more complex programming and asset preparation was required.

Perhaps the most successful attempts in DVD to supersede the format's inherent modal experience came through the use of the format's multiangle feature. Pioneered by companies like MX Entertainment and Technicolor in the US, and the BBC and The Yard in the UK, titles began to appear on the market that would allow the user to make interactive selections, activating alternate video and audio material without interrupting playback. Although the feature was most often used for concert videos such as *U2: Live from Boston* (2001), *Rolling Stones: Four Flicks*, and *Rush: Rush in Rio* (2003), its debut was on a special edition title produced by the BBC for their *Walking with Beasts* television series. Unlike the concert videos, which primarily used the technique for providing an on-screen, intuitive user interface for accessing alternate video angles, the BBC used the feature to create fully interactive pop-up menus over video that provided access to audio stream selection and alternate, synchronized video content. At any time throughout the 30-minute episode, the user could switch among four related video programs. The primary video, an episode of the television series, served as a starting point. The alternate programs provided specific information related to the main program, such as details about the flora of the time or creatures being shown. Unfortunately, even this ground-breaking title was not able to fully overcome the constraints of the DVD format. Due to the nature of multiangle video, a half-hour program consumes virtually the entire disc's capacity. In addition, the changes from one menu to another menu created non-seamless pauses in the video playback — not as bad as jumping completely out to a separate menu, but still rather disruptive to the video program. Nevertheless, these titles have paved the way for the next generation's new model for interactivity.

The Seamless User Experience

Dubbing it *the seamless user experience*, the studios have all gravitated toward a methodology based on the convenience of being able to watch a movie and, without interrupting the film (without even a glitch in the audio or video), use thematic pop-up menus that provide access to special features, audio and subtitle selections, and alternate scenes or camera angles, as well as to launch into highly interactive applications like video games that may appear to be part of the film itself.

Imagine, for example, putting a mystery film into a next-generation player. Instead of going directly to a menu screen, playback launches directly into the feature film (with requisite logos and warning cards). During the opening credits, a stylized menu appears in the lower third of the screen, offering the option of watching just the film or receiving periodic, interactive prompts about additional bonus features on the disc. You select the latter option and the movie continues to play effortlessly. A minute or two later, a prompt appears to let you know that a director's commentary is available, which will start in 10 seconds. Would you like to hear it? As the counter ticks down the seconds, you activate the option and the director himself appears in the lower left corner of the screen, superimposed over the movie. When the timer hits zero, the director says hello, introduces himself, and proceeds to tell you about how he made the film, sharing interesting points and highlighting portions of the screen as the movie plays. Later in the film, an option appears for selecting a game in which you have to identify "who dunnit." You launch the feature and suddenly find yourself in control of an interactive magnifying glass that you can move around the screen during the film and select items for closer consideration. As you collect these "clues" you receive additional insight about each character. In time, between the sequence of events in the film and the clues you've found, you decide you know who the guilty party is and make your selection. At the conclusion of the film you find out that your selection was not only correct but made in record time!

The hope, of course, is that this new paradigm for interactivity provides a compelling new way in which fans can interact with the content. At the same time, a simple movie-watching experience is also maintained for those who just want to see the film. After all, it's the film itself that ultimately drives the purchase of the content, but many argue that it's the look and feel of the content that sells the platform.

Target Applications

Beyond the fundamental approach of how content should be accessed, a wide range of advanced applications were considered for the platform, including (but certainly not limited to):

- Ability to select objects in the feature video, adding them to a wish list for later purchase.

- Pop-up trivia questions tied to events taking place in the film.

- The film's script in a scrolling window next to or over the video. As the video plays, the script scrolls, remaining in sync with the video.

- Multi-player games in which two users can play together simultaneously using multiple controllers, such as game pads.

- Director's commentary in which the director appears on-screen, overlayed on the video, and is able to point out key areas in the video with graphical indicators (like a chalk board).

- Scrapbook feature in which the user is able to take "snapshots" of the film by simply pressing a button while the movie plays.

- Coloring book application in which the user can chose a scene from the movie to bring up an image for coloring with interactive paint brushes and colors.

These are a very small set of the types of applications that were envisioned for these new formats, all of which are designed around the feature itself.

Creating the Paradigm

A next-generation player, although far more advanced than today's DVD player, will still be a *constrained device*, with specific limits on memory capacity, processing speed, and bandwidth. In part, this is due to the realities of trying to build a reasonably priced player, because a $10,000 player won't sell very well. However, it is also due to the need for interoperability across all of the different players that will be manufactured.

In order to ensure that a piece of authored content operates consistently on all players, *minimum requirements* have to be put on the players to allow files of a certain size to be handled and to be able to handle a certain amount of graphics in real-time. These minimum requirements, therefore, become the constraints on content since a producer will generally want the material to work on all players, and staying within the minimum requirements of the players is the only way that can be guaranteed.

This concept of building applications within the constraints of the platform is often quite different than what is actually done in today's software development community. With the ever-growing and changing PC market, there is no single reference specification, and therefore no guarantee of cross-compatibility. As a result, application developers must define the platform for which they will build their application. These requirements are often listed on the product packaging so that the consumer can make sure that the software will work in their PC. If their system does not meet the minimum requirements, then they know that the software will probably not work, and they may chose to upgrade their system so it will. This could be thought of as a *best effort* approach, in which the PC is only expected to make its best effort to run the software. If the PC doesn't meet the minimum requirements, then its best effort is not likely to be good enough. This allows users to enhance their PCs and grow their capabilities at their own pace, but it also creates an extremely complex market with frequent incompatibility problems and general market confusion. However, because people have grown up with this model for computers, expectations for software seem to be much lower than for other types of products. For example, you wouldn't expect your car to crash (literally or figuratively) every few uses; in fact, that would be completely unacceptable — and yet one usually does expect software to crash from time to time (we said *expect*, not *like*.) Let's put it this way, if you knew that your car's steering wheel or brakes would lock up once every few hundred miles and could only be fixed by closing all the windows, powering down, and then restarting, you would probably sue the manufacturer for creating faulty automobiles. However, you don't see software users up in arms about applications that crash every few months and require you to close all the windows and restart the machine.

DVD players, including the next-generation devices, tend to fall on the automobile side of the equation — they have to actually work. And so we return to constraints. The trick for content producers is to think imaginatively about how to get interesting new content to look different from all the other content on the market, while still ensuring it will work within the minimum system constraints. It is also important that the device be designed from the start to handle the target application and, in the case of the next-generation player, the seamless user experience.

Designing for Interactivity

Both the HD DVD and BD formats have, in their own ways, been designed for the seamless user experience. Both facilitate this by defining an independent interactive graphics plane that may overlay the video. This graphics plane can be controlled by authoring and filled with pop-up graphics of all sorts, including those necessary for overlaying menus onto video with *alpha blending* capabilities so that portions of the graphics may appear translucent or completely transparent. In addition, the formats each provide a means for *caching* application resources, such as image data, so that it can be accessed seamlessly during disc playback. (Caching, or pre-loading of content, is necessary to prevent the disc pickup head from having to seek to another part of the disc to read the data, thus interrupting the main feature playback.) The formats also define a timing mechanism that allows the applications to trigger different functions at specific points during the movie playback, such as showing the option for a director's commentary or video game that is synchronized to the movie. By building these features into the formats, the designers help to ensure that the interaction model will operate smoothly even on these constrained devices.

Additional Features

In addition to the seamless user experience, several other key features have been specified in the new standards that will have a significant effect on how people interact with the content. For example, both HD DVD and BD have defined Internet connectivity as a key feature in the player, as well as local, persistent storage and multiple types of input devices.

Internet Connected Players

Although the player itself is not required to be attached to the Internet, several studios plan to create exciting new forms of content that will make it worth the effort to try. In addition to providing self-updating content (e.g., cast & crew biographies that are always current) and downloadable supplemental content (such as audio and subtitle streams that were not available when the disc was released), it is expected that Internet-based events will be scheduled in which live video feeds of directors and producers will provide a chance to interact directly with the filmmakers. This use of networked content can make a next-generation disc seem boundless, but it also introduces new interaction characteristics that DVD never had to address.

Some key differences between networked content and traditional disc-based material center around access time and quality of service. Expectations for disc-based content have been established by the DVD experience. For instance, each jump from one piece of content to another, though usually non-seamless, completes in one or two seconds. This, however, is rarely the case for access to network content, which may take quite some time to identify the server, locate the content, and download enough of it into the player's buffers to start playback. Likewise, there is always the possibility that something fails along the way — the connection to the player could be faulty, the outbound connection to the Internet may go down, or the server itself may be down (or just overloaded with incoming requests). As a result, any applications that utilize networked content must be designed to handle such situations.

In general, there can be a delay between action and response of about 10 seconds before a person thinks the system is broken. On paper that may sound like a very short period of time, but imagine pressing a doorbell and having to count 1-2-3...up to 10 before you finally hear the doorbell sound. (Ten seconds seems like a long time now, doesn't it?) With networked content, it can easily take 10 seconds or longer to begin playback, so it is imperative that the system be able to provide feedback that the operation is underway. Likewise, because the delays associated with network access tend to be somewhat random in duration, it becomes very important that the feedback provided is informative enough that the user will realize what is happening. For example, take the standard computer approach of having an hourglass cursor appear to indicate that the computer is busy doing something. This does provide immediate feedback to the user, but it's not very informative. After a few seconds, the user starts to wonder if the system has broken down. On the other hand, if a progress bar is used to show how much content has actually downloaded, then the user can clearly tell if the content is actually coming or if the whole system hung up at the start, before any of the material had actually downloaded.

Not only will next-generation content need to provide clear feedback for network operations, but it will also have to be built with the expectation that the networked content may not be accessible. For example, imagine a photo gallery with images that are periodically updated online. If the feature is authored on the disc so that you can go to the photo gallery even when there is no network connection, the experience will be pretty bad if there are no photos to be seen. One way to address this would be to simply dissallow the feature, or even hide it altogether, when the server is unreachable. However, that can lead to confusion when the viewer sees the feature available on the same disc in a friend's player (who has a network connection). A third alternative, which perhaps offers the best experience for the non-networked user, is to have a collection of photos for the gallery already stored on disc. Those with a network connection can download additional photos, but this way even those without a connection can have a satisfying experience. A fourth and final approach is to have all network access occur in the background while the movie is playing. Since network transfers require relatively low bandwidth, processing time or memory, next-generation players should be able to perform background transfers relatively easily. For instance, one might download the latest movie trailer to local storage while the movie plays so that when the movie is finished, the download will have completed and the trailer may be played immediately following the film (making it a trailer in a truer sense of the word).

Another way in which network-connected players may affect the interaction model is through the use of *hooks*, which can be authored into a disc prior to their scheduled release such that the disc automatically checks the network server for any new applications or content that it should download. Such hooks can be placed at key locations within the playback sequence, such as immediately when the disc is first loaded, just before the film plays, just after the film plays, etc. Likewise, the hooks can be designed so that if the server doesn't have any content for them, they just continue playing normally, but if the server does have available content, it will download and launch. In this way, one can "future proof" a disc so that new content and applications can be added any time after release of the product. For instance, let's say a movie is released on disc with one of these hooks. Two years later, it is decided that a sequel for the movie will be made. Using the hook on the disc, a simple notification (or full-blown movie trailer) could be made to appear whenever the original film's disc is played in a networked player, thus notifying the viewers who would most likely want to know about a sequel.

Local Storage

Another key feature of the next-generation discs is local storage. Both HD DVD and BD mandate that all next-generation players have some amount of persistent memory or local storage. This can range from built-in 128 MB flash memory up to a multi-gigabyte local hard drive. Removable flash memory and network storage are options that players can support in addition to the built-in storage. The purpose of this local storage is to allow information and content to be saved between sessions. For example, with DVD, as soon as you eject a disc from the player, all memory associated with that disc is reset. As a result, there is no way to distinguish the first time a disc is played from the 100th time. If you were playing a DVD-Video game and just reached your highest score, it would be forgotten by the player the moment the disc was removed. However, with local storage, the disc will be able to save your high score to the player so that it is remembered the next time you access that disc.

Local storage can have other interesting uses as well, many of which affect how viewers will interact with their discs. For example, multi-disc box sets of DVDs are very popular now, and are expected to remain so for television series and movie sequels on the next-generation formats. Whenever it is necessary to eject one disc and load another, the local storage can be used to save information about what to do when the new disc is loaded. For example, one might implement a quick-launch option in which the menu for a multi-disc television series lists all of the episodes across all the discs. When an episode is selected that exists on another disc, the menu can ask the user to insert the appropriate disc, while it also saves which episode to launch in the local storage. When the other disc is inserted, it can be programmed to automatically check if it should jump immediately to the saved episode number, quickly bypassing all the other steps like logo sequences, warning cards, or other episodes.

Local storage can also be used to store disc status information. A producer creates a disc with many different elements and it may be helpful for the user to keep track of which elements have already been viewed, even across multiple sessions. For example, imagine an exercise program in which the disc helps you set up an ideal workout session with appropriate time assigned to different exercises, mapped out over three month times. With each session, the disc can present the appropriate exercises for that particular session, and can even be used to help track your performance (weight, strength, etc.) throughout the course of the program. Wouldn't that make a new DVD player worth purchasing?

Types of User Input

Perhaps one of the greatest challenges producers of next-generation content will be dealing with is the increased complexity of user input. Many of the applications that are expected to debut will require keyboard input, such as entering a name and address for e-commerce transactions. However, most players will not ship with a physical keyboard. Therefore, the application may need to provide some sort of virtual keyboard for use when a real keyboard is not available. Likewise, some applications will work much better with a mouse, but most players will probably not come with a mouse or trackball, so again, the user with the standard IR remote control must be accommodated. Game controllers are the third likely input device type to be supported, but again, few players will come with game controllers. The fact that next-generation players will support a wide range of input devices is a tremendous benefit and will allow for more creative applications. However, it will also complicate interoperability issues since the application developer may not know what types of controllers each user may have.

Another challenge that content producers will face is *inter-application navigation*. The new formats each make it possible to have multiple applications active and displayed simultaneously, each with its own interactive user interface. Due to the complexities of navigating among different interactive elements (even within a single application), the specifications do not generally provide for automatic inter-application navigation. As a result, an author must remain constantly aware of what applications will be active at any one time and determine how to navigate among them (or always just use one application).

One of the more forward-thinking features of the new formats is their support for a *generic device framework*. This allows new types of controllers to be created and used after the launch of the format. By supporting a method of acquiring raw data input from a controller, applications can be written to support new and innovative types of input devices, such as specialized kiosk controllers, new game pads, or some other type of input device that no one has yet conceived. Unfortunately, because of how specifications are written, we won't know for certain what will be possible until players actually start to appear on the market.

Chapter 14
DVD in Home, Business, and Education

From its introduction, the primary focus for the marketing of DVD has been home entertainment. Even though there were seven times as many DVD PCs as home DVD players in the year 2000, there were about 100 times as many video titles as computer software titles. As the technology matured, business and education have adapted DVD to serve a variety of purposes.

The effectiveness of video has long been recognized for instruction and learning, and the advantages of DVD-Video are sufficient in many cases to warrant the complete replacement of existing video equipment with DVD equipment. DVD is also a natural fit for marketing and communications.

Although some of this chapter addresses how to use a DVD player at home — how to get the best picture and sound, viewing distance — that information is also very useful for non-home use of DVD. This chapter continues with how DVD can be a boon to business and an aid to education, and concludes with disc care and handling information.

How to Get the Best Picture and Sound

Now that digital connections are possible between a DVD player and a display, it requires new purchases of digital-capable devices. As noted in Chapter 10, digital connections provide a surety of copyright protection that is not possible with analog signals. However, if you do not have an all-digital environment, analog connections provide a very high level of picture quality and aural enjoyment.

Whenever possible, use component video connections. If this is not an option, an s-video connection is better than a composite video connection. The improvement from composite to s-video is much greater than the improvement from s-video to component. Further, a progressive scan DVD player or a DVD-equipped computer, when connected to a progressive scan display, can provide a significant improvement over component video.

Beyond the connection scheme that you choose, one of the most important steps to getting better picture and sound is to adjust the television display properly. Turn the sharpness control on the TV all the way down. Video from DVD is much clearer than from traditional analog sources. The TV's sharpness feature adds an artificial high-frequency boost. If the sharpness control is not turned down, it exaggerates the high frequencies and causes distortion, just as the treble control set too high for a CD causes it to sound harsh. An overly sharp display can create a shimmering or ringing effect. Reduce the brightness control, as it is usually set too high as well. Many DVD players output video with a black-level setup of 0 IRE (Japanese standard) rather than 7.5 IRE (U.S. standard). On TVs that are not adjusted properly, this may cause some blotchiness in dark scenes. DVD video has exceptional color fidelity, so muddy or washed-out colors almost always indicate a problem in the display, not in the DVD player or disc.

If you get audio hum or noisy video, it is probably caused by interference or a ground loop. Interference can be reduced with an adequately shielded cable. The shorter the cable, the better is the result. You may be able to isolate the source of the interference by turning off all equipment except the pieces you are testing. Try moving things farther apart or plugging them into a different circuit. Wrap your entire house in tinfoil. If those steps do not remove the hum or noise, make sure all equipment is plugged into the same outlet. If nothing works, then it may be time to get a new TV, as the culprit may be the TV's power supply.

It may be hard to believe, but televisions are not necessarily adjusted properly in the factory. Get your TV professionally calibrated, or calibrate it yourself. A correctly calibrated TV is adjusted to proper color temperature, visual convergence, and so on, resulting in accurate colors and skin tones, straight lines, and a more accurate video reproduction than is generally provided by a television when it comes out of the box. Organizations such as the Imaging Science Foundation (ISF) train technicians to calibrate televisions using special equipment. They usually charge between $175 and $600. The accompanying DVD demonstration disc includes extensive test and optimization resources that may be used to aid in setting up your viewing environment.

Another option is to use Joe Kane's *Video Essentials* DVD, updated for DVD and based on the Video Essentials laserdisc. The *AVIA* and *The Ultimate DVD: Platinum* discs also provide instructions, test pictures, and other resources to calibrate audio and video systems.

Connect the DVD player to a good sound system. This may sound like a strange way to improve the picture, but numerous tests have shown that when viewers are presented with identical pictures and two different quality levels of audio, they perceive the picture that is accompanied by high-quality audio to be better than the picture associated with low-quality audio.

Viewing Distance

A great deal of research has gone into human visual acuity and how it relates to viewing distance, information size, and display resolution. Viewing distance is usually measured in display screen heights. In general, the best viewing distance is about 3 to 5 times the screen height. Some industry recommendations vary from 2 to 10 times the screen height (the lower the number, the more likely it's the optometric industry). If you sit too close to the screen, you will see video scanline structures and pixels. If you sit too far away, you will lose visual detail and will not engage enough of your field of view to draw you into the experience. The optimal viewing experience is to fill at least 35 degrees of your field of vision while staying beyond the limit of picture resolution. The THX recommendation is a 36 degree viewing angle, whereas Fox video recommends a 45 degree viewing angle.[1] There is disagreement about the dimensions of the average human field of view, especially as it relates to watching

---

[1] The vertical visual angle is calculated as arctan (height/2/distance) × 2. The horizontal viewing angle is arctan (width/2/distance) × 2. Dividing height or width by 2 and multiplying the final result by 2 centers the viewpoint in the middle of the screen.

video, but the vertical range is about 60 to 90 degrees, whereas the horizontal angle varies from 100 to 150 degrees, depending on the individual. Viewing angle fields broader than about 40 degrees, however, can make the viewer uncomfortable after a period of time because human vision tends to focus toward the center.

Often the resolution of the display determines the distance at which viewers naturally sit. Psychophysical studies have shown that human viewers tend to position themselves relative to a scene so that the smallest detail of interest subtends an angle of about 1 minute of arc, which is the limit of angular discrimination for normal vision. For the 480 visible lines of 525-line (NTSC) television, this produces a viewing distance of about 7 times picture height, with a resulting horizontal viewing angle of about 11 degrees for a 4:3 picture and 15 degrees for a 16:9 picture. To deliver the optimal one-pixel-per-arc-minute viewing experience at DVD resolution requires a viewing distance of 7 times screen height for NTSC, 6 times screen height for PAL.

THX Certification

Lucasfilm THX (the letters come from Tomlinson Holman experiment) was developed in 1983 with the goal of ensuring that theatergoers experience the quality picture and sound presentation as intended by the filmmaker. THX technologies and standards were then extended to the living room. The Home THX program works with leading manufacturers to incorporate proprietary designs into certified home theater components such as receivers, speakers, and laserdisc players. The THX Digital Mastering program offers studios the certification of picture and audio quality during film-to-video transfer, mastering, and replication of laserdiscs and tapes. Lucasfilm has adapted these two programs for DVD.

THX is not an audio format like Dolby Digital or DTS. It is a certification and quality control program that applies to sound systems and acoustics in theaters, home equipment, and digital mastering processes. Certification is performed by THX of software and hardware.

Software Certification

A large part of the THX software certification process for DVD is the same as it was for laserdisc. This includes calibrating the video and sound equipment, monitoring the transfer from film to video master tape, and adding a special THX test signal in the vertical blanking interval. The same video master is generally used for laserdisc, videotape, and DVD.

The new and critical aspect of THX certification for DVD involves the MPEG-2 video encoding process. Part of the task is simply understanding the potential problems and paying attention to the many details such as video noise, picture detail, color balance, black level, white level, and correct audio level. The MPEG encoding process attempts to reduce redundant information; the lower the entropy of the signal, the better the compression. Entropy can be increased by many characteristics of the source video such as instability (weave), random noise (film grain), edge sharpness, etc. These cause the encoder to require a higher bit rate and may affect the quality if the bit rate is higher than can be allowed. For example, the first DVD movie to be THX-certified — *Twister* — contained considerable hand-held camera work. The shakiness of the picture required a higher bit rate and careful attention to difficult sequences to make sure it was encoded cleanly.

The THX staff also works to request the best possible transfer, using D1 or D5 tapes with minimal image processing and noise reduction, and will request an audio remix if needed. THX does not interfere with the choice of aspect ratio or anamorphic/nonanamorphic mode (although many DVD fans point out that only anamorphic mode can provide the highest picture quality for widescreen video). THX specifies multipass variable-bit-rate encoding with manual correction of trouble spots and places limits on the number of audio tracks (to keep the video data rate above 5 Mbps). The final steps of THX certification are a check of the physical stampers and thorough testing of check discs and final replicated discs.

The primary difficulty is that few tools exist for objectively measuring the output of an MPEG encoder. No machines or software algorithms can simulate human visual response. Therefore, THX relies on trained viewers to monitor the output from a reference decoder connected to the MPEG encoder. The viewers are trained to look for digital compression artifacts such as macroblocking (visible squares) and "mosquitoes" (fuzzy dots around sharp edges). The encoding process is adjusted and repeated as many times as needed to eliminate problem areas. In some cases where the nature of the transfer or print is the cause of encoding problems, a new video transfer or even a different print may be requested.

THX verifies that a preliminary check disc plays as expected on a wide range of players. One advantage of DVD is that there is little chance for errors to be introduced after the digital premaster is created. The mastering and replication operations deal with a bitstream, not an analog video signal that might be degraded by improperly adjusted equipment.

Before the widespread adoption of multichannel theater audio for listening environments, many of the older movies were released in stereo even though they had four- or six-track masters. To make the most of DVD's capabilities with these older films, a new sound track is mixed from the original multitrack masters. Although now a standard for both video and audio, THX was developed initially for audio, and this is still its forte. THX engineers supervise the mixing process, test the equipment, and even adjust the equalization where necessary to make up for the difference between the prevailing standards when the program material was made.

Although DVD supports 5.1-channel Dolby Digital audio, the vast majority of viewers will hear the downmixed audio on a standard stereo system or a system based on Dolby Pro Logic surround sound. Therefore, the audio encoding process is monitored with a decoder connected to a Dolby Pro Logic processor. Trained listeners audition the result in a calibrated THX listening room, checking primarily that the dialogue — which was designed originally for a discrete center speaker — remains audible, even when played on only two speakers or on multiple speakers through Pro Logic decoding. Older action movies require that the audio mix be adjusted slightly to boost the dialogue or reduce the sound effects. This is a delicate process, since the balance and accuracy of the 5.1-channel version must be preserved along with the artistic decisions of the original audio editor. In a few cases, when no amount of tweaking can achieve a satisfactory result, a separate Dolby Surround track is added to the disc.

Hardware Certification

THX continues to develop criteria for THX-certified DVD players. These players are held to a high level of performance, which can be measured using test signals in the same way

THX-certified laserdisc players were once tested. The most critical component is the analog video encoder. The numerical output of MPEG decoders varies little from player to player. On the other hand, the circuits that produce the composite and s-video signals have the most significant effect on the visual signal quality and therefore must be given the closest attention.

The original THX specification for home theater equipment based on studio reference equipment is now called *THX Ultra*. Whereas, *THX Select* is a simplified specification designed for audio systems in the average living room, allowing for smaller speakers and more flexibility in speaker design.

THX-certified amplifiers enhance Dolby Pro Logic and Dolby Digital: bass management sends low-frequency signals from all channels (or front channels for Dolby Surround) to the subwoofer; front channel re-equalization compensates for the high-frequency boost in theater mixes designed for speakers behind the screen; rear channels are timbre matched and are decorrelated when both channels have the same audio; low frequencies are emphasized; and signal processing is used to adjust for the distance of each speaker from the viewing location.

THX's quality assurance and certification programs for audio and video help bring out the best of DVD. Many people feel that THX engineers raised the quality of laserdisc close to the level of studio master tapes that was once considered to be the standard. THX is likewise working to help DVD live up to its potential and approach the quality of today's digital studio masters.

The Appeal of DVD

DVD has many advantages over other media, including videotape, print, CD-ROM, and the Internet. For example, in the case of CD-ROM multimedia, the capacity of DVD to carry large amounts of realistic, full-screen video makes it more compelling, effective, and entertaining. Other benefits include:

Low cost. Production and replication costs of DVD-Video and DVD-ROM have quickly dropped below that of videotape and CD-ROM, especially when cost calculations take the larger capacity into account. Corporate and government databases that were filling dozens or hundreds of CD-ROMs are now being put on fewer DVD-ROMs, with one DVD-18 taking the place of 25 CD-ROMs. Businesses that were spending millions of dollars on videotapes have begun to reduce the cost of duplication and inventory by a factor of four or more now that DVD players have become widespread and disc production costs decreased. At this point, it has become cost-effective to equip entire groups of recipients with "bare bones" players or with DVD-ROM upgrades for their computers, simply to reap the benefits of the medium.

Simple, inexpensive, reliable distribution. Five-inch discs are easier and cheaper to mail than tapes or books. Optical discs are not susceptible to damage from magnetic fields, x-rays, or even cosmic rays, which can damage tapes or magnetic discs in transit. One DVD is easier to store than videotapes, multiple CDs, or multiple audiotapes. Production is quicker, logistics are simpler, and inventory is streamlined.

Ubiquity. An apparent drawback to DVD had been the lack of an installed base of players. Regardless of how long it took DVD-Video players to begin showing up in place of

VCRs, most businesses have at least one DVD-equipped computer. As DVD-ROM continues its dynastic march and overtakes CD-ROM, it has greatly amplified the audience for DVD-based material. Various forecasts indicate that between 2002 and 2010 the number of installations capable of playing DVD-Video will exceed the number of VCRs. This will occur in the business world much sooner than in the home.

High capacity. A double-sided, dual-layer DVD-Video disc can hold more than eight hours of video, and over 28 hours if compressed at VHS videotape quality. Eight hours of video would require two bulky and expensive videocassettes or would take more than 25 hours to transmit over the Internet with a high-speed T1 connection. Many hours of video and hundreds of gigabytes of data can be sent anywhere in the world by slipping a few discs into an overnight express mailer.

Self-contained ease of use. DVD-ROM programs obviously can include integrated instruction. The features of DVD-Video are also sufficient to provide instruction, tutorials, and pop-up help. A disc can start with a menu of programs, how-to sections, and background information. Rather than being tied to a linear taped presentation, the viewer can select appropriate material, instantly repeat any piece, or jump from section to section. Unlike previous commercial media such as videotape, audiotape, and laserdisc, DVD-Video discs need no ancillary material for training or user education — everything explaining how to use the disc can be put on the disc itself.

Portability. Video presentations no longer require a VCR. A portable DVD-Video player (the size of a portable CD player) can be slipped into a briefcase and hooked up to any television or video monitor. One-on-one or small-group presentations can be done using a portable player with an integrated LCD video screen or a laptop DVD computer. Large-audience presentations can be done with a video projector and a portable player or laptop computer. There are now portable video projectors with built-in DVD players. Notebook computers with DVD-ROM drives and audio/video decoding capabilities can be used for both DVD-Video and DVD-ROM multimedia presentations. Beyond presentations, portable players and laptops can be used for training and learning in any location. DVD players can even be rented in airports for trips.

Desktop production. Desktop video editing and recordable DVD are doing to the video industry what desktop publishing and laser printers did to the print industry. Video production can be done from beginning to end with inexpensive desktop equipment. A complete setup will become remarkably affordable within several years of the introduction of DVD: A digital video camera (under $300) can be plugged into a digital video editing computer (under $1500), and the final product can be assembled and recorded onto a DVD-R disc with DVD authoring software and recording hardware (less than $100). This constitutes an entire video production studio on a desktop for under $2,000.

Mixed media. DVD bridges the gap between many different information sources. A single disc can contain all the information normally provided by such disparate sources as videotapes, newspapers, computer databases, audiotapes, printed directories, and information kiosks. Training videos can be accompanied with printable manuals, product demonstrations can include spec sheets and order forms, databases can include Internet links for updated information, product catalogs can include video demonstrations, and so on. DVD is the perfect medium for this because it provides high-quality video (unlike CD-ROM); searchable, dynamic text (unlike paper); hours and hours of random-access audio (unlike tapes); and a rapidly growing base of devices to read it all.

The Appeal of DVD-Video

The conveniences of DVD-Video, which in the home are enjoyable but not essential, are translated in the office into efficiency and effectiveness. DVD-Video also brings computers into the picture. Unlike in the home, where the value of being able to play a video disc on a computer is questionable, the usefulness of computers doing double duty as video players is clearly apparent.

The natural inclination when working with computers is to take advantage of their additional features, such as keyboard entry, graphic interactivity, and etc., but in many cases this is counterproductive and inefficient. Simpler may be better.

The integration of DVD-Video features into computer multimedia is not straightforward. Most authoring environments and delivery systems still do not widely support DVD-Video content (see Chapter 15 for details). Even after the wrinkles are smoothed out and the learning curve is eased, developing a computer-based multimedia project may be more difficult and expensive than developing a similar project using only DVD-Video.

There are certain advantages to standard DVD-Video over DVD-ROM based multimedia:

- *Easier development at a lower cost*. For simple titles, fewer programming and design requirements exist. Creating a set of menu screens and related video can be done with low-end, low-cost DVD-Video authoring packages.

- *Easier for the customer*. The limited interface is simple to learn and is usually accessible using a remote control. Hooking a player to a TV is much simpler than getting a multimedia computer to work.

- *Familiar interface*. Menus and remote controls are similar.

- *Larger audience and no cross-platform complications*. Macintosh computers, Windows computers, workstations, DVD players, and even video game consoles can all play standard DVD-Video discs.

Certain kinds of programs lend themselves better to DVD-Video, including programs with large amounts of video, programs intended for users who may not be comfortable with computers, programs with still pictures accompanied by extensive audio, and so on. Here are a few examples of material well-suited for DVD-Video:

- Employee orientation and sensitivity training

- Press kits, corporate reports or newletters

- Emergency response training or information systems

- Product demonstrations and catalogs

- Information kiosks — product/service searches, traveler's aid, and way-finding

- Product training

- Sales and marketing tools
- Educational learning kits
- Video tours or video brochures
- Video "billboards"
- Video greeting/holiday "cards"
- Testing, including licenses and professional certification
- Video portfolios (ads, promo spots, demo reels)
- Trade show demo discs
- Point-of-sale displays
- Ambient video and music
- Video "business cards"
- Video "yearbooks"
- Lecture support resources
- Repair and maintenance manuals
- Medical informed consent information
- Patient information systems, home health special needs instructions
- Language translation assistance

DVD-Video also has its disadvantages when compared with other media (the disadvantages of DVD-Video compared with DVD-ROM are covered in the next section). Following are a few examples of material that may not be appropriate for DVD-Video: (Of course, the best of both formats may be combined on a hybrid disc)

- Documentation (Not searchable, not easily read on TV)
- Databases (DVD-Video is not well suited for large amounts of text)
- Productivity applications (Word processing, checkbook balancing, etc. cannot be supported by DVD players)
- Network applications (DVD data rate is higher than what an average network can support)
- Text entry
- Constantly changing databases

The Appeal of DVD-ROM

DVD-ROM has its own advantages over DVD-Video. Because a DVD-ROM can contain any type of computer data and software, the disc content possibilities are practically endless. And, DVD-ROM is appealing because of its increased capacity. There is little question that it has superseded CD-ROM as the medium of choice for computers. Some of the advantages of DVD-ROM over DVD-Video include:

- *Flexibility*. Any application or content may be used. The only limitation is the target computer platform.
- *Compatibility*. Existing software can be put onto DVD-ROM with little or no change.
- *Familiar development tools*. There is no need to switch from a programming language or multimedia authoring package already in use, and most support the improved audio/video capabilities of DVD.
- *Memory*. Unlike DVD players, computers can store information such as preferences, scores, updates, and annotations.
- *Interface*. Computers allow keyboard entry and point-and-click graphic interface.
- *Connectivity*. Computers can be connected to networks, the Internet, hard-disk drives, and so on.

Additional advantages and applications of DVD-ROM are covered in Chapter 15.

Sales and Marketing

DVD-Video is an excellent sales and presentation tool. A complete sales presentation system can be contained in a portable DVD-Video player. For example, home sales presentations can be enhanced greatly by professional video supplements provided on DVD. The presenter simply plugs a portable player into the customer's TV (or puts a disc in the customer's player). Unlike videotape, which must be watched in a linear manner, a DVD can contain different segments for different scenarios, with answers to common questions, and so on. The need to train sales and marketing representatives is reduced by having them rely on carefully prepared presentations that are called up as needed.

Advertising representatives can make presentations to clients using a portable DVD player rather than lugging around a computer and/or a videotape deck. Hundreds of 30-second ad spots, along with DVD-Video based presentation slideshows, can be put on a single disc. If the client already has a DVD player, the rep needs nothing more than a disc and some breath mints. Alternatively, the DVD-Video disc can be integrated with a laptop computer. A custom PowerPoint presentation can include ad spots that play instantly in full-screen mode from the disc.

A point-of-information station or a trade show video presentation is vastly improved with DVD. The disc can be set to loop forever — customers will not be lost due to a black screen after the tape runs out or while it rewinds. If someone is working the exhibit, he or she can respond to customer questions by quickly jumping to appropriate sections of the disc.

Product catalogs with thousands of photographs and video vignettes can be put on a single disc for a fraction of the cost of printed catalogs. Of course, the video catalog cannot be read at the kitchen table — at least not until thin DVD "videopad" players become available. Environmental resource waste from printed catalogs is becoming a big concern. Environmentally conscious companies can replace tons of paper with polycarbonate discs. By producing discs that connect back to the company via the Internet, the life of the discs can be extended with updated prices, product information, new promotions, and other supplements.

With the large capacity of DVD, companies can band together to put multiple catalogs and product videos on a single disc. This type of cooperative disc lowers the cost barrier for small companies and provides more for the recipient to choose from. It also allows companies take advantage of relationships between services and products to do cross-promotions.

A DVD can contain literally hundreds of hours of audio, any part of which can be accessed in seconds, making it the perfect vehicle for instructional audio programs.

Communications

Companies spend billions of dollars a year producing printed information, much of which requires unwieldy indexes and other reference material merely to make it accessible. In addition, much of it becomes out of date in a very short time. Companies are learning to use CD-ROM and the Internet, but for very large publications or those which benefit from a graphic or video ingredient, DVD-ROM and DVD-Video provide an extremely cost-effective means of distribution coupled with improved access to the content.

DVD is also a high-impact business-to-business communications tool. It can be used as a standalone, "no instructions needed" communication device, or it can back up an in-person presentation. The message can include high-quality corporate videos, advertising clips, interviews with company officers, video introductions, dynamic video press releases, visual instruction manuals, and documentaries of corporate events such as new office openings and seminars.

Companies can send free DVD discs in the mail to targeted audiences. Unlike VHS tapes, which people often ignore because they do not want to have to sit through the whole thing, video on DVDs can be broken into small, easily digested segments that the viewer can select from on-screen menus. This revolutionizes the way businesses can communicate with their customers and with other businesses. Unlike CD-ROMs, the discs can be played in both set-top players and computers, so traveling executives might be tempted to pop a free DVD into their DVD laptop computer while flying or when sitting in a hotel room, for instance.

Effective video-supported presentation demands on-the-fly, instant, context-sensitive access to any point in the footage. Linear videotapes are woefully inadequate for this task. By supporting multiple language tracks, a single DVD can lower geographic and cultural barriers. Whether distributed to individuals for playback on PCs, shown to groups via settop players or portable PCs connected to video projectors, or mounted in a network server for remote viewing, DVD helps get the message to every recipient.

Training and Business Education

Now that sufficient numbers of DVD players and DVD-equipped computers are established in businesses, there is no question that DVD has become a leading delivery format for business training. The Internet has certain advantages such as low cost and timeliness, but the demand for high-quality multimedia far exceeds the capabilities of the Internet for the near future. DVD is better suited to deliver such multimedia and can be integrated easily with the Internet to provide the best of both worlds.

DVD-Video is strictly defined and fully supported by DVD players and by DVD-Video navigation software on most DVD-equipped computers. As with all authoring systems, the ease of development is inversely proportional to the flexibility of the tools (Figure 14.1).

Figure 14.1 The Authoring Environment Spectrum: Utility vs. Ease of Use

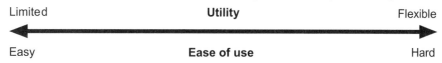

As the parameters of the system are constrained, the complexity of the task is reduced. Since DVD-Video is quite constrained, it is relatively easy to develop for it, given the proper tools. Those considering video training programs should decide if the features of DVD-Video meet their requirements. A simple product containing mostly menus, pictures, and movies may be developed in less time and for less money than with a complex computer authoring system.

DVD-Video works so well for certain corporate education and training applications that the content will sell the hardware. Player purchases will prove to be an insignificant part of the deal. With player prices under $50, companies spending millions of dollars on video-based education and training programs should not think twice about equipping their employees or laboratories with players. Many companies that have specialized systems such as study kiosks or CD-i players are switching to DVD players.

DVD-ROM, on the other hand, covers the entire spectrum from custom-programmed software to fill-in-the blank lesson templates. Practically any authoring or software development system can produce material to be delivered on DVD-ROM. The main advantages of DVD-ROM over other media are space, data transfer speed, and cost per byte. A significant advantage of DVD-ROM over CD-ROM is responsiveness. DVD-ROM access times and transfer rates are much better than CD-ROM platforms. DVD-ROM drive transfer rates are similar to hard disks, and even though its access rates are slower, in most cases DVD-ROM's higher capacity and lower cost more than compensate.

Industrial Applications

DVD discs are being used increasingly in specialized applications. Custom discs can be produced on DVD-R that are unique and proprietary, not designed for sale. Simple installations can use inexpensive, off-the-shelf players. More demanding installations may use professional DVD players, which are more reliable and can be connected to specialized hardware such as multiplayer controllers, video synchronizers, video walls, touch screens, custom input devices, lighting controllers, robotic controllers, etc. Cheap kiosks can be made from a PC motherboard, a DVD-ROM drive, and an input device such as a trackball or touch screen.

A few examples of industrial applications include:

- Kiosks and public learning stations
- Point-of-purchase displays
- Museums
- Video walls and public exhibits
- Vocational skills training
- Corporate presentations and communications
- House video in a store, bar, or dance club
- Theme park and amusement park exhibits
- Closed-circuit television
- Video simulation and video-based training
- Tourism video on buses, trains, and boats
- Hotel video channels
- Media literacy

Classroom Education

Laserdiscs were a success in education almost from day one. Teachers quickly saw the advantage of rapid access to thousands of pictures and high-impact motion video sequences. They began investing in laserdisc players and discs after seeing the effectiveness of laserdisc-based instruction in the classroom. By 1998, 20 years after their debut, more than 250,000 laserdisc players were still found in schools in the United States. Many are still in use, and most are enhanced with laser bar code technology to provide quick and easy access to video by scanning a bar code printed in a textbook or on a student worksheet. Computer multimedia is now replacing laserdisc in the classroom, but the ease of popping in a laserdisc and pressing play, or scanning a few bar codes, may never be matched by computers with their complicated cables, and software setups and daunting troubleshooting requirements.

DVD is poised to take over from laserdisc, but in a different way. DVD-Video players will only trickle into schools, but DVD computers, will proliferate. These computers will be able to run existing CD-ROM programs and new DVD-ROM programs as well as play DVD-Video discs.

Educational publishers were slow to embrace DVD-Video. They did not see a large market, and many of them were hurt by the mass flocking of teachers to the Internet as the new source of free educational technology. Educational video titles require a large amount of work to develop and must be designed to meet curriculum standards. Until more discs intended specifically for education are produced, other titles will help fill the void. Documentaries, historical dramas, newsreel archives, and even popular movies, TV shows, and "edutainment" programs can be adapted and repurposed for use in the classroom. The particular advantage of DVD for this application is random access. Teachers can quickly jump to any desired video segment and can skip sections during playback. If a computer is used to play the disc, it can be programmed to show particular sections in a specific order.

As DVD eventually becomes a mainstream vehicle for delivering video, a large base of educational content will build up. It will be able to take advantage of DVD-Video features such as random access to hundreds of video segments, subpicture overlays to enhance video presentation, on-screen quizzes, multiple-scenario presentations, adaptation to learner needs, multiple languages, tailored commentary tracks, and much more.

Care and Feeding of Discs

Because DVDs are read by a laser, they are resistant — to a point — to fingerprints, dust, smudges, and scratches. However, surface contaminants and scratches can cause data errors. On a video player, the effect of data errors ranges from minor video artifacts to frame skipping to complete unplayability. Therefore, it is a good idea to take care of your discs. In general, treat them the same way as you would a CD.

Your player cannot be harmed by a scratched or dirty disc unless there are globs of nasty substances on it that might actually hit the lens. Still, it is best to keep your discs clean, which also will keep the inside of your player clean. Never attempt to play a cracked disc because it could shatter and damage the player. It probably does not hurt to leave a disc in the player (even if it is paused and still spinning), but leaving the player running unattended for long periods of time is not advisable.

In general, there is no need to clean the lens on your player, since the air moved by the rotating disc keeps it clean. However, if you commonly use a lens cleaning disc in your CD player, you may want to do the same with your DVD player. It is best to only use a cleaning disc designed for DVD players because there are minor differences in lens positioning.

There is no need for periodic alignment of the pickup head. Sometimes the laser can drift out of alignment, especially after rough handling of the player, but this is not a regular maintenance item.

Handling and Storage

Handle discs only by the hub or outer edge. Do not touch the shiny surface with your popcorn-greasy fingers. Store the disc in a protective case when not in use. Do not bend the disc when taking it out of the case, and be careful not to scratch the disc when placing it in the case or in the player tray. Make certain that the disc is seated properly in the player tray before you close it.

Keep discs away from radiators/heaters, hot equipment surfaces, direct sunlight (near a window or in a car during hot weather), pets, small children, and other destructive forces. Magnetic fields have no effect on DVDs. The DVD specification recommends that discs be stored at a temperature between 20 and 50°C (4 and 122°F) with less than 15°C (59°F) variation per hour at a relative humidity of 5 to 90 percent.

Coloring the outside edge of a DVD with a green marker (or any other color) makes no difference in video or audio quality. Data is read based on pit interference at one-quarter of the laser wavelength, a distance of less than 165 nanometers. A bit of dye that, on average, is more than 3 million times farther away is not going to affect anything.

Cleaning and Repairing DVDs

If you notice problems when playing a disc, you may be able to correct them with a simple cleaning. Do not use strong cleaners, abrasives, solvents, or acids. With a soft, lint-free cloth, wipe gently in only a radial direction (in a straight line between the hub and the rim). Since the data is arranged circularly on the disc, the microscratches you create when cleaning the disc will cross more error correction blocks and be less likely to cause unrecoverable errors. Do not use canned or compressed air, which can be very cold from rapid expansion and may stress the disc thermally.

For stubborn dirt or gummy adhesive, use water, water with mild soap, or isopropyl alcohol. As a last resort, try peanut oil. Let it sit for about a minute before wiping it off. Commercial products are available to clean discs, and they provide some protection from dust, fingerprints, and scratches. Cleaning products labeled for use on CDs work as well as those that say they are for DVDs.

If you continue to have problems after cleaning the disc, you may need to attempt to repair one or more scratches. Sometimes even hairline scratches can cause errors if they happen to cover an entire ECC block. Examine the disc, keeping in mind that the laser reads from the bottom. There are essentially two methods of repairing scratches: (1) fill or coat the scratch with an optical material or (2) polish down the scratch. Many commercial products do one or both of these, or you may wish to buy polishing compounds or toothpaste and do it yourself. The trick is to polish out the scratch without causing new ones. A mess of small polishing scratches can cause more damage than a big scratch. As with cleaning, polish only in the radial direction.

Libraries, rental shops, and other venues that need to clean many discs may want to invest in a commercial polishing machine that can restore a disc to pristine condition, even after an amazing amount of abuse. Keep in mind that the data layer on a DVD is only half as deep as on a CD, so a DVD can only be repolished about half as many times.

Chapter 15
DVD on Computers

Home entertainment devices and computers used to be very separate items, with little in common other than screens and power cords. Then came CD-ROM, multimedia PCs, the Internet, and DVDs, and the world began hearing the term "convergence." DVD and next-generation formats enrich both sides of the equation by bringing them closer together. Lamentably, each side also brings a new set of problems and old, bad habits.

Just as replicated and recordable CD-ROMs are used for purposes as diverse as databases, multimedia, software distribution, backups, document archiving, publication, photographic libraries, and so on, DVD-ROM and the next-generation disc formats are being used for all of that and more.[1] It's when audio and video features enter the equation that things become confusing.

Understanding how optical discs relate to computers can become very complicated, very quickly. On one hand, there's DVD-ROM, HD DVD-ROM, and BD-ROM, which can be thought of as nothing more than improved versions of CD-ROM. On the other hand are the application formats such as DVD-Video, HD DVD-Video, and BD-J, which can stand completely on their own or be tightly integrated with computers. The details become very tangled between the two hands. This chapter attempts to sort things out and keep all the fingers in the right place.

NOTE
Many technical details of the information in this chapter are explained in Chapters 7, 8, and 9 (physical formats and application details).

DVD-Video Sets the Standard

Ironically, the MPEG-2 video and the Dolby Digital audio features of DVD-Video were originally selected for home video players, but because of the influence of DVD they have become de facto video and audio standards on computers as well. Most computer video graphics cards now include additional circuitry to accelerate decoding and displaying MPEG video, and many audio card manufacturers include multichannel features and digital audio output for Dolby Digital and DTS. Companies that specialize in simulating multispeaker surround sound from two computer speakers jumped on the DVD bandwagon and ensured that their systems work with Dolby Digital audio tracks. Operating systems developers such as Microsoft and Apple have also built support for MPEG-2 and Dolby Digital into their core multimedia software layers.

---

[1]This chapter uses the term DVD-ROM to refer to discs and content intended specifically for use on computers, as opposed to DVD-Video or DVD-Audio content. Of course, DVD-Video and DVD-Audio are applications of DVD-ROM, and audio and video are merely specific kinds of data, but there is no simple and unambiguous way to refer to computer content in particular.

Multimedia: Out of the Frying Pan . . .

Although the primary use of CD-ROM is for applications and business data, the most high-profile use has been for multimedia, which is computer software that combines text, graphics, audio, motion video, and more. To battle-weary multimedia CD-ROM developers, DVD either seems like a breath of fresh air or more of "been there, done that." Multimedia CD-ROM products are expensive and difficult to produce. A huge disparity of computer capabilities exists among customers, as well as a frustrating inconsistency of support for audio and video playback, especially on the popular Windows platforms. Hundreds of multimedia CD-ROM titles are being produced each year, though far fewer than in the 1990s. Only a handful of them make money, in part because customers have been burned too many times by CD-ROMs that don't work on their computers.

A Slow Start

Within a year or two after DVD was introduced, there were about 10 times as many DVD computers as settop DVD players. In some countries, where DVD-Video was slow in getting started, DVD computers outnumbered DVD players by a factor of 100. Most people expected this ratio to persist or to grow even larger, but the opposite happened.

DVD-Video sprinted to a huge success, while the DVD-ROM industry limped along. By the end of 2000, DVD computers outnumbered settop players in the US about five to one. By 2005, there were slightly more DVD players worldwide than computers. Just as content protection and a lack of titles delayed the introduction of DVD players, compatibility problems, a lack of titles, and fitful customer demand delayed DVD-ROM in the computer industry. DVD-ROM was slow to take off for several reasons:

- No pressing need: Unlike CD-ROM, which was a quantum leap beyond floppy disks, DVD doesn't offer enough to sway many publishers. Most multimedia titles and software applications fit on one or two CDs. It was simply not worth it for publishers to move some of their titles to DVD, especially because they knew that their CDs will work in DVD-ROM drives. It was hard enough to make money in the CD multimedia business without having to worry about DVD.

- No killer application: The video and audio quality are better on DVD, but many game and animation designers focus on 3D engines instead of canned video. A number of games appeared on DVD, but many of them were merely multi-CD versions packed onto a single DVD or slightly modified versions that used higher data rates with PC codecs for moderately improved video quality. Even using MPEG-2 DVD video for transition scenes has its drawbacks, because the high-quality video shows the flaws in the graphics used for the rest of the game. Video and audio can be used for other purposes, such as video help files, but it takes more work to create the video than many developers are willing to give.

- Lack of a consistent development target: Microsoft was slow in developing with a unified solution, and in the meantime, decoder makers each created their own proprietary and incompatible solutions. Even those that used the MCI standard were inconsistent in their implementation. Once Microsoft released DirectShow in 1998, the situation should have improved quickly, but decoder vendors and OEMs did not wholeheartedly support it. Two years after DirectShow was released, about 70 percent of the installed base of DVD PCs worked with DirectShow. No developer wanted to have to write 10 versions of their code to support proprietary decoder implementations for the remaining 30 percent. Apple was even slower to provide QuickTime support for DVD. Apple had a player to play movies, but didn't release a platform for developers to directly access DVD-Video features.

■ No interest in bundling: In the heyday of CD multimedia it was almost de rigueur that new PCs come with a pile of discs containing games, encyclopedias, and other applications. It's now almost impossible to get a deal to bundle DVDs with new PCs. OEM margins are tight, and OEMs don't get much differentiation from their competitors by bundling DVD titles. Even though most early CD-ROM bundles were low-quality titles that customers never used more than once, the bundling deals helped to jump-start the CD-ROM publishing industry. This never happened for DVD.

In spite of stumbling at the starting gate, DVD has inexorably infiltrated the PC market. Most computer makers now ship DVD drives in more than 50 percent of their systems.

DVD-ROM for Computers

The simplest implementation of DVD on a computer is a DVD-ROM drive for reading computer data files. Support for the audio, video, and navigation features of DVD-Video is provided as an additional software or hardware layer, independent of the drive (see the following section, "DVD-Video for Computers").

A DVD-ROM drive (sometimes referred to as a DVD logical unit) is supported by the operating system as a random-access, block-oriented input device. It may be considered simply a large, read-only data storage device. The writable DVD variations are random-access, block-oriented input/output devices, although DVD-RAM is the most suitable format for random access and multiple rewrites.

Features

A single-speed DVD-ROM drive transfers data at 11.08 Mbps (1.321 MB/s)[2] which is just over nine times the 1.229 Mbps (150 KB/s) data rate of a single-speed CD-ROM drive (see Table 15.1). Although 40X and faster CD-ROM drives achieve a higher data transfer rate than 1X DVD-ROM drives, they push the limits of CD technology and begin to suffer from spin instability and read errors. A single-speed DVD-ROM drive spins a little faster than a single-speed CD-ROM drive, but it reads data much faster. A double-speed DVD-ROM drive transfers data at 22.16 Mbps (2.642 MB/s), which is equivalent to an 18X CD-ROM

---

[2]The oft-used figure of 1.385 refers to millions of bytes per second, not megabytes per second.

drive. Most drives include read-ahead RAM buffers to enable burst data rates of over 100 Mbps (12 MB/s) as long as the connection between the drive and the computer is fast enough. By 2005, most DVD-ROM drives ran at 16X speeds, which seems to be the practical limit of speed increases. DVD-ROM average seek times are around 100 ms, with average access times of about 150 ms.

Table 15.1 DVD Drive Speeds

DVD Drive Speed	Data Rate	Equivalent CD Rate	Actual CD Speed of Drive
1X	11.08 Mbps (1.32 MB/s)	9X	8X to 18X
2X	22.16 Mbps (2.64 MB/s)	18X	20X to 24X
4X	44.32 Mbps (5.28 MB/s)	36X	24X to 32X
5X	55.40 Mbps (6.60 MB/s)	45X	24X to 32X
6X	66.48 Mbps (7.93 MB/s)	54X	24X to 32X
8X	88.64 Mbps (10.57 MB/s)	72X	32X to 40X
10X	110.80 Mbps (13.21 MB/s)	90X	32X to 40X
16X	177.28 Mbps (21.13 MB/s)	144X	32X to 40X
20X	221.6 Mbps (26.42 MB/s)	180X	40X+
24X	265.92 Mbps (31.70 MB/s)	216X	40X+
32X	354.56 Mbps (42.27 MB/s)	288X	40X+

Single-speed blue laser formats transfer data at 32 Mbps (3.8 MB/s), which is roughly 25 times faster than single-speed CD-ROM and three times faster than single-speed DVD-ROM.

Almost all optical disc drives include an analog stereo output for reproducing audio from CDs. For internal drives, the analog audio signals are usually mixed into the computer's own audio circuitry by connecting the analog output to the audio card or motherboard. The computer can also read the digital audio from the CD directly. For external drives, the output can be connected to a pair of headphones or speakers. The audio feature is only for audio CDs, not DVDs or next-generation discs. Almost no DVD-ROM drives include the Dolby Digital or MPEG audio decoders necessary to directly play audio or video from a disc. Because of cost, complexity, and content protection issues, the audio/video decoding and playback systems are in the computer rather than in the drive.

Compatibility

All DVD-ROM drives are able to read CDs. Most drives can read variations of the CD format — CD-DA, CD-ROM, CD-ROM XA, Video CD, and Enhanced CD. For Video CD and CD-i, specific software is required to make use of the data after it's read from the disc. Many first-generation DVD-ROM drives were not able to read CD-Rs, but almost all current DVD-ROM drives have no problem reading CD-R discs. Likewise, first-generation

DVD-R drives could not write to CD-R and CD-RW discs, but now, essentially all current DVD-R/RW and DVD+R/RW drives can write to CD-R and CD-RW discs. Refer to Chapter 6 for more details on compatibility.

Interface

Optical drives are available with E-IDE/ATAPI or SCSI-2 interfaces. Some extend the IDE interface to use USB or IEEE 1394 connections. Some also support the newer serial ATA (SATA) connection. The MMC (Mt. Fuji) standard extends the ATAPI and SCSI command sets to provide support for reading from CD, DVD, HD DVD, and BD media, including physical format information, copyright information, regional management, decryption authentication, burst-cutting area (BCA), disc identification, and manufacturing information (see Table 15.2). Because the MMC extensions use the existing ATA and SCSI protocols, standard CD-ROM software drivers usually support DVD-ROM, HD DVD-ROM, and BD-ROM discs. On the other hand, the recordable formats usually require specialized software applications.

Table 15.2 DVD ATAPI/SCSI Interface Command Information

Physical

DVD book type and version	DVD-ROM, DVD-RAM, DVD-R, DVD-RW, DVD+R, DVD+RW; 0.9, 1.0, 2.0, etc.
Disc size	120 mm or 80 mm
Minimum read rate	2.52 Mbps, 5.04 Mbps, or 10.08 Mbps
Number of layers	One or two
Track path	PTP or OTP
Layer type	Read/write
Linear density	0.267 μm/bit, 0.293 μm/bit, 0.409 to 0.435 μm/bit, 0.280 to 0.291 μm/bit, or 0.353 μm/bit
Track density	0.74 μm/track, 0.80 μm/track, 0.615 μm/track
BCA present	Yes or no

Copyright

Protection system	None, CSS/CPPM, or CPRM
Region management	Eight bits, one for each region, set = playable, cleared = not playable

Authentication

Disc key	2,048-byte authentication key
Media key block	24,576-byte key block (CPRM)

BCA (optional)

BCA information	12 to 188 bytes

Manufacturing

Manufacturing info	2,048 bytes of manufacturing information from lead-in area

Disk Format and I/O Drivers

DVD and the blue laser formats are designed around the UDF file system. DVD includes the ISO 9660 *bridge* format for backward-compatibility (refer to the File Systems section of Chapter 9 for details). At the lowest level, an optical disc is simply a block-oriented data storage medium and may be formatted for almost any file system: Apple HFS, Windows NTFS, and so on. However, because UDF provides explicit cross-system support, standardization via UDF is strongly recommended. Technically, the UDF bridge format is required on all DVD discs by the DVD file system specification books.

Volume size and file size limitations of older OSs are quickly revealed by DVD (see Table 15.3). Sometimes large files or files stored on a disc above four gigabytes are not readable. Limitations on the size of files or volumes that are accessible depend partly on the operating system, partly on the file system used to read DVD volumes (ISO 9660 or UDF), and partly on the underlying native file system (FAT16, FAT32, NTFS, HFS, and so on). The standard Windows autorun.inf file or Apple QuickTime autostart feature can be used to create DVD products that automatically begin execution when they are inserted into a drive.

Table 15.3 Operating System File and Volume Size Limitations

OS	Native File System[a]	Maximum File Size	Maximum Volume Size	File System Used to Read DVD
Windows 95	FAT16	Two gigabytes (2×2^{30})	Two gigabytes (2×2^{30})	ISO 9660
Windows NT 4.0	NTFS	Two terabytes (2×2^{40})	16 exabytes (16×2^{60})	ISO 9660
Windows 98, Millennium	FAT32	Four gigabytes (4×2^{30})	Eight terabytes (8×2^{40})	UDF
Windows 2000, XP, Vista	NTFS	Two terabytes (2×2^{40})	16 exabytes (16×2^{60})	UDF
Mac OS older than 8.1	HFS	Four gigabytes (4×2^{30})		ISO 9660
Mac OS 8.1	HFS Plus	Two gigabytes (2×2^{30})	Two terabytes (2×2^{40})	UDF
Mac OS 9, X	HFS Plus	Two terabytes (2×2^{40})	Two terabytes (2×2^{40})	UDF

[a]Some later releases of operating systems can use newer file systems. For example, Windows 95 OSR supports FAT32.

DVD drives are supported for basic data-reading operations in any Microsoft Windows operating system with CD-ROM extensions and ATAPI/SCSI drivers. Even DOS and Windows 3.x can usually read data from DVD drives. Only newer versions of Windows support DVD-Video, and no version of Windows had native support for the blue-laser video application formats when the formats were finally released.

Almost any Macintosh computer can support DVD-ROM and blue laser drives, and older Macs can connect to external DVD drives using the built-in SCSI port. DVD-ROM volumes with the UDF bridge format are recognized by the existing ISO 9660 file system extension. Full support for UDF discs also has been added with a file system extension in Mac OS 8.1. Support for reading and writing DVD-RAM, using UDF 1.5, was added in Mac OS 8.6.

DVD-Video for Computers

From a certain point of view, DVD movies on a computer may seem to be an oxymoron. Why take an audio-visual experience originally designed for maximum sensory impact on a large screen with a theatrical sound system and reproduce it on a small computer screen with midget computer speakers? The answer is that computers are steadily becoming more like entertainment systems, even to the point that some home computers are designed to be installed in the living room. Not all computers are going to sprout remote controls and channel tuners, but there is much more to DVD-Video than movies, including educational videos, business presentations, training films, product demonstrations, computer multimedia, DVD/Internet combinations, etc.

Computer hardware and software companies are avidly exploiting the convergence of PC and TV, or at least the development of TVs to the point that they have PC-like features and operating systems, or the development of PCs to where they can be sold as replacements for TVs. This has motivated the hardware and software companies to cooperate amazingly well with the entertainment and consumer electronics industries. Computer companies also recognize that the issues of protecting artistic ownership rights in a digital environment must be dealt at some point, and DVD became the first battleground. Earlier skirmishes had occurred with DAT and MiniDisc, but those had little to do with computers. Since then, the importance of entertainment to the computer industry has increased significantly. TVs and PCs won't merge overnight, and they will always be differentiated, but their blending will continue, with DVD and its successors squarely in the center.

In the short term, DVD-Video for computers is important for customer perception and as a source for content. Many customers do not understand the difference between DVD-ROM and DVD-Video. Making sure that DVD in all its variations is supported by a DVD computer greatly alleviates customer confusion and dissatisfaction.

The DVD-Video Computer

Not just any computer can play a DVD-Video disc. A DVD drive is required, of course, but a drive alone is insufficient; specific hardware or software is needed. For video playback using only software, the computer must be fast enough and have enough display horsepower to keep up with full-motion, full-screen video. When DVD was first available, specialized decoding hardware was often provided, but since then computers have gotten fast enough to handle everything in software.

Drive speed and video memory have little to do with playback quality. DVD-Video is designed for 1X playback, so a faster drive provides nothing more than possibly smoother scanning and faster searching. Higher speeds only make a difference when reading computer data, such as when playing a multimedia game or when using a database. Decoder software design, and video card quality (especially of video scaling) are the keys to video quality.

DVD-Video Playback

Essentially any computer faster than a 500-MHz Pentium III can play DVD-Video discs using only software. A common approach is to use the minimum amount of specialized hardware. Certain MPEG decoding tasks such as motion compensation, inverse discrete cosine transform (IDCT), inverse variable length coding (IVLC), and even subpicture decoding can be performed by additional circuitry on a video graphics chip. This improves the performance of software decoders and frees up the central processor and memory bus to perform the rest of the decoding as well as other simultaneous tasks. This is called *hardware decode acceleration*, *hardware motion comp*, or *hardware assist*. Some card markers also call it hardware decode, even though they don't do all the decoding in hardware.

Most modern graphics cards take this approach. The cards provide hardware-accelerated DVD-Video playback by dedicating a small portion of the card circuitry to handle MPEG decoding tasks. This usually involves dedicating a few thousand gates on a chip (which has hundreds of thousands of gates) and essentially provides a hardware speedup of about 35 percent for free. Microsoft defined the DirectX VA (video acceleration) API to standardize the hardware acceleration of software decoding in Windows 2000. The API was extensively improved and incorporated into DirectX 8 in late 2000 for general use in Windows operating systems.

Because MPEG video uses $Y'C_bC_r$ (as opposed to RGB) colorspace with 4:2:0 sampling and interlaced scanning, a graphics card with native support of these formats is advantageous. All modern graphics cards provide hardware colorspace conversion ($Y'C_bC_r$ to RGB). PCI graphics cards can usually handle the load, but Advanced Graphics Port (AGP) graphics cards provide better DVD-Video performance.

Decoders

Decoder hardware or software is required for video and audio, along with specialized drivers for DVD I/O. A specialized driver is needed for CSS authentication, for example. A complication is that the multimedia content of DVD is streaming data. Unlike traditional computer data, which can be read into memory in its entirety or in small chunks, audio and video data is continuous and time-critical. In the case of DVD, a constant stream of data must be: fed from the DVD-ROM drive; split into separate streams; fed to decryptors, decoders, and other processors; combined with graphics and other computer-generated video; possibly combined with computer-generated audio; and then fed to the video display and to the audio hardware. This data stream, which can be monstrous when decompressed, may pass through the CPU and the bus many times on its trip from the disc to the display and speakers. To make this process more efficient, optimized low-level operating support is essential. Some operating systems, such as Apple's MacOS with QuickTime, already support streaming data, while Microsoft Windows supports streaming data in the newer OS versions that include DirectShow.

Microsoft Windows Architecture

Microsoft did not use its aging MCI interface to support DVD-Video, and instead moved to the DirectShow architecture and streaming class drivers based on the Win32 Driver Model (WDM). DirectShow is the streaming multimedia component of DirectX, the core audio/video technology set for Windows. DirectShow provides standardized compatibility and an abstraction from the underlying decoders. A developer who writes a DVD application using the DirectShow API does not need to worry whether the system uses a hardware decoder or a software decoder, because the interface is the same. DirectShow is essentially a virtual machine that handles hardware/software decoding, WDM drivers, demultiplexing, DVD navigation, hardware acceleration, video overlay, colorkey, VPE, CSS, APS, regional management, and so on.

The WDM streaming class driver interconnects device drivers to handle multiple data streams. This driver, in turn, handles issues such as synchronization, direct memory access (DMA), internal and external bus access, and separate bus paths (such as sending the video display data over a specialized channel to avoid overwhelming the system bus). Microsoft has implemented a generic design that enables hardware manufacturers to develop a single minidriver to interface their component to the WDM streaming class driver. For example, an MPEG-2 video decoder card maker, a Dolby Digital audio decoder card vendor, or an IEEE 1394/FireWire interface card developer has to write only one minidriver that conforms to a single driver model in order to operate with each other (see Figure 15.1). Software decoders are implemented as DirectShow filters, which are code components designed according to the DirectShow COM API.

Support for DVD-Video is available in Windows 95, Windows 98, Windows 2000, Windows Me, Windows XP, and newer operating systems. Microsoft DirectShow is not available for Windows NT 4.0, so no OS-level support exists for DVD-Video in Windows NT, although third-party applications are available for playing DVD-Video.

Windows does not include an MPEG-2 or Dolby Digital decoder. These must be provided by third parties. Almost all Windows PC systems with a built-in DVD-ROM drive include a full DVD player and decoder package provided by the PC manufacturer.

DVD-Video playback and navigation is supported by DirectShow versions 5.2 and later. The DirectDraw API also supports video display. Because full frame-rate decoded video has too high a bandwidth to be passed back and forth on most current implementations of the PCI bus, it often must travel on a separate path to the display adapter. Intel has designed the AGP architecture to support this. Microsoft's DirectDraw hardware abstraction layer (HAL) and video port extension (VPE) provide hardware-independent support for AGP and similar parallel bus or parallel memory access architectures.

Regionalization is supported by the Windows operating system, which gets the region from the drive. The decoder driver may also include its own regional code. If the disc does not match the code in the decoder driver and the code stored by the OS, the system will refuse to play the disc.

The Windows operating system also supports content protection. Windows and DirectShow do not provide a decryption module, but act as an agent for facilitating the exchange of authentication keys between the DVD-ROM drive and the hardware or software decryption module. Video display adapter cards are required to implement Macrovision on composite video output to receive decoded video from the system.

Figure 15.1 Microsoft Windows DirectShow DVD Architecture

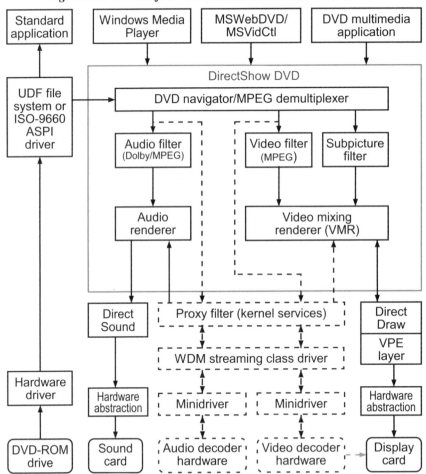

 TIP

For details on regional management and content protection, see Chapter 5.

Apple Macintosh Architecture

Apple was very slow to support DVD, which was surprising given the synergy between DVD and Apple's artistic-minded customers. During the period when Apple was concentrating on rebuilding market share and stock value, it unfortunately ignored DVD. Only in 1999, when iMacs were released with DVD-ROM drives and high-end G4s were released with DVD-RAM, drives did Apple begin to again pay attention to the technology that so suitably matched its user profile.

Early developers of DVD hardware and software players for Macintoshes went out of business. Apple finally developed its own player software for Mac OS 9 that at first relied on hardware cards, then later was implemented in software with hardware decode acceleration in the graphics card.

The stream-oriented architecture of QuickTime works well for media types such as MPEG in Mac OS as well as on Microsoft Windows computers, but the interactive features of DVD-Video don't fit so easily into the QuickTime design. MPEG-1 software for playback was provided by the QuickTime MPEG extension, which was released in February 1997. Apple and other hardware and software vendors developed codecs for MPEG-2 video, Dolby Digital audio, and other DVD-Video streams to work with QuickTime, but after many releases of QuickTime (up to version 7 in 2005), there was still no API for applications to control DVD-Video playback other than the rudimentary Apple DVD SDK for HTML released in 2003.

DVD-Video Playback Details

Hardware video decoders can be built on specialized cards that use a VGA *pass-through* system for analog overlay. The pass-through system takes the output of the VGA display adapter and keys the video into pixels in a certain color range. *Hardware overlay* is also used for both hardware and software decoders, where the video is sent to the display adapter via a special bus: the VPE. The display adapter then overlays the video into the display using a color key. The key color is often magenta, which may flash now and then. The overlay approach is primarily designed to compensate for video hardware that is not powerful enough to deal with memory-mapped video at 30 frames per second. As video cards become more powerful and system busses become faster, the video port and overlay approach will give way to memory-mapped surfaces. The advantage will be that the video can be manipulated like other graphics surfaces, with geometric deformations, surface mapping, and so on. For example, MPEG video can be wrapped onto a spinning object or reflected onto a simulated window.

Because computer pixels are square, while DVD pixels are rectangular, the video must be scaled to the proper aspect ratio. The exact number of pixels is not as important as the displayed picture aspect ratio. Optimal square-pixel resolutions without scaling the video are 720×540 (no horizontal scaling) or 640×480 for NTSC with no vertical scaling or 768×576 for PAL with no vertical scaling. For full-screen viewing, the entire 1.33 screen should be used

$(800 \times 600, 1024 \times 768,$ etc.). On modern graphics cards with high-quality scaling hardware, video that is scaled up to fit the entire screen should look very close to windowed video in quality. Optimal square-pixel resolutions for anamorphic video without scaling the video are 720×405 (no horizontal scaling) or 854×480 for NTSC (no vertical scaling) or 1024×576 (no vertical scaling) for PAL.

Video often looks better if the display refresh rate is adjusted to match the source frame rate of material on the disc: 60 Hz for NTSC interlaced video (2×30), 72 Hz for film-sourced video (3×24), 75 Hz for PAL interlaced video (3×25). This avoids temporal aliasing such as tearing and jerkiness. A good sweet spot, still out of reach of most monitors, is 120 Hz $(4 \times 30$ and $5 \times 24)$.

DVD By Any Other Name: Application Types

The confusing part about the relationship between DVD and computers is determining where the domain of DVD-Video ends and that of computers begins. The only constant is the DVD-ROM drive itself. Some computer upgrade kits include a drive and an entire DVD player on a card. Other upgrade kits or DVD-Video-enabled computers include complete DVD audio and video decoding hardware; others include only small amounts of extra graphics circuitry to accelerate DVD-Video playback. Some systems rely entirely on software, others contain only a DVD-ROM drive and are incapable of playing video or audio. It's clear that a text database on DVD-ROM, for example, has nothing to do with the DVD-Video standard. It's also fairly clear that a blockbuster Hollywood movie that can be viewed on a computer monitor relies heavily on the DVD-Video standard. But what if the disc containing the blockbuster movie includes additional computer-only bonuses such as a screen saver or a computer game based on the movie?

The following classifications of DVD products in Table 15.4 may help clarify the variety of options. Note that the classifications cover both DVD-ROM and DVD-Video.

Table 15.4 DVD Product Classifications for Computers

Classification	Content	Ice Cream Analogy	Level
Pure	DVD-Video only	Vanilla	2
Bonus	DVD-Video plus computer supplements	Sprinkles	2,1
Augmented	DVD-Video plus computer enhancements or framework	Carmel swirl	2,1
Split	Independent DVD-Video and computer versions	Neapolitan	2,1
Multimedia	DVD-ROM for computer, any format of audio and video	Sundae	0
Data	DVD-ROM for computer data or applications only	Frozen yogurt	0

Pure DVD-Video

A pure DVD-Video disc is designed entirely for use on a standard DVD-Video player. It can also be played on a DVD-Video-enabled computer as well, with no differences other than the tiny computer screen and the chair that's not as comfortable as the living room couch. In this case, the computer serves merely as an expensive DVD-Video player. The computer's DVD-Video navigation software is all that is needed or used.

The target audience for this type of product is movie and video viewers. Note that the flexibility of the DVD-Video format enables a pure DVD-Video to contain still images, text screens, additional footage, supplementary audio, and other enhancements generally associated with special editions. Examples of pure DVD-Videos include movies, product demo videos, and instructional videos.

Computer Bonus DVD-Video

Just as Enhanced CDs contain both music to be played on a CD audio player and additional goodies to be used on a computer, bonus DVD-Video products contain extra material designed for a computer. The disc can be played normally on a DVD-Video player, while the video part of the disc may also be played with the computer's DVD-Video navigation software. The primary audience is similar to that of pure DVD-Video. The difference is that the disc contains computer software or data that the producer thinks will make the disc more appealing: movie scripts, searchable databases of supplemental information, still pictures in computer format, instruction manuals in electronic form, short video clips in computer video format, screen savers, icons, computer games, Internet Web links, and so on.

Computer-augmented DVD-Video

An augmented DVD product may be played on a DVD-Video player, but it is amplified and changed when played on a computer. The viewer might have more control over the sequence of the video, or the video might respond to the actions of the viewer, for example interludes in the video may offer a game or puzzle, or perhaps the video is subsumed entirely by a computer-based game or environment in which the video plays a supplemental or reward role. In this case, the computer might present video segments in a different order or generate additional graphics and create a composite image.

For instance, a murder mystery could be played and enjoyed in a straightforward way on a DVD-Video player and perhaps even offer multiple endings. But when the same disc is put into a computer, the viewer could choose to become a character in the drama. If the viewer chooses to play the role of the detective, the viewer can interrogate characters, search for records, collect and manipulate objects, answer questions, and become an interactive participant controlling the environment and outcome of the game.

Split DVD-Video/DVD-ROM

This type of disc may be used in a player or in a computer, but no overlap exists between the two. The computer version is usually related to the video player version, but no content is shared. It's basically a pure DVD-Video product on one part of the disc and a multimedia DVD-ROM on the other part (see the following section).

Multimedia DVD-ROM

A multimedia DVD-ROM product cannot be played in a DVD-Video player. It's intended for use in a computer only, like a traditional multimedia CD-ROM. The disc may contain audio and video in the same MPEG and Dolby Digital format as a DVD-Video disc, but it can be played back only on a computer. The disc may also contain video at higher resolutions and different aspect ratios than those supported by the DVD-Video format.

The multimedia DVD-ROM is the most visible application of DVD technology on computers. Current CD-ROM multimedia producers will gravitate toward DVD-ROM depending on their needs and the changing platforms of their customers. Multimedia programs requiring more than one CD are early candidates. As soon as the installed base of DVD-ROM drives in a given market reaches critical mass, associated multimedia publishers can switch their existing content from CD-ROM to DVD-ROM. They may want to take advantage of the increased capacity and higher minimum data rate of DVD-ROM to include more content and to improve the quality of audio and video.

The multimedia DVD-ROM could even become the primary application of DVD at home if major computer companies such as Microsoft, HP, and Intel succeed with their plans to capture the home market with a computer-based, multimedia, HDTV entertainment system.

Data Storage DVD-ROM

With all the audio and video hype swirling around DVD-ROM, it's easy to overlook it as a simple data storage solution. DVD-ROM can hold the same data as hard disks, floppy disks, and CD-ROMs, but in vastly greater quantity. Text documents, graphical documents, databases, catalogs, software applications, etc. can all be stored on DVD-ROM — the bigger, the better. By contrast, most CD-ROMs are not multimedia products, and multidisc sets are common.

DVD Production Levels

DVD development can also be classified according to the way the content is produced and used and what it demands from the PC. Three classifications exist for DVD development: level 0, level 1, and level 2.

Level 0 — Level 0 discs contain data and software only; they do not require MPEG-2 decoding. The content of a level 0 disc has computer data, perhaps even multimedia, but no dependence on MPEG-2 video. Traditional computer software development tools are used as well. No special features are needed in the computer to play back level 0 discs other than a DVD drive. Obviously, level 0 discs don't play in standard settop DVD players.

Level 1 — Level 1 uses only MPEG-2 video files with no DVD-Video requirements, whereas, a level 2 disc contains only ISO MPEG-2 program stream files. A level 1 disc does not use anything from the DVD-Video (or DVD-Audio) specification, the advantage being that the DVD authoring process is unnecessary. Once the video is encoded as an MPEG-2 file, it can be played back directly by the application software. This makes testing and development easier, especially because meaningful filenames may be used (such as intro.mpg and bouncingball.mpg), rather than cryptic DVD-Video filenames (such as vts_01_01.vob).

The drawbacks are that level 1 discs don't play in settop players. Also, because many PC DVD decoders are designed specifically for DVD playback and not for general MPEG-2 playback, compatibility problems occur with older computers, especially those using hardware decoders. Jumping around within files is also tricky because it requires the proper byte offsets to be predetermined at development time or that a slow timecode search be used, which demands that the decoder/playback software scan through the file to find the right position.

It's possible to produce video files with a DVD authoring system and play the resulting .vob files as MPEG-2 files. This works reasonably well, as long as the video is sequential. If interleaving is used for multiple camera angles or multistory branching, each interleaved block will be played in sequence, resulting in fragmented scenes.

Level 1 discs require a computer with an MPEG-2 decoder that has the capability to play individual MPEG-2 files. Level 1 development is supported by both Microsoft DirectShow and Apple QuickTime.

Level 2 — Level 2 discs rely fully on the DVD-Video specification. The video and audio content is created with a DVD authoring system, so the disc can play in a settop DVD player. The computer uses the DVD-Video presentation and navigation structures to present the content. Instead of specifying a filename to play a particular video segment, the application specifies a title and chapter, or a title and timecode.[3] Because the features of DVD-Video are used with time maps and other navigation information, access is very fast and accurate. Features such as angles, subtitles, multiple audio tracks, and seamless branching may be used, as long as they are authored on the disc. Care must be taken during authoring to make sure that all entry points on the disc are accessible. Either a chapter point must be at each place on the disc that the application would want to jump to or the video must be authored as a one-sequential-PGC title so that timecode searches can be used.

Level 2 discs require a computer with an MPEG-2 decoder and DVD navigator software that has an API to provide application control over DVD playback. Level 2 development is supported by Microsoft DirectShow. As of late 2005, Apple QuickTime does not support level 2.

---

[3]Timecode, when used in DVD playback, refers to an hours:minutes:seconds:frames representation of a time offset from the start of the disc, where 00:00:00:00 is the start time of the disc. The DVD-Video format does not directly support SMPTE timecode entry or guarantee accurate disc access to content via timecode entry.

WebDVD

Too many pundits and predictors would have us believe that DVD-quality television and movies will be streaming over the Internet any day now. The truth is that streaming video on demand is only for those who aren't very demanding, at least for another half-decade or so. Although more than 60 percent of US Internet-connected households are expected to have a broadband connection by the end of 2005, how broad is broadband? Roughly 44 percent of these high-speed households use DSL modems, which literally requires 24 hours or longer to download a typical DVD movie. The other half of the 60 percent have cable modems, which are faster, but not nearly fast enough for real-time delivery of DVD video. The inescapable fact is that broadband Internet simply is not developing at a rate that will make the delivery of DVD-quality video feasible any time soon. It will eventually happen, but not before 2010 for the typical household, and by that time most consumers will want high-definition, which raises the bar too high yet again. The development of true broadband will happen even slower in countries with less-developed Internet infrastructures.

The idea of TV shows and movies delivered to living rooms "live" over the Internet is appealing, but it's far from reality. Consider that in 1978, a laserdisc connected to a computer could provide interactive, full-screen, full-motion video. Admittedly, the video was often on a separate screen, but it was a good start for interactive multimedia. Since then, many supposed advances in technology have actually taken us in the wrong direction from that promising beginning. When QuickTime appeared on the scene in the late '80s, many people were so impressed that digital video from a CD-ROM worked at all that they didn't seem to notice that it filled a tiny fraction of the screen at less than 10 frames per second. The technology slowly improved, especially with the introduction of MPEG-1 and Video CD, but the quality of video on a computer was still below that of TV. Eventually the Internet arrived, which was painstakingly slow in sending full-screen graphics, let alone full-screen video. The newest player in this strange back-and-forth game is DVD. Finally, a mainstream technology exists with a fast-enough data rate and a high-enough capacity for high-quality video and audio to be completely integrated into a computer, but now that we're used to the instant gratification of e-mail and Internet search engines, a static, physical medium such as DVD seems somewhat limiting. So, why can't we have the best of both worlds?

Unlike the Internet, with its slow data rates, tiny, fuzzy video, long download times, and lost data packets, DVD has very high data rates for high-quality video and audio, guaranteed availability (as long as the correct disc is in the drive), and fast access to enormous amounts of data. DVDs can be distributed for free in magazines, at store displays, or in the mail. A DVD-9 sent via express mail can deliver in 24 hours, anywhere in the world, eight gigabytes of content that would literally take two weeks to download on a 56k modem. Until true broadband Internet capabilities are available and affordable to everyone who wants them, one of the best ways a Website can provide a high-impact experience is to depend on a local DVD that supplements HTML pages with hours of video and days of audio.

WebDVD, also known as connected DVD, Internet DVD, online DVD, enhanced DVD, and many other names, is the simple but powerful concept of enhancing DVD-Video or DVD-ROM with Internet technology. From a reverse point of view, it can be seen as enhancing Web pages with video, audio, and data from a DVD. WebDVD combines the best of

DVD with the best of the Internet. It's the marriage of two of today's hottest technologies, which complement each other beautifully. It's not a new idea; CD-ROMs have been linked to the Internet for years, but the character of DVD elevates this to a new level. DVD makes PCs true audio/video machines, with better-than-TV video and better-than-CD audio, and the Internet links them together, creating connections between creators and consumers and communities.

Until the Internet becomes a true broadband highway for average home viewers, DVD has an interesting role to play. When a DVD disc is placed in the local drive of a computer or an Internet-connected settop box, it can supply the quality video that the Web sorely lacks. For example, an educational Website can play detailed instructional videos from the local DVD drive, or a movie fan site can pull up favorite clips. A director presenting a live on-line chat can pull up behind-the-scenes footage hidden on the disc inserted in each viewer's drive, delivering it simultaneously to thousands of viewers with a "virtual bandwidth" that would bring any Web server to its knees. A Web-enhanced DVD movie can connect to the Internet for promotional deals, viewer discussion forums, multiplayer games, and more. A marketing disc, perhaps distributed in a magazine, can entice viewers with gorgeous product videos backed up by Web pages for on-the-spot purchases. The possibilities are endless, as long as the customer already has the disc or can be easily supplied with one.

When a DVD is designed to connect to the Internet, it becomes a different kind of product. Even after the disc leaves the hands of the developer, it can be improved or refreshed, and what would otherwise be a static medium becomes dynamic and renewable, extending the life of the product. WebDVD titles can take advantage of the timeliness of the Internet with automatic software updates, supplemental content as new information becomes available, updated links to related Web pages, current news and events, special offers of associated services and products, and even communities of customers who participate in online chats or discussion groups and provide feedback about the content of the disc. For example, the discographies included on many DVDs are often out of date by the time you watch the film. A WebDVD discography is never out of date.

Web content, such as HTML pages viewed in a Web browser, can be greatly enhanced with embedded audio and video assets, including MPEG-2 video and multichannel Dolby Digital audio played from local DVD storage. Conversely, DVD-Video titles can be enhanced with supplements and dynamic information from the Internet. Such DVD-Video titles will play normally on standalone DVD players, but when played on a DVD-enabled PC or a Web-enabled settop DVD player, they will be enriched with additions delivered via the Web. Here are some examples of such additions:

- A movie with built-in Web links or Web wrappers. Thousands of Hollywood movies have been enhanced for use in a computer. These discs play normally in a DVD-Video player, but when put into the DVD-ROM drive in a PC, much more content becomes available, such as links to information about the movie, an active screenplay that can play associated scenes, multimedia annotations, interactive games, special product discounts for Web purchases, and more. Some enhancements are intriguing, some frivolous, and some are downright lame, but none have yet come close to taking advantage of the power of a PC to take the movie experience to a different level.

■ A custom interface for the PC-based use of a DVD. HTML pages on the disc or on a remote Website can control the DVD, giving instant access to various sections of the disc, controlling angles, changing audio, and more. This type of custom interface can be quick, simple, and much easier to use than the on-disc menus that must be accessed by a standard DVD remote control or through the DVD player software interface. Hundreds of HTML development tools are at hand to quickly and easily create specialized interfaces in thousands of different styles.

■ A content pool for dynamic front ends. Advertising companies and corporate video development companies can use WebDVD for sales and marketing presentations. Hundreds or thousands of video and audio clips can be collected on the DVD: product demos, customer testimonials, tutorials, sales presentations, and so on. Custom HTML pages can be created to organize and present selected clips and other information from the disc. A new DVD can be shipped to offices and sales reps every few months or once a year. Between disc deliveries, different creative teams can put together new PowerPoint presentations and custom HTML pages. The resulting presentations, perfectly tailored to specific needs or clients, can be quickly e-mailed or downloaded anywhere in the world, even though they include full-screen video. The high-bandwidth content pulled from the DVD doesn't need to be downloaded. Urgent deadlines can be easily met with new pages that can be dashed out over e-mail, as long as the necessary video or audio is on the disc.

■ HTML-based reference material on DVD. Corporations, government agencies, and other institutions can use DVD to compile massive databases of text, pictures, graphics, audio, and video available to users and customers. Because the interface is done with HTML, large Websites can be easily moved onto a DVD. New content can be put on a Website between DVD releases, and it can be integrated into the existing body of HTML content for the next version of the disc. The DVD-based content is available in situations where no Internet connection exists, such as when traveling or for presentations.

■ Commentaries, annotations, and analyses. Anyone who loves (or hates) a movie enough that they want to share their opinion with the rest of the world can use WebDVD to create their own personal tour of the DVD. They can put their analysis on a Website with clickable text or buttons that play associated clips. Visitors to the Website need only put the proper disc into their DVD-ROM drive. Such a WebDVD site could even include a custom audio track, streamed over the Internet and synchronized to the DVD, supplying a running commentary (either the serious kind or the *Mystery Science Theater 3000* kind). Because the person viewing the Web page owns their copy of the disc, there should be no copyright worries.

■ A student of Shakespeare can create a Website that enhances any standard DVD version of one of the bard's plays. As the video runs, notes can appear on the side explaining the historical context, what is happening, what the actors are saying, etc. An alternative design might give the viewer the option to stop the video at any point and bring up background information. Web-based annotation works well with many kinds of DVDs, such as documentaries, historical dramas, or even TV shows. A WebDVD page that explained all the inside-jokes and references in episodes of *The Simpsons* could be a huge hit. Teacher's lectures, educational programs, and all manner of instructional video can be enhanced with pop-up annotations and ancillary information to enrich the learning process.

- Customized advertising and commerce. A DVD-based department store catalog could offer an interactive showcase of all the merchandise, complete with audio and video. Because the disc is connected to the store via the Internet, the consumer can get current prices, check availability, order merchandise, communicate with a shopping consultant, pay bills, and so on. The advantage of the DVD component is that the customer does not have to wait for slow downloads of pictures and even slower downloads of low-quality video. If the DVD is sent free in the mail, it creates a push to get the customer to visit the Website.

- Sharing high-bandwidth content. Until high-speed Internet connections are available between research groups and far-flung corporate divisions, WebDVD is an excellent tool for sharing large collections of data such as 3D visualizations, rendered graphics, and video simulations. The video and data can be put on a recordable DVD and shipped anywhere in the world. Recipients need only put the disc in their DVD PC and connect to a Web page that provides playback control, indexes, notes, and even collaborative discussions.

WebDVD Applications

WebDVD can be implemented easily on a computer. Traditional settop DVD-Video players can be enhanced with WebDVD capabilities, just as Internet appliances can be enhanced with DVD drives. As the convergence continues, other devices such as satellite receivers and cable boxes will also implement both Web and DVD features.

Still, most WebDVD development centers on Web browsers, specifically Microsoft Internet Explorer, which provides the best development environment for DVD. In addition, WebDVD applications can be created using PowerPoint, Macromedia Director, ToolBook, Visual Basic, and other multimedia tools. When references are made in this chapter to Web pages controlling playback, keep in mind that the same thing can often be done in other environments as well.

The continuum of WebDVD possibilities runs from a simple startup menu to a sophisticated application that displays video combined with synchronized content from the Internet. Some examples include the following:

- A menu that appears on the screen when the disc is inserted. The simplest "surf or show" menu would have two options: connect to a Website or play the disc.

- Hyperlink buttons that appear in player interfaces or over motion video when associated Web content is available.

- HTML information that appears in a separate window, synchronized to the video.
- An HTML-based application to control DVD playback in an embedded window or in full-screen mode.
- DVD-enhanced Web pages, where high-quality DVD content is played from the local drive in place of low-quality streaming Web content.
- Web-updated DVD pages, where new or time-sensitive information on the DVD is superseded and/or augmented by content from the Web.
- A custom application with a window playing DVD-Video, a Web-based sidebar synchronized with the video, and a chat window to interact with people viewing the same content.

Enabling WebDVD

HTML with associated scripting languages and related technologies has become the easiest and most widely supported computer development environment. More authoring systems, utilities, support services, and reference resources are available for HTML than for traditional languages such as C++, Visual Basic, or Macromedia Director. In addition, more people are versed in HTML than in any programming language. Millions of people, from children to grandmothers, know how to create a Web page. It's easy to add an object to play DVD-Video in a window on a Web page. Ironically, some of the most interesting applications of WebDVD don't even use an Internet connection, they simply rely on the ease of HTML authoring to create interactive DVDs for PCs.

The key is that a WebDVD platform must provide a way to integrate a window on the HTML page that shows video from the DVD. The best approach enables the video to be fully integrated into the page, with HTML objects overlaid on top of the video. Full-screen mode, where the video takes up the entire window, with HTML-based controls and information boxes that can pop up over the video, is essential for full-fledged WebDVD applications. At the other end of the spectrum, some WebDVD platforms enable only limited control of video playback within a separate window. Other applications play the video and simply link to URLs within a separate browser window. In most cases, DVD playback control is accomplished using a script language such as ECMAScript (or JavaScript) to call methods and access properties defined by the DVD API of the platform. In some cases, declarative DVD control is also possible, using markup tags such as those defined for SMIL.

WebDVD pages may be jazzed up with any other Web technology supported by the browser, such as CSS, DHTML, XML, CGI, and so on. Even proprietary or platform-specific formats, such as 360-degree picture viewers, ASP, PHP, Flash, Shockwave, Real Networks audio, Microsoft Windows Media Audio, and so on, can be used in conjunction with the DVD playing in a window on the same HTML page. Simply put, anything that works in a Web browser can be enhanced with DVD. This is a key point: the hard part of WebDVD development is not the DVD portion; it's everything else that goes with it. Too many people get hung up trying to figure out what they can and can't do with WebDVD. The answer is that if you can do it in a Web browser, you can add DVD to it.

TIP

The golden rule of WebDVD: Anything you can do with the Internet or HTML, you can combine with DVD.

Creating WebDVD

The preliminary step in creating a WebDVD is to determine which parts of the project are specific to DVD and which are not. Elements and processes that are not specific to WebDVD include page layout, script and application programming, customer registration, e-commerce and online sales, the general mechanics of conditional access and payment, user authentication, searches, and printing. These can usually be accomplished with tools that have no WebDVD components and by developers who know nothing about DVD. Only a few tasks are specific to WebDVD (see Table 15.5).

Table 15.5 WebDVD Needs

DVD-quality video and audio	Use existing disc or produce new content.
PC control of video and audio	Use WebDVD features of authoring system or write scripts or generate SMIL/HTML + TIME tags.
Synchronization of PC content	Use WebDVD features of authoring system or write scripts based on time event and list of timecode or on ATVEF trigger events.
Unlocking of content on disc	Use simple tricks to hide the content on disc or use conditional access and encryption tools designed to work with DVD-Video content.
Linking to URLs	Put HTML pages on disc. Minimize links from disc to Web because destinations may move or disappear.
Multiple platforms	Write generic control scripts that don't depend on a specific WebDVD object and API.

Essentially, the HTML content of a WebDVD may reside in four places:

- On the DVD, designed to work without a connection to the Internet.

- On the DVD, designed to connect to the Internet, pulling content from the Internet or linking to other HTML pages.

- On the Web, designed to be activated by the DVD. A single link on the DVD jumps to the Web page, which then controls the disc and provides additional content and links to other Web pages.

- On the Web, intended to work with a DVD. The HTML page requires that the disc be inserted in the local drive so that the HTML page can control the disc.

Making and Accessing the Audio and Video

Video content needs to be identified or created. If the WebDVD project uses an existing disc, the only task is to identify the titles, chapters, and timecodes needed to access parts of the disc. If the project requires a new disc, the video and audio content needs to be created. You must decide whether to author the content as DVD-Video or simply put MPEG-2 or other multimedia files on a DVD-ROM. In general, authoring a DVD-Video disc is the best approach. The DVD-Video content will play in all DVD platforms, even those with no WebDVD support, and DVD-Video-specific features such as subpictures and camera angles can be used. DVD-Video volumes are more compatible with computers and with settop WebDVD players than generic MPEG-2 files.

Access to the DVD-Video zone of the disc is accomplished with the WebDVD API of the target platform, which must provide a way to jump to specific titles and chapters, and preferably to specific timecodes as well. For simple WebDVD titles intended only to show related HTML pages as the video plays, an authoring system with WebDVD support may be all you need; you can type in URLs that are to be associated with each chapter. For more complicated WebDVD titles, you will need to write scripts and possibly create lists or databases that link the video to the Web or the Web to the video. In either case, a title number plus a chapter number, or a title number plus a timecode (hours: minutes: seconds: frames), will be the link to the video. A URL (such as http://dvddemystified.com/webdvd), possibly with a script to be executed on the page, is the link to the Web. URLs may be relative (no http://domain.com/ at the front), in which case the HTML pages come from the DVD itself.

If a new disc is created, it should be authored with WebDVD in mind. User operation controls (UOPs) should be used sparingly. Disabling the fast-forward, next, and menu operations to lock the viewer into introductory logos, legal warnings, and ads will also prevent control by HTML pages. Scripts on the page won't be able to jump to desired locations on the disc, while the necessary user operations are blocked. Commands and GPRMs should be used judiciously. A disc that depends on certain values being placed in GPRMs may behave in unexpected ways when the HTML page that controls it skips over sections that set or check values. Video access points should be identified ahead of time. That is, every place in the video to which you might want to have the HTML page jump to should be marked as a chapter during DVD-Video authoring because chapter access is extremely fast, easy, and accurate. It's possible to jump to an arbitrary point in a video sequence using timecodes, but only if the title is authored in one sequential PGC form. Most players can only jump to the I frame nearest a timecode, so the granularity of access is usually about 0.5 seconds.

Designing the Interface

A common approach to a WebDVD interface design is to create links from the video to the Internet or to the PC world. For anything other than very simple titles, this is a dead-end approach, primarily because the DVD-Video specification has no provisions for jumping outside its limited universe. A much more flexible approach is to design the disc so that the computer takes control and wraps the DVD-Video inside its own much larger universe. Menus and other features authored on the disc can still be used, but the PC can provide additional control and content. The autorun or autostart mechanism can be used to launch an HTML page when the disc is inserted in the drive, then the page can take over with its own menus and windows or it can play the video in full-screen mode to mimic normal disc playback, perhaps placing a small icon in the corner that the user may click to gain access to enhanced content. Because the PC can do a better job of implementing a user interface, providing graphical enhancement and controlling the video in a more interactive way, a PC-centric approach works better than a DVD-Video-centric approach.

Synchronization

Fancy WebDVD titles may synchronize PC-based content with the video or audio playing from the disc in a variety of ways. The most straightforward method requires that the DVD navigator component generate a time signal or a recurring time interval event. The page can repeatedly poll the current time position or respond to the time event. A list of synchronization points can be compared to the current playback time position. When a synchronization point is passed, the page can show text, activate a link, start audio playback, or begin an animation.

Connections

Most WebDVD titles connect from the disc to the Internet, or from one Web page to another. WebDVD authors must make the commitment to maintain Web-based content and links. Along with the ability to extend the reach of a DVD comes the responsibility to keep the extensions working.

Internet links change with alarming frequency. A year after producing a WebDVD title, you may discover that half the links to other sites no longer work. The best way to deal with this is to make sure that the links permanently stored on the disc point only to sites under your control. You can then use passthrough mechanisms to redirect the user to other sites. If the sites move or disappear, the central Web site can be updated as needed. Some WebDVD tool providers, such as InterActual, offer a redirection service. A simpler approach than redirection is to simply make sure that all pages that link to other sites are stored on your Web site instead of the disc. These pages can then be updated whenever needed to maintain working connections to the rest of the Internet.

Conditional Access

Hiding or locking the content on a disc is done for many reasons and in many ways. You might want to play special hidden tracks as an incentive for customers to visit your Web site, or you might want to charge customers for access to value-added content. Or, you might simply have extra video clips on the disc that only make sense when played through a custom PC interface.

The simplest approach is to put video clips in titles that aren't accessible through any of the menus on the disc. Users who know how to use the title search feature of their player will be able to play the clips, but most users won't even know to look for them, let alone play them. For added security, extras may be authored in a separate DVD-Video volume that is stored in a subdirectory of the disc. That is, the .VOB and .IFO files that would normally be stored in the VIDEO_TS root directory are stored in a different directory, possibly a few levels down in other directories. A standard DVD player will never see the secondary volume, but a PC application can be designed to read the DVD-Video content from its non-standard location. Another option for securing access to video is to add a precommand to each restricted section to check for a special value in a GPRM. The PC will store the special value in the GPRM, possibly after a transaction or authorization process.

WebDVD for Windows

Microsoft provides free basic components for WebDVD development as part of the Windows operating system. DirectShow, part of the DirectX set of core audio/video technologies, provides a standardized framework for DVD decoders and DVD applications. C++ programs can write directly to the DirectShow API, but most WebDVD developers will use the Windows Media Player or the MSWebDVD of MSVidCtrl object. Each is an ActiveX control that can be added to any HTML page to create a scriptable DVD window. The controls are built on top of DirectShow, so they require that a DirectShow-compatible DVD decoder be installed in the computer. Because ActiveX support is required, Windows Media Player and MSWebDVD don't work in Netscape (unless it has since been updated to support ActiveX). Also, because of design flaws in Windows Media Player versions 5.2 to 7.x, the DVD features work only in Internet Explorer, but not in PowerPoint, Visual Basic, and other ActiveX hosts.

The scriptable DVD APIs for Windows Media Player and MSWebDVD provide access to essentially all the features of the DVD-Video format. These include title search and play, chapter search and play, timecode search and play, operation of onscreen menu buttons, audio track selection, video angle selection, subpicture selection, scan, slow, step, etc. Various properties enable scripts to determine the status of playback, to query for features available on the disc, to respond to events, and so on. The feature set is sufficient to the point that a complete DVD-Video player can be implemented in HTML, even though most WebDVD applications use only a few basic commands for playing video clips.

Adding DVD to a Web page using Windows Media Player or MSWebDVD is far simpler than many Web page creation tasks, but it does require scripting. For those looking for easier but possibly more limited solutions, various options are available. InterActual Player (formerly called PCFriendly), the software used for most PC-enhanced movies, is available to developers directly from Sonic Solutions. Sonic also provides tools such as eDVD to make it easy to add Web features to any authored DVD-Video. Sonic, Visible Light, and others make plug-ins for HTML-authoring systems that simplify the process of creating WebDVD pages. Some DVD-Video authoring tools include basic WebDVD features. Also, a number of Xtras add DVD playback to Macromedia Director projects. The sample disc that comes with this book includes demos of various WebDVD tools.

WebDVD for Macintosh

As of 2005, Apple QuickTime doesn't yet provide control of DVD-Video. Until this happens, it's difficult to develop WebDVD applications that work on Macs. Apple has implemented JavaScript control of the Mac OS DVD player application, which at least provides the rudimentary capability for HTML pages to control DVD playback. Sonic's InterActual Player software, and Apple's related DVD SDK, supports scriptable control of Apple DVD playback, providing the first glimmer of cross-platform WebDVD tools.

No plug-ins exist for Macintosh Web browsers, and without QuickTime support, they would have to be written to talk directly to specific hardware or software DVD players, thus limiting future compatibility.

WebDVD for the Rest of the World

In December 2000, the DVD Forum created a new ad-hoc group (AH1-12) chaired by InterActual Technologies to develop a specification for WebDVD. The idea was to have a single format for consumer DVD players as well as PCs so that there was more incentive for Hollywood studios and others to produce enhanced titles. Members of the ad-hoc group spent years developing what became known as the eNav specification (for enhanced navigation). The format uses XHTML, ECMAScript, and other Web-oriented technologies to add highly interactive control of the DVD-Video portion of discs. But, as the specification neared completion in mid 2003, the next-generation formats were coming to fruition. Many companies, particularly the studios, felt that it was better to wait and incorporate eNav into a single new high-definition format. Thus eNav died, but from its ashes arose the interactive capabilities of the next-generation formats. Most of the advanced interactivity of HD DVD-Video and BD-J grew from work done on the eNav specification.

Why Not WebDVD?

WebDVD doesn't make sense in certain applications. For example, video on demand is intended to provide video without requiring a physical copy. Video or audio collections that change frequently or have recurring additions are difficult to keep up to date on distributed media. Large libraries of video that can't fit on a single disc are generally not appropriate for WebDVD, unless DVD jukeboxes can support them. And WebDVD obviously requires DVD PCs or WebDVD players.

Sending data over the Internet is generally perceived to be free. In spite of the costs that are hidden within access fees, subscriptions, and hundreds of charges, specific data transactions usually have no direct charges. DVDs cost money to produce. Even though replicated or recordable discs cost well under a dollar each, DVD cannot compete with the virtual free ride of the Internet.

Why WebDVD?

As the Internet inexorably entangles itself into everything we do, Web-enhanced products will become the norm. Customers will be disappointed if the DVDs they buy don't have Web features. Home and business Internet users will become increasingly frustrated that the quality of video and audio on their PC is so inferior to their TV and DVD player. The convergent evolution of settop devices such as cable TV boxes, game consoles, personal video recorders, and WebTV will lead to the integration of the Internet into TVs, DVD players, and every kind of information and entertainment device. The advantages of WebDVD — content, that can be customized and updated, dynamic data integration, customer-publisher communication, e-commerce, and more — make it very appealing to developers and customers alike. Today's Web-enhanced DVD titles are just the tip of the iceberg, given the unlimited potential that is unlocked by the marriage of DVD and Internet.

Besides computer geeks who do it just because they can, a surprising number of people watch movies on a PC. For example, those with limited living space, such as college students, may make a PC double as a TV. Airplane travelers with laptop computers can watch their own movies without getting a permanent kink in their neck from craning to see the television screen in the center aisle. Even if you have a home theater with a standard DVD-Video player, there's no reason you can't enjoy the movie from your comfy chair, and then later pop it into your computer to see what else is available. One out of five copies of Web-enhanced movies are viewed on a PC. Keep in mind also that WebDVD goes beyond movies. It works even better for personal improvement videos, documentaries, education, business training, product info, and so forth.

The fact is that eventually we will all watch movies on a computer, we just won't realize it. Today, a computer is one of the best solutions for progressive DVD display and for digital television and it was the first solution for high-definition DVD. The trend begun by WebTV and accelerated by TiVo continues, until the day when every TV and settop box is smarter and more connected than the fastest PCs available when DVD was born.

Traditional linear storytelling will always be an important part of the entertainment mix, but interactive cinema will continue to grow. DVD, and especially WebDVD, is the first technology that makes Hollywood-quality interactive cinema possible for the average producer and available for the average consumer. Until Internet bandwidth catches up, WebDVD provides portable broadband. This is a proven, practical, and functional way to deliver high-quality interactive video to the worldwide market.

Chapter 16
Production Essentials

This chapter covers the basics of producing DVD-Video titles and DVD-ROM titles. It is not a detailed production guide, but rather an overview of the production processes which can serve as a guide and checklist, as well as an introduction for those who want to understand what's involved. Some aspects of DVD authoring that are confusing, misunderstood, or overlooked are also covered.

General DVD Production

Creating a DVD is a tricky process. DVD-Video and DVD-Audio production demand knowledge and training, especially for highly interactive projects. Even DVD-ROM production is surprisingly more difficult than CD-ROM production. This is partly because DVD has more capabilities and more flexibility and thus more potential for mistakes, and partly because the tools for DVD haven't gone through a long period of refinement. The tools will undoubtedly improve and the process will become easier, but in the meantime, do not assume that making a DVD is a walk in the park.

Special utility software is needed for DVD-ROM formatting and for writing DVD-Rs. Specialized production systems are required for DVD-Video and DVD-Audio. Movie studios, in conjunction with DVD hardware partners, developed proprietary production systems to create content before the introduction of DVD. At the introduction, one authoring system was commercially available, and only for SGI workstations: Scenarist DVD, created by Daikin and distributed by Sonic Solutions. A few compression and premastering systems were commercially available from companies including Daikin/Sonic Solutions, Minerva, Innovacom, and Zapex. Commercial production and compression systems ranged from $80,000 to $250,000. Following the introduction of DVD, dozens of new production systems began to be developed. After a few years, the tools ranged from barebones authoring applications priced at less than $75 to full-featured authoring packages that cost tens of thousands of dollars, but could do far more than the original $100,000 systems.

Producing a DVD is significantly different from producing a videotape, laserdisc, audio CD, or CD-ROM. There are basically three stages: authoring (including encoding), formatting (premastering), and replication (mastering). DVD production costs are not much higher than for other media, unless the extra features of DVD-Video or DVD-Audio such as multiple sound tracks, camera angles, synchronized lyrics, and so on are employed.

Authoring costs are proportionately the most expensive part of DVD production. Video, audio, and subpictures must be encoded, menus have to be laid out and integrated with audio and video, and control information has to be created. This all has to be multiplexed into a single data stream and finally laid down in a low-level format.

In comparison, videotapes don't have significant authoring or formatting costs, aside from video editing, but they are more expensive to replicate. The same is true for laserdiscs, which also include premastering and mastering costs. CDs require formatting and mastering, but cost less than DVDs. Since DVD production is based mostly on the same equipment used for CD production, mastering and replication costs will eventually drop to CD levels. Double-sided or dual-layer DVDs cost slightly more to replicate because they are more difficult to produce.

Project Examples

DVD projects can take a variety of forms. The list in Table 16.1 is by no means exhaustive, but illustrates the ways DVD can go far beyond a portion of linear video transferred from tape.

Table 16.1 Project Examples

Project	Example
Video and audio without menus, graphics, or options	Archived video
Video and audio with a single menu to choose selections	Home videos
Video and audio with a main menu and submenus with supplements	Simple movie, educational disc
Audio with a few menus and stills	Music albums (DVD-Video or DVD-Audio)
Multichannel audio and music videos with menus and supplemental video	Music video albums (DVD-Video or DVD-Audio)
Main menu and submenus for video format, audio language, subtitles, angles, chapters, and so on	Special edition movie
Any of the above with computer content or Web connectivity on hybrid disc	PC-enhanced movie
DVD-ROM with data or applications	Auto parts database, computer game, and application suite

CD or DVD?

For computer applications, CD and DVD do not differ greatly other than in their capacity. If audio and video are involved, then CD — particularly Video CD — may be an option if video quality is not paramount. For short video that can be played on computers, DVD-formatted content can be put on a CD.

For those considering moving from CD to DVD, the financial break-even point is at two or three discs. Although it may be cheaper to produce and replicate two CDs compared to one DVD, cost savings for DVD also come from simpler packaging, lower shipping and inventory costs, etc.

DVD-5 or DVD-9 or DVD-10?

The disc type is generally mandated by the size of the content. Single-sided DVD-9 discs and double-sided DVD-10 discs are similar in capacity. Although DVD-10 discs hold slightly more and are slightly cheaper, DVD-9 is almost always a better choice, since it allows a full label and doesn't require flipping. Rumors of production problems and compatibility issues are just that — rumors. Early production yield problems were solved long ago, and minor dual-layer compatibility problems affect an insignificant number of readers.

DVD-Video or DVD-ROM or Both?

A DVD-Video disc plays in any DVD system: DVD players, DVD computers, DVD game consoles, and so on. It has more limitations, making project design and production easier in some ways. In general, DVD-Video discs require little or no tech support. Placing computer applications on DVD-ROM, however, provides more flexibility, and it has better support for text display, data manipulation, searching, and interactivity. The best approach is to combine both, using the best qualities of each. Of course, this option requires more work. See Chapter 15 for more on computer use of DVD and enhancing DVD-Video and DVD-Audio discs with Internet technology.

DVD-Video and DVD-Audio Production

DVD-Video and DVD-Audio production require specialized equipment and software. Melding the computer world with the audio/video world, in ways that are unfamiliar to many, is a complex process. Computer graphics artists, accustomed to square pixels and an RGB palette, must learn about non-square video pixels, video-safe colors, overscan, and interlace artifacts. Video engineers, accustomed to headroom and waveforms, have to deal with $Y'C_bC_r$ colorspace, sampling, compression, and bit rates. Audio engineers also must deal with issues such as multichannel mixing, apportioning the LFE channel, downmixing, sampling frequencies and word sizes, and synchronizing multiple audio tracks to video.

DVD production work is done in a variety of ways and places. Each business model is optimized for the market it serves, with attendant advantages and disadvantages. Smaller businesses may provide services for only part of the DVD production process, such as menu design, subtitles, or audio formatting, while large service houses can provide a one-stop shopping experience for an entire project. Several types of DVD production environments are available:

- Movie/music studios or subsidiaries develop their own products from content they create.

- Film/video/audio post-production houses produce DVD as a sideline to complement core business.

- Replication facilities operate authoring services to generate more disc manufacturing business and can provide one-stop shopping.

- DVD shops focus primarily on encoding and authoring.

- Specialty DVD shops focus on market niches such as government, museums, or weddings.

- Corporate A/V departments develop discs for internal training or external marketing and may provide production services to other companies.

- Individual or small-group projects involve one person or a few people with a vision, such as independent filmmakers.

A large studio has experience creating hundreds of titles. They have a staff with experience in what works and what doesn't. They have the financial wherewithal to stay in business and can eat cost overruns if something goes wrong. On the downside, they operate in factory mode and are not very flexible. Most creative people don't function well within this environment. Internal titles or big customers come first, potentially sidelining smaller projects. Costs may be higher, but the quality generally matches the price.

A traditional film, video, and audio post facility has the production background and experience to make DVD work, even though DVD is much more complicated than film transfers and editing. If problems with the assets occur, such as bad tapes, or missing pieces, a professional post house usually has the equipment to make repairs on the spot. Artists and creative people tend to be on hand, and good service is an expected way of life. The downside may be that DVD is not the prime focus of the facility.

Replication facilities that offer DVD premastering services can handle most elements of a job from input to packaged product. Disc manufacturers need titles to feed the production lines, and they will do plenty to keep the production lines going, including authoring and premastering, often at very competitive rates. Replicators can afford to use compression and authoring as a loss leader in order to make money in replication. However, a sideline authoring service may not carefully focus on quality and creativity. The user of such packaged services is usually a price-conscious customer, likely to switch suppliers for the next cheaper deal.

The dedicated DVD shop tends to be a small but focused group, originally from a creative multimedia background. The personnel skill sets are well suited and dedicated to all things DVD, and good service is the norm. Often, the president is also the person that runs the machines, does the shipping, and makes the coffee. DVD-only companies are more likely to try new techniques. They offer more value for the dollar, since they have to work harder just to survive. The disadvantage is they may be undercapitalized and unable to keep up with changing conditions and technology. If a project goes bad, it could seriously affect their bottom line. Some will grow to be big players, but many will fall away.

Specialty shops are the best at what they do in a limited role, and they know enough to stay in their niche and be successful. The disadvantage is competition with all-in-one operations. Producers must establish multiple relationships with suppliers and providers of related services. Technology can change and make niche markets obsolete.

Corporate audio/video departments often evolve DVD production capabilities from existing services. In some cases, the DVD side of things may be awkwardly grafted onto existing structures, or in other cases, the DVD group may be well positioned to understand the needs of corporate video and know how to apply unique features of DVD.

Sometimes the best titles spring from the creative drive of one person. This person may do it alone, finding the cheapest production tools possible, or they may beg, borrow, or barter services from others. The most difficult part is finding a distributor to put the creator's labor of love into the marketplace.

Tasks and Skills

Many skills and processes are needed for DVD production. The larger the facility, the more specialized each position is. No facility is likely to have all facets down to perfection, but many have multiple areas of expertise or relationships with other service providers. Small DVD production shops rely on outside facilities to do much of the production work.

The following list covers the essential job positions of DVD production. These could be combined into a few positions or spread across multiple people. Rare is the person who excels at all tasks of DVD production, although it's a good idea to cross-train personnel so they can assist in other areas as needed.

- A project designer works with clients and is responsible for layout, design, and general budget, and therefore must understand the steps of the production process.

- A project manager deals with bottlenecks and should be familiar with all other production tasks, works with clients for testing and signoff and is responsible for schedule, coordination, budget details and asset management.

- A colorist transfers film to video, is responsible for color correction, and often works with the director or director of photography (DP).

- A video editor is responsible for assembling, editing, and preparing all the video assets for encoding; uses video editing systems, usually digital video editing software; may need to restore and clean up footage; may do compositing of video and computer graphics; should be familiar with video encoding and DVD authoring.

- An audio editor is responsible for assembling, mixing, editing, and preparing all the audio assets for encoding; uses audio editing systems, usually software; is often required to sweeten or clean up audio; must conform audio to match edited video, especially when multiple languages are involved; and should be familiar with audio encoding and DVD authoring.

- A video compressionist must understand MPEG-2 compression, preferably VBR; needs to be familiar with DVNR equipment; should be familiar with colorist and video editor jobs; have working knowledge of DVD authoring process; and must be familiar with professional videotape equipment.

- An audio compressionist must understand multichannel audio and principles of perceptual audio encoding; must have in-depth knowledge of details of Dolby Digital encoding, and DTS if needed; must be familiar with professional digital and analog audio recording equipment; should be familiar with audio editor job; and have working knowledge of DVD authoring process.

- A graphic artist is responsible for graphic design, including menus and stills; may also create subpictures; may be responsible for user interface design; may also work on disc label and package design; uses graphics packages such as Adobe Photoshop; and must understand nuances of digital video and analog video.

■ A 3D animator produces animation sequences for menus and transitions; uses graphics packages such as Adobe After Effects; must understand nuances of digital video and analog video.

■ An author uses specialized DVD authoring applications; must have detailed knowledge of DVD specification and navigation design. Advanced authoring requires programming knowledge.

■ A tester is responsible for quality control (QC) of product; reviews encoded audio, video, and subtitles; tests navigation; and should be familiar with digital video and audio.

The Production Process

The DVD video and audio production process can be divided into basic steps, which are shown in roughly chronological order in Table 16.2. The QC and testing steps are spread throughout most of a project. Proper project design and careful asset management are key to the successful creation of a DVD title.

Table 16.2 The DVD Production Process

Project planning	Develop schedule and milestones, storyboard video, lay out disc, design navigation, budget bits.
Asset preparation	Collect, create, capture, process, edit, and encode video, audio, graphics (menus, subpictures, stills), and data. Check source assets, digitizing, and encoding.
Authoring	Import, synchronize, and link together assets; create additional content, define and describe content; simulate and check compliance with DVD specification.
Formatting	Multiplex and create volume image. Emulate, check compliance, run check discs, and compare checksums. Write DLT or DVD-R.
Replication	Create master image, record glass master, mold discs, bond, print, and QC final copies.
Packaging and distribution	Insert discs and printed material, insert source tags, shrink wrap, box, and ship.

Authoring refers to the process of designing and creating the content of a DVD-Video title. The more complex and interactive the disc, the more authoring is required. The authoring process is similar to other creative processes such as making a movie, writing a book, or building an electronic circuit. It may include creating an outline, designing a flowchart, writing a script, sketching storyboards, filming video, recording and mixing audio, taking photographs, creating graphics and animations, designing a user interface, laying out menus, defining and linking menu buttons, writing captions, creating subpicture graphics, encoding audio and video, and then assembling, organizing, synchronizing, and testing the material. Authoring is done on an authoring system, which may be integrated with encoding systems and a premastering system.

Authoring sometimes refers only to the specific step of arranging the assets using an authoring system. Encoding and authoring are sometimes referred to as premastering, although premastering more accurately refers to the final step of multiplexing files, formatting a UDF bridge image, and writing it to DLT in preparation for mastering and replication.

Production Decisions

During the production process, many decisions between different options need to be made. The choices made depend on the particular project and on personal preference.

Service Bureau or Home Brew?

You must decide whether to do some, most or all of the work yourself, or to outsource the project to a DVD production house. (Unless, of course, you are a DVD production house.) The advantages of producing a title yourself are that you have full control of the project and can make changes as needed. If you intend to produce a number of DVDs, you can build a base for long-term production. Service bureaus have the advantage of knowledge and experience, and they own most of the production equipment needed to produce a DVD. They have trained personnel and can usually produce a disc quickly, if necessary. They have tools for testing and may even have a collection of players and DVD PCs for testing and quality control. They also usually have connections with replicators.

To find an appropriate service facility for your DVD project, check with other media producers for referrals, comments and recommendations, or look in trade magazines and features for knowledgeable DVD personnel and service bureaus. The better companies are well known and are often featured in news articles. Advertisements display the companies with marketing budgets, but often the best service providers remain so busy that they don't need to advertise for more business.

With a list of several companies in hand, do a comparative analysis between them. Request company information, including background, personnel, equipment, DVD experience, production capacity, and other available services that can help you with your project. Ask for a demonstration or sample discs. Once you choose a facility, develop a good relationship and communicate regularly during projects.

VBR or CBR?

Variable bit rate (VBR) compression enables better quality with longer playing times, since it uses disc capacity more efficiently. Good VBR encoding is somewhat of a black art and requires an experienced compressionist. *Constant bit rate* (CBR) compression is cheaper and faster because it doesn't require multiple passes. For projects without a lot of video, high data rate CBR can achieve the same quality as VBR.

NTSC or PAL?

DVDs in 525/60 (NTSC) format will play in more than 95 percent of all DVD players, since PAL and SECAM players play NTSC discs. DVDs in 625/50 (PAL/SECAM) format provide better resolution, but will not play in most NTSC players. However, performing standards conversions from one format to the other may result in significant degradation of the video quality, so it's usually best to remain with the video format of the content.

PCM or Dolby Digital or MPEG Audio or DTS?

Dolby Digital audio format gives the widest coverage. Every DVD player, including computers, can play Dolby Digital audio tracks. Every player is also required to play PCM audio, but PCM doesn't leave much room for video. DTS audio is a secondary format that appeals to an audiophile niche. Most players and computers must be connected to a DTS decoder, since they can't natively decode DTS. MPEG audio is not widely used.

Multiregion and Multilanguage Issues

Before you begin designing a DVD-Video disc, ask about the distributions needs for the project. Will it stay within a single geographical region or will it be distributed worldwide? Do you want to restrict the regions in which it will play? Unless you expect special exclusive regional distribution arrangements, there is no reason to restrict the regions of a disc. If you make an all-region disc, it can be distributed anywhere in the world. If distribution is to be restricted to one or a few regions, this may cause a problem later if you want to broaden distribution; you'll have to reformat and remaster the disc.

If you want to set region codes on the discs, you must also CSS encode the contents. Region code restrictions in players are required only in conjunction with CSS. It's possible to set the region flags on the disc without using CSS, but players are then not required to honor the region settings.[1]

If you decide to use regional management, then you need to decide which regions. Again, broadening the scope of the disc early may be a lifesaver later, when marketing plans become more ambitious. Table A.1 lists what regions apply to which countries. You must also decide menu languages, and choose which audio languages and which subtitle languages to put on the disc. Carefully check the audio and subtitle tracks to make sure they match the video you are using. In many cases, an existing audio dub or subtitle set may be from a different cut of the movie, and will have to be modified or redone to match the cut you are using. Audio from film usually has to be sped up (approximately 4 percent) for PAL, so make sure the audio properly matches the video.

If you cover multiple regions, or even if you only cover a single region such as Europe/Japan, you may need to consider certification. Each country has its own requirements for classifications and certification. A disc with a "children's" rating in one country

---

[1]As a matter of practice, if region codes are defined for the content, players will behave accordingly, whether or not there is any CSS encrypted material on the disc. While it is true that the CSS license is what compels manufacturers to support regional management, once it's in the player, it's more of a burden to turn it off for non-CSS protected discs than to just leave it active.

may end up with a "teen" rating in another country. If you have to cut scenes to get a particular classification, you will have to deal with conforming the audio and any subpictures. Using DVD-Video parental management features usually will not work. The classification boards rate a work according to all the versions on the disc, regardless of how you have restricted access to them.

You must also decide if you want a single package (often called a SKU, *stock-keeping unit*) or multiple packages for multiple regions. In some cases, creating multiple SKUs is easier, since you can concentrate on making each version appropriate for its region rather than trying to make a single version work everywhere. For example, you might want to make three versions of a disc for release in Europe: one for the UK and Ireland, one for Germany and Scandinavia, and one for central and southern Europe (France, Spain, Italy, and so on). Multiple SKUs can also help with legal restrictions concerning how soon the home video version can be sold or rented after theatrical release.

Regardless of the number of regions, if you expect to use more than one language on the disc, especially for menus, keep all the graphic and video elements as language-neutral as possible. Keep text separate from graphics so it can be easily changed for other languages. Use subpictures, including forced subpictures, to put text on video, rather than permanently setting the text into the video. Keep in mind that American English is different from British English regarding usage and spelling. Likewise there are differences between American Spanish and Castillian Spanish, between Canadian French and continental French, between Brazilian Portuguese and Portugal Portuguese, and so on. In all cases, check to ensure that you have proper rights for the content. A particular work may have only been licensed for distribution in one country, or a foreign language audio dub may be restricted for use in certain countries.

Supplemental Material

One of the great appeals of DVD is that a disc can hold an incredible variety of added content to enhance the featured program. Being able to watch a movie a second time and listen to the comments of the director, writer or actors gives viewers a deepened appreciation for the movie and for the filmmaking process. Adding more video, audio, graphics, and text information can greatly enrich a disc. Anyone producing a DVD should consider how the product can be enhanced with supplements; not just because it's neat or expected, but to truly improve the viewing experience. The following list provides some examples of added content for DVD-Video and DVD-Audio discs. Some types of added content, especially text, and information that can become out of date, may be better provided as computer-based or Web-based supplement, and are marked with an asterisk (*):

- Multiple languages and subtitles, menus in other languages

- Biographies: information about actors, directors, writers, and crew, describing careers, influences, and so on*

- Filmographies: information about actors, directors, writers, and crew describing previous films, acting experience, awards, and so on*

- Commentaries: audio or subtitle commentaries from people who worked on the film, people who have studied the film, people with historical information, and so on
- Song list: a chapter index to all the songs on the soundtrack
- Backgrounders: historical context, timeline, important details, related works*
- "Making of" documentary
- Production notes and photos, behind-the-scenes footage, lobby cards, memorabilia
- Live production details, such as synchronized storyboards, screenplay, or production notes*
- Trivia quiz or game
- "Easter eggs:" hidden features that are found by selecting disguised buttons on menus
- Related titles: "If you liked this disc, you should see...," lists or previews of discs with similar characteristics or with the same actors
- Trailers or ads for other titles
- Bibliographies: where to find more information on subject matter*
- Web links: Websites for the movie, actors, studio, producer, fans, and the like*
- Excerpts for computer use: screenplay, documents, scanned photographs, sound bites, and so on for use on a computer*
- Related computer applications: screen savers, games, quizzes, digitized comic books, novelizations, and so on*
- Additional content more easily programmed or accessed through ROM on a computer
- Product catalog and ordering system to sell related goods such as other DVD titles, action figures, etc*
- Newspaper and magazine articles or reviews*

Scheduling and Asset Management

The secret to a smooth and successful project is a carefully monitored production schedule. The schedule enables you to allocate resources and determine bottlenecks. Project planning software is very helpful. In many cases you will start with a rough schedule for laying out the project, then produce an updated schedule based on the project plan.

Throughout the production process, keep track of the assets. Make a list of each asset, when it's due, when it arrived, what format the source is in, when it was digitized and encoded and by whom, what the filename is, when the client signed off on each stage of asset conversion or production, and so on. Clarify who owns the copyright for new content that may be created during production.

Project Design

A project design document is the blueprint from which everyone will work; it is the important first step for creating a disc. Decide up front the scope and character of your title. Make a plan and try to stick to it. Authoring is a creative process, but if you continually change the project as it goes along, you will waste time and money. Determine the level of interactivity you want. The more interactive, the more complex the layout, and thus the more time you need to spend at this early stage.

Laying out the disc generally involves the following:

- **Flowchart**: Diagram each part of the title. Decide what appears when the disc is first inserted. Does the disc go straight to the video or to a menu? Diagram each menu and how it links to other menus and video segments. For a complex title, unless you create a clean design with clear relationships between the various hierarchical levels, your viewers will become lost when trying to navigate the disc. Each box on the flowchart should indicate how the viewer will move to other boxes and what keys or menu buttons they will use to do so (see Figure 16.1).

- **Storyboard**: Draw pictures of the menus and the video segments. You may want to draw a storyboard picture for each chapter point. Make sure that transitions between menus and from menus to video are smooth.

- **Prototype**: Create rough versions of the menus in PhotoShop or PowerPoint or another graphics program. Simulate jumping from one menu to another by using layers or by opening new files. Bring in other people to test the simulation. Make sure the experience is smooth, clear, and easy.

- **UOP controls**: For every menu and video segment, determine what the user will be able to do. Start with all user operations enabled, and only disable them for a good reason. Do not needlessly frustrate your viewers by restricting their freedom. Specify user operation results.

- **If you have a dual-layer disc, choose PTP or OTP (RSDL)**: For sequential playback across layers, you will need RSDL. In this case, start thinking about where the layer switch will be. For two versions of a movie, such as pan & scan and widescreen, a PTP layout will usually give you more flexibility.

The layout process serves as a preflight check to make sure you have accounted for all the pieces and that they work together. Good layout documents provide a roadmap to be used by the entire team throughout the remainder of the project: a checklist for coordinating assets and schedules, a framework for menus and transitions, a guide for bit budgeting, an outline for the authoring program, a checklist for testing, and so on.

Many authoring systems attempt to provide layout and storyboarding tools in the application. Some succeed better than others. In general, at least a basic layout should be done on paper or in a separate program (or in Pioneer's *DVDesigner* layout tool) before jumping into the authoring program, especially since the authoring stations are often key pieces of expensive equipment that are in constant demand.

Figure 16.1 Sample Flowchart

DVD Demystified 3 Sample Disc Flowchart

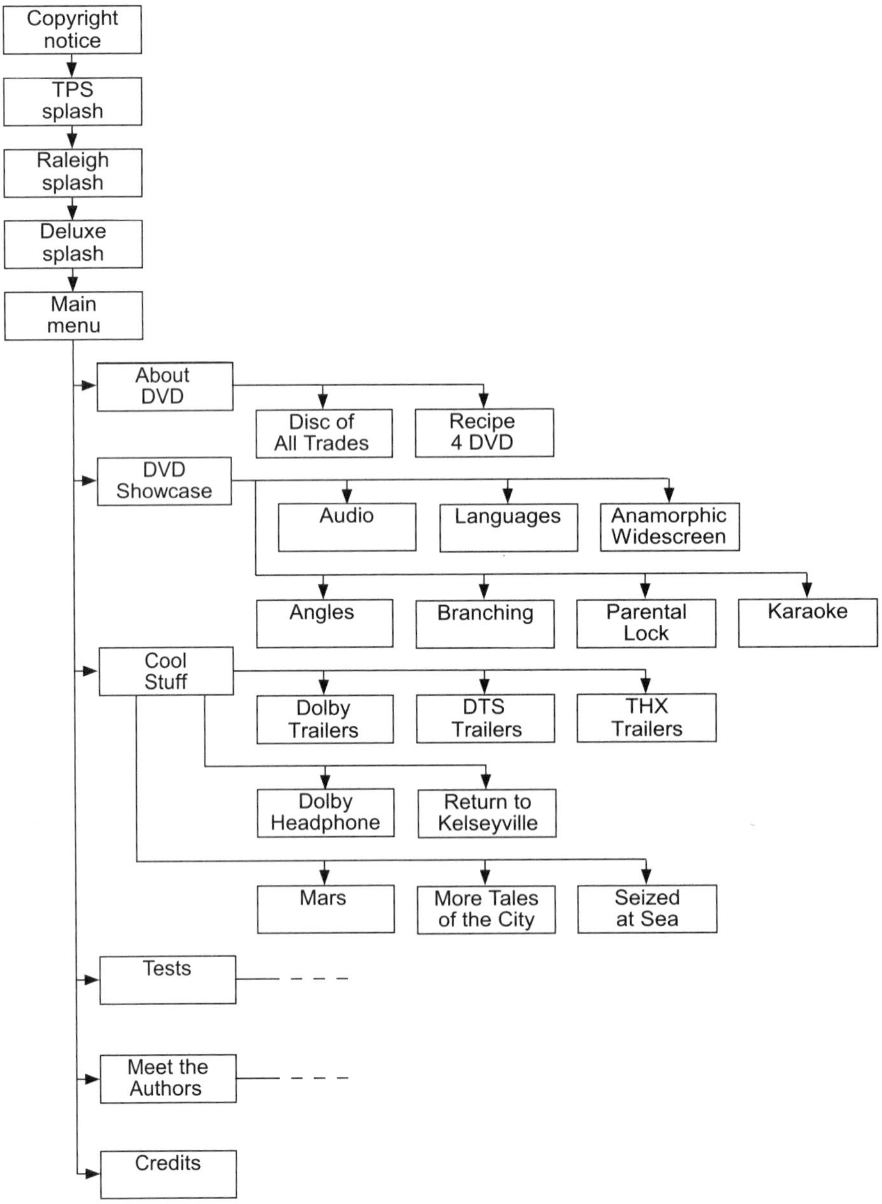

Menu Design

Menus make the DVD; they set the tone and define the experience of using a disc. They are also fun to design. The possibilities are endless, so you must make many decisions up front. Do you want still menus or moving menus? If you have a moving menu, consider carefully how to design the video so that it loops smoothly from the end, back to the beginning. If a great deal of movement occurs, consider starting with a still and ending with the same still. Fade the audio down just before the end, and fade it back up at the beginning.

Do you want live transitions? That is, do you want a video or animated chunk that appears at the beginning of the main menu, or between menus and video segments? Make sure the transitions are smooth and appropriate. Even if you don't use motion transitions, jumping from one design and color scheme to a completely different one can be very jarring. Likewise, jumping from one type of audio in the menu to a completely different type of audio in the video program can be grating.

Do you want background audio in the menu? Be very careful with this. One of the most obnoxious features of DVD discs are menus with audio that is too loud and too intrusive. Drop the audio to a pleasingly low level; then cut the level in half. Loop the video and audio, playing it over and over at least 20 times to discover just how annoying your creation can be.

Do you want multilanguage menus? If so, keep the text separate from the graphics and video so that it can be easily changed for each language version of the menu. If you are creating the original backgrounds in English, leave 30 percent more space for translations.

Will you use subpicture highlights or action buttons? Action buttons jump to a different menu, optionally with motion, when the button is highlighted. This approach lets you use full-color graphics for highlighting rather than one-color subpicture highlights. You can even do fancy motion highlighting such as flaming or spinning buttons, but it's much slower on many players. It will also cause problems on PCs, since moving a mouse over the button will usually not cause the action highlight effect. If you use action buttons, be sure to test them on more than one computer.

Do you want to idle out? That is, do you want the menu to automatically jump somewhere else after a particular amount of time? If so, choose a long timeout period. Make sure the menu doesn't time out while the viewer is still reading it.

Menu Creation

Menus are produced with a still or motion video background and a subpicture highlight overlay. The actual "art" of the buttons is usually in the video background, while the highlights are used to color or surround the currently selected button (see Figure 16.2). One of the tricky parts of menu creation is implementing the subpicture highlights. There are essentially three ways to create highlights: as a separate color-coded graphic, as layers in a Photoshop file, or directly in the authoring system.

Figure 16.2 Menu Subpicture Highlight Overlay

Menu video graphic   **Subpicture graphic**

Four highlight pixel types exist (see Chapter 9 for details). The background pixel type is usually transparent for the video to show through the pixel. When using separate graphics, the background pixels are usually mapped to white by the authoring system. The other three pixel types can be used for highlight effects. These are usually mapped to blue, red, and black in separate graphics. The four pixels types each have one set of color and transparency when the button is highlighted, and a separate set when the button is activated. This means you can use the same shape for highlighting and activating, but with different colors, or you can use different shapes by having the activation pixels be transparent when the button is highlighted, and the highlight pixels be transparent when the button is activated. Since activation lasts less than a second, you may decide it's not worth creating separate graphic designs.

The upshot is that you can use three colors (with three transparency levels) for button highlighting. But, unless you do some clever antialiasing or you stick with rectangular designs, your highlight graphics will look rough and jaggy.

If you use the auto action feature of DVD to create "action buttons" or "24 bit rollover buttons," you must make a separate graphic for each highlighted button. That is, if the menu has six buttons, then you have to make six graphics, each one with a different button in its highlight state. Each version is actually a different menu page that is automatically jumped to when the user changes the button selection.

When you lay out a menu, consider interbutton navigation. The user has four directional arrow keys to use in jumping from button to button. Try to lay out the menu so that it's obvious which button will be selected when a particular arrow key is pressed. At the same time, don't forget that your disc inevitably will be played in a computer with a mouse interface, so don't rely on the user moving sequentially from button to button. In general, it's best to have button links wrap around, thus when the user presses the down key on the bottom button, the top button should be selected. When the user presses the up arrow key on the top button, the bottom button should be selected, likewise for left and right movement. The reasoning is that it's better to have a keypress do something than nothing. The brief confusion experienced by novice users when the arrow key functions wrap around is preferable to the angry frustration of users who expect arrows to wrap when they don't.

Tips and Tricks

The following tips will assist you in the process of creating a menu:

- The number one mistake DVD menu designers make is failing to differentiate between a highlighted and non-highlighted button. For example, the viewer sees a screen with two buttons; one is yellow, one is green. Which one is highlighted? If the user presses an arrow key, the yellow button turns green and the green button turns yellow. The user still can't tell which one is highlighted. This can even be confusing with more than two buttons if some of the buttons are different colors in their unhighlighted state. In many cases, it's better to put a box or a circle around a button, or have an icon or pointer appear next to it, rather than only changing the color or contrast.

- Make sure the highlighted color is different enough from the unhighlighted color so that button selections are visible. Do not rely on menu simulation in the computer because colors can end up appearing much more similar on a TV.

- Each menu should prehighlight the button that the user is most likely to select. For example, if the main menu is shown before the feature, the "play" button on the menu should be preselected so the user can simply press "Enter" or "Play" to start. This is especially important since the "Play" key on most remote controls does the same thing as the "Enter" key when in a menu. Beginning DVD users will intuitively press the "Play" key when they reach the first menu.

- Author menus so that the button that takes the viewer out of the menu is highlighted upon return. If a video segment or another menu jumps to a menu, the button that returns the viewer to where they just came from should be highlighted.

- Design your menus for numerical access. Most remote controls enable the user to press the 1 through 9 keys on a menu to directly choose a button. Arrange the buttons in proper numerical order. Be careful with invisible buttons. On a disc with several menus and complex navigation, consider putting a number next to each button.

- An introductory video sequence for the main menu will become a flaming irritant if the user has to sit through it every time they return to the menu. Use the intro for initial play only, then stifle it for later menu accesses.

- Motion transitions between menus can be enjoyable, but keep them short and sweet; never make transitions longer than two seconds. When creating the transitions, play them over and over and imagine yourself pressing a menu button each time the video loops. If it seems like a long time between button presses, then it is too long a time.

- If you have video in 16:9 anamorphic form, you will also want to make the menus work in 4:3 mode. One approach is to make all your menus in widescreen form, then use the built-in pan & scan feature to have the player crop the sides. Make sure that no button art is outside the area that will be cropped. You will need to make two sets of button highlight graphics, as explained in the subpicture section.

■ Make sure the button highlight/selection rectangle is over the top of the button art in the background. If you put the rectangle somewhere else on the screen, the disc will still work with arrow keys, but the button will not work when a computer user clicks it.

■ Double check that the subpicture overlay aligns with the underlying background. If they don't match, the highlight will jump or be offset. See the graphics preparation section for details on making graphics match motion video.

■ Buttons are subpictures, and subpictures go away when the user presses fast forward or rewind. To avoid this problem with motion menus or with buttons over the top of video, disable the FF/REW user operations or refresh the subpictures every few seconds.

■ Temper creativity with good user interface design. Weigh functionality over aesthetics and experimentation. Frustrated users will head for the eject button and will never see your avant-garde designs.

Navigation Design

Navigation is one of the hardest aspects of DVD production to get right. In addition to the confusion of "Title" key, "Menu" key, and the inconsistency of navigation on different discs, it's difficult to balance creativity and playfulness against ease of use.

When you create the flowchart, identify the effect of important menu keys for each menu. Clearly indicate where the viewer will go when they press "Top" ("Title"), "Menu," or "Return" ("GoUp"). If you have several menus, use a hierarchical design that the user will be able to intuitively understand. If you use more than three levels in your hierarchy, try combining menus to flatten the structure. However, try not to have more than seven to ten buttons on each menu in order to avoid overwhelming the user with options. Keep all selectable features and options as close to the main menu page as possible, with a minimal number of button selections. Make it as easy to navigate back up through menus as to navigate in the down direction.

Understand the difference between the "Top" ("Title") key and the "Menu" key. "Top" means "take me to the very top, to the table of contents." "Menu" means "take me to the most appropriate menu for where I am." For multilevel menu structures, the "Menu" key can be used to go up one level, or the "Return" key can accomplish this feature. A well-designed disc will program the "Return" ("GoUp") function to act somewhat like the back button on a Web browser. Each time the viewer presses "Return," they move up one level until they reach the top (see Figure 16.3). Every remote should include the three basic navigation keys: "Top" ("Title"), "Menu," and "Return." Take advantage of the intended function of each (see Figure 16.4).[2]

---

[2]Unfortunately, some player interfaces, especially on computers, leave out the Top (Title) key or the Return key. However, the incompetence of some player designers should not cause you to castrate your disc. The more discs there are that properly use the navigation keys, the sooner players will be fixed to provide them all. In the meantime, you can overcome the problem by providing on-screen buttons for navigation. See the following paragraph.

Figure 16.3 Example Menu Structure

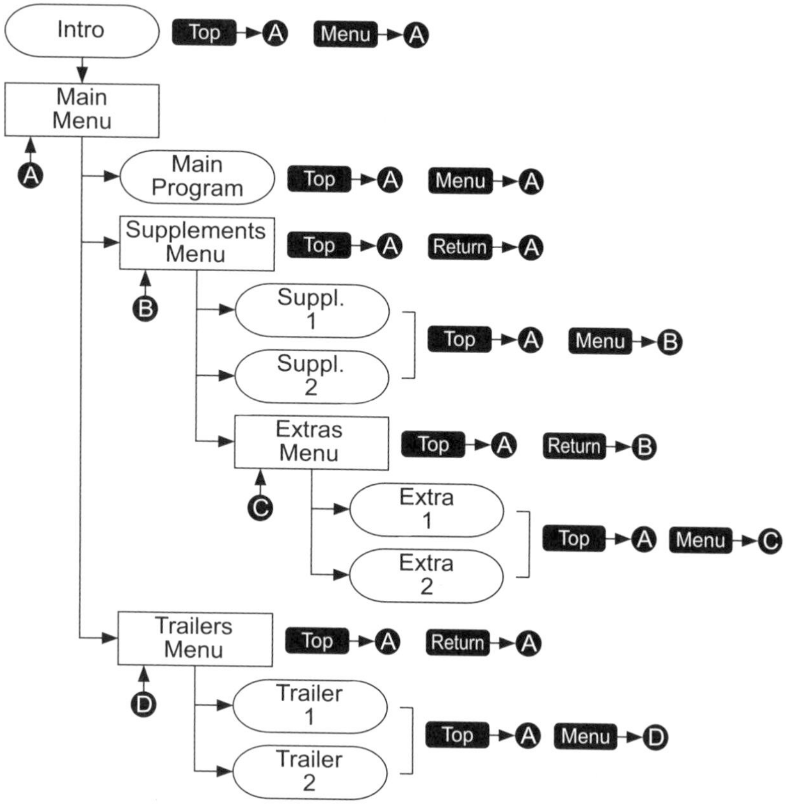

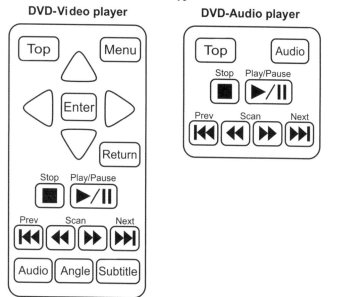

Figure 16.4 Basic Archetypal Remote Control

Always enable navigation with on-screen buttons as an alternative to remote control keys. Assume that the viewer has only the directional arrow keys and the "Enter" key, or assume that the viewer has only a mouse. Make sure that buttons labeled "back" or "main menu" or something similar are included to enable the viewer to navigate through all the menus without relying on any other keys.

Figure out where each chapter point will be. Anywhere the viewer might want to jump to should be a chapter point. Identify the chapter points by timecode on the storyboard or flowchart. Do this before the video is encoded, since each chapter entry point in the video must be encoded as an I-frame at a GOP header.

TIP

Rule of thumb: There should be no more than 10 minutes between chapters.

Tips and Tricks

The following tips may assist your efforts to make user navigation easier and clearer:

- If you only have one menu, make both the "Top" ("Title") key and "Menu" key go to it. Most discs do not have multiple titles on them, so the "Title" key should act like the "Menu" key.

- Never disable the "Top" ("Title") or "Menu" keys. The viewer should always be able to leave the current video segment by pressing either of these keys.

- Do not author the disc so that pressing the "Top" ("Title") key starts over and forces the viewer to watch logos and FBI warnings again.

- Make sure the "Next" and "Prev" keys always work. At minimum, they should move through chapters.

- Consider placing a "How to use this disc" segment on every disc you make. It can explain the basics of navigation using the arrow keys, "Enter," "Top" ("Title"), "Menu," and "Return." Keep in mind that the "Top" button may be labeled "Title" or "Guide" or something else, and that the "Menu" button may be labeled "DVD Menu" or something else.

Balancing the Bit Budget

Bit budgeting is a critical step before you encode the audio and video. You must determine the data rate for each segment of the disc. If you underestimate the bit budget, your assets won't fit and you will have to re-encode and re-author. If you overestimate the bit budget, you will waste space on your disc that could have been used for better quality encoding. However, be conservative — using less than the entire disc still works, but exceeding its capacity, even by a small amount, forces you to rework the project.

The idea is to list all the audio and video assets and determine how many of the total bits available on the disc can be allocated to each. Part of the early bit budgeting process involves balancing program length and video quality. Two axes of control exist: data rate and capacity. Within the resulting two-dimensional space, you must balance title length, picture quality, number of audio tracks, quality of audio tracks, number of camera angles, amount of additional footage for seamless branching, and other details. The data rate (in megabits per second) multiplied by the playing time (in seconds) gives the size (in megabits). At a desired level of quality, a range of data rates will determine the size of the program. If the size is too large, reduce the amount of video, reduce the data rate (and thus the quality), or move up to a bigger capacity disc. Once you have determined the disc size and the total playing time, maximize the data rate to fill the disc, which will provide the best quality within the other constraints. See Figure 16.5 to get an idea of needed disc sizes. A DVD-5 holds 1 to 9 hours of video or 2 to 160 hours of audio. A DVD-9 holds 2 to 16 hours video or 3 to 300 hours of audio. Of course, the video quality at high-end playing times is much lower than what is expected from DVD.

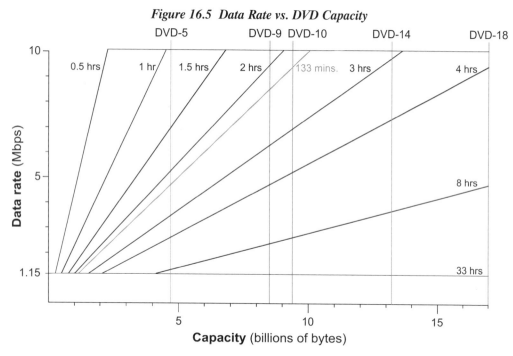

Figure 16.5 Data Rate vs. DVD Capacity

The easiest way to make a bit budget is to use a spreadsheet. One is included on the sample disc that comes with this book. For simple projects with only one video segment, use the *DVDCalc* spreadsheet on the sample disc. Some authoring programs will calculate bit budgets for you when you use their layout features.

To simplify calculations, keep track of sizes in megabits, rather than megabytes (see Table 16.3). Allow an overhead of 7 percent for control data and backup files which are added during formatting and multiplexing. This also allows a bit of breathing room to make sure everything fits on the disc.

Table 16.3 Disc Capacities for Bit Budgeting

	DVD-5	DVD-9	DVD-10	DVD-14	DVD-18
Billions of bytes	4.7	8.54	9.4	13.24	17.08
Megabits	37,600	68,320	75,200	105,920	136,640
7% overhead	2,632	4,782	5,264	7,414	9,565
Adjusted capacity	34,968	63,538	69,936	98,506	127,075

In addition to fitting the assets into the total space on the disc, also make sure that the combined data rate of the streams in a video program do not exceed the maximum data rate (or *instantaneous bit rate*) of 10.08 Mbps. Subtract the audio data rates from the maximum to get the remaining data rate for video. Lowering the maximum data rate reduces the ability of a variable bit rate encoder to allocate extra bits when needed to maintain quality.

Example A typical bit budgeting process for a disc with a single main video program, plus a few ancillary segments, goes as follows (see Table 16.4).

Table 16.4 Sample Bit Budget

Element	Time	Data rate	Size
English audio track (5.1)	110 minutes	0.448 Mbps	$110 \times 60 \times 0.448 = 2,957$ Mbits
English audio track (2.0)	110 minutes	0.192 Mbps	$110 \times 60 \times 0.192 = 1,267$ Mbits
Spanish audio track (5.1)	110 minutes	0.448 Mbps	$110 \times 60 \times 0.448 = 2,957$ Mbits
4 subpicture tracks	110 minutes	0.040 Mbps	$4 \times 110 \times 60 \times 0.04 = 1,056$ Mbits
Subtotal (movie audio and subpicture tracks)	—	1.248 Mbps	—
2 motion menus and 3 transitions	48 seconds	8 Mbps (incl. audio)	$48 \times 8 = 384$ Mbits
Intro logos	32 seconds	8 Mbps (incl. audio)	$32 \times 8 = 192$ Mbits
Movie preview	2 minutes	4.5 Mbps	$2 \times 60 \times 4.5 = 540$ Mbits
Audio for preview	2 minutes	0.192 Mbps	$2 \times 60 \times 0.192 = 23$ Mbits
Interview	30 minutes	4.5 Mbps	$30 \times 60 \times 4.5 = 8100$ Mbits
Audio for interview	30 minutes	0.192 Mbps	$30 \times 60 \times 0.192 = 350$ Mbits
Total (non-movie elements)	—	—	17,826 Mbits
Space remaining for movie	—	—	$63,538^a - 17,826 = 45,712$ Mbits
Movie	110 minutes	Avg: $45,712/(110 \times 60) = 6.92$ Mbps Max: $10.08^b - 1.248 = 8.832$ Mbps	

[a]Adjusted capacity for DVD-9 from Table 12.3.

[b]Maximum combined data rate allowed for DVD.

Calculate the total bit rate and size of the main program's audio tracks and subpicture tracks. If there are additional video segments such as animated logos, motion menus, and previews, calculate their sizes (including audio and subpictures), and add them to the total size of the main program's audio and subpicture tracks. Still logos and copyright warnings should be authored as timed stills, which have no impact on disc space. Subtract the total size of ancillary material from the disc capacity minus overhead. This gives remaining capacity for the main program. Calculate the average video bit rate for the main program by dividing the remaining capacity by the running time. Also calculate the maximum video bit rate by subtracting the combined audio and subpicture bit rate from 10.08. These numbers will be fed into the encoder when compressing the video for the main program. Note that minor elements such as subpicture tracks and motion menus are usually so small that they often can be omitted from calculations. If they were left off of the example in Table 16.4, the average data rate would come out to 4.17 Mbps, which is close enough.

An alternative bit budgeting approach is to divide the disc capacity (in bits) by the total running time of all segments (in seconds) to get the average data rate (in Mbps). Adjust up or down for quality of the different programs as needed. For each program, subtract the combined data rate of its audio and subpicture streams to get video rate. It's a good idea to follow up by summing the sizes of each segment (including audio) to check that the total does not exceed disc capacity.

Tips and Tricks

The following list of tips will help you in balancing the bit budget:

- Don't forget motion menus and menu transitions, as well as logo sequences.

- Don't forget to leave room for any added PC content. A few programs or HTML files are not worth worrying about, but large multimedia applications and installer packages need to be accounted for.

- Each camera angle increases the data; two camera angles double it, three angles triple it, and so on (including the corresponding audio and subtitles). Reduce the maximum data rate according to the number of angles (see Table 9.31).

- If you have two or more programs that contain much of the same data, consider using the seamless branching feature to combine the duplicated segments (but only if you have an authoring tool that supports the feature).

- Still images take up so little space that you don't need to include them in your calculations unless you have dozens or more of them. Unless you are including more than four subpicture streams, you may wish to omit subpictures from the calculations, since the space they take up is negligible.

Asset Preparation

A typical project requires dozens of assets from a variety of sources. A complicated project may require thousands (see Table 16.5). Each asset must be properly prepared before it can be fed into the DVD authoring process. The following sections cover the details for preparing each type of asset.

Table 16.5 Typical Project Assets

Type	Assets	Tasks
Video	Source video tapes (movie, trailer, supplements)	Digitize, clean up, encode
	Already encoded video (logos, computer graphics)	Check
Audio	Original language source tape	Digitize, conform
	Foreign language source tapes	Digitize, check length, conform
	Commentary or other supplemental audio sources	Digitize, conform
	Already encoded audio	Check synchronization
Graphics	Menu graphics	Create/integrate graphics, create button highlights
	Supplemental graphics	Create/integrate graphics, create button highlights
	Production stills and other photos	Digitize, correct color, create button highlights
Subtitles	Text files	Add timecodes
	Graphic files	Create timecode-filename mapping file

Preparing Video Assets

The final quality of DVD video is primarily dependent on three things: the condition and quality of the source material, the visual character of the material, and the data rate allocated to the video during encoding. The better the quality of the source, the better job the encoder can do. Always use the best quality source and insist on a digital copy whenever possible, such as D1, D5, or Digital Betacam. Beta SP will do in a pinch, but in many ways it is inferior to DVD's capabilities.

If the video is of poor quality or contains noise, use *digital video noise reduction* (DVNR) to improve it. Noise is random, high-entropy information, which is antithetical to MPEG-2 encoding. Efficient encoding depends on reducing redundant information, which is obscured by noise. Noise can come from grainy film, dust, scratches, video snow, tape dropouts, cross-color from video decoders, satellite impulse noise, and other sources. Complex video, including high detail and high noise levels, can be dealt with in two ways: increase the data rate or decrease the complexity. The DVNR process compares the video across multiple frames and removes random noise. After DVNR, the video compresses better, resulting in better quality at the same data rate.

If possible, keep the production path short and digital. A frequent irony of the process is that video is edited on a computer, then output to tape in order to be encoded back into a computer file. If you use a nonlinear digital editing system (NLE), look into options for directly outputting DVD-compliant MPEG-2 streams. Anticipate the need for aspect ratio conversion. Make sure the various sources are in the proper aspect ratio (4:3 or 16:9), especially for assets that need to be edited together.

Make sure all the video on a disc (or on a single side) uses the same video standard. Be prepared to convert from PAL to NTSC or vice versa. Good-quality conversion is difficult.

Unless you will be converting hundreds of clips, it's better to find a video production house with the equipment and expertise to do it correctly. Be aware that the audio must be conformed to match.

You may need to choose between variable bit rate encoding (VBR) or constant bit rate encoding (CBR). VBR doesn't directly improve the quality, it only means that quality can be maintained at lower average bit rates. For best quality of long-playing video, VBR is the only choice. For short video segments, CBR set at or near the same data rate of VBR will produce similar results with less work and lower cost. For VBR encoding, set the average and the maximum bit rate as determined in the bit budgeting step.

Tips and Tricks

The following tips will help you in preparing video assets:

- The display rate must be 25 frames per second for 625/50 (PAL/SECAM) video or 29.97 frames per second for 525/60 (NTSC) video, not 30 frames per second. This is especially important when using NLE systems that allow many different frame rates. Many authoring tools will reject streams with non-compliant frame rates. Note that the display frame rate can be different from the source frame rate. Twenty-four frame-per-second film can be encoded to display at 29.97 frames per second.

- Ensure that the encoder can produce DVD-compliant streams, with proper GOP size, display rates, and so on. Also look for an encoder that can do *segment-based re-encoding*, which enables you to re-encode small segments as needed, rather than reprocess hours of video to correct a small problem.

- If you are working with film transferred to video, make sure the encoder can do inverse telecine.

- Chapters and programs must start at a sequence header or GOP header, and the GOP must be closed, not open. Each cell must also start at a sequence header. If you have very precise points for chapter marks, make sure they are encoded properly, otherwise the authoring software may move the chapter point to the nearest I frame.

- In general, the easiest way to put still video on a disc is to author each still as a menu.

- Static video, where the picture doesn't change over time, such as FBI warnings, logos, and so on, should be authored as a timed still. Not only does this look better, with no movement or variation when played back, it can save space on the disc.

- Make sure the video does not have burned-in subtitles. Use the DVD subpicture feature instead, particularly for multilingual discs.

- If you want to pack a large amount of video on a disc, don't try to encode to MPEG-2 under 2 Mbps. Use MPEG-1 between 1 and 1.86 Mbps; it will look better than MPEG-2 at the same data rates. You could also use "half D1" resolution (352×480) at rates of about 1.5 to 3 Mbps, which can look surprisingly good. Be aware that although half D1 video is mandatory in the DVD-Video specification, it will cause problems on a few players. Before encoding the video, use heavy digital video noise reduction (DVNR) and liberal application of blurring filters in video processing programs. These steps reduce the high-frequency detail in the video, which makes MPEG encoding more efficient. Also try reducing the color depth from 24 bits to something between 16 bits and 23 bits. Do not use a dither process, which increases high-frequency detail. This can help with encoding efficiency and can reduce posterization artifacts.

■ If video quality is a top consideration, create or request your content in progressive format if at all possible: shoot on film or use a progressive digital video camera. If your video assets come from an interlaced source, consider converting to progressive form before encoding. Depending on the nature of the video, progressive conversion may cause artifacts that unacceptably degrade the video. For some types of interlaced content, however, a high-quality progressive conversion will result in a DVD that looks significantly better when played on progressive DVD players and computers. See Chapter 3 for more on progressive video.

Preparing Animation and Composited Video

Video produced on a computer, or camera-source video that is composited with computer graphics, has its own set of issues. The most important guideline to follow is to avoid too much detail, especially vertical detail, which wreaks havoc on interlaced TVs. Avoid thin horizontal lines; use the anti-aliasing and motion blur features of your animation or editing software whenever possible. Use a low-pass or blur filter, but don't overdo it because too much blurring reduces resolution on progressive-scan players and computers.

For 525-line (NTSC) video, render animations at 720×540 and scale down to 720×480 before encoding. For 625-line (PAL) video, render animations at 768×576 and scale down to 720×576 before encoding, or if your software supports it, render with D1 or DV pixel geometry.

If at all possible, do not use the "print to video" or similar TV video rendering feature of your software, which will almost always result in interlaced output. Instead, render at 24 fps progressive or 30 fps progressive, then encode for 29.97 fps display. This requires an encoder that can handle progressive video input files and perform "synthetic telecine" to add flags to the MPEG-2 stream for 2-3 pulldown in the player (see the following graphics preparation section for more details).

Preparing Audio Assets

Contrary to most expectations, audio is usually just as complex as video, and often more complex. You must deal with multiple tracks, synchronization, audio levels across the disc, and so on. As with video, the better quality the source, the better the results after encoding.

Audio assets usually come on *modular digital multitrack* (MDM) formats such as Tascam DA-88, or you may receive PCM (WAV) files on a CD-R, or already-encoded Dolby Digital or DTS files. With a multitrack tape source, make sure the track-to-channel assignments are correct. Dolby has specified that the sequence is L, R, C, LFE, Ls, Rs, (Lt, Rt), but not all production houses follow this standard.

Capture your audio mixes at the highest sample rate and word size possible. If someone else is doing the audio mixing, request that they do the same. As recording and mastering studios acquire the new equipment needed to work with high-resolution audio, 96 kHz 24 bit audio will become commonplace. For DVD-Audio or for audio-oriented DVD-Video, it's usually not a good idea to add video to a 96 kHz 24 bit PCM track due to little remaining headroom.

When converting from NTSC to PAL, or transferring from film to PAL, the audio must be conformed to match the changed video length with a 4 percent speedup, and preferably a pitch shift to restore proper pitch. If the audio is from a CD, it must be upsampled from 44.1 kHz to 48 kHz for DVD-Video. Many audio files produced on PCs are also at 44.1 kHz. DVD-Audio supports 44.1 kHz sampling, unless a Dolby Digital version is being added for compatibility, in which case the audio must be converted to 48 kHz before encoding.

Do all necessary processing before encoding: clean up and sweeten the audio, re-mix and equalize, adjust phase, adjust speed and pitch, upsample or downsample, and so on. Audio levels can be adjusted before encoding, or for Dolby Digital, they can be adjusted during encoding by using the dialog normalization feature. If the mix was done for the theater, re-equalize for the home environment. Establish consistent audio levels and equalization throughout all audio on the disc, including menus and supplements. This is especially important with multiple audio tracks from various sources, since the viewer will be able to jump at will between them and will be bothered by inconsistencies in level or harmonic content. Resist the temptation of locking out the audio key on the remote control, which will only annoy your viewers. Instead, get all sources in PCM format so they can be matched and harmonized before being encoded.

Tips and Tricks

The following tips will assist you in preparing the audio assets:

- Make sure timecode is included with the audio so that it can be synchronized with the video (but make sure the timecode for the audio actually matches that of the video).

- It's almost always a good idea to mute the audio for a second or so at the beginning of each segment. This ensures that the player (or receiver) doesn't clip the audio as it begins the decoding process.

- If you are doing 5.1-channel Dolby Digital encoding, or using multichannel DVD-Audio PCM tracks, test the downmix.

- If you do your own encoding, request a copy of the *Dolby Digital Professional Encoding Manual* from Dolby Labs.

The Zen of Subwoofers

If you understand the proper use of the LFE channel, you belong to an elite minority. As Dolby Labs has pointed out over and over, apparently with little success, LFE does not equal subwoofer. All five channels of Dolby Digital are full range, which means they are just as capable of carrying bass as the LFE channel. Most theaters have full-range speakers, and don't need separate subwoofers since most of the speakers have a built-in subwoofer. In the home, it's the responsibility of the receiver or the subwoofer crossover to allocate bass frequencies to the subwoofer speaker. This is not the responsibility of the DVD player (unless it has a built-in 5.1 channel decoder) or of the engineer mixing the audio in the studio.

The LFE channel is heavily overused in most audio mixes. A case in point is a certain famous movie released on laserdisc with Dolby Digital 5.1 tracks in which the booming bass footsteps of large dinosaurs were placed exclusively in the LFE channel. When downmixed for two channel audio systems, the bass was discarded by the decoder, leaving mincing, tip-toeing dinosaurs; this was not the fault of downmixing or the decoder. It could have been fixed by a separate two channel mix, but it also could have been fixed with a proper 5.1 mix or a proper 5.0 mix. Had the footsteps been left in the main five channels, which are all full range, the bass effects would have come through properly on all home systems.

Dolby Digital decoders ignore the LFE channel when downmixing. This is because the typical stereo or four channel surround system doesn't have a subwoofer. Again, this does not mean that the LFE channel represents the subwoofer, only that the extra "oomph" intended to be supplied in the LFE channel is inappropriate and could muddy the audio. Dolby Digital decoders all have built-in bass management, which directs low-frequency signals from all six channels to the subwoofer, if one exists. Therefore, the LFE channel should be reserved only for added emphasis to very low frequency effects such as explosions and jets.[3]

As illustrated by the tiptoeing dinosaurs, moving low-frequency audio to the LFE channel does not make the soundtrack better in the home environment. This mistaken approach is what has caused complaints about the LFE channel being omitted in downmixing. Nothing important in a film soundtrack, including low-frequency audio, should ever be moved out of the main five channels. Isolating low-frequency audio in the LFE channel does not remove any sort of burden from the center and main speakers since the built-in bass management already does this.

It's possible to mix all "subwoofer sounds" in Dolby Digital 5.0 without an LFE channel at all. Some theatrical releases are mixed this way. Such a mix will play the same in 5.1 channel home theaters as in cinemas, since the decoder and receiver automatically route the bass to the speakers that can best reproduce it. The .1 channel is present only for an extra kick that the audio engineer might decide should only appear in full 5.1 audio systems.

Slipping Synchronization

An occasional problem with DVD is lack of proper synchronization between the audio and

---

[3]Roger Dressler, Technical Director for Dolby, has remarked that the LFE channel should be renamed as the "explosions and special effects" channel in order to avoid confusion.

the video (see Chapter 12 for more detail). Players are partly to blame for the problem, but steps can be taken during production to keep synchronization tight.

Check the timecodes on the audio to make sure they are correct. A handy trick for testing audio sync, especially when no timecodes exist and it has to be aligned by hand, is to bump the audio track forward several frames, check playback, bump it back several frames from the original position, and check playback again. If moving it one direction had little or no effect, while moving it in the other direction had a very big effect, chances are that the original position was not optimal.[4] Make sure the master print was not incorrectly dubbed before being transferred to video.

If sync steadily worsens, check to see if non-drop frame timecode is being used on the audio master. The difference between 29.97 and 30 fps will slowly move the audio out of sync with the video. Also make sure the film chain in the telecine machine is run at the proper video speed. If sync suddenly changes, suspect a slip during reel change in the telecine process or an edit in the NLE system independent of the audio.

Preparing Subpictures

Subpictures are used for two things: graphic overlays (usually subtitles) and menu highlighting. (Menus were in a previous section.) The possible uses for subpicture overlays are endless. They can expand and alter existing video, emphasize or de-emphasize regions on the screen, cover over or censor, tint the video a different color, add a logo bug, and so on.

A handy feature of subpictures is that they can be forced to appear under program control. This is especially useful for multilanguage discs because the video can be kept clean of all subtitles, but they can be made to appear in the appropriate language during normal playback of the disc.

Subpictures can be used to create limited animation overlaying the video. A subpicture rate of 15 per second is the practical limit for many players, although anything faster than 2 per second will cause problems on some players. Lowering the maximum data rate for the other streams can help avoid problems. Likewise, using simple subpictures that encode more efficiently than large, complex ones can make all the difference.

Effects such as fade, wipe, and crawl can be performed by the player. A good authoring system will be able to add the subpicture display commands needed to create these effects.

Tips and Tricks

The following tips will assist you in preparing subpictures:

- Subpictures cannot cross chapter or program points.

- The size of a subpicture frame is limited, but because subpictures are RLE compressed, it's difficult to determine the size ahead of time. It's easiest to let the authoring software flag subpictures that are too big.

---

[4]When using this trick, keep in mind that synchronization problems in audio delayed behind the video are less perceptible than in audio that precedes the video.

Subpictures for Anamorphic Video

Subpictures accompanying anamorphic video are not cropped or scaled by the player. This means that separate subpictures must be created for each allowed display mode (widescreen, letterbox, and pan & scan). The basic steps are as follows:

- Start out with a 16:9 overlay in square-pixel 854×480 resolution for NTSC, 1024×576 for PAL.

- For widescreen mode, resize the subpicture to 720×480 for NTSC or 720×576 for PAL. This squeezes in the horizontal direction to make an anamorphic version of the subpicture.

- For letterbox mode, add transparent (white) pixels to the top and bottom to make the image 854×640 for NTSC or 1024×768 for PAL. This accounts for the letterbox mattes that are added by the player. Resize the image to 720×480 for NTSC or 720×576 for PAL.

- For pan & scan mode, crop the sides of the picture to make it 720×540 for NTSC or 768×576 for PAL. You can do this in *Photoshop* by changing the canvas size. Then resize the image to 720×480 for NTSC or 720×576 for PAL.

Subtitles

Subtitling, particularly creating foreign language subtitles, is a tricky job best given to a subtitling house. Provide the subtitling service with a master copy that has clear audio and *burned-in timecode* (BITC) that exactly matches the DVD video master. This is critical for frame-accurate subtitling work; timing is very important.

You may request that the subtitle source be provided in text form rather than graphic form, as text is much easier to change, but text subtitles must then be turned into graphics.

If you are creating your own subtitles, consider the following:

- Don't use serif fonts. Sans serif fonts are easier to read on a video screen, especially since the serifs tend to create single-pixel horizontal lines that produce interlaced twittering.

- Use a drop shadow or an outline, either black or translucent (using the subpicture transparency feature).

- Each new subtitle should appear at the scene change, not at the dialog start. This provides more on-screen time, and studies show that it reduces reading fatigue, even though it's annoying to hearing users who have turned on the subtitles to accompany the audio.

- Center the subtitles. If there is more than one line, left-justify the lines within the centered position (or right-justify the lines for right-to-left writing systems). Justifying the lines makes them easier to scan.

- Try to limit titles to two lines. Keep the first line shorter, which covers less video and clues in the eye that there's another line.

- If there is more than one speaker, indicate speaker change with a hyphen at the beginning of the line.

- Display subtitles for no less than one second and no longer than seven seconds.

- Keep in mind that jumping into the middle of a scene, or using fast forward or rewind, will not refresh a subtitle. Subtitles with long periods will take a long time to appear. To avoid this problem, you may wish to repeat the same subtitle at intervals of five to ten seconds.

- Proofread! With thousands of subtititles, especially in foreign languages, it's easy for misspellings to occur.

Preparing Graphics

Graphics are generally used for menus. They are also used for slideshow sequences such as photo galleries and information pages, although these are usually implemented as menus. Graphics can be tricky, since the traditional tools for creating computer graphics are not always well suited for television graphics.

Colors

Graphics should be created in 24 bit mode. Choose RGB colorspace (or YUV colorspace, if available) rather than CMYK or other print-oriented colorspace. The file format (TIFF, BMP, etc.) doesn't really matter, as long as it's one that your authoring system can import. Avoid JPEG file format, since JPEG compression can cause artifacts that may be compounded by MPEG compression. If you must use JPEG, choose the highest quality level.

Use NTSC-safe colors. Saturated colors, especially bright reds and yellows, will bleed on an NTSC television. Keep the saturation of all colors below 90 percent. Another way to accomplish the same thing is to keep RGB values below 230. Some graphics programs have an NTSC color filter that will tone down colors that are out of the NTSC-safe gamut. Use the filter on your graphics until you get a feel for the colors. Certain color combinations don't mesh in NTSC colorspace. For example, light blue next to peach will usually bloom badly. The only way to avoid problems like this is to check the graphics on an NTSC monitor.

Image Dimensions

The pixel density of the image doesn't matter. Whether you use 72 dpi or 96 dpi or 300 dpi, what matters is the final pixel dimensions of the output. Pixel geometry does matter. Most computer graphics programs use square pixels. DVD is based on the ITU-R BT.601 formats that do not use square pixels. In some cases you don't need to worry about the difference, but in other cases, such as when the graphics have squares or circles, you need to account for pixel aspect ratio changes or else you will get rectangles and ovals. There's a 9 percent vertical distortion for NTSC, and a 9 percent horizontal distortion for PAL.[5]

To account for pixel geometry differences, create 525 line (NTSC) video graphics at 720×540 and then scale them to 720×480 before encoding. Create 625 line (PAL) video

---

[5]You may be wondering why the distortion isn't 12.5 percent for NTSC (540/480) and 6.7 percent for PAL (768/720). Technically, these are the correct differences in raw pixel dimensions, but the pixel aspect ratios for MPEG-2 video also take into account the horizontal overscan ratio. So, the visual distortion is 9 percent (see Table 9.20).

graphics at 768×576 and scale them to 720×576 before encoding. Do not constrain proportions when scaling. The resulting picture will look slightly squashed; this is normal. Some authoring software will automatically scale the video down from larger sizes. If so (and if the scaling algorithm produces good quality results), this will save a step. Try to avoid scaling up from 640×480, since this will make the picture fuzzy.

For 16:9 anamorphic video, create 525 line (NTSC) video graphics at 854×480 or 960×540, then scale to 720×480. The 960×540 size is useful if you plan on cropping the graphics to 4:3. Create 625 line (PAL) video graphics at 1024×576, then scale to 720×576.

If your graphics production software gives you the option of using video pixel aspect ratios, consider using it, but keep in mind that images on the computer screen may appear distorted. Choose DV, which matches DVD. If DV is not an option, then choose D1, but be aware that you may have to crop six lines to match DVD picture dimensions.[6] Some graphics software will also let you set the picture aspect ratio. Choose 4:3 (1.33) or 16:9 (1.78).

One way to avoid all this is to create the graphics at native DVD resolution. If you're working with text or with images that look okay when slightly distorted, then don't bother with resizing. This has the advantage of avoiding artifacts caused by downscaling.

If you need to match graphics to video, e.g., for menu transitions or motion menu subpicture overlays, then you may need to alter the scaling process noted above for 525 line (NTSC) video. The exact details depend on the video source and the MPEG-2 video encoder. D1 tape video has pixel dimensions of 720×486. To conform to MPEG-2 dimensions of 720×480, some encoders scale and some encoders crop the extra six lines; however, no consistent approach exists. Some crop three from the top and three from the bottom, others crop six from the bottom, some crop two and four, and so on.[7] You will need to do some research or testing of your encoder to find out what it does. Then, instead of scaling to 720×480, scale to 720×486 and crop the picture the same way your video encoder does. This will ensure that everything lines up properly when transitioning from graphics to video or vice versa. Another approach is to render graphics as one second stills at the end of the motion video. After encoding, you can extract the first I-frame of the still sequence and use it for the menu graphic. Since it was encoded along with the video, it will match perfectly.

Safe Areas

Most televisions employ overscan, which covers the edges of the picture. Overscan can hide as much as 10 percent of the picture. Two common "safe areas" are used to account for this. The *action-safe* area defines a 5 percent boundary as a guideline for the area where action or important video content should be kept. The *title-safe* area defines a 10 percent boundary as a guideline for the area in which text and other vital information should be kept (see Figure 16.6). The DVD Demystified sample disc includes safe area templates. Some programs have built in video safe templates. In *After Effects*, for example, press the apostrophe key to turn them on and off. Most modern TVs don't have more than 5 percent overscan, but older models may approach 10 percent. Computers, LCD or plasma screens, and many video projectors have little or no overscan.

---

[6]DV and D1 video formats use the same pixel aspect ratio, but D1 uses 486 lines for 525 line (NTSC) video, while DV uses 480.

[7]Cropping an odd number of lines (1, 3, or 5) is a bad thing for video, since it changes field dominance and can cause interlacing artifacts. For stills, field dominance is irrelevant.

Figure 16.6 Video Safe Areas

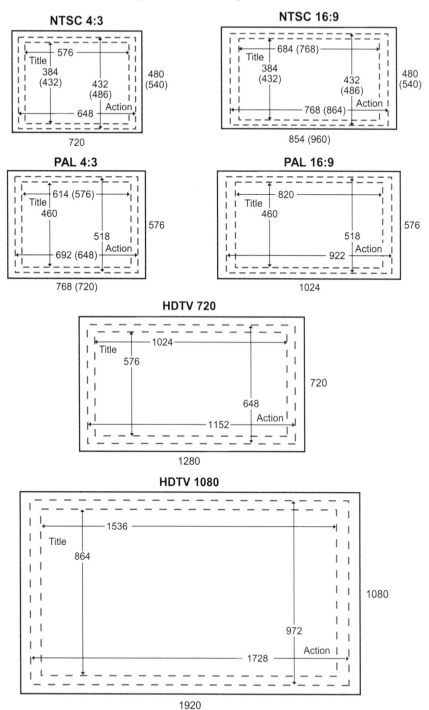

Video Artifacts

Because there are one billion interlaced televisions in the world that your DVD video might be displayed on, you must consider interlace artifacts. The fundamental problem is that since only every other horizontal scan line is displayed at a time, thin horizontal lines disappear and reappear 30 times a second. This interlaced *twitter* effect (also called *flicker*) can be especially bad with computer-generated graphics. In addition to interlace artifacts, chroma crawl and color crosstalk from composite video signals make thin lines or sharp color transitions problematic, despite their orientation.

To reduce these effects, use the antialiasing and feathering features of your graphics software whenever possible. Make sure that all horizontal lines are at least two pixels thick. To be even more careful, keep the thickness of horizontal lines at multiples of two pixels.[8] Use gradual color transitions instead of sharp contrasts between dark and light colors. To adapt existing graphics, or as a final pass before encoding, apply a blur filter to the entire image or to offending areas. For example, a Gaussian blur of 0.5 to 1 has little visual effect but helps to reduce video artifacts. Also, because blurring reduces high-frequency detail, the MPEG compression will be more efficient, possibly resulting in higher quality.

Text is also affected by interlacing and other video artifacts. Small point sizes produce single-pixel lines, which are bad, especially the horizontal lines created by serifs. Small text sizes are also difficult to read from standard viewing distances. The smallest object that normal human vision can discern subtends one minute of arc on the retina. Studies have shown that for legibility, the height of a lower case character must subtend at least nine or ten minutes of arc; more as the viewer moves off axis. ANSI standards recommend 20 to 22 minutes of arc, with a minimum of 16 minutes. (See Chapter 14 for more information on viewing distances and resolution.) Angular measurements are independent of screen size. This means, for example, that at a viewing distance of five feet, 21 minutes of arc equates to 0.37 inches, while at a viewing distance of 15 feet, 21 minutes of arc results in 1.10 inches. As a rule, stick with 24 point or larger sizes. 14 point, bold text should be the absolute minimum. Use antialiased text whenever possible.

Tips and Tricks

Here are a few tips that will help you in preparing graphics. Table 16.6 also provides some guidelines to follow.

- Proofread all graphics before importing them into the authoring system. Don't rely on the graphic artist.

- If you work with clients, have them sign off on every graphic and menu page before the start of authoring.

---

[8]This advice only applies when you are working at native NTSC video resolution of 480 lines. If you create graphics at higher resolutions, the number of pixels will change when the graphic is scaled down. This doesn't apply to 625 line (PAL) video, since it's generally scaled horizontally, not vertically.

Table 16.6 Video Graphics Checklist

	525-line video (NTSC)	625-line video (PAL/SECAM)
▪ Pixel geometry	Create at 720×540, resize to 720×480. (For 16:9 anamorphic, create at 854×480 or 960×540, resize to 720×480.) Or create at 720×480 if vertical distortion is not a problem.	Create at 768×576, resize to 720×576. (For 16:9 anamorphic, create at 1024×576, resize to 720×576.) Or create at 720×576 if horizontal distortion is not a problem.
▪ Colors	No bright reds or yellows. Keep saturation below 90 percent or RGB values below 230.	Not an issue.
▪ Signal Range	Check for below black (< 16 in Rec 601 Video) or above white (> 235 in Rec 601 Video) and make sure it comes out okay on DVD players.	
▪ Safe areas	Keep text and important detail about 70 pixels from the sides, and about 50 pixels from the top and bottom.	Keep text and important detail about 70 pixels from the sides, and about 56 pixels from the top and bottom.
▪ Small detail	Avoid single-pixel lines, especially horizontal. Use antialising and blurring. Use feathering and color gradations instead of sharp transitions.	
▪ Text	14 points minimum. Avoid serifs. Antialias.	

Putting It All Together (Authoring)

Once all the assets are prepared and proofed, bring them into the authoring system and knit them together. A variety of authoring paradigms are available. Some systems stick close to the DVD specification, while others try to hide terminology and details. Most provide an onscreen flowchart, timeline, or combination of the two. Large production environments may benefit from networked authoring workstations, where each station focuses on a specific task (encoding, menu creation, asset integration, simulation, formatting, and so on) and the project data flows from station to station over a high-speed network.

At the authoring stage, specify basic parameters of the volume such as video format and disc size. Import audio and video, synchronize them, and add chapter points. Then, apportion titles and title sets. Create menus by importing graphics and subpictures, link buttons together for directional highlighting, and link buttons to video segments or to other menus. Those are the essential authoring tasks. More complex projects require much more detailed work.

Once everything is imported and connected, simulate the final product. Good authoring systems can show menus, video, and audio on-the-fly, giving a reasonable facsimile of how the disc will look in a player. Take advantage of the simulation feature to test navigation and overall layout before going further.

The authoring system is responsible for verifying compliance with the DVD specification. The better the authoring system, the better a job it will do of warning you when you have violated the spec. Better authoring systems will also provide information or warnings from the "reality spec" — they will warn you about things that are allowed by the spec but may cause problems on certain players. Some authoring systems also implement behind-the-scenes workarounds to avoid known player deficiencies.

Region coding and CSS or CPPM flags may be added in the authoring system. In general, the actual encryption is done by the replicator. This is the best option since the replicator will have already executed a license and have been given sets of disc keys and media key blocks.

Tips and Tricks

The following tips will help you with the authoring process:

- Make sure you have more hard disk storage space than you think you could possibly need; DVD assets take up huge amounts of space. The first time you have to suspend a project because of a bottleneck and start work on a different project, you will either be driven mad by the amount of time it takes to offload a project to tape, or you will be glad you have the disk space needed to work on another project on the same station.

- If you must do a layer change in the middle of the video, look for slow-moving pictures, a still, or a fade to black. The secret to a smooth transition is low data rate, which gives more time to the DVD player to refocus on the second layer before the buffer underflows. At the encoding stage, make the data rate as low as possible for a few seconds going into the layer change, as well as a few seconds coming out of the layer change.

Formatting and Output

Once the authoring process is complete and everything works in simulation, it's time to create the final disc image. The authoring system will multiplex all the assets together, adding navigation and control information according to the DVD specification, and it will write out a special set of .VOB, .AOB, .IFO, and .BUP files in a VIDEO_TS directory (for DVD-Video or DVD-Audio discs with video) or an AUDIO_TS directory (for DVD-Audio-only discs). You will often end up going through this step more than once after finding a problem during emulation or when testing with a DVD-R disc.

The data files can be written to a hard drive. This is useful for emulation testing, where you use a computer software player to test for problems that don't show up when simulating the disc in the authoring system. The data files can be written to DVD-R (or other writable DVD media) using the UDF bridge format. This creates a fully functional DVD disc that can be tested on standard players and other computers.

Once everything is working perfectly, the output can be directed to *digital linear tape* (DLT). The disc image is written as an ANSI tape file, along with a DDP control file that includes additional information to be used by the replicator. All DVD replicators accept compact DLT type III or type IV in DDP 2.0 format. Some replicators will also accept DVD-Rs (which they usually turn around and write to DLT). Discs to be protected with CSS or CPPM must be submitted on DLT, since DVD-Rs have only 2048 byte sectors without extra header information. A dual layer disc takes two DLTs. Only one layer can be written per tape. A new option is supported by some replicators to allow electronic delivery of DDP file sets. However, this is usually restricted to those who have access to private, secure networks.

Testing and Quality Control

Testing is the most important factor in the success of a project. If you work with clients, quality control is vital to your reputation. Testing should never be left to the end; it must be an ongoing process from the beginning.

The first testing stage is source asset checking. Never assume that your supplier will get it right. QC all incoming assets, at least until you have verified which suppliers you can rely on to always get it right. Although it seems redundant and a drain on time and resources, it will save you money and time in the long run. Screen all incoming assets for quality and accuracy. View every video tape and encoded video file. Audition all audio tapes and audio files. View every graphic and subpicture file, checking for spelling mistakes and missing elements. If you can't check foreign language subtitles, make sure the supplier guarantees accuracy and takes responsibility for costs associated with reauthoring the disc in case of subtitle errors. Check timecodes and synchronization between audio and video and between subtitles and video.

Review video and audio during and after compression. Check encoded audio against the master to spot sync, level, and equalization problems. Check all text before it goes to the graphic artist. Provide text to the graphic artist as text files that they copy and paste (or otherwise import) into the graphics application. Do not let them type text — they always misspell. If you work with clients, have the client proof and sign off on all the prepared assets (video, audio, and graphics) before authoring. Check navigation, layout, and general functionality with the simulation features of the authoring system. Don't worry about audio and video quality checking at this point, since simulation is not accurate enough.

After outputting formatted files, do emulation testing with computer software players. At this point, check whether the computer has reliable playback, and QC video and audio, especially synchronization. Since the computer displays the video in progressive mode, this is a good time to check for unanticipated effects of progressive display. Also be sure to check menu functionality, since the mouse-based user interface changes the way things work. Emulation testing can catch errors before you spend the time and expense of writing a DVD-R for further testing.

After emulation testing, output to DVD-R for testing on standard players and on other PCs. Test as many players as you can, and test with as many of the popular PC hardware and software decoders as you can. However, keep in mind that DVD-Rs don't play properly in all players, and can even cause spurious errors. Reserve some of this *matrix testing* time for replicated check discs. Consider sending the DVD-R to a verification service or running it through verification software or hardware. If the video has Closed Captions, connect a player to a TV with a Closed Caption decoder and make sure they appear correctly.

When you are sure everything works, send the DLT off to the replicator and request check discs. Check discs are a short replication run from the stamping master. A check disc is essentially the same as a final mass-produced disc. If you approve the check disc, the replicator can use the existing master for the full replication run. Most replicators provide one round of check discs free of charge as part of their replication service. However, they may charge a fee if you find errors and have to send in a new DLT for another set of check discs. Unless you have thoroughly tested with DVD-Rs and are comfortable with the potential of having thousands of shiny coasters at your disposal, don't skip the check disc step.

Tips and Tricks

The following tips may help your testing and quality control be more effective and efficient:

- Test on a cheap TV. Most of the production process is done with high-end, high-quality video equipment, but many of your customers will watch your work on a cheap, old television. Make sure that details of video and graphics, menu highlights, and so on are not lost on the average viewer.

- It's almost impossible to test all the permutations of a highly interactive disc; try to break things down into testable subsections.

- If you have computer software that can mount a formatted DVD image file from DLT or a hard drive, mount the volume for more thorough emulation. This will find file system errors that normal emulation will not find.

- If you put computer applications or large amounts of computer data on the disc, use a checksum program to verify the accuracy of the original data against DVD-Rs, DLTs, check discs, and final discs. A file-compare program such as windiff is also useful for verifying filenames and data integrity. Be careful when testing DVD-Rs, since some DVD-ROM drives have problems reading DVD-Rs and will report errors that will not be present on a check disc or the final replicated copies.

- If you put Macintosh files on the disc, make sure that the icons display properly and that the resource forks are still intact (use ResEdit or similar utility). Check that the filenames are correct when played on a Windows PC.

Table 16.7 is also provided to help you test QC issues on your project.

 NOTE

Anyone serious about producing DVD-Video must invest in three things: an expensive, high-quality monitor; a cheap TV; and a DVD PC.

Table 16.7 Testing Checklist

▪ Intro sequence	Does the disc do what's intended when inserted? Are the intros annoyingly longer than you realized?
▪ Menus	Check arrow movement in all four directions from each button. Test that Top (Title), Menu, and Return functions work. Check motion loops and idle outs. Make sure menu transitions are not annoyingly long and menu audio is not too loud or annoying.
▪ User operations	Which functions work? Which are locked out? Are they supposed to be locked out?
▪ Video	Verify quality. Check chapter search and Previous/Next functions. Check any angles, branching, and layer switch. Test on interlaced and progressive equipment.
▪ Audio	Make sure languages match the language name displayed by the player. Test multichannel playback and stereo downmix. Carefully look for audio sync problems.
▪ Subpictures	Check that subtitles appear at proper times. Make sure languages match the language name displayed by the player.
▪ Closed Captions	Verify that Closed Captions can be decoded by a TV.
▪ Parental control	Verify that disc or section lockout works. Try the disc at each parental level, and also with no parental level set.
▪ DVD-ROM filenames	Compare ISO 9660 and UDF filenames. Make sure long filenames were not truncated and that special characters are intact.
▪ DVD-ROM data integrity	Verify readability of all files. (If there are errors with a DVD-R, try a different drive.)

Replication, Duplication, and Distribution

At this point, things are mostly out of your hands. Your DLT, representing weeks or months or years of your life, is in the hands of someone else. Most replicators have excellent processes in place and will take good care of your project.

The replicator will use the DLT to cut a glass master and make stamping masters. Some replicators first verify the DLT formatting and check for specification compliance. The replicator will send you check discs made from the master. Once you approve the check disc, then thousands or millions of copies of your disc will be churned out.

If you want to serialize the discs or put other custom information in the BCA, a small number of replicators have the necessary equipment. If you need to make less than a hundred discs, you may want to use DVD-R duplication instead. You can burn DVDs by hand, buy a DVD-R bulk-duplication system, or have a service bureau duplicate the discs.

Most replicators can do *source tagging*: inserting *electronic article surveillance* (EAS) labels into packaging. The two most popular source tagging technologies are acousto-magnetic tags from Sensormatic and RF tags from Checkpoint Systems.

Disc Labeling

Many options for disc labeling are available, such as silkscreen printing, offset printing, pit art, and holograms. The area of a disc available for label printing normally extends from the 46 mm to 116 mm diameter points (the data area runs from 48 mm to 116 mm). For special applications, the print area can start at 34 mm, just past the clamping area. The BCA area is at 44.6 mm to 47 mm, so discs that use the BCA must start the label at about 48 mm. A DVD-10, DVD-14, or DVD-18, with data on both sides, has a narrow ring from only 39.5 to 43.5 mm for printing.

Silkscreening applies layers of colored ink to the label side of the disc. If more than two colors are used, or if the disc has large ink coverage areas, care must be taken that the disc does not warp as the ink dries. Offset web printing causes fewer warping problems from ink shrinkage than screen printing, but it is more expensive.

Pit art is produced by creating a stamper that embosses a graphic design onto the back substrate. The black print of the source graphic will appear as a mirrored surface, while the white areas of the graphic will appear as a frosted surface. The pit art area extends from 40 mm diameter to 118 mm. The source graphic is a one-bit monochrome image, usually 2048×2048 pixels.

Package Design

Packaging is the spokesperson for your disc. Make it reflect the contents of the disc and grab the attention of the customer. Even with Internet shopping, the disc jacket influences the buyer's impression of the work.

Design with the collector in mind. Part of the reason people buy DVDs rather than renting them or taping them (or waiting for the network broadcast) is because they enjoy collecting. Use quality artwork on the package. Produce an informative and enjoyable insert. Consider including little goodies in the package that relate to the disc.

Explain what's on the disc. Do not assume that the customer is familiar with the contents. Include a list of titles, chapters, extras, and so on. Use standard text and icons to indicate the format of the contents. Clearly identify number and type of audio tracks, aspect ratio, disc region, video format, running time, and so on (see Figure 16.7). Sometimes it's a good idea to delay package design until the end, since details of aspect ratio, soundtracks, extras, and so on may change during production.

Tips and Tricks

The following advice will help you to avoid problems during the replication, duplication and distribution processes:

- If you need a rush job, schedule replication early, since large orders (especially things such as *Star Wars* discs or Microsoft operating systems) can sometimes swallow all available replication capacity. Establishing a relationship with one or two replicators will help when you have special demands.

- If you want to use CSS, CPPM, or Macrovision make sure the replicator is appropriately licensed.

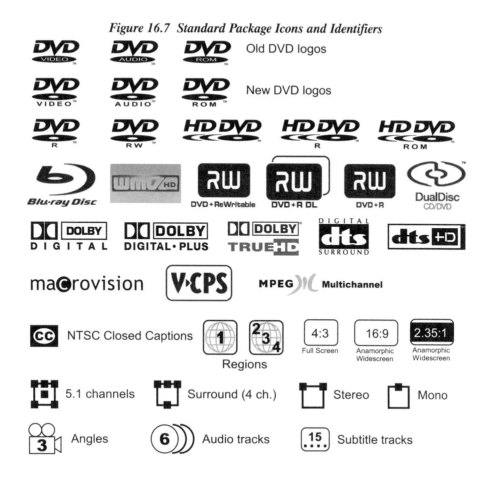

Figure 16.7 Standard Package Icons and Identifiers

Production Maxims

For your edification and enjoyment, here is a small collection of maxims, reminders, and rules of thumb — many of them learned the hard way — collected from various DVD authors:

- Players display time (and sometimes chapters) only for one_sequential_PGC titles.

- Corollary: Timecode search, particularly on computers, only works with one_sequential_PGC titles.

- Corollary: DVD players that use barcodes require timecode, and thus one_sequential_PGC titles.

- Playback between PGCs is non-seamless. Cell commands are executed in a non-seamless manner.

- Audio will mute during a non-seamless break.

- A layer change (on an RSDL disc) does not have to be at a chapter or program boundary, only at a cell boundary.

- All titles in a title set must have the same video aspect ratio (4:3 or 16:9) and audio format.

- Programs are usually the same as chapters (PTTs), but programs that are not chapters behave in special ways. To be safe, always put a chapter mark on every program.

- Next/previous commands can only be used in the title domain. The return (go up) command can be used in all domains.

- Entry points (chapters, programs, and cells) must start at a GOP header and must be at least 0.4 seconds apart.

- Corollary: An I frame is not necessarily a GOP header. To create an entry point, you must force a GOP header, not just an I frame.

- Subpictures cannot cross an entry point (chapters, programs, or cell).

- GPRMs are reset to zero on title search or stop.

- Cell commands are executed after the cell is presented.

- If you want to play continuously from chapter to chapter, do not put a next program command at the end of the program, since the player will interrupt playback to execute the command. There is no need to add any commands, since playback will automatically go to the next chapter.

- Never assume that anyone, even the client, knows what the DLT contains. After authoring and writing to a DLT, fill out the replicator's project sheet personally to avoid inaccuracies.

- Any amount of QC on a replicated check disc is worth more than the cost of paying for a disc run that failed for the simplest problem. If there will not be enough time or budget to go through check disc testing, then you, above all others, *need* to test the check disc.

- Last, but not least, no disc is ever *easy*, *simple* or *trivial*, and saying that it will be usually guarantees that it won't be.

DVD-ROM Production

Many details of producing a DVD for computers are covered in Chapter 15. This section goes over a few of the basics. Also, see the testing section earlier in this chapter.

File Systems and Filenames

DVD-Video discs are specified to use the UDF bridge format, which includes ISO 9660 and Micro UDF, but a DVD-ROM may be formatted with any file system: full UDF, ISO

9660 only, Windows NTFS, Apple HFS, and so on. ISO 9660 Level 1 limits filenames to eight characters plus a three character extension. Directory names are limited to eight characters with no extension. ISO 9660 Level 3 limits filenames to 30 characters (not counting the period between filename and extension), although it's possible to store up to 221 characters as the file identifier. Directory names are limited to 31 characters, but 221 are possible. UDF supports filenames of up to 255 characters using the Unicode character set, which supports characters from almost every language. However, an operating system has no guarantee that it will be able to correctly display a given Unicode character.

To provide better compatibility with operating systems such as Windows 95, DOS, and Unix that are unable to read the UDF file system on DVD volumes, Microsoft recommends that DVD image formatting software, including DVD-Video and DVD-Audio authoring systems, implement the Joliet extensions to allow long filenames within the ISO 9660 file system.[9] Long filename support is especially important to publishers putting legacy software onto DVD-ROM volumes.

The Joliet extensions to ISO 9660 provide for filenames longer than 30 characters and for the Unicode character set in file and directory names. The use of Joliet extensions does not violate the DVD specification or the Micro UDF specification. Standardized directory names of VIDEO_TS and AUDIO_TS and the standardized names of files within these directories follow the 8.3 filename limitations, but these limitations do not apply outside of the video zone and audio zone.

Joliet applies only to the ISO 9660 section of the UDF bridge format and is independent of UDF, which has its own provisions for long file names. Windows 98 and Windows 2000 use UDF and do not read the ISO 9660 portion of UDF bridge DVDs. The intent behind using Joliet extensions with DVD volumes is to make them behave more like today's CD-ROMs when used with an OS that recognizes Joliet but does not recognize UDF.

The UDF file system also supports Macintosh resource fork, including file creator and file type information. The authoring system or DVD-ROM formatting software must be able to recognize Macintosh file information in order to preserve it for UDF formatting. On a Windows-based authoring system, software that supports Macintosh files, such as *PC MacLAN* from Miramar Systems, may be needed.

It's possible to make a bootable DVD-ROM. The Micro UDF spec specifically forbids boot records, but most computers don't look for UDF boot records, they look for ISO 9660 boot records according to the El Torito specification. Very few authoring systems and DVD-ROM formatting utilities can create a bootable disc. If you need this capability you may even have to produce a DVD-ROM image and modify it by hand to create the boot sector before writing the image disc or DLT.

Bit Budgeting

Bit budgeting on DVD-ROM is a breeze compared to DVD-Video or DVD-Audio. You just add up all the file sizes and see if they fit on a disc; however, a few potential pitfalls might arise. Don't forget the difference between billions of bytes and gigabytes. The operating sys-

[9]Unfortunately, what Microsoft does not mention is that there are some DVD-Video players that cannot read discs that contain Joliet extensions, and therefore will not play the disc.

tem will report file sizes in gigabytes, but DVD capacities are usually given in billions of bytes. Don't forget to leave some room for the directory entries.

When adding up file sizes, use the actual file size, not the "size on disk" value, since the space taken on the hard disk is dependent on the block size. Likewise, space taken up by each file on a DVD can be more than the actual file size. For a very tight bit budget, adjust the size of each file by the DVD sector size: divide the file size in bytes by 2048, round up, then multiply by 2048. This will give the sector-adjusted file size, which represents the true amount of space the file occupies on the DVD.

Creating a DVD-9 is not as simple as it might seem. You need two DLTs, but simply dividing your project in half and creating two disc images will not work. A DVD-9 has a single UDF bridge directory section located on the first layer that references files on both layers. If you use formatting software to create two images, you will get two volumes, each with its own directory section. The first layer will work fine, but the second layer will be inaccessible. Use a DVD authoring application or a DVD-ROM formatting utility that knows how to create DVD-9 images, preferably one that lets you specify where the layer break occurs.

A/V File Formats

When creating a multimedia DVD for playback on PCs, you have essentially three choices for the A/V format: DVD-Video files (.vob), MPEG files (.mpg), or computer-oriented media files (.avi, .mov, .mp3, et cetera). The advantages of using DVD-Video content are that the disc will also play in standard DVD players, the features of DVD-Video such as angles, seamless branching, fast chapter access, and subpictures are available, and the video will be more compatible with PC decoders, which are oriented more toward DVD than MPEG-2. The advantages of using MPEG-2 program stream files are that you don't have to use a DVD-Video authoring system, and you can easily create and test playback. You can also make each video clip a file with a meaningful name rather than trying to find the right access point in the middle of a .vob file.

The advantages of using other formats are that the development tools are simple, widely available, and often inexpensive. Since the data rate of most DVD drives is at least equivalent to an 18x CD-ROM drive, high data rates can be used with codes such as Cinepak, Sorensen, and Windows Media Video (MPEG-4) to provide surprisingly good video. It may also be much easier to create discs that work on both Windows and Macintosh computers by using tools such as Macromedia *Director*, Click2Learn *Toolbook*, or even HTML authoring applications combined with multimedia playback plugins.

Hybrid Discs

If you are adding DVD-ROM content to a DVD-Video or DVD-Audio disc (or adding DVD-Video or DVD-Audio content to a DVD-ROM disc), be mindful of a few things. Data in the DVD-Audio and DVD-Video zones must come first on the disc and must be contiguous. If you have a set of VOB and IFO files in a VIDEO_TS directory or a set of AOB and IFO files in an AUDIO_TS directory and you feed the files into a DVD-ROM formatting application, it must recognize the DVD-Video or DVD-Audio files and physically place them on the disc in the proper way. The IFO files include references to sector offsets, which requires that the IFO and AOB/VOB files be in proper sequential order on the disc. If not, the disc will fail to play on many DVD players. It will play fine in software players on PCs, since they ignore physical placement of data on a disc. If you add more than a few DVD-ROM files to a DVD-Video or DVD-Audio disc, put them in a subdirectory. Too many extra files in the root directory will cause some DVD players to fail.

Impact of the new formats

As the next-generation formats (UMD™, HD DVD, and BD) are released, there will be some significant changes made to the disc content production process, while some aspects of the process will remain unchanged. In general, the basic concepts of planning, design, asset preparation, compression, authoring and testing will also apply for the new formats, but there are changes in specific details for some of the production steps. There are, however, some general issues that will be new, which are explained below.

A Lot More Elements

Unlike DVD-Video, in which the smallest element is one full screen graphic or a subpicture, the new formats allow a presentation to be composed of multiple layers with different graphic and video elements. A DVD-Video menu (see Figure 16.8) could consist of a still image or a video background, with a subpicture highlight.

Figure 16.8 Standard DVD-Video Menu Composition

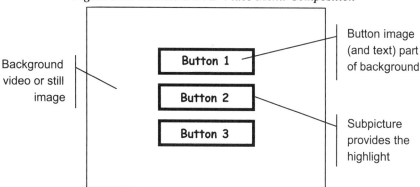

With the new formats, a menu may consist of multiple elements — video, animations, graphics, subpictures, and audio — that create the full presentation. Each button may be a separate graphic object, with different images or effects applied for each state (normal, selected and activated). Animation sequences and other graphic elements may be blended with the background image or video to create advanced visual effects. Rendered text, sound effects and secondary picture-in-picture video may also be used in some formats to provide a dynamic user experience (see Figure 16.9).

Figure 16.9 Next-Generation Disc Formats Menu Composition

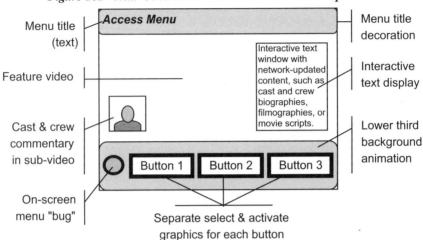

Just think of all the exciting new features that could be added and the many extra elements they would be required. And then think of all the extra planning and coordination that will be needed!

Design vs. Authoring

Because of this significant difference in how the user interface for menus and other interactive features are constructed, the title layout and programming becomes much more closely tied to the design. In standard DVD, an author, armed with a small amount of documentation, could easily understand what the designer intended and it was quite simple to separate the design and authoring activities. One person (or company) could do the former and a separate person (or company) could do the latter.

However, when dealing with a more complex design in which multiple elements must be positioned with exacting accuracy, and individual behaviors must be programmed for each element (behaviors that may, at times, cause elements to conflict with one another), it becomes far more difficult to separate the design process from the authoring process.

The title designer must provide documentation that thoroughly explains the intended complex element construction, and work closely with the author to insure that the design is executed properly. Either that, or the designer has to expect that the author will simply do his or her best with what there is to go on.

When Authoring Becomes Programming

Although the HD DVD and BD formats include simplified modes for DVD-like navigation, most of the compelling new features fall in the area of enhanced content. Like UMD, the enhanced navigation mode of each of these formats requires that the content be *programmed* far beyond what could ever be done with DVD-Video. In the case of HD DVD *advanced content* and UMD, even the most basic applications require significant amounts of script programming combined with complex XML markup code. Blu-ray Disc takes it up a notch, requiring full Java application development to achieve the same result. Although there are a few tools on the market designed to help make programming (whether it's XML, ECMAScript, or Java) a bit easier, it still requires a fundamental knowledge of software constructs and methodologies. It's not authoring anymore — it's programming!

Chapter 17
DVD and Beyond

In the fall of 2005, as this book was being finished, DVD was almost 10 years old. It had succeeded beyond many expectations but also failed to meet others. Given the wide range of speculation and the variety of ways DVD can be used, this is no surprise. This chapter examines past and present forecasts, and looks ahead to possibilities for DVD over the next decade or so.

Questions such as "Will DVD succeed?" and "When will DVDs outsell videotapes?" have been answered, but many other questions remain: When will DVD recorders replace VCRs? What will be the effect of hard-disk-based personal video recorders (PVRs)? When will the next-generation disc formats finally appear? Will there be a format war and how long will it last? Will we ever need another audio format? What comes after blue laser? Will the Internet obviate the need for physical media? This chapter touches on these and other questions.

The Prediction Gallery: 1996

The variables in predicting the path of DVD are extremely complex, but this has not deterred many people from making predictions ranging from the sensible to the outrageous.

The DVD format was announced in December 1995. In the following 11 months until the appearance of the first DVD players in Japan, the only people making money from DVD were journalists and analysts. Considering the disparity between their forecasts and reality, we humbly submit, some of them were making more money than they deserved. It's interesting to look back and see who had the most accurate auguries. Here is a potpourri of prognostications made in 1996 (the reality check follows the list of predictions):

Philips: 25 million DVD-ROM drives worldwide by the year 2000 (10 percent of projected 250 million optical drives).

Pioneer: 500,000 DVD-ROM drives sold in 1997, 54 million sold in 2000.

Toshiba: 120 million DVD-ROM drives in 2000 (80 percent penetration of 150 million PCs). Toshiba projects that it will no longer make CD-ROM drives in the year 2000.

International Data Corporation (IDC): 10 million DVD-ROM drives sold in 1997, 70 million sold in 2000 (surpassing CD-ROM), and 118 million sold in 2001. Over 13 percent of all software will be available on DVD-ROM in 1998. DVD recordable drives will make up more than 90 percent of the combined CD/DVD recordable market in 2001, with DVD drive units installed in 95 percent of computer systems with fixed storage. IDC later revised its figures to 3.7 million in 1997, 10.8 million in 1998, 36.5 million in 1999, and 95 million in the year 2000.

AMI: An installed base of seven million DVD-ROM drives by 2000.

Intel: 70 million DVD-ROM drives by 1999 (sales will surpass CD-ROM drives in 1998).

InfoTech: 1.1 million DVD-Video players in 1997, with 10 million by 2000. 1.2 million DVD-ROM drives in 1997, with 39 million by 2000. 250,000 interactive (video game or cable TV settop) DVD players in 1997, with six million by 2000. InfoTech revised its DVD-Video predictions in January of 1997 to a slightly lower 820,000 units in the first year, with an installed base of 80 million by 2005.

Dataquest: One-year sales of 15 million DVD-Video players in 2000, with a DVD optical drive market of $35 million by 1996 and $4.1 billion by 2000.

Paul Kagan Associates: 12 million DVD-Video players in the US after five years. Total DVD-Video business of $6 billion per year by the year 2000.

Simba: Annual DVD software sales of up to $35 million by 1997 and $100 million by 1999.

Freeman Associates: DVD-ROM drive sales of 89 million by 2001, accompanied by zero sales of CD-ROM drives.

Electronic Industry Association of Japan (EIAJ): A combined US and Japanese DVD-Video and DVD-ROM market of $23.6 billion by 2005, with a DVD-ROM drive in about 80 percent of all new PCs.

It is enlightening to compare the success of DVD with the growth rate of previous consumer products (see Figure 17.1 and Table 17.1).

Figure 17.1 Consumer Electronics Success Rates

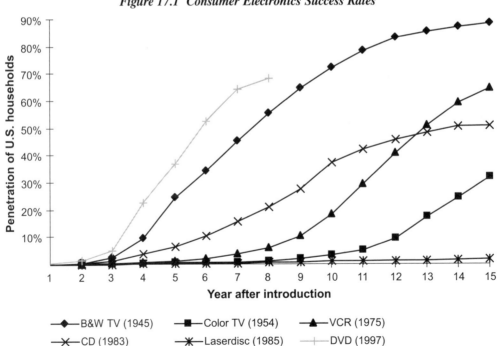

Table 17.1 Technology Penetration Rates

Format	Years to Reach 50 Percent of US Homes
Telephone	71
Radio	10
B&W TV	9
Color TV	17
Cable TV	23
Cellular phone	14
VCR	11
CD	11
PC	19
DVD	7

Sources: US Census, CEA, IDC, JPR, Technology Futures

Four-year Reality Check: 2000

The actual number of DVD players sold to dealers in the US in 1997 was 349,000. First-year sales of DVD players more than doubled that of VCRs from 1975 to 1977 and totaled more than 12 times the first-year sales of CD players in 1983. Just over one million DVD players were sold in 1998. More than four million DVD players were sold in 1999 for a US total of about 5.5 million at the end of 1999. DVD reached 16 percent penetration of home theater systems in 1999. In October 2000, the 10 millionth US DVD player was shipped. In three-and-a-half years since its US introduction, DVD players achieved the mark that VCRs took eight years to reach and CD players took eight years to match. Sales of DVD players in the US in 2000 hit 9.8 million (worth roughly $2 billion), reaching a US grand total of over 14 million DVD players. About 40 million discs shipped in the US in 1998, about 100 million discs shipped in 1999, and approximately 227 million shipped in 2000, representing more than $4 billion in retail revenue. In 2000, DVDs accounted for about 32 percent of home movie sales in the US. In Europe, 270,000 players shipped to dealers in 1998, 1.6 million in 1999, and approximately four million in 2000, for a total of just under six million. About 20 million DVD-Video discs shipped in Europe in 1999, with shipments of over 50 million in 2000. In Japan, 240,000 DVD players were sold in 1999, 390 thousand were sold in 1999, and about 800,000 were sold in 2000. Sony sold over two million PlayStation® 2 consoles since its March 2000 introduction, essentially tripling the number of DVD players in Japan. Approximately 18 million DVD players were sold worldwide in 2000, bringing the total to around 26 million. An additional 56 million DVD PCs and 10 million Playstation 2 units brought the grand total of DVD playback devices on the planet to roughly 91 million.

For comparison, an installed base of about 700 million CD players and about 240 million CD-ROM drives existed worldwide at the end of 2000. Roughly 47 million new CD players were sold in 2000, equaling sales of $4.5 billion. Total audio shipment revenues in 1999 passed the $8 billion mark for the first time since 1995. Audio system revenues in 1999 were $2.1 billion. By the end of 2000, about 600 million VCRs and 1.3 billion televisions were owned worldwide. Over 93 percent of American households had at least one VCR. Nearly 23 million new VCRs were sold in the US in 1999, compared to just under 24 million in 2000. American consumers spent $9 billion to rent about four billion videotapes in 2000. They spent $500 million renting DVDs. The pay-per-view market had grown to $1 billion (after 15 years). About 40 million TVs were sold in 2000. More than 90,000 digital TVs were sold in 1999, followed by about 600,000 in 2000. About 40 million Video CD players were in homes, mostly in Asia, by the end of 2000. Worldwide sales of digital cameras hit 2 million units in 1999 and 11 million in 2000.

The Prediction Gallery: 2000

Here's a sampling of expectations at the beginning of the new millennium. The following forecasts were made in 2000:

DVD Forum: $16 billion in DVD-ROM/RAM drive sales in 2001, $4 billion in DVD audio players, and $8 billion in DVD video players/recorders.

Technicolor: The number of PCs equipped with DVD-ROM drives will reach more than 130 million units worldwide by 2001.

Jon Peddie and Associates: DVD player sales in the US in 2004 will total nearly 20 million units.

IRMA: Thirteen million players will ship in the US in 2000, with 31 million sold in 2003. Two hundred and eighteen million DVD-Video titles will ship in 2000, with 360 million units in 2001. DVD-ROM and DVD-Audio will ship a combined 70 million discs in 2001.

Screen Digest: Western Europe will have 5.4 million DVD players by the end of 2000, in 3.5 percent of TV households. The total will rise to 47 million by 2003, or about 30 percent of TV households. DVD spending will rise from $450 million in title purchases and $18 million in rentals ($470 million total) in 1999 to $4.9 billion in sales and $1 billion in rentals ($5.8 billion total) in 2003. VHS spending will fall from $6.1 billion in 1999 to $3 billion in 2003.

Baskerville Communications: Worldwide spending on DVDs will surpass spending on videos by 2003. By 2010, 625 million DVD players will be in homes worldwide (55 percent of TV households), with spending on DVDs at $64.6 billion versus $2.5 billion for VHS. Also by 2010, China will have the highest number of homes with DVD players at 127 million.

Sony: Ten million PlayStation 2 units will be shipped worldwide by the end of March 2001: four million in Japan, three million in the US, and three million in Europe.

InfoTech: Approximately 25 million DVD video recorders will be sold by 2005.

Disk/Trend: DVD-ROM drive sales will surpass CD drive sales in 2001, with 60.3 million DVD units sold, versus 56.8 million CD units. By 2002, consumers will buy 92.8 million DVD drives, compared to 30.3 million CD drives.

Hewlett Packard: More than 140 million DVD-ROM drives and DVD video players will be in use by the end of 2001.

Understanding and Solutions: The attach rate of DVD-ROM PCs in Western Europe will hit 9 percent in 2000, 20 percent in 2001, 37 percent in 2002, and 58 percent in 2003.

Strategy Analytics: Global DVD hardware sales (players and drives) will reach 46 million units in 2000, including 21 million in the US and 17 million in Europe. The average price of a DVD player in 2001 will fall to $200 in the US and $270 in Europe. Fourteen percent of US homes and 5 percent of European homes will own at least one TV-based DVD player by the end of 2000. By 2002, 58 percent of US homes will own at least one DVD device (a player, game console, or PC). DVD PCs, which accounted for 75 percent of the installed base at the beginning of 2000, will fall to 59 percent by 2002 as TV-based DVD becomes more widespread. Video titles accounted for over 90 percent of the software market in 1999. By 2005, their share will have fallen to 43 percent, while DVD-ROM will account for 28 percent and games formats will total 24 percent. Worldwide DVD-Video disc shipments in 2000 will reach nearly 400 million units, growing to 2.3 billion by 2005, worth $44 billion. Including other DVD formats (games, ROM, and audio), shipments will reach 3.7 billion by 2005, worth nearly $100 billion.

Dataquest: DVD drives outsold CD-RW in 1999 (16.2 million versus 12.5 million), but CD-RW will take the lead in 2000 with 28.7 million CD-RW units shipped compared with 22.6 million DVD drives. By 2002, DVD will overtake CD-RW, and by 2004 a predicted 105 million DVD drives will be shipped, versus 28 million CD-RW drives.

Paul Kagan Associates: Hard-disk-based PVR sales in the US will reach one million in 2000, 12 million in 2004, and 40 million by 2008.

Forrester Research: Fourteen million PVRs will be in homes by 2004. About 80 percent of US homes could have PVRs by 2010.

Five-year Reality Check: 2005

Near the end of 2005 over 50,000 video titles are available on DVD in the US,[1] not including an estimated 30,000 "adult" titles. US consumers spent more than $21 billion buying and renting DVDs in 2004. 1.5 billion discs were shipped to retailers in the US in 2004, and another 770 million have shipped by mid-2005, adding up to over 5 billion DVDs shipped in the US since the birth of the format. More than 140 million DVD players were in homes in

[1]According to DVD Release Report.

the US,[2] with 4 out of 5 households owning a DVD player before the end of 2005, closing in on the 98 percent penetration of color TV and the 94 percent penetration of VCRs. Almost half of US homes had two or more DVD players. Given the saturation of the marketplace, retail revenues for DVD players began to fall in 2005 as growth slowed. Some studios reported lower-than-expected sales of DVD titles — probably because their predictions didn't take into account that nearly everyone already owned a DVD player and there weren't tens of millions of new player owners looking for discs. DVDs had become the leading source of theatrical revenue, with box office accounting for 22 percent, television accounting for 27 percent, and DVD/videotape accounting for 51 percent, according to Kagan Associates.

In comparison, about 18 million digital cameras were sold in the US in 2004, compared to over 63 million sold worldwide in 2004, and over 50 percent of US homes had digital cameras before the end of 2005. According to the Consumer Electronics Association, US households in 2005 owned an average of 100 music CDs, more than 40 DVD movies, 16 video games, and three televisions. Roughly 13 percent of US households owned a digital high-definition television (HDTV), about 10 percent owned a digital video recorder (DVR), and approximately 15 percent owned a portable MP3 player. Approximately 7 million portable music (MP3) players were sold in the US in 2004.

According to Screen Digest, DVD video player/recorder penetration in Western Europe passed 50 percent of TV households at the end of 2004, six years after its official European launch in 1998. By comparison, the VCR achieved a similar level of European penetration shortly before the end of 1990. The total number of DVD players worldwide was estimated to be just over 300 million by the end of 2005, close to 700 million if DVD PCs and DVD game consoles are included.

Over 17 billion optical discs were replicated worldwide in 2004: 5.3 billion audio CDs, 4.2 billion CD-ROMs, 2.8 billion Video CDs, 4.1 billion DVD-Videos, 750 million DVD-ROMs, and 44 million DVD-Audio discs. More than 10.4 billion recordable CDs were manufactured worldwide in 2004 (CD-R and CD-RW) compared to 1.9 billion recordable DVDs (DVD-RW, DVD+RW, DVD-R, DVD+R, and DVD-RAM), while blank VHS tapes dropped to 280 million.

DVD-Audio succeeded, barely, on the coattails of DVD-Video. If DVD-Audio had to stand on its own it would have fallen flat on its face. Sales of DVD-Audio and SACD players and discs were a small fraction of DVD-Video players and discs. According to RIAA figures from 2004, 350,000 DVD-Audio discs were sold in the US (down 20 percent from 2003) and 790,000 SACD discs were sold (down 39 percent from 2003, but still a reversal from 2003, when DVD-Audio discs outsold SACDs five to one), compared to 1.3 million LPs and almost 766 million audio CDs.

The DVD PC market slowly displaced the CD-ROM PC market, with more than half of new PCs equipped with DVD-ROM drive, a combo CD-RW/DVD-ROM drive, or a DVD recordable drive. CD media will continue to be widely used, however, but CD-ROM drives will suffer the fate of floppy drives (remember those?).

---

[2]According to the Digital Entertainment Group, counting settop and portable DVD players, DVD recorders, home-theater-in-a-box systems, TV/DVD and DVD/VCR combination players, but not counting DVD PCs or DVD game consoles.

The Death of VCRs

VCRs and VHS tapes will eventually disappear, it's just a question of when.[3] Analog videotape technology will be replaced by digital technologies such as DVD players and recorders, digital video recorders (DVRs), and non-physical delivery methods such as video on demand (VOD) over cable, satellite, and Internet. VCR revenues in the US peaked in 1996 and prerecorded VHS sales peaked in 1999. VCR sales growth in 2000 was less than 5 percent above 1999, compared to over 100 percent sales growth of DVD players. Worldwide VCR sales fell from less than 25 million in 2003 to less than 17 million in 2004 and approximately 12 million in 2005. Informa Media Group expects that worldwide DVD homes will reach 473 million by 2009, surpassing 443 million VCR homes, which will already be in decline by that time (VCR penetration began to decline in the US in 2004). Buena Vista Home Entertainment expects that revenue from US VHS releases will drop from $4.6 billion in 2004 to $1.7 billion in 2007. Cambridge Associates shows that worldwide demand for blank VHS tapes dropped from roughly 1.5 billion in 1999 to around 900 million in 2004. Informa predicts that by 2010, 31 per cent of global TV households, or 334 million homes, will have a DVD recorder. Japan Electronics and Information Technology Industries Association (JEITA) reports that 4 million DVD video recorders were sold in Japan in 2004 (representing 56 percent of combined sales of players and recorders) compared to 1.85 million VCRs, down 38 percent from 2003 sales.

The displacement of VCRs by DVD recorders can be compared to the displacement of cassette tapes by CDs. It is interesting to note that it took until 2000, 17 years after the introduction of CD audio, for prerecorded cassette tapes to begin disappearing completely from music store shelves. If this historical perspective is applied to DVD, we might expect VHS tapes to begin fading away before 2014.

Other technologies, such as hard-disk-based video recorders, will compete with DVD recorders to become the VCR of the future, but the final outcome will be a melding of both technologies: digital recorders that combine hard disks for time shifting and DVD drives for playing movies and making long-term copies of recordings from the hard disk onto recordable DVDs.

Eventually, broadband Internet connections will reach the point where DVD-quality video can be streamed in real time to living rooms and offices on demand. Before then, high-definition video will reach mainstream status, which will require another bump in connected bandwidth. Once again, DVD (more specifically blue-laser DVD) will fill the need for the delivery, storage, and archiving of high-bandwidth, high-quality content. Techniques such as downloading video overnight to a hard drive will help overcome bandwidth deficiencies, but even when high-definition video or digital cinema are delivered in real time over the Internet, there will still be a need for making permanent physical copies.

---

[3]And, no, we are not going to hazard a guess as to when VHS and VCRs will disappear, for fear that at some time in the future someone else may do to us what we've just done to others. Thank you very much. (See footnote 7)

Hybrid DVD Systems

The DVD format is being combined with other products and systems in intriguingly diverse ways. Next-generation players will add Internet connections, and home Internet-TV appliances may add DVD-Video players or DVD-ROM drives. Discs will appear for free in the mail to be viewed or purchased by having the player make a quick phone call to authorize the charge to an account or credit card. Cable/satellite settop boxes are beginning to include DVD player or recorder units, as are digital video recorders (DVRs). Scientific-Atlanta discovered in 2004 that the feature more than 60 percent of cable subscribers wanted most in their DVR was the ability to record to DVD.

Both home and commercial video game systems will continue to include DVD-Video and DVD-ROM units. Public information kiosks will incorporate more DVD technology. Car navigation systems will continue to take advantage of DVD storage capacity for maps and graphics and even pipe DVD movies to a monitor in the back seat. As MPEG video and Dolby Digital audio become more common, the digital decoder will move into the television or stereo fields, making the player an inexpensive transport that, along with other devices such as a digital satellite receiver, digital cable box, digital TV tuner, or video phone connection, would simply feed compressed signals to the video display. Digital TVs with built-in DVD players could become common. Video editing applications on PCs, or even video-editing features of cameras and settop boxes, enable users to record digital pictures and camcorder footage to be shared with others.

Electronic sell-through of DVDs will take off in 2006, where titles will be downloadable to PC or PVR hard drives for recording to DVD on demand. Essentially, some of the millions of DVDs formerly replicated in factories and shipped to consumers will be burned in hundreds of in-store kiosks and in millions of consumers' homes.

The living room of the cost-conscious consumer may be the primary setting for splicing DVD into other formats. These possibilities are explored further in the section entitled "The Changing Face of Home Entertainment," which follows later in this chapter.

The UMD Explosion

By August of 2005 Sony had shipped about 6 million PlayStation® Portable (PSP™) hand-held game players. In spite of widespread incredulity that anyone would want to watch blockbuster movies on a tiny screen, over 8 million UMD™ movie titles have been sold, compared to about 9 million UMD game titles. Over 200 UMD movie titles are expected to be available in the US before the end of 2005. Sony has predicted 130 million UMDs will be sold by 2008, with over 60 percent being movies.

Growth of Next-generation DVD

The Consumer Electronics Association (CEA) said that 7.3 million digital television (DTV) products were sold in 2004, followed by 3.8 million in the first half of 2005, and projected that 15 million DTVs would be sold in 2005 for a cumulative total of 36 million.[4] InfoTech predicted that as many as 1 billion blue-laser devices (consumer players/recorders, PC drives, and so on) could be sold over the lifetime of the format, which it predicts will peak around 2015, assuming the formats merge. Research and Markets expected shipments of blue-laser recorders and players to reach 4 million in 2008, not including video game consoles. The Trojan horse is Sony's PlayStation® 3, scheduled for release in early 2006, which will play Blu-ray movies. If it does as well as its predecessor (PlayStation 2) it will put about 9 million Blu-ray players into the market in one year, around 50 million players in three years, and around 90 million players in five years.

The underlying question is: how compelling is a next-generation format? DVD is a tough act to follow because it was such a vast improvement over VHS tape. DVD heralded the switch from analog to digital video, added multichannel audio, quick access to programs with chapters, menus and interactivity, multiple audio tracks and subtitle, and much more. Next-generation formats offer the same. For instance, picture quality is better, but DVDs already look pretty good on HDTV sets. Interactivity is significantly improved, but consumers aren't exactly rioting in the streets demanding more interactive titles. The promise of fresh new content delivered to Internet-connected players is probably the most interesting new feature. Is it enough to get consumers to make the switch to new players, and more importantly, start changing their collections over to HD format? It depends, in part, on the draw of HDTV. After someone pays a lot of money for a high-definition display, they will want to watch high-definition content to feel like they're getting their money's worth.

Some people expect that the blue-laser formats will be adopted even faster than DVD, but the specter of a format war invites a less utopian scenario where consumers shun the confusion of multiple formats and gravitate to alternative sources such as pay-per-view, video on demand, and the Internet. It is painful but not inconceivable to envision a scenario where DVD is the last significantly successful packaged media format.

The outlook is a little brighter on the computer side, where demand for increased storage never abates. Major computer manufacturers such as Apple, Dell, and HP support the Blu-ray format, making it the likely winner in the PC space. However, if the entertainment market grows too slowly to drive down hardware costs and keep attention focused on the Blu-ray format, the Blu-ray PC market will not blossom like the PC DVD market did.

Beyond Next-generation DVD

In 2005, before blue-laser formats reached the marketplace, variations such as four layers were already being demonstrated. Even more layers had been successfully implemented in research labs. In 1999, Constellation 3D Inc. began touting fluorescent multilayer disc (FMD) technology, which stores data in pits on multiple layers of fluorescent material. When a laser excites the material, it emits light that can be read more easily than the reflected coherent light of a laser. In theory, hundreds of data layers can be used to create very high-

---

[4]For reference, in 1999 the CEA predicted the first 10 million DTV units would be sold by 2003, the next 10 million by 2005, and 10.8 million would be sold in 2006 alone. At the beginning of 2005 the CEA predicted sales of 20 million for the year, but in April revised the prediction to 15 million.

capacity volumetric storage media. However, FMD was a case study of how technology that is too removed from mainstream and not supported by a sufficient number of leading companies has little chance of hitting the big time. Constellation 3D ran out of money in 2002.

Another example of an intriguing technology that failed to catch on is multilevel or grayscale recording, as promoted by Calimetrics.[5] In Sisyphusian fashion it was put forward for CDs, then DVDs, then blue-laser discs, but was rebuffed each time.

One clear trend is that discs will stay at 12 cm (5 inch) and smaller sizes. Rumors about laserdisc-sized 30 cm (12 inch) discs faded away after the turn of the millennium when it became clear that advances in technology could pack plenty of data on the much more convenient 12 cm size. However, because of the insatiable demand for higher capacities, it's not likely that smaller formats will dominate, although 8 cm (3 inch) discs will flourish in products such as camcorders.

The most likely successor technology to blue-laser formats, other than another refinement with ultraviolet lasers, is holographic storage. Optical storage technology takes a quantum leap in capacity when laser holography is used to record data.[6] Holographic storage reads and writes millions of bits simultaneously in a few milliseconds by arranging them in "pages" that are stored on 12 inch discs in holographic interference patterns. Both InPhase and the Holographic Versatile Disc (HVD) Alliance (a group of six primarily Japanese companies, including CMC Magnetics, Fujifilm, and Optware) made announcements in February 2005 of plans for holographic discs holding up to 1.6 trillion bytes (about 180 times the capacity of a dual-layer DVD), with data transfer rates of over one gigabit per second (about six times the speed of 16x DVD). Initial versions could hold about 200 billion bytes, and because holographic technology is laser based, it's not a stretch to make the drives read CDs and DVDs.

Other future possibilities include diamond-based x-ray lasers, charged particle beam recorders, near-field scanning microscopy, and any number of creative new technologies for manipulating matter in strange ways to store vast amounts of data.

The Death of CD-ROM

As evidenced in the preceding Prediction Galleries, the range of dates given for the disappearance of CD-ROM would make a fine office pool, but all agree that CD-ROM will become obsolete. This will not happen overnight, to be sure, since CD-ROM had a 100 million-unit lead on DVD-ROM when it was introduced.[7] According to Taiwan's Industrial Technology Research Institute (ITRI), global shipments of CD-RW drives are expected to decline by 10 million units per year over a period of three years: from 60 million units in 2003 to 40 million units in 2005 and to 30 million units in 2006. International Data Corp. (IDC) estimated 2005 demand for CD-RWs at 19 percent, lower than demand for DVD-RWs at 28 percent and DVD-ROMs at 27 percent.

---

[5]Calimetrics was acquired in November 2003 by LSI Logic.

[6]This refers to using holographic technology to store data, not to three-dimensional holographic video.

[7]The second edition of this book, in 2000, said it would be a pretty sure bet that the last CD-ROM drive production lines would have changed over before 2005. It's easy to underestimate the staying power of established technologies.

The insatiable demand for storage space is propelling DVD-ROM into the computer mainstream. Because DVD-ROM drives can read CDs, the only real barriers are support and price. Prices will also drop close to CD-ROM levels due to economies of scale and because most of the mechanisms and components are similar to existing CD technology.

The Interregnum of CD-RW

Rewritable CD (CD-RW), finalized at the end of 1996 and introduced shortly after DVD-ROM, had a tough row to hoe. It finally brought the Holy Grail of erasability to the basic CD format, but only by being incompatible with all existing CD drives and players. Had the makers of DVD-RAM been aggressive and seized the opportunity, they could have taken over the writable market, especially if they had introduced DVD-RAM drives that could write to CD-R discs. Instead, DVD-RAM drives remained high-priced specialty items, while CD-RW drives sold better than expected. This was not because CD-RW itself was a success, but because CD-R media became astoundingly cheap. For every CD-RW disc sold in 2000, about 100 CD-R discs were sold. In 1999, 1.7 billion CD-Rs were produced worldwide to meet a demand between 1.3 and 1.4 billion, while CD-RW demand was just over 15 million worldwide. In 2004 about 17 billion discs were produced worldwide, with CD-Rs representing 83 percent, CD-RWs taking 2 percent, DVD±Rs representing 14 percent, and DVD±RWs making up only 1 percent. People bought CD-RW drives so they could record CD-Rs, specifically music CD-Rs that would play in existing CD audio players. In 2000, CD-RW drives outsold DVD drives 28.7 million to 22.6 million. But since DVD recordable drives that also write to CD-R and CD-RW discs dropped below $50, they began to seriously encroach on the CD-R/RW drive market. Combo drives that play DVDs but record CD-R/RWs may also continue to be successful.

Standards, Anyone?

Computer standards often develop like meandering cowpaths that later become so well traveled that they end up being paved into meandering roads. DVD offers both promise and hardship: a new format that can be done well or a new jungle through which too many crooked paths can be blazed.

Take, for instance, DVD's predecessor, MPEG-1 video on CD, which was a standards nightmare. Each hardware and software implementation of MPEG-1 was slightly different, with companies designing their own MCI interfaces and data formats. The market for MPEG-1 video hardware decoding, which at first looked so promising as an alternative to quarter-screen, low-resolution video codecs, took a nosedive before it ever got off the ground. Software publishers discovered that their movies might play fine on one card but not on another. The OpenMPEG Consortium was created in an attempt to resolve the problems but did not receive sufficient industry support and was never able to accomplish much.

Some of the members of the OpenMPEG Consortium, by now sadder and wiser, formed the OpenDVD Consortium with more pragmatic goals. No attempt was made to define a standard; and instead, the organization provided a common meeting ground for computer

DVD developers and served as a clearinghouse for information. Unfortunately, the OpenDVD Consortium barely got off the ground before it faded away.

The Interactive Multimedia Association (IMA), with slightly loftier goals, established numerous technical working groups (TWGs) in December of 1996 to deal with issues such as standardized architectures, cross-platform interactive media formats, hybrid Internet-DVD applications, data interchange formats for DVD authoring and production, technical safeguards, and compliance testing. The IMA was later absorbed by the Software Publishers Association (SPA), which lost interest in DVD as it went through its own metamorphoses and name changes. The DVD Association (DVDA), established in 1999, also formed working groups for standardizing DVD applications and production processes, but never accomplished much.

Microsoft's DirectShow provides the broadest standard for DVD application development, in spite of poor support from computer makers and decoder software developers. Sonic's InterActual DVD application development software, which builds on top of DirectShow but also supports other proprietary decoders, along with other platforms such as Apple Mac OS, has become the de facto standard, partly because it's used by most of the Hollywood studios for PC-enhanced titles.

The Changing Face of Home Entertainment

The year 1996 presaged the digital revolution of home entertainment. Direct broadcast digital satellite (DBS) systems were introduced and purchased by the millions. DVD was unveiled, though it did not manage to appear anywhere but in Japan before the end of the year. In December of 1996, the FCC approved the US DTV standard[8] that ushered in 1997, the Year of Waiting for HDTV.[9]

DBS and DVD each have one foot in the past and one foot in the future. They use digital signal recording and compression methods to squeeze the most quality into their limited transmission and storage capacities, but they convert the signal to analog format for display on conventional televisions. As digital televisions slowly push their creaky predecessors out of living rooms and offices, the pure digital signal of DVD and similar technologies will be ready and waiting for them. Video from existing DVDs and players will look better than ever on widescreen digital televisions. New-generation DVD players will enhance venerable NSTC or PAL video by generating progressive-scan video, upscaling it, and supplying it via digital connectors, leaving behind the flicker and flutter of interlaced video and enabling television personalities everywhere to wear pinstriped fabrics and sit in front of miniblinds.

As digital television spreads, we'll witness the computerization of television. This does not mean that every television will sprout a keyboard, a mouse, and system crash error messages.

---

[8]DTV is a set of broadcast standards for digital HDTV transmission in the US. Other countries have developed different HDTV standards, such as DVB in Europe.

[9]Followed by 1998, the Year No One Could Afford HDTV. Followed by 1999, the Year No One Paid Much Attention to HDTV. Followed by 2000, the Year of Wondering If the DTV Format Would Be Changed. Followed by 2001, the Year People Finally Started Buying HDTVs.

It does mean, however, that as consumer electronics companies and home computer makers endeavor to make their products more suitable to the basic entertainment needs, communication needs, and productivity needs of home users, features from computers such as onscreen menus, pointers, multiple windows, and even an Internet connection will become commonplace. DVD is the perfect format for this environment because it is able to carry content for both the television, the computer, and all the variations in between.

In the meantime, the traditional television is getting better with each passing year. The term "big-screen TV" applies to ever-larger sizes. Thirty-five-inch sets are now the price of 25 inch sets a few years ago, and of 19 inch sets a few years before that. Video projectors are at the magic $2000 price point, where they can compete with direct-view and rear-projection big-screen TVs. Widescreen popularity in Europe and in Japan, where over 60 percent of new television purchases are widescreen, is helping to drive prices down. The biggest impediment was a lack of movies and other video sources in widescreen format, but DVD has filled this hole with its native support for wide aspect ratio displays.

Convergence will happen with systems other than the computer and TV. The prime candidates are cable settop boxes, satellite receivers, video game consoles, Internet TV boxes, telephones, cellular phones, hand-held computers, and, of course, DVD players (see Figure 17.2). Engineers are designing new boxes that combine the features of digital satellite receivers and DVD players, since they share many of the same MPEG-2 digital video components. Digital cable boxes will combine the features of cable TV tuners and DVD players and will top it off with an Internet connection. Hard-disk based recorders are selling well and will more commonly add DVD recording capabilities once prices dip low enough. Program listings from the Internet can be shown on the TV screen, for instance, so that shows can be marked for recording with the click of a remote control, and the system automatically compensates for delays caused by ball games or newsbreaks. Video game consoles include DVD drives for game play, giving them the capability to double as DVD movie players.

In short, because of the flexibility and interconnectability of inexpensive digital electronics, all the independent pieces we have grown accustomed to can be tossed in a big hat, shaken around, and pulled back out as new, mixed-and-matched systems with more features at lower prices. Customers will benefit from, and soon come to expect, personalization, freedom of location, and freedom of time. Our electronic devices will enable us to access anything, anytime, in the way we want.

Even movie theaters are being changed by digital technology. Digital cinema trials have shown that film can and should be replaced by digital files. The main issue is how to deploy a new system and who will pay for it. Digital movie files are often delivered on DVD-ROM discs but will eventually be sent over Internet connections. International release dates have been steadily getting closer to US release dates as the old model of movie distribution has been forced to change by Internet advertising that hits all audiences simultaneously. Movies that used to take six months to creep into theaters around the world now make a worldwide release in only one or two months.

Figure 17.2 Media Convergence in the Digital Age

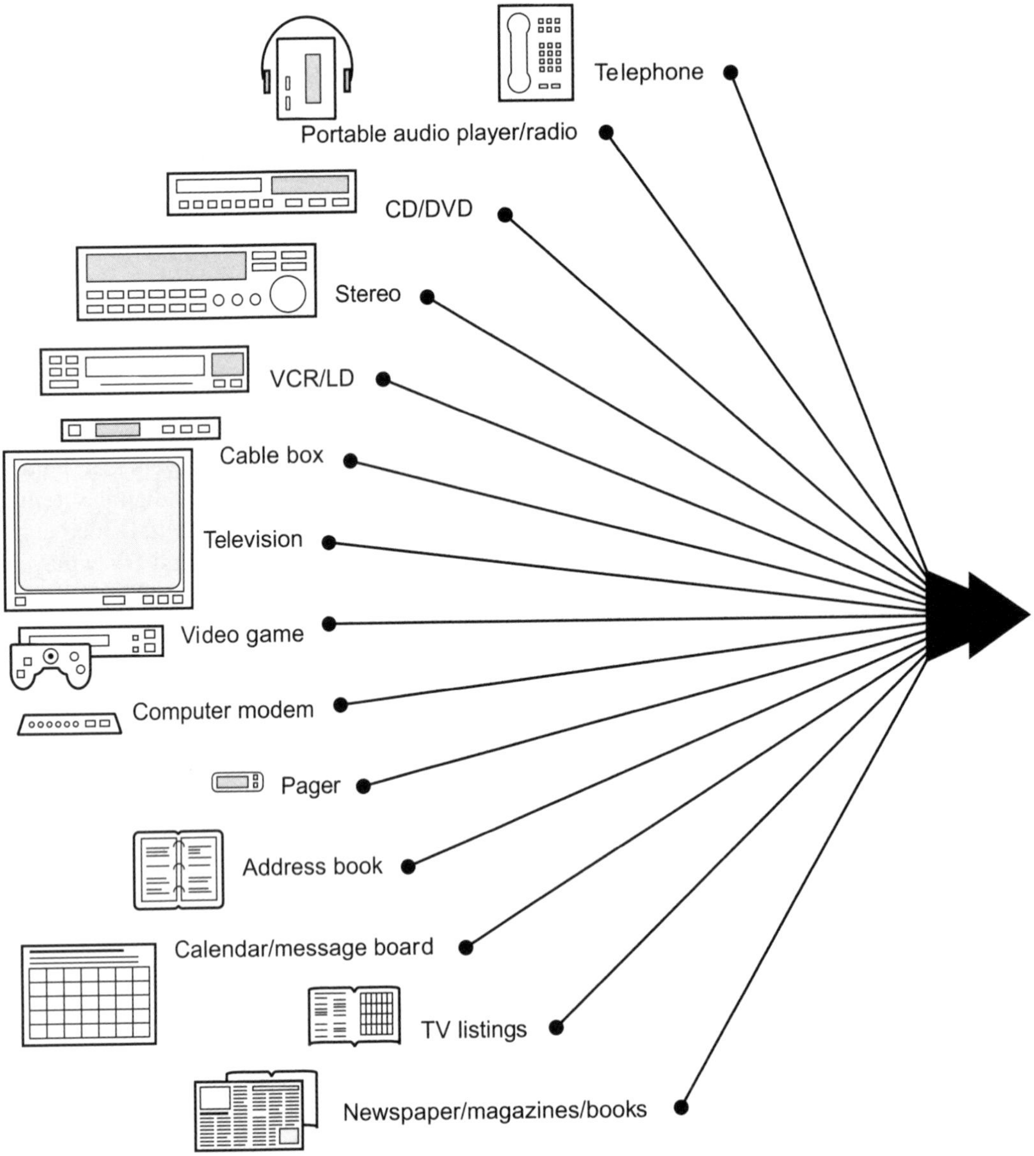

Telephone

Portable audio player/radio

CD/DVD

Stereo

VCR/LD

Cable box

Television

Video game

Computer modem

Pager

Address book

Calendar/message board

TV listings

Newspaper/magazines/books

continues

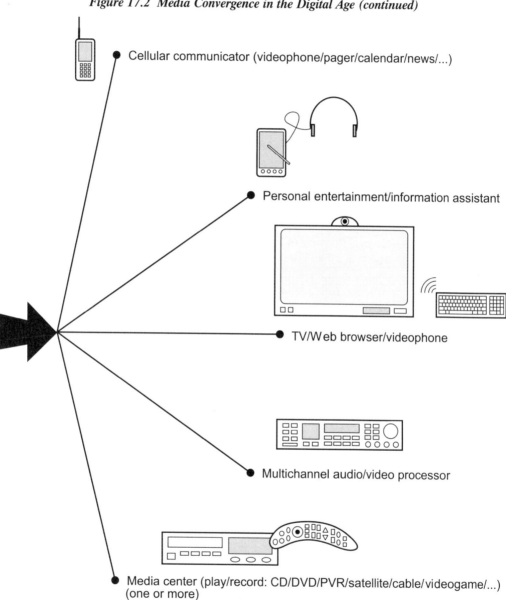

Figure 17.2 Media Convergence in the Digital Age (continued)

Cellular communicator (videophone/pager/calendar/news/...)

Personal entertainment/information assistant

TV/Web browser/videophone

Multichannel audio/video processor

Media center (play/record: CD/DVD/PVR/satellite/cable/videogame/...)
(one or more)

The Far Horizon

In 2004 two very rich men predicted the imminent death of DVD. In July 2004, speaking to Germany's *Bild* magazine, Bill Gates said, "The entertainment of the future will definitely not be on a DVD player, that technology will be completely gone within 10 years at the most." And, in an August 2004 Internet blog, Mark Cuban said, "Which is the better way to deliver a movie or movies? On a DVD with a boring, lifeless future, or hard drives?" Perhaps too much money puts one out of touch with reality, because they're both wrong. Bill Gates is not wrong about the eventual disappearance of packaged media, but he is wrong about the timeframe.[10]

If you look back at predictions of the death of CD from Internet-delivered songs, it's easy to see how people get carried away by new technology and new paradigms, underestimating the inertia of established technologies. Five years after Napster launched the music download craze, online services accounted for less than 5 percent of music sales and CDs were still going strong. Eventually the demand for physical media will decline, but it will require at least an astonishing increase in bandwidth from the completely insufficient typical rate of 2 or 3 Mbps, and it may require a new generation of consumers who are conditioned to be comfortable with ephemeral delivery of content rather than ownership of tangible property. This might happen as soon as 2020, but certainly not before 2010 or even 2015, as some people predict.

In the very long term, the Internet will merge with cable TV, broadcast TV, radio, telephones, satellites, and eventually even newspapers and magazines. The Internet will take over the communications world. News, movies, music, advertising, education, games, financial transactions, e-mail, and most other forms of information will be delivered via this giant network. Internet bandwidth, currently lagging far behind the load being demanded of it, will eventually catch up, as did other systems, such as intercontinental telephone networks and communications satellites. Discrete media such as DVD will then be relegated to niches such as software backups, archiving, time-shift "taping," and collector's editions of movies. Why go to a software store to buy a DVD-ROM or to a movie rental store to rent a DVD when you can have it delivered right to your computer or your TV or even your digital video recorder? In the intervening years, however, DVD in all its permutations and generations promises to be the definitive medium for both computers and home entertainment.

---

[10]Whereas, Cuban seems to ignore the fact that millions of copies of DVDs can be mass produced for about 50 cents each, and even cheap hard disks would cost hundreds of times more. Even putting hundreds of movies on a single hard drive would never make it economical to ship the drives to thousands or millions of customers.

Appendix A - Reference Data

Figure A.1 DVD-Video Conversion Formulas

Formula		Example
Time (hours) =	$\dfrac{\text{space (G bytes)} \times 8000}{\text{data rate (Mbps)} \times 3600}$	$\dfrac{4.7}{5.5} \times 2.22 = 1.9$ hours
Data rate (Mbps) =	$\dfrac{\text{space (G bytes)} \times 8000}{\text{time (hours)} \times 3600}$	$\dfrac{8.5}{3} \times 2.22 = 6.3$ Mbps
Space (G bytes) =	$\dfrac{\text{data rate (Mbps)} \times \text{time (hours)} \times 3600}{8000}$	$5 \times 2 \times 0.45 = 4.5$ G bytes

Note: For gigabytes (instead of billions of bytes) replace 8000 with 8590 (or replace 2.22 with 2.39, and 0.45 with 0.42)

Figure A.2 Data Rate vs. DVD Playing Time

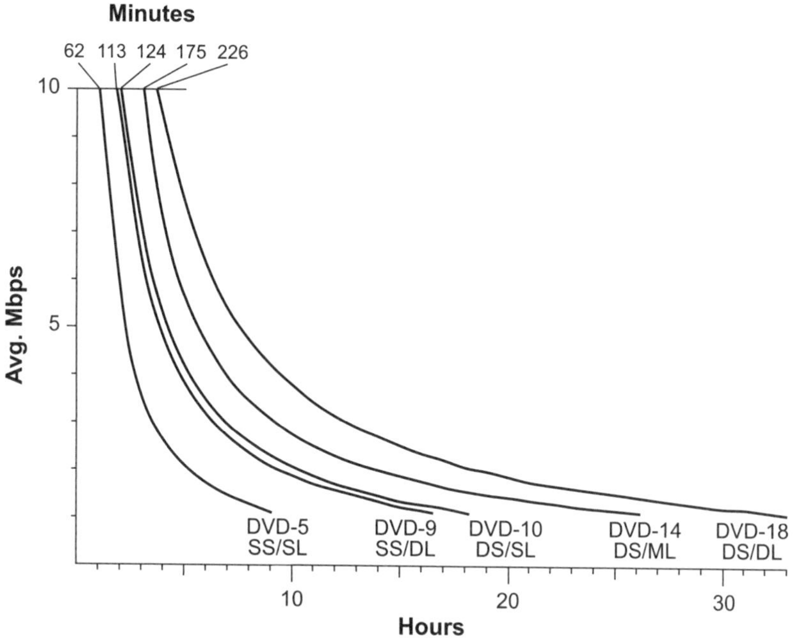

Figure A.3 Relative Sizes of DVD-Video Data Streams

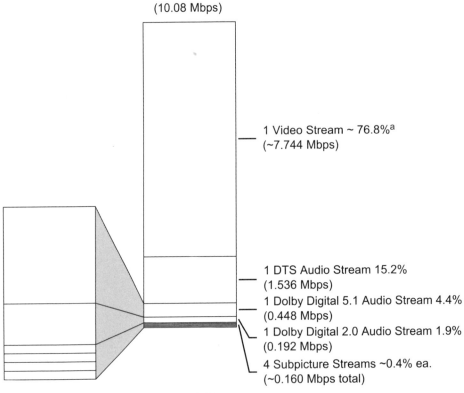

Maximum data rate
(10.08 Mbps)

1 Video Stream ~ 76.8%[a]
(~7.744 Mbps)

1 DTS Audio Stream 15.2%
(1.536 Mbps)

1 Dolby Digital 5.1 Audio Stream 4.4%
(0.448 Mbps)

1 Dolby Digital 2.0 Audio Stream 1.9%
(0.192 Mbps)

4 Subpicture Streams ~0.4% ea.
(~0.160 Mbps total)

[a] Represents the peak video data rate available.
Average VBR video data will be lower.

Figure A.4 Data Rate vs. DVD / HD DVD / BD / UMD Capacities

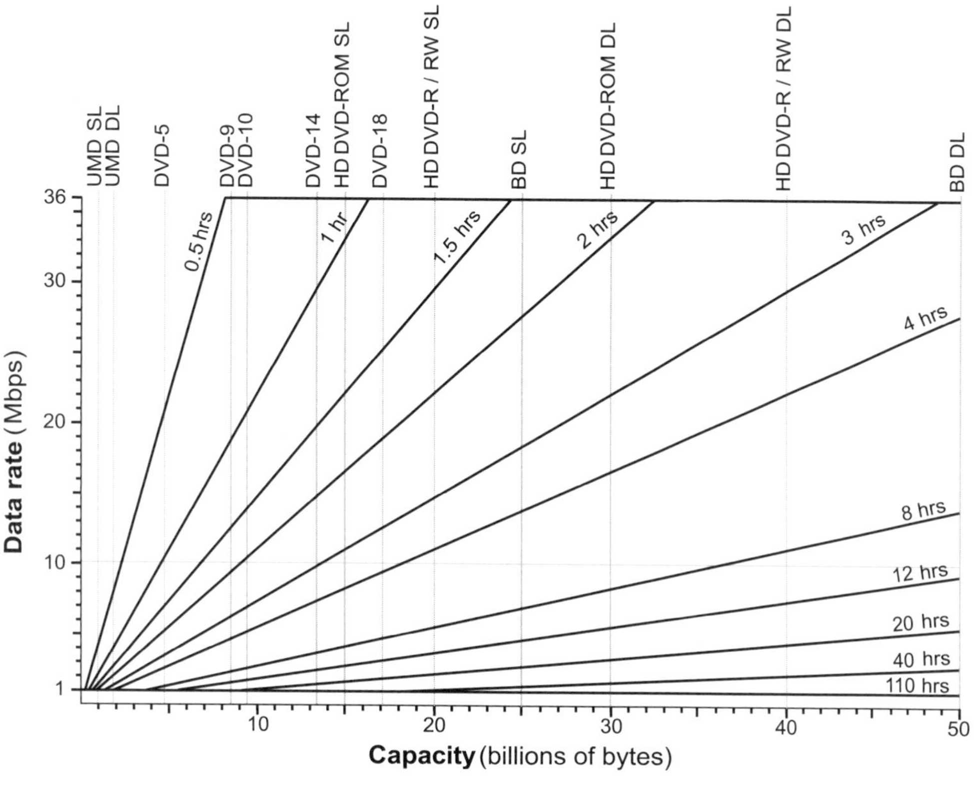

Table A.1 ISO 3116 Country Codes and DVD Regions

Country	ISO codes			DVD region
Afghanistan	AF	AFG	004	5
Albania	AL	ALB	008	2
Algeria	DZ	DZA	012	5
American Samoa	AS	ASM	016	1
Andorra	AD	AND	020	2
Angola	AO	AGO	024	5
Anguilla	AI	AIA	660	4
Antarctica	AQ	ATA	010	?
Antigua and Barbuda	AG	ATG	028	4
Argentina	AR	ARG	032	4
Armenia	AM	ARM	051	5
Aruba	AW	ABW	533	4
Australia	AU	AUS	036	4
Austria	AT	AUT	040	2
Azerbaijan	AZ	AZE	031	5
Bahamas	BS	BHS	044	4
Bahrain	BH	BHR	048	2
Bangladesh	BD	BGD	050	5
Barbados	BB	BRB	052	4
Belarus	BY	BLR	112	5
Belgium	BE	BEL	056	2
Belize	BZ	BLZ	084	4
Benin	BJ	BEN	204	5
Bermuda	BM	BMU	060	1
Bhutan	BT	BTN	064	5
Bolivia	BO	BOL	068	4
Bosnia and Herzegovina	BA	BIH	070	2
Botswana	BW	BWA	072	5
Bouvet Island	BV	BVT	074	?
Brazil	BR	BRA	076	4
British Indian Ocean Territory	IO	IOT	086	5

continues

Table A.1 ISO 3116 Country Codes and DVD Regions (continued)

Country	ISO codes			DVD region
Brunei Darussalam	BN	BRN	096	3
Bulgaria	BG	BGR	100	2
Burkina Faso	BF	BFA	854	5
Burundi	BI	BDI	108	5
Cambodia	KH	KHM	116	3
Cameroon	CM	CMR	120	5
Canada	CA	CAN	124	1
Cape Verde	CV	CPV	132	5
Cayman Islands	KY	CYM	136	4
Central African Republic	CF	CAF	140	5
Chad	TD	TCD	148	5
Chile	CL	CHL	152	4
China	CN	CHN	156	6
Christmas Island	CX	CXR	162	4
Cocos (Keeling) Islands	CC	CCK	166	4
Colombia	CO	COL	170	4
Comoros	KM	COM	174	5
Congo	CG	COG	178	5
Congo, the Democratic Republic of the	CD	COD	180	5
Cook Islands	CK	COK	184	4
Costa Rica	CR	CRI	188	4
Cote D'Ivoire	CI	CIV	384	5
Croatia (Hrvatska)	HR	HRV	191	2
Cuba	CU	CUB	192	4
Cyprus	CY	CYP	196	2
Czech Republic	CZ	CZE	203	2
Denmark	DK	DNK	208	2
Djibouti	DJ	DJI	262	5
Dominica	DM	DMA	212	4
Dominican Republic	DO	DOM	214	4
East Timor	TP	TMP	626	3

continues

Table A.1 ISO 3116 Country Codes and DVD Regions (continued)

Country	ISO codes			DVD region
Ecuador	EC	ECU	218	4
Egypt	EG	EGY	818	2
El Salvador	SV	SLV	222	4
Equatorial Guinea	GQ	GNQ	226	5
Eritrea	ER	ERI	232	5
Estonia	EE	EST	233	5
Ethiopia	ET	ETH	231	5
Falkland Islands (Malvinas)	FK	FLK	238	4
Faroe Islands	FO	FRO	234	2
Fiji	FJ	FJI	242	4
Finland	FI	FIN	246	2
France	FR	FRA	250	2
French Guiana	GF	GUF	254	4
French Polynesia	PF	PYF	258	4
French Southern Territories	TF	ATF	260	?
Gabon	GA	GAB	266	5
Gambia	GM	GMB	270	5
Georgia	GE	GEO	268	?
Germany	DE	DEU	276	2
Ghana	GH	GHA	288	5
Gibraltar	GI	GIB	292	2
Greece	GR	GRC	300	2
Greenland	GL	GRL	304	2
Grenada	GD	GRD	308	4
Guadeloupe	GP	GLP	312	4
Guam	GU	GUM	316	4
Guatemala	GT	GTM	320	4
Guinea	GN	GIN	324	5
Guinea-Bissau	GW	GNB	624	5
Guyana	GY	GUY	328	4
Haiti	HT	HTI	332	4

continues

Table A.1 ISO 3116 Country Codes and DVD Regions (continued)

Country	ISO codes			DVD region
Heard and McDonald Islands	HM	HMD	334	4
Holy City (Vatican City State)	VA	VAT	336	2
Honduras	HN	HND	340	4
Hong Kong	HK	HKG	344	3
Hungary	HU	HUN	348	2
Iceland	IS	ISL	352	2
India	IN	IND	356	5
Indonesia	ID	IDN	360	3
Iran, Islamic Republic of	IR	IRN	364	2
Iraq	IQ	IRQ	368	2
Ireland	IE	IRL	372	2
Israel	IL	ISR	376	2
Italy	IT	ITA	380	2
Jamaica	JM	JAM	388	4
Japan	JP	JPN	392	2
Jordan	JO	JOR	400	2
Kazakhstan	KZ	KAZ	398	5
Kenya	KE	KEN	404	5
Kiribati	KI	KIR	296	4
Korea, Democratic People's Republic of	KP	PRK	408	5
Korea, Republic of	KR	KOR	410	3
Kuwait	KW	KWT	414	2
Kyrgyzstan	KG	KGZ	417	5
Lao People's Democratic Republic	LA	LAO	418	3
Latvia	LV	LVA	428	5
Lebanon	LB	LBN	422	2
Lesotho	LS	LSO	426	2
Liberia	LR	LBR	430	5
Libyan Arab Jamahiriya	LY	LBY	434	5
Liechtenstein	LI	LIE	438	2

continues

Table A.1 ISO 3116 Country Codes and DVD Regions (continued)

Country	ISO codes			DVD region
Lithuania	LT	LTU	440	5
Luxembourg	LU	LUX	442	2
Macau	MO	MAC	446	3
Macedonia, the Former Yugoslav Republic of	MK	MKD	807	2
Madagascar	MG	MDG	450	5
Malawi	MW	MWI	454	5
Malaysia	MY	MYS	458	3
Maldives	MV	MDV	462	5
Mali	ML	MLI	466	5
Malta	MT	MLT	470	2
Marshall Islands	MH	MHL	584	4
Martinique	MQ	MTQ	474	4
Mauritania	MR	MRT	478	5
Mauritius	MU	MUS	480	5
Mayotte	YT	MYT	175	5
Mexico	MX	MEX	484	4
Micronesia, Federated States of	FM	FSM	583	4
Moldova, Republic of	MD	MDA	498	5
Monaco	MC	MCO	492	2
Mongolia	MN	MNG	496	5
Montserrat	MS	MSR	500	4
Morocco	MA	MAR	504	5
Mozambique	MZ	MOZ	508	5
Myanmar	MM	MMR	104	3
Namibia	NA	NAM	516	5
Nauru	NR	NRU	520	4
Nepal	NP	NPL	524	5
Netherlands	NL	NLD	528	2
Netherlands Antilles	AN	ANT	530	4
New Caledonia	NC	NCL	540	4

continues

Table A.1 ISO 3116 Country Codes and DVD Regions (continued)

Country	ISO codes			DVD region
New Zealand	NZ	NZL	554	4
Nicaragua	NI	NIC	558	4
Niger	NE	NER	562	5
Nigeria	NG	NGA	566	5
Niue	NU	NIU	570	4
Norfolk Island	NF	NFK	574	4
Northern Mariana Islands	MP	MNP	580	4
Norway	NO	NOR	578	2
Oman	OM	OMN	512	2
Pakistan	PK	PAK	586	5
Palau	PW	PLW	585	4
Panama	PA	PAN	591	4
Papua New Guinea	PG	PNG	598	4
Paraguay	PY	PRY	600	4
Peru	PE	PER	604	4
Philippines	PH	PHL	608	3
Pitcairn	PN	PCN	612	4
Poland	PL	POL	616	2
Portugal	PT	PRT	620	2
Puerto Rico	PR	PRI	630	1
Qatar	QA	QAT	634	2
Reunion	RE	REU	638	5
Romania	RO	ROM	642	2
Russian Federation	RU	RUS	643	5
Rwanda	RW	RWA	646	5
Saint Kitts and Nevis	KN	KNA	659	4
Saint Lucia	LC	LCA	662	4
Saint Vincent and the Grenadines	VC	VCT	670	4
Samoa	WS	WSM	882	4
San Marino	SM	SMR	674	2

continues

Table A.1 ISO 3116 Country Codes and DVD Regions (continued)

Country	ISO codes			DVD region
Sao Tome and Principe	ST	STP	678	5
Saudi Arabia	SA	SAU	682	2
Senegal	SN	SEN	686	5
Seychelles	SC	SYC	690	5
Sierra Leone	SL	SLE	694	5
Singapore	SG	SGP	702	3
Slovakia (Slovak Republic)	SK	SVK	703	2
Slovenia	SI	SVN	705	2
Solomon Islands	SB	SLB	090	4
Somalia	SO	SOM	706	5
South Africa	ZA	ZAF	710	2
South Georgia and the South Sandwich Islands	GS	SGS	239	4
Spain	ES	ESP	724	2
Sri Lanka	LK	LKA	144	5
St. Helena	SH	SHN	654	5
St. Pierre and Miquelon	PM	SPM	666	1
Sudan	SD	SDN	736	5
Suriname	SR	SUR	740	4
Svalbard and Jan Mayen Islands	SJ	SJM	744	2
Swaziland	SZ	SWZ	748	2
Sweden	SE	SWE	752	2
Switzerland	CH	CHE	756	2
Syrian Arab Republic	SY	SYR	760	2
Taiwan	TW	TWN	158	3
Tajikistan	TJ	TJK	762	5
Tanzania, United Republic of	TZ	TZA	834	5
Thailand	TH	THA	764	3
Togo	TG	TGO	768	5
Tokelau	TK	TKL	772	4
Tonga	TO	TON	776	4

continues

Table A.1 ISO 3116 Country Codes and DVD Regions (continued)

Country	ISO codes			DVD region
Trinidad and Tobago	TT	TTO	780	4
Tunisia	TN	TUN	788	5
Turkey	TR	TUR	792	2
Turkmenistan	TM	TKM	795	5
Turks and Caicos Islands	TC	TCA	796	4
Tuvalu	TV	TUV	798	4
Uganda	UG	UGA	800	5
Ukraine	UA	UKR	804	5
United Arab Emirates	AE	ARE	784	2
United Kingdom	GB	GBR	826	2
United States	US	USA	840	1
United States' Minor Outlying Islands	UM	UMI	581	1
Uruguay	UY	URY	858	4
Uzbekistan	UZ	UZB	860	5
Vanuatu	VU	VUT	548	4
Venezuela	VE	VEN	862	4
Vietnam	VN	VNM	704	3
Virgin Islands (British)	VG	VGB	092	4
Virgin Islands (U.S.)	VI	VIR	850	1
Wallis and Futuna Islands	WF	WLF	876	4
Western Sahara	EH	ESH	732	5
Yemen	YE	YEM	887	2
Yugoslavia	YU	YUG	891	2
Zambia	ZM	ZMB	894	5
Zimbabwe	ZW	ZWE	716	5

TABLE A.2 ISO 639 Language Codes

Language	Code	Hex	Dec
Abkhazian	ab	6162	24930
Afar	aa	6161	24929
Afrikaans	af	6166	24934
Albanian	sq	7371	29553
Amharic	am	616D	24941
Arabic	ar	6172	24946
Armenian	hy	6879	26745
Assamese	as	6173	24947
Avestan[1]	ae	6165	24933
Aymara	ay	6179	24953
Azerbaijani	az	617A	24954
Bashkir	ba	6261	25185
Basque	eu	6575	25973
Bengali; Bangla	bn	626E	25198
Bhutani	dz	647A	25722
Bihari	bh	6268	25192
Bislama	bi	6269	25193
Bosnian[1]	bs	6273	25203
Breton	br	6272	25202
Bulgarian	bg	6267	25191
Burmese	my	6D79	28025
Byelorussian	be	6265	25189
Cambodian	km	6B6D	27501
Catalan	ca	6361	25441
Chamorro[1]	ch	6368	25448
Chechen[1]	ce	6365	25445
Chichewa; Nyanja[1]	ny	6E79	28281
Chinese	zh	7A68	31336
Church Slavic[1]	cu	6375	25461
Chuvash[1]	cv	6376	25462
Cornish[1]	kw	6B77	27511
Corsican	co	636F	25455

continues

Table A.2 ISO 639 Language Codes (continued)

Language	Code	Hex	Dec
Croatian	hr	6872	26738
Czech	cs	6373	25459
Danish	da	6461	25697
Dutch	nl	6E6C	28268
English	en	656E	25966
Esperanto	eo	656F	25967
Estonian	et	6574	25972
Faeroese	fo	666F	26223
Fiji	fj	666A	26218
Finnish	fi	6669	26217
French	fr	6672	26226
Frisian	fy	6679	26233
Galician	gl	676C	26476
Georgian	ka	6B61	27489
German	de	6465	25701
Greek	el	656C	25964
Greenlandic	kl	6B6C	27500
Guarani	gn	676E	26478
Gujarati	gu	6775	26485
Hausa	ha	6861	26721
Hebrew[2]	iw	6977	26999
Herero[1]	hz	687A	26746
Hindi	hi	6869	26729
Hiri Motu[1]	ho	686F	26735
Hungarian	hu	6875	26741
Icelandic	is	6973	26995
Indonesian[3]	in	696E	26990
Interlingua	ia	6961	26977
Interlingue	ie	6965	26981
Inuktitut[1]	iu	6975	26997
Inupiak	ik	696B	26987

continues

Appendix A   A-13

Table A.2 ISO 639 Language Codes (continued)

Language	Code	Hex	Dec
Irish	ga	6761	26465
Italian	it	6974	26996
Japanese	ja	6A61	27233
Javanese	jw	6A77	27255
Kannada	kn	6B6E	27502
Kashmiri	ks	6B73	27507
Kazakh	kk	6B6B	27499
Kikuyu[1]	ki	6B69	27497
Kinyarwanda	rw	7277	29303
Kirghiz	ky	6B79	27513
Kirundi	rn	726E	29294
Komi[1]	kv	6B76	27510
Korean	ko	6B6F	27503
Kuanyama[1]	kj	6B6A	27498
Kurdish	ku	6B75	27509
Laothian	lo	6C6F	27759
Latin	la	6C61	27745
Latvian, Lettish	lv	6C76	27766
Letzeburgesch[1]	lb	6C62	27746
Lingala	ln	6C6E	27758
Lithuanian	lt	6C74	27764
Macedonian	mk	6D6B	28011
Malagasy	mg	6D67	28007
Malay	ms	6D73	28019
Malayalam	ml	6D6C	28012
Maltese	mt	6D74	28020
Manx[1]	gv	6776	26486
Maori	mi	6D69	28009
Marathi	mr	6D72	28018
Marshall[1]	mh	6D68	28008
Moldavian	mo	6D6F	28015

continues

Table A.2 ISO 639 Language Codes (continued)

Language	Code	Hex	Dec
Mongolian	mn	6D6E	28014
Nauru	na	6E61	28257
Navajo[1]	nv	6E76	28278
Ndebele, North[1]	nd	6E64	28260
Ndebele, South[1]	nr	6E72	28274
Ndonga[1]	ng	6E67	28263
Nepali	ne	6E65	28261
Northern Sami[1]	se	7365	29541
Norwegian	no	6E6F	28271
Norwegian Bokmål[1]	nb	6E62	28258
Norwegian Nynorsk[1]	nn	6E6E	28270
Occitan, Provençal	oc	6F63	28515
Oriya	or	6F72	28530
Ossetian; Ossetic[1]	os	6F73	28531
Oromo (Afan)	om	6F6D	28525
Pali[1]	pi	7069	28777
Pashto, Pushto	ps	7073	28787
Persian	fa	6661	26209
Polish	pl	706C	28780
Portuguese	pt	7074	28788
Punjabi	pa	7061	28769
Quechua	qu	7175	29045
Rhaeto-Romance	rm	726D	29293
Romanian	ro	726F	29295
Russian	ru	7275	29301
Samoan	sm	736D	29549
Sangro	sg	7367	29543
Sanskrit	sa	7361	29537
Sardinian[1]	sc	7363	29539
Scots Gaelic	gd	6764	26468
Serbian	sr	7372	29554

continues

Table A.2 ISO 639 Language Codes (continued)

Language	Code	Hex	Dec
Serbo-Croatian[4]	sh	7368	29544
Sesotho	st	7374	29556
Setswana	tn	746E	29806
Shona	sn	736E	29550
Sindhi	sd	7364	29540
Singhalese	si	7369	29545
Siswati	ss	7373	29555
Slovak	sk	736B	29547
Slovenian	sl	736C	29548
Somali	so	736F	29551
Spanish	es	6573	25971
Sundanese	su	7375	29557
Swahili	sw	7377	29559
Swedish	sv	7376	29558
Tagalog	tl	746C	29804
Tahitian[1]	ty	7479	29817
Tajik	tg	7467	29799
Tamil	ta	7461	29793
Tatar	tt	7474	29812
Telugu	te	7465	29797
Thai	th	7468	29800
Tibetan	bo	626F	25199
Tigrinya	ti	7469	29801
Tonga	to	746F	29807
Tsonga	ts	7473	29811
Turkish	tr	7472	29810
Turkmen	tk	746B	29803
Twi	tw	7477	29815
Uighur[1]	ug	7567	30055
Ukrainian	uk	756B	30059
Urdu	ur	7572	30066

continues

Table A.2 ISO 639 Language Codes (continued)

Language	Code	Hex	Dec
Uzbek	uz	757A	30074
Vietnamese	vi	7669	30313
Volapuk	vo	766F	30319
Welsh	cy	6379	25465
Wolof	wo	776F	30575
Xhosa	xh	7868	30824
Yiddish[5]	ji	6A69	27241
Yoruba	yo	796F	31087
Zhuang[1]	za	7A61	31329
Zulu	zu	7A75	31349

Note:

The DVD specification refers to ISO 639:1988, which has since been updated. Because the normative reference is to the 1988 version, it is recommended that old codes be used in disc production. It is recommended that players recognize old codes and new codes.

[1]Added after original publication.

[2]Hebrew was changed from iw to he after the original publication.

[3]Indonesian was changed from in to id after the original publication.

[4]Serbo-Croatian was deprecated after the original publication in favor of Bosnian (bs), Croatian (hr), and Serbian (sr).

[5]Yiddish was changed from ji to yi after the original publication.

Table A.3 DVD/CD Capacities

Type	Sides/layers[a]	Billions of bytes[b]	Gigabytes[b]	Typical hours[c]	Min. to max. hours[d]	Typical audio hours[e]	Min. to max. audio hours[f]
12-cm size							
DVD-ROM (DVD-5)	SS/SL	4.7	4.37	2.25	1 to 9	4.5	1.7 to 164
DVD-R(G) 2.0							1.7 to 163
DVD-R(A) 2.0							
DVD-RAM 2.0							
DVD-RW 1.0							
DVD+RW 2.0							
DVD-ROM (DVD-9)	SS/DL	8.54	7.95	4	1.9 to 16.5	8.3	3.1 to 296
DVD-R DL (DVD-R9)							
DVD+R DL (DVD+R9)							
DVD-ROM (DVD-10)	DS/SL	9.4	8.75	4.5	2.1 to 18.1	9.1	3.4 to 326
DVD-R(G) 2.0							
DVD-RAM 2.0							
DVD-RW 1.0							
DVD+RW 2.0							
DVD-ROM (DVD-14)	DS/ML	13.24	12.33	6.25	2.9 to 25.5	12.8	4.8 to 459
DVD-ROM (DVD-18)	DS/DL	17.08	15.91	8	3.8 to 33	16.5	6.2 to 593
DVD-R 1.0	SS/SL	3.95	3.67	1.75	0.9 to 7.6	3.8	1.4 to 137
DVD-RAM 1.0	SS/SL	2.58	2.4	1.25	0.6 to 4.9	2.5	0.9 to 89
DVD-RAM 1.0	DS/SL	5.16	4.8	2.5	1.1 to 9.9	5	1.9 to 179
CD-ROM[g]	SS/SL	0.682	0.635	0.25[g]	0.2 to 1.3	0.7	0.2 to 23
DDCD-ROM	SS/SL	1.36	1.28	0.59	0.3 to 2.6	1.3	0.5 to 47

continues

Table A.3 DVD/CD Capacities (continued)

Type	Sides/layers[a]	Billions of bytes[b]	Gigabytes[b]	Typical hours[c]	Min. to max. hours[d]	Typical audio hours[e]	Min. to max. audio hours[f]
8-cm size							
DVD-ROM	SS/SL	1.46	1.36	0.75	0.3 to 2.8	1.4	0.5 to 50
DVD-ROM	SS/DL	2.65	2.47	1.25	0.6 to 5.1	2.6	1 to 92
DVD-ROM	DS/SL	2.92	2.72	1.5	0.6 to 5.6	2.8	1.1 to 101
DVD-ROM	DS/ML	4.12	3.83	2	0.9 to 7.9	4	1.5 to 143
DVD-ROM	DS/DL	5.31	4.95	2.5	1.2 to 10.2	5.1	1.9 to 184
DVD-RAM 2.0	SS/SL	1.46	1.36	0.75	0.3 to 2.8	1.4	0.5 to 50
DVD-RAM 2.0	DS/SL	2.92	2.72	1.5	0.6 to 5.6	2.8	1.1 to 101
CD-ROM[g]	SS/SL	0.194	0.18	0.07[h]	0 to 0.3	0.2	0.1 to 6
DDCD-ROM	SS/SL	0.388	0.36	0.14[h]	0.1 to 0.7	0.4	0.1 to 13

[a]DVD-14 (and the corresponding eight-cm size) has one layer on one side and two layers on the other.

[b]Reference capacities in billions of bytes (10^9) and gigabytes (2^{30}). Actual capacities can be slightly larger if the track pitch is reduced.

[c]Approximate video playback time, given an average data rate of 4.7 Mbps. Actual playing times can be much longer or shorter (see next column).

[d]Minimum video playback time at the highest data rate of 10.08 Mbps. Maximum playback time at the MPEG-1 data rate of 1.15 Mbps.

[e]Typical audio-only playback time at the two-channel MLP audio rate of 96 kHz and 24 bits (2.3 Mbps).

[f]Minimum audio-only playback time at the highest single-stream PCM audio rate of 6.144 Mbps. Maximum audio-only playback time at the lowest Dolby Digital or MPEG-2 data rate of 64 kbps.

[g]Mode 1, 74 minutes (333,000 sectors) or 21 minutes (94,500 sectors). Audio/video times are for comparison only.

[h]Assuming that the data from the CD is transferred at a typical DVD video data rate, about four times faster than a single-speed CD-ROM drive.

Table A.4 DVD, HD DVD, BD, and CD Characteristics Comparison

	DVD	HD DVD	BD	CD
Thickness	1.2 mm (2×0.6)	1.2 mm (2×0.6)	1.2 mm (0.1+1.1)	1.2 mm
Mass (12 cm)	13 to 20 g	14 to 20 g	12 to 17 g	14 to 33 g
Diameter	120 or 80 mm	120 mm	120 mm	120 or 80 mm
Spindle hole diameter	15 mm	15 mm	15 mm	15 mm
Lead-in diameter	45.2 to 48 mm	46.6 mm	44 to 44.4 mm	46 to 50 mm
Data diameter (12 cm)	48 to 116 mm	48.2 mm	39.8 to 48 mm	50 to 116 mm
Data diameter (8 cm)	48 to 76 mm	-	-	50 to 76 mm
Lead-out diameter	70 to 117 mm	115.78 mm	116 mm	76 to 117 mm
Outer guardband diameter (12 cm)	117 to 120 mm	115.78 mm to 120 mm	116 to 120 mm	117 to 120 mm
Outer guardband diameter (8 cm)	77 to 80 mm	-	-	77 to 80 mm
Reflectivity (full)	45% to 85%	-	35 to 70% (DL),12 to 28% (DL)	70% min.
Readout wavelength	650 or 635 nm	405 nm	405 nm	780 nm
Numerical aperture	0.60	0.65	0.85	0.38 to 0.45
Focus depth	0.47 μm	-	-	1 (±2 μm)
Track pitch	0.74 μm	0.4 μm	0.32 μm	1.6 mm (1.1 μm[a])

continues

Table A.4 DVD, HD DVD, BD and CD Characteristics Comparison (continued)

	DVD	HD DVD	BD	CD
Pit length	0.400 to 1.866 μm (SL), 0.440 to 2.054 μm (DL)[b]	0.204 to 1.020μm	0.149 to 0.695 μm	0.833 to 3.054 μm (1.2 m/s), 0.972 to 3.560 μm (1.4 m/s); [0.623to2.284 μm[a](0.90m/s)]
Pit width	0.3 μm	-	-	0.6 μm
Pit depth	0.16 μm	0.10 μm	0.10 μm	0.11 μm
Data bit length	0.2667 μm(SL), 0.2934 μm(DL)	0.154 μm	0.11175 μm	0.6 μm(1.2 m/s), 0.7 μm(1.4 m/s)
Channel bit length	0.1333 μm(SL), 0.1467 μm(DL)	-	0.745 μm	0.3 μm
Modulation	8/16	ETM, RLL(1,10) (8:12)	1-7 PP[e], RLL(1.7) (2:3)	8/14 (8/17 w/merge bits)
Error correction	RS-PC	RS-PC	LDC and BIS picket	CIRC (CIRC7[a])
Error correction overhead	13%	-	17%	23%/34%[c]
Bit error rate	10^{-15}	-	2×10^{-4}	10^{-14}
Correctable error (1 layer)	6 mm (SL), 6.5 mm (DL)	7.1 mm	7.0 mm	2.5 mm
Speed (rotational)[d]	570 to 1600 rpm	-	-	200 to 500 rpm
Speed (scanning)[d]	3.49 m/s (SL), 3.84 m/s (DL)	-	4.917 m/s	1.2 to 1.4 m/s (0.90 m/s[a])
Channel data rate[d]	26.15625 Mbps	64.8	66.0 Mbps	4.3218 Mbps (8.6436 Mbps[a])
User data rate[d]	11.08 Mbps	36.55	35.965 Mbps	1.41 Mbps/1.23 Mbps[c]
User data: channel data	2048:4836 bytes	2048:3631	2048:3758	2352:7203/2048:7203[c]
Format overhead	136 percent	77%	83%	206 percent/252 percent[c]
Capacity	1.4 to 8.0 GB per side	15 to 30 GB per side	25 to 50 GB per side	0.783/0.635 GB[c]

[a]Double-density CD
[b]SL 5 single layer, DL 5 dual layer
[c]CD-DA / CD-ROM Mode 1.
[d]Reference value for a single-speed drive.
[e]PP=Parity preserve//Prohibit RMTR. RMTR=Repeated minimum transition run length.

Table A.5 Video resolution

Format		VHS (1.33)	VHS (1.78)	VHS (2.35)	LD (1.33)	LD (1.78)	LD (1.85)	LD (2.35)	VCD (1.33)	VCD (1.78)	VCD (2.35)
NTSC	TVL	250	250	250	425	425	425	425	264	264	264
	H pixels	333	333	333	567	567	567	567	352	352	352
	V pixels	480	360	272	480	360	346	272	240	180	136
	Total pixels	159,840	119,880	90,576	272,160	204,120	196,182	154,224	84,480	63,360	47,872
PAL	TVL	240	240	240	450	450	450	450	264	264	264
	H pixels	320	320	320	600	600	600	600	352	352	352
	V pixels	576	432	327	576	432	415	327	288	216	163
	Total pixels	184,320	138,240	104,640	345,600	259,200	249,000	196,200	101,376	76,032	57,376

Format		DVD (1.33/1.78)	DVD (1.85)	DVD (2.35)	DTV3 (1.33)	DTV3 (1.78)	DTV3 (2.35)	DTV4 (1.78)	DTV4 (2.35)
NTSC	TVL	540/405	540/405	540/405	720	720	720	1,080	1,080
	H pixels	720	720	720	1,280	1,280	1,280	1,920	1,920
	V pixels	480	461	363	960	720	545	1080	817
	Total pixels	345,600	331,920	261,360	1,228,800	921,600	697,600	2,073,600	1,568,640
PAL	TVL	540	540	720					
	H pixels	720	720	720					
	V pixels	576	554	436					
	Total pixels	414,720	398,880	313,920					

continues

Table A.5 Video resolution (continued)

Notes:

1. DTV is neither PAL nor NTSC. The values are placed in the NTSC rows for convenience.

2. Wide aspect ratios (1.78 and 2.35) for VHS, LD, and VCD assume a letterboxed picture. For comparison, letterboxed 1.66 aspect ratio resolution is about 7 percent higher than 1.78. Letterbox is also assumed for DVD and DTV at a 2.35 aspect ratio. DVD's native aspect ratio is 1.33; it uses anamorphic mode for 1.78. DTV's native aspect ratio is 1.78.

3. The very rare 1.78 anamorphic LD has the same pixel count as 1.33 LD. Anamorphic LD letterboxed to 2.35 has almost the same pixel count as 1.78 LD (567×363). The almost non-existent 1.78 anamorphic VHS has the same pixel count as 1.33 VHS. Anamorphic VHS letterboxed to 2.35 has almost the same pixel count as 1.78 VHS (333×363). No commercial 2.35 anamorphic format exists and no corresponding stretch mode exists on widescreen TVs.

4. TVL is lines of horizontal resolution per picture height. For analog formats, the value is derived from the actual horizontal pixel count adjusted for the aspect ratio. DVD's horizontal resolution is lower for 1.78 because the pixels are wider. Pixels for VHS and LD are approximations based on TVL and scan lines.

5. Resolutions refer to the medium, not the display. If a DVD player performs automatic letterboxing on a 1.85 movie (stored in 1.78), the displayed vertical resolution on a standard 1.33 TV is the same as from a letterboxed LD (360 lines).

Table A.6 Player and Media Compatibility

Disc	DVD-Video Player	DVD-ROM Drive	DVD/LD Player	LD Player	CD Player	CD-ROM Drive	Video CD Player
DVD-Video	yes	depends[1]	yes	no	no	no	no
DVD-ROM[2]	no	yes	no	no	no	no	no
LD	no	no	yes	yes	no	no	no
Audio CD	yes	yes	yes	yes[3]	yes	yes	yes
CD-ROM[4]	no	yes	no	no	no	yes	no
CD-R[5]	few[6]	some[6]	few[6]	yes[3]	yes	yes	yes
CD-RW[5]	yes	yes	yes	no	yes	some[7]	no
CDV	part[8]	part[8]	usually[3]	usually[3]	part[8]	part[8]	part[8]
Video CD	some[9]	depends[1]	some[9]	no	no	depends[1]	yes
Photo CD	no	depends[6,10]	no	no	no	depends[10]	no
CD-i	no	depends[11]	no	no	no	depends[11]	no

[1]Computer requires hardware or software to decode and display audio/video.
[2]DVD-ROM containing data other than standard DVD-Video files.
[3]Most newer LD players can play audio from a CD and both audio and video from a CDV.
[4]CD-ROM containing data other than standard CD digital audio.
[5]CD-R/RW containing CD digital audio data.
[6]DVD units require an additional laser tuned for CD-R readout wavelength.
[7]Only MultiRead CD-ROM drives can read CD-RW discs.
[8]CD digital audio part of disc only (no video).
[9]Not all DVD players can play Video CDs.
[10]Computer requires software to read and display Photo CD graphic files.
[11]Computer requires hardware or emulation software to run CD-i programs.

Table A.7 Comparison of MMCD, SD, and DVD[a]

	MMCD	SD	DVD
Diameter	120 mm	120 mm	120 mm
Thickness	1.2 mm	2×0.6 mm	2×0.6 mm
Sides	1	1 or 2	1 or 2
Layers	1 or 2	1 or 2	1 or 2
Data area (diameter)	46 to 116 mm	48 to 116 mm	48 to 116 mm
Min. pit length	0.451 μm	0.400 μm	0.400 μm
Track pitch	0.84 μm	0.74 μm	0.74 μm
Scanning velocity	4.0 m/s	3.27 m/s	3.49 m/s
Laser wavelength	635 nm	650 nm	650 or 635 nm
Numerical aperture	0.52	0.60	0.60
Modulation	8/16	8/15	8/16
Channel data rate	26.6 Mbps	24.54 Mbps	26.16 Mbps
Max. User data rate	11.2 Mbps	10.08 Mbps	11.08 Mbps
Avg. User data rate	3.7 Mbps	4.7 Mbps	4.7 Mbps
Capacity (single layer)	3.7 G bytes	5.0 G bytes	4.7 G bytes
Capacity (dual layer)	7.4 G bytes	9.0 G bytes	8.54 G bytes
Sector size	2048 bytes	2048 bytes	2048 bytes
Error correction	CIRC+	RS-PC	RS-PC
Stated playing time	135 minutes	140 minutes	133 minutes
Video encoding	MPEG-2 VBR	MPEG-2 VBR	MPEG-2 VBR
Audio encoding	MPEG-2 Layer II	AC-3	AC-3, MPEG-2, PCM, etc.

[a]MMCD and SC were precursor formats to DVD. See Chapter 1.

Table A.8 Architectural Layers of BD and HD DVD

Book	BD	HD DVD
Content protection	AACS + SPDC	AACS
Interactive application	BD-J	iHD
Movie application	HDMV	HD DVD-Video
Recording application	BDAV	HD DVD-VR
Codecs	MPEG-2, VC-1, AVC, PCM,	
File system	UDF 2.5/2.6	UDF 2.5
Data storage	BD modulation	PRML, ETM
Physical	BD-R/RE/ROM	DVD, HD DVD-ROM/R/RW

Table A.9 DVD Stream Data Rates

	Minimum (Mbps)	Typical (Mbps)	Maximum (Mbps)
Video			
MPEG-2 video	1.500[a]	4.500	9.800
MPEG-1 video	0.900[a]	1.150	1.856
Audio			
PCM (DVD-Video)	0.768	1.536	6.144
MLP/PCM (DVD-Audio)	n/a	6.900	9.600
Dolby Digital 1.0	0.064	0.096	0.448
Dolby Digital 2.0	0.128	0.192	0.448
Dolby Digital 5.1	0.256	0.384 or 0.448	0.448
DTS 5.1	0.768	0.768 or 1.509	1.524
MPEG-1 audio	0.064	0.192	0.384
MPEG-2 audio	0.064	0.384	0.912
Subpicture	n/a	0.040	3.360

[a]Not an absolute limit but a practical limit below which video quality is too poor

Table A.10 Stream Data Rates for Next-Generation Technologies

	Rates for UMD (Mbps)			Rates for HD DVD (Mbps)			Rates for BD (Mbps)		
	Min.	Typical	Max.	Min.	Typical	Max.	Min.	Typical	Max.
Video									
MPEG-4 AVC video[a]	0.50[b]	1.50	10.00	4.00[b]	14.00	29.40	4.00[b]	16.00	40.00
SMPTE VC-1 video				4.00[b]	16.00	29.40	4.00[b]	18.00	40.00
MPEG-2 video				8.00[b]	24.00	29.40	8.00[b]	24.00	40.00
Audio									
ATRAC3 2.0 audio	0.096	0.128	0.128						
Dolby Digital Plus 1.0 audio[c]				0.032	0.064	0.504	0.032	0.064	0.512
Dolby Digital Plus 2.0 audio[c]				0.032	0.128	0.504	0.032	0.128	0.512
Dolby Digital 5.1 audio				0.384	0.448	0.448	0.384	0.448	0.640
Dolby Digital Plus 5.1 audio				0.256	0.448	3.024	n/a	n/a	n/a
Dolby Digital Plus 7.1 audio				0.512	1.024	3.024	0.640	1.024	1.664
DTS 5.1 audio[d]				0.768	1.509	1.524	0.768	1.509	1.524
DTS-HD 2.0 (LBR)[c]				0.048	0.064	0.384	0.048	0.064	0.256
DTS-HD 5.1 (LBR)[c]				n/a	n/a	n/a	0.192	0.256	0.512
DTS-HD 7.1 (CBR) audio				1.509	1.509	3.000	1.509	1.509	3.000
DTS-HD 5.1 Lossless 24-bit audio				0.800[b]	4.000[e]	18.00	0.800[b]	4.000[e]	24.50
Dolby TrueHD 5.1 Lossless audio				n/a	3.900[e]	18.00	0.800[b]	3.900[e,f]	18.00

[a]For UMD, video refers to primary video at SD resolution. For HD DVD and BD, video refers to primary video at HD resolution.

[b]Not an absolute limit but a practical limit below which presentation quality is too poor

[c]HD DVD and BD both support streamed secondary audio at low data rates, as shown here.

[d]DTS core rates now include 768, 960, 1152, 1344 and 1509 kbps.

[e]Note that the compressed rate of DTS-HD and Dolby TrueHD is heavily dependent on the source material. Values shown here represent movie source material, which can typically be more highly compressed than music.

[f]For BD discs, Dolby TrueHD streams are required to be accompanied by an additional Dolby Digital audio stream (typically 448 kbps), which is not reflected here.

Appendix B
Standards Related to DVD

DVD is based on or has borrowed from dozens of standards developed over the years by many organizations. Most of the standards in this appendix are listed as normative references in the DVD format specification books.

Physical and Device Interface Standards

Disc Format:

- *ECMA 267: 120 mm DVD - Read-Only Disc* (DVD-ROM part 1)
  — Equivalent to *ISO/IEC 16448*

- *ECMA 268: 80 mm DVD - Read-Only Disc* (DVD-ROM part 1)
  — Equivalent to *ISO/IEC 16449*

- *ECMA-279: 80 mm (1.23 Gbytes per side) and 120 mm (3.95 Gbytes per side) DVD-Recordable Disc* (DVD-R 1.0)
  — Equivalent to *ISO/IEC 20563*

- *ECMA-272: 120 mm DVD Rewritable Disc* (DVD-RAM 1.0)
  — Equivalent to *ISO/IEC 16824*

- *ECMA-273: Case for 120 mm DVD-RAM Discs*
  — Equivalent to *ISO/IEC 16825*

- *ECMA-274: Data Interchange on 120 mm Optical Disc using +RW Format - Capacity: 3.0 Gbytes and 6.0 Gbytes* (DVD+RW 1.0)
  — Equivalent to *ISO/IEC 16969*

- *ECMA-330: 120 mm (4.7 Gbytes per side) and 80 mm (1.46 Gbytes per side) DVD Rewritable Disc* (DVD-RAM 2.0)
  — Equivalent to *ISO/IEC 17592*

- *ECMA-331: Cases for 120 mm and 80 mm DVD-RAM Discs*
  — Equivalent to *ISO/IEC 17594*

- *ECMA-337: Data Interchange on 120 mm and 80 mm - Optical Disc using +RW Format - Capacity: 4,7 and 1,46 Gbytes per side* (DVD+RW 2.0)
  — Equivalent to *ISO/IEC 17341*

- *ECMA-338: 80 mm (1.46 Gbytes per side) and 120 mm (4.70 Gbytes per side) DVD Re-recordable Disc* (DVD-RW)
  — Equivalent to *ISO/IEC 17342*

- *ECMA-349: Data Interchange on 120 mm and 80 mm Optical Disc using +R Format - Capacity: 4.7 and 1.46 Gbytes per Side* (DVD+R)

Device Interface:

- *INF 8090 ATAPI/SCSI* (MMC3 [Draft ANSI NCITS T10 standard]; Mt. Fuji; SFF 8090i)
- *ANSI X3.131-1994: Information Systems-Small Computer Systems Interface-2* (SCSI-2)
- *ANSI X3.277-1996: Information Technology-SCSI-3 Fast-20*
- *ANSI X3.221-1994: Information Systems-AT Attachment Interface for Disk Drives* (ATA/IDE)
- *ANSI X3.279-1996: Information Technology-AT Attachment Interface with Extensions (ATA-2)* (EIDE)
- *ANSI X3.298-1997: Information Technology - AT Attachment-3 Interface (ATA-3)*
- *ANSI NCITS 317-1998: AT Attachment with Packet Interface Extension (ATA/ATAPI-4)* (EIDE; Ultra DMA 33)
- *ANSI NCITS 340-2000: Information Technology - AT Attachment with Packet Interface - 5 (ATA/ATAPI-5)* (EIDE; Ultra DMA 66)
- *IEC 60856: Prerecorded optical reflective videodisc system (PAL)*
- *IEC 60857: Prerecorded optical reflective videodisc system (NTSC)*
- *IEEE 1394-1995 IEEE Standard for a High Performance Serial Bus* (FireWire)

System Standards

File System:

- *OSTA Universal Disc Format Specification Revision 1.02: 1996* (OSTA UDF Compliant Domain of ISO/IEC 13346:1995 *Volume and file structure of write-once and rewritable media using non-sequential recording for information interchange)* — *ISO/IEC 13346* is equivalent to *ECMA 167*
- *OSTA Universal Disc Format Specification Revision 2.50: 2003* (OSTA UDF Compliant Domain of *ECMA 167 3rd edition*)
- *OSTA Universal Disc Format Specification Revision 2.60: 2005* "OSTA UDF Compliant Domain" of *ECMA 167 3rd edition*
- *ISO 9660:1988 Information processing-Volume and file structure of CD-ROM for information interchange* — Equivalent to *ECMA 119, 2d edition, 1987*
- *ECMA TR/71 DVD Read-Only Disk File System Specifications* (UDF Bridge)
- *Joliet CD-ROM Recording Specification, ISO 9660:1988 Extensions for Unicode* (Microsoft)

MPEG-2 System:

- *ISO/IEC 13818-3:2000 Information technology-Generic coding of moving pictures and associated audio information: Systems* (ITU-T H.222.0) (program streams for DVD and HD DVD, transport streams for BD)

CD:

- *IEC 60908 (1987-09) Compact disc digital audio system* (Red Book)

CD-ROM:

- *ISO/IEC 10149:1995 Information technology-Data interchange on read-only 120 mm optical data discs (CD-ROM)* (Yellow Book) (Note: Equivalent to *ECMA 130, 2nd Edition, June 1996*)
- Philips/Sony Orange Book part-II Recordable Compact Disc System
- Philips/Sony Orange Book part-III Recordable Compact Disc System
- *IEC 61104: Compact Disc Video System, 12 cm* (CDV Single).

Video Standards

MPEG-1 Video:

- *ISO/IEC 11172-2:1993 Information technology-Coding of moving pictures and associated audio for digital storage media at up to about 1.5 Mbit/s-Part 2: Video*

MPEG-2 Video:

- *ISO/IEC 13818-2:1996 Information technology-Generic coding of moving pictures and associated audio information: Video*
  — Equivalent to *ITU-T H.262*
- *SMPTE RP202-2000: Video Alignment for MPEG-2 Coding*

MPEG-4 AVC:

- *ISO/IEC 14496-10:2004 Information technology-Coding of audio-visual objects-Part 10: Advanced Video Coding*
  — Equivalent to *ITU-T H.264: Advanced video coding for generic audiovisual services: 2005-03*

VC-1 :

- *SMPTE 421M VC-1 Compressed Video Bitstream Format and Decoding Process*
- *SMPTE RP227 VC-1 Bitstream Transport Encodings*
- *SMPTE RP 228 VC-1 Decoder and Bitstream Conformance*

Source Video:

- *ITU-R BT.601-5 Studio encoding parameters of digital television for standard 4:3 and widescreen 16:9 aspect ratios*
- *ITU-R BT.709-5 Parameter values for the HDTV* standards for production and international programme exchange (2002-04)* (1080-line video)

NTSC Video:

- *SMPTE 170M-1994 Television-Composite Analog Video Signal-NTSC for Studio Applications*
- *ITU-R BT.470-4 Television Systems*

PAL Video:

- *ITU-R BT.470-4 Television Systems*

DTV:

- *ATSC A/53C (2004)* with Amendment 1 (2004) and Corrigendum 1 (2005)
- *SMPTE 274M-2005 Television - 1920 x 1080 Image Sample Structure, Digital Representation and Digital Timing Reference Sequences for Multiple Picture Rates*
- *SMPTE 293M-1996 Television - 720x483 active line at 59.94 Hz progressive scan production - digital representation*
- *SMPTE 296M-2001 Television - 1280 x 720 Progressive Image Sample Structure - Analog and Digital Representation and Analog Interface*

Additional Video Signals:

- *CEA-708-B Digital Television (DTV) Closed Captioning (1999)*
- *ETS 300 294 Edition 2:1995-12 Television Systems; 625-Line Television: Wide Screen Signaling (WSS)*
- *EN 300 468 (V1.3.1, 1998-02): Digital Video Broadcasting (DVB): Specification for Service Information (SI) in DVB systems*
- *ITU-R BT.1119-1 Widescreen signaling for broadcasting. Signaling for widescreen and other enhanced television parameters*
- *IEC 61880 (1998-01) Video systems (525/60) - Video and accompanied data using the vertical blanking interval - Analogue interface* (CGMS-A; NTSC line 20; PAL/SECAM/YUV line 21)
- *EIA/CEA-608-B Line 21 Data Services* (Closed Captions, XDS, CGMS-A; NTSC line 21; YUV line 21)
- *EIA/IS 702 Copy Generation Management System (Analog)* (CGMS-A data in XDS. Superseded by *EIA/CEA-608-B*
- *EIA-708-B Digital Television (DTV) Closed Captioning (December 1999)*
- *EIA-744-A Transport of Content Advisory Information using Extended Data Service (XDS) (1998)*

- *EIA-770.1-A Analog 525 Line Component Video Interface - Three Channels*
- *EIA-770.2-A   Standard Definition TV Analog Component Video Interface*
- *EIA-770.3-A   High Definition TV Analog Component Video Interface*
- *EIA-775-A   DTV 1394 Interface Specification*
- *EIA/CEA-805 Data Services on the Component Video Interfaces* (CGMS-A for pro-gressive-scan component video, including YPbPr, RGB, and VGA)
- *ETS 300294* (PAL/SECAM/YUV CGMS-A)
- *EIA-608 Recommended Practice For Line 21 Data Service* (NTSC Closed Captions)
- *EIA-746 Transport Of Internet Uniform Resource Locator (URL) Information Using Text-2 (T-2) Service* (TV links; ATVEF triggers)
- *ETS 300 294 Edition 2:1995-12* (Film/camera mode)
- *ITU-R BT.1119-1 Widescreen signaling for broadcasting. Signaling for widescreen and other enhanced television parameters* (PAL CGMS-A)
- *IEC 61880 Video systems (525/60) - Video and accompanied data using the vertical blanking interval - Analogue interface (1998-01)* (NTSC VBI line 20 CGMS-A)
- *IEC 61880-2 Video systems (525/60) - Video and accompanied data using the vertical blanking interval - Analogue interface - Part 2: 525 progressive scan system (2002-09)* (520p line 41 CGMS-A)
- *JEITA CPR-1204 (NTSC widescreen signaling and CGMS-A, progressive-scan; former-ly EIAJ CPX-1204)*
- *SMPTE 259M-1997 Television - 10-Bit 4:2:2 Component and 4fsc Composite Digital Signals - Serial Digital Interface* (SDI)
- *SMPTE 292M-1998 Television - Bit-Serial Digital Interface for High-Definition Television Systems (HD-SDI)*

Audio Standards

Dolby Digital Audio (AC-3):

- *ATSC A/52 1995*
  — Revision B adds *Enhanced AC-3* (Dolby Digital Plus)
- *ETSI TS 102 366 V1.1.1 (2005-02) Digital Audio Compression (AC-3, Enhanced AC-3) Standard*

MPEG-1 Audio:

- *ISO/IEC 11172-3:1993 Information technology-Coding of moving pictures and associ-ated audio for digital storage media at up to about 1,5 Mbit/s-Part 3: Audio*

MPEG-2 Audio Extensions:

- *ISO/IEC 13818-3:1995 Information technology-Generic coding of moving pictures and associated audio information-Part 3: Audio*

Digital Audio Interface:

- *IEC 60958-1: Digital audio interface - Part 1: 1999 - General*
- *IEC 60958-2: Digital audio interface-Part 2: 1997 - Software information delivery mode*
- *IEC 60958-3: Digital audio interface - Part 3: 2000 - Consumer applications* (S/PDIF, "type II" consumer-use version of AES3)
- *IEC 60958-4: Digital audio interface - Part 4: 2000 - Professional applications*
- *AES3-1992: AES Recommended practice for digital audio engineering - Serial transmission format for two-channel linearly represented digital audio data* (formerly AES/EBU; complement of IEC 60958)
- *IEC 61937-1 Interfaces For Non-Linear PCM Encoded Audio Bitstreams Applying IEC 60958 - Part 1: Non-Linear PCM Encoded Audio Bitstreams For Consumer Applications (obsoletes ATSC A/52 Annex B: AC-3 Data Stream in IEC 958 Interface)*
- *EIAJ CO-1201*
- *EIAJ CP-340* (optical digital audio; "Toslink")

Recording Codes:

- *ISO 3901:1986 Documentation-International Standard Recording Code (ISRC)*

Other Standards

Language Codes:

- *ISO 639:1988 Code for the representation of names of languages* (see Table A.1)

Country codes:

- *ISO 3166:1993 Codes for the representation of names of countries* (see Table A.2)

Text information:

- *BIG5 - Institute for Information Industry, "Chinese Coded Character Set in Computer" (March 1984)*
- *GB18030-2000 - Information technology - Chinese ideograms coded character set for information interchange - Extension for the basic set*
- *GB-2312 - "Coding of Chinese Ideogram Set for Information Interchange Basic Set", GB 2312-80*
- *ISO/IEC 646:1991 Information technology-ISO 7-bit coded character set for information interchange*
- *ISO 8859-1:1987 Information processing-8-bit single-byte coded graphic character sets-Part 1: Latin alphabet No. 1*
- *ISO 8859-2:1987 Information processing-8-bit single-byte coded graphic character sets-Part 2: Latin alphabet No. 2*

- *ISO/IEC 2022:1994 Information technology-Character code structure and extension techniques*
- *ISO/IEC 10646-1:1993, Information technology - Universal Multiple Octet Coded Character Set (UCS) - Part 1: Architecture and Basic Multilingual Plane with Amendment 1, 2, 3, 4, 5, 6, and 7*
- *JIS, Shift-JIS*, and others
- *RFC2279 - UTF-8, a transformation format of ISO 10646*
- *RFC2781 - UTF-16, an encoding of ISO 10646*
- *JIS X0208:1997 Appendix 1 (Shift JIS)*
- *KS C 5601-1987 - Korea Industrial Standards Association, "Code for Information Interchange (Hangul and Hanja)," Korean Industrial Standard, 1987, Ref. No. KS C 5601-1987*

Digital A/V Interface:

- *IEC 61883 Standard for Digital Interface for Consumer Electronic Audio/Video Equipment* (transport protocol for IEEE 1394)
- *1394 Trade Association Audio/Video Control Digital Interface Command Set (AV/C)* (control protocol for IEEE 1394).

Interactive applications:

- *ECMA-262 ECMAScript Language Specification (December 1999)*
- *ECMA-327 ECMAScript 3rd Edition Compact Profile (June 2001)*
- *ITU-T.81 Information technology - Digital compression and coding of continuous-tone still images - Requirements and guidelines (1992-09)* (JPEG)
  — Equivalent to *ISO/IEC 10918-1*
- *JPEG File Interchange Format Version 1.02*
- *W3C Recommendation: Extensible Markup Language (XML) 1.1 (3rd Edition, 04 February 2004, edited in place 15 April 2004)*
- *W3C Recommendation: XHTML 1.0 The Extensible HyperText Markup Language (2nd Edition, 26 January 2000, revised 1 August 2002)*

Appendix C
References and Information Sources

For an up-to-date list of references and information sources, plus lists of companies serving the DVD industry, visit dvddemystified.com and the DVD FAQ (dvddemystified.com/dvdfaq.html).

Recommended References

Benson, K. Blair. *Television Engineering Handbook: Featuring HDTV Systems* (revised ed.). McGraw-Hill, 1992. ISBN: 007004788X.

Dunn, Julian. *Sample Clock Jitter and Real-Time Audio over the IEEE1394 High-Performance Serial Bus*. Preprint 4920, 106th AES Convention, Munich, May 1999.

Dunn, Julian, and Ian Dennis. *The Diagnosis and Solution of Jitter-Related Problems in Digital Audio*. Preprint 3868, 96th AES Convention, Amsterdam. February 1994.

Haskell, Barry G., Atul Puri, and Arun N. Netravali. *Digital Video: An Introduction to MPEG-2*. Chapman & Hall, 1996. ISBN: 0412084112.

Jack, Keith. *Video Demystified* (2d ed.). Hightext Publications, 1996. ISBN: 187870723X.

LaBarge, Ralph. *DVD Authoring and Production*. CMP Books, 2001. ISBN: 1-57820-082-2

Mitchell, Joan L., William B. Pennebaker, and Chad E. Fogg. *MPEG Video: Compression Standard*. Chapman & Hall, 1996. ISBN: 0412087715.

Negroponte, Nicholas and Marty Asher. *Being Digital*. Vintage Books, 1996. ISBN: 0679762906.

Pohlmann, Ken C. *Principles of Digital Audio* (3d ed.). McGraw-Hill, 1995. ISBN: 0070504695.

Poynton, Charles A. *Digital Video and HDTV: Pixels, Pictures, and Perception*. John Wiley & Sons, 2001. ISBN: 0471384895.
— *A Technical Introduction to Digital Video*. John Wiley & Sons, 1996. ISBN: 047112253X.

Solari, Stephen J. Digital Video and Audio Compression. McGraw-Hill, 1997. ISBN: 0070595380.

Watkinson, John. *The Art of Digital Audio* (2d ed.) Butterworth-Heinemann, 1994. ISBN: 0240513207.
— *Compression in Video and Audio*. Focal Press, 1995. ISBN: 0240513940.
— *An Introduction to Digital Audio*. Focal Press, 1994. ISBN: 0240513789.

DVD Information and Licensing

General DVD Information
DVD Forum
www.dvdforum.org
Tokyo, Japan
181-3-5777-2881, fax 181-3-5777-2882

DVD Specification and Logo
DVD Format/Logo Licensing Corporation (DVD FLLC)
www.dvdfllc.co.jp
Tokyo, Japan
181-3-5777-2881, fax 181-3-5777-2882

Patent Licensing (DVD: Hitachi/Matsushita/Mitsubishi/Time Warner/Toshiba/Victor Pool)
Toshiba Corporation
DVD Business Promotion and Support
1-1 Shibaura 1-chome,
Minato-ku, Tokyo 105-01
Japan
181-3-3457-2473, fax 181-3-5444-9430

Patent Licensing (DVD: Philips/Pioneer/Sony Pool)
Philips Standards and Licensing
www.licensing.philips.com
Eindhoven, The Netherlands
Fax 131-40-2732113

Patent Licensing (DVD)
Thomson Multimedia
Director Licensing
46 Quai Alphonse Le Gallo
92648 Boulogne Cedex
France
33 1 4186 5284, fax 33 1 4186 5637

Patent Licensing (Optical Disc)
Discovision Associates
2355 Main Street, Suite 200
Irvine, CA 92614
949-660-5000, fax 949-660-1801

Patent Licensing (MPEG)

MPEG LA, LLC
www.mpegla.com
Denver, Colorado
303-331-1880, fax 303-331-1879

Patent Licensing (Dolby Digital and MLP Audio)

Dolby Laboratories Licensing Corporation
www.dolby.com
San Francisco, CA
415-558-0200, fax 415-863-1373

Patent Licensing (CD and DVD Packaging)

Business Development Europe (BDE) (inside EU)
International Standards & Licensing (IS&L) (outside EU)

Copy Protection Licensing

Macrovision Corporation
www.macrovision.com
Sunnyvale, California
408-743-8600, fax 408-743-8610

Copy Protection Licensing

License Management International, LLC (LMI)
Umbrella entity for:
    DVD Copy Control Association, LLC (CCA); licenses CSS
    Digital Transmission Licensing Administrator, LLC (DTLA); licenses DTCP
    4C Entity, LLC; licenses CPPM, CPRM, and 4C/Verance watermark
www.lmicp.com
Morgan Hill, CA
408-776-2014, fax 408-779-9291

Copy Protection Licensing

Digital Content Protection, LLC; licenses HDCP
www.digital-cp.com

Newsletters and Industry Analyses

Adams Media Research
Market research
tomadams@ix.netcom.com
Carmel Valley, CA
408-659-3070, fax 408-659-4330

The CD-Info Company (CDIC)
Industry directories, newsletters, and other publications
www.cd-info.com
Huntsville, AL
205-650-0406, fax 205-882-7393

Cahners In-stat Group
www.instat.com
Newton, MA
617-630-3900

Centris
www.centris.com
Santa Monica, CA
877-723-6874, fax 310-264-8776

Computer Economics
www.computereconomics.com
Carlsbad, CA
800-326-8100, fax 760-431-1126

Corbell Publishing
www.corbell.com
Marina del Rey, CA
310-574-5337, fax 310-574-5383

Dataquest
Market research
www.dataquest.com
San Jose, CA
408-468-8000, fax 408-954-1780

DVD Intelligence
www.dvdintelligence.com

Ernst & Young
www.ey.com

Home Recording Rights Coalition
www.hrrc.org
Washington, DC
800-282-8273

InfoTech
Market research
www.infotechresearch.com
802-763-2097, fax 802-763-2098

International Data Corporation (IDC)
Market research
www.idcresearch.com
Framingham, MA
508-872-8200, fax 508-935-4015

Jon Peddie Associates (JPA)
www.jpa.com
Mill Valley, CA
415-331-6800, fax 415-331-6211

Knowledge Industry Publications, Inc. (KIPI)
Newsletters, magazines, conferences
www.kipinet.com
White Plains, NY
800-800-5474

Market Vision
Market research
www.webcom.com/newmedia
Santa Cruz, CA
408-426-4400, fax 408-426-4411

Paul Kagan Associates
Market research
Carmel, CA
408-624-1536

SIMBA Information Inc.
Market research, newsletters
www.simbanet.com
Wilton, CT 06907
203-358-0234, fax 203-358-5824

Strategy Analytics
Bedfordshire, UK
144 (0)1582 405678, fax: 144 (0)1582 454828

Magazines

Digital Video Magazine
www.dv.com

DVD Report
www.kipinet.com/dvd

eventDV (formerly EMedia)
www.eventdv.net

Medialine News (formerly Replication News)

One to One

Standards Organizations

Audio Engineering Society (AES)/AES Standards Committee (AESSC)
www.aes.org
New York, NY

American National Standards Institute (ANSI)
www.ansi.org
New York, NY

Commission Internationale de l'Éclairage/International Commission on Illumination (CIE)
ciecb@ping.at
Vienna, Austria

Deutsches Institut für Normung/German Institute for Standardization (DIN)
www.din.de
Berlin, Germany

European Telecommunications Standards Institute (ETSI)
www.etsi.fr
Cedex, France

European Broadcasting Union (EBU)
www.ebu.ch

European Computer Manufacturers Association (ECMA)
www.ecma-international.org
Genève, Switzerland

International Electrotechnical Commission (IEC)
www.iec.ch
Genève, Switzerland

International Organization for Standardization (ISO)
www.iso.ch
Genève, Switzerland

International Telecommunication Union (ITU)
www.itu.int
Genève, Switzerland

National Committee for Information Technology Standards (NCITS)
(Formerly the Accredited Standards Committee X3, Information Technology)
www.ncits.org
Washington, DC

Optical Storage Technology Association (OSTA)
www.osta.org
Santa Barbara, CA

Society of Motion Picture & Television Engineers (SMPTE)
www.smpte.org
White Plains, NY

Other Related Organizations

Acoustic Renaissance for Audio (ARA)
www.meridian.co.uk/ara

Business Software Alliance (BSA)
www.bsa.org
Washington, DC

Computer and Business Equipment Manufacturer's Association (CBEMA)
Washington, DC

Consumer Electronics Association (CEA)
CEA represents U.S. manufacturers of audio, video, consumer information, accessories, mobile electronics, and multimedia products.
www.ce.org
Arlington, VA, fax

The DVD Association
www.dvda.org

Electronic Industries Association (EIA)
A trade association representing all facets of electronics manufacturing.
www.eia.org
Arlington, VA

Information Technology Industry Council (ITI)
www.itic.org
Washington, DC

Motion Picture Association of America (MPAA)
The MPAA serves as the advocate of the American motion picture, home video, and television production industries.
www.mpaa.org

Recording Industry Association Of America (RIAA)
www.riaa.com
Washington, DC

SFF (Small Form Factor) Committee
Saratoga, CA

Video Software Dealers Association (VSDA)
www.vsda.org
Encino, CA

Glossary

1080i 1080 lines of interlaced video (540 lines per field). This usually refers to a 1920×1080 resolution in a 1.78 aspect ratio.

1080p 1080 lines of progressive video (1080 lines per frame). This usually refers to a 1920×1080 resolution in a 1.78 aspect ratio.

2-2 pulldown The process of transferring 24-frame-per-second film to video by repeating each film frame as two video fields. (See Chapter 3, Technology Primer, for details.) When 24-fps film is converted via a 2-2 pulldown to 25-fps 625/50 (PAL) video, the film runs four percent faster than normal.

2-3 pulldown The process of converting 24-frame-per-second film to video by repeating one film frame as three fields, and then the next film frame as two fields. (See Chapter 3 for details.)

3-2 pulldown An uncommon variation of 2-3 pulldown, where the first film frame is repeated for three fields instead of two. Most people mean 2-3 pulldown when they say 3-2 pulldown.

4:1:1 The component digital video format with one C_b sample and one C_r sample for every four Y′ samples. This uses 4:1 horizontal downsampling with no vertical downsampling. Chroma is sampled on every line, but only for every four luma pixels (one pixel in a 1×4 grid). This amounts to a subsampling of chroma by a factor of two compared to luma (and by a factor of four for a single C_b or C_r component). DVD uses 4:2:0 sampling, not 4:1:1 sampling.

4:2:0 The component digital video format used by DVD, with one C_b sample and one C_r sample for every four Y′ samples (one pixel in a 2×2 grid). This uses 2:1 horizontal downsampling and 2:1 vertical downsampling. C_b and C_r are sampled on every other line, in between the scan lines, with one set of chroma samples for each two luma samples on a line. This amounts to a subsampling of chroma by a factor of two, compared to luma (and by a factor of four for a single C_b or C_r component).

4:2:2 The component digital video format commonly used for studio recordings, with one C_b sample and one C_r sample for every two Y′ samples (one pixel in a 1×2 grid). This uses 2:1 horizontal downsampling with no vertical downsampling. This allocates the same number of samples to the chroma signal as to the luma signal. The input to MPEG-2 encoders used for DVD is typically in 4:2:2 format, but the video is subsampled to 4:2:0 before being encoded and stored.

4:4:4 A component digital video format for high-end studio recordings, where Y′, C_b, and C_r are sampled equally.

480i 480 lines of interlaced video (240 lines per field). This usually refers to 720×480 (or 704×480) resolution.

480p 480 lines of progressive video (480 lines per frame). 480p60 refers to 60 frames per second, 480p30 refers to 30 frames per second, and 480p24 refers to 24 frames per second (film source). This usually refers to 720×480 (or 704×480) resolution.

4C The four-company entity consisting of IBM, Intel, Matsushita, and Toshiba.

5C The five-company entity that consists of IBM, Intel, Matsushita, Toshiba, and Sony.

525/60 The scanning system of 525 lines per frame and 60 interlaced fields (30 frames) per second. This is used by the NTSC television standard.

625/50 The scanning system of 625 lines per frame and 50 interlaced fields (25 frames) per second. This is used by PAL and SECAM television standards.

720p 720 lines of progressive video (720 lines per frame). This offers a higher definition than standard DVD (480i or 480p). 720p60 refers to 60 frames per second, 720p30 refers to 30 frames per second, and 720p24 refers to 24 frames per second (film source). This usually refers to a 1280×720 resolution in a 1.78 aspect ratio.

8/16 modulation The form of modulation block code used by DVD to store channel data on the disc. See modulation.

A

AAC Advanced audio coder. An audio-encoding standard for MPEG-2 that is not backward-compatible with MPEG-1 audio.

AC Alternating current. An electric current that regularly reverses direction. It has been adopted as a video term for a signal of non-zero frequency. Compare this to DC.

AC-3 The former name of the Dolby Digital audio-coding system, which is still technically referred to as AC-3 in standards documents. AC-3 is the successor to Dolby's AC-1 and AC-2 audio coding techniques.

Access restriction A function that restricts presentation of a Cell when accessed with a Forward/Backward Scan or Search user operation.

access time The time it takes for a drive to access a data track and begin transferring data. In an optical jukebox, the time it takes to locate a specific disk, insert it in an optical drive, and begin transferring data to the host system.

ActiveMovie The former name for Microsoft's DirectShow technology.

ADPCM Adaptive differential pulse code modulation. A compression technique that encodes the difference between one sample and the next. Variations are lossy and lossless.

AES The Audio Engineering Society.

AES/EBU A digital audio signal transmission standard for professional use, defined by the Audio Engineering Society and the European Broadcasting Union. Sony/Philips digital interface (S/P DIF) is the consumer adaptation of this standard.

AGC Automatic gain control. A circuit designed to boost the amplitude of a signal to provide adequate levels for recording. See Macrovision.

aliasing A distortion (artifact) in the reproduction of digital audio or video that results when the signal frequency is more than twice the sampling frequency. The resolution is insufficient to distinguish between alternate reconstructions of the waveform, thus admitting additional noise that was not present in the original signal.

AMGM_VOBS The Video Object Set for Audio Manager Menu.

analog A signal of (theoretically) infinitely variable levels. Compare this to digital.

angle In DVD-Video this is a specific view of a scene, usually recorded from a certain camera angle. Different angles can be chosen while viewing the scene.

angle block A group of up to nine angles, all exactly the same length.

angle cell One scene recorded from a specific point of view.

Anamorphic A widescreen cinema theatrical release film format.

ANSI American National Standards Institute (see Appendix C, References and Information Sources).

AOTT_AOBS Audio Object Set for Audio-Only Title.

apocryphal Of questionable authorship or authenticity; erroneous or fictitious. One of the authors of this book is fond of saying that the oft-cited 133-minute limit of DVD-Video is apocryphal.

application format A specification for storing information in a particular way to enable a particular use.

artifact An unnatural effect not present in the original video or audio, produced by an external agent or action. Artifacts can be caused by many factors, including digital compression, film-to-video transfer, transmission errors, data readout errors, electrical interference, analog signal noise, and analog signal crosstalk. Most artifacts attributed to the digital compression of DVD are in fact from other sources. Digital compression artifacts always occur in the same place and in the same way. Possible MPEG artifacts are mosquitoes, blocking, and video noise.

aspect ratio The width-to-height ratio of an image. A 4:3 aspect ratio means the horizontal size is a third wider than the vertical size. The standard television ratio is 4:3 (or 1.33:1). The widescreen DVD and HTDV aspect ratio is 16:9 (or 1.78:1). Common film aspect ratios are 1.85:1 and 2.35:1. Aspect ratios normalized to a height of one are often abbreviated by leaving off the :1.

Assistant vocal A leading guide function for singing a song, as in the vocal part of Karaoke songs in Karaoke-equipped DVD Video players. It is also called the "guide vocal".

ASV Audio Still Video. A still picture on a DVD-Audio disc.

ASVOBS Audio Still Video Object Set.

ATAPI Advanced Technology Attachment (ATA) Packet Interface. An interface between a computer and its internal peripherals such as DVD-ROM drives. ATAPI provides the command set for controlling devices connected via an IDE interface. ATAPI is part of the Enhanced IDE (E-IDE) interface, also known as ATA-2. ATAPI was extended for use in DVD-ROM drives by the SFF 8090 specification.

ATSC The Advanced Television Systems Committee. In 1978, the Federal Communications Commission (FCC) empaneled the Advisory Committee on Advanced Television Service (ACATS) as an investigatory and advisory committee to develop information that would assist the FCC in establishing an advanced broadcast television (ATV) standard for the U.S. This committee created a subcommittee, the ATSC, to explore the need for and to coordinate development of the documentation of Advanced Television Systems. In 1993, the ATSC recommended that efforts be limited to a digital television system (DTV), and in September 1995 issued its recommendation for a DTV standard, which was approved with the exclusion of compression format constraints (picture resolution, frame rate, and frame sequence).

ATV Advanced television with significantly better video and audio than standard TV. Sometimes used interchangeably with HDTV, but more accurately encompasses any improved television system, including those beyond HDTV. ATV is also sometimes used interchangeably with the final recommended standard of the ATSC, which is more correctly called DTV.

Audio coding mode The method by which audio is digitally encoded, such as PCM, AC3, DTS, MPEG Audio or SDDS.

authoring For DVD-Video, authoring refers to the process of designing, creating, collecting, formatting, and encoding material. For DVD-ROM, authoring usually refers to using a specialized program to produce multimedia software.

autoplay or **automatic playback** A feature of DVD players that automatically begins playback of a disc if so encoded.

Autostop or **automatic picture stop** The function that stops the playback at a preset point in a scene of a Title, and displays a still picture.

B

bandwidth Strictly speaking, this is the range of frequencies (or the difference between the highest and the lowest frequency) carried by a circuit or signal. Loosely speaking, this is the amount of information carried in a signal. Technically, bandwidth does not apply to digital information; the term data rate is more accurate.

BCA Burst cutting area. A circular section near the center of a DVD disc where ID codes and manufacturing information may be inscribed in bar-code format.

birefringence An optical phenomenon where light is transmitted at slightly different speeds depending on the angle of incidence. Also refers to light scattering due to different refractions created by impurities, defects, or stresses within the media substrate.

bit A binary digit. The smallest representation of digital data: zero/one, off/on, no/yes. Eight bits make one byte.

bitmap An image made of a two-dimensional grid of pixels. Each frame of digital video can be considered a bitmap, although some color information is usually shared by more than one pixel.

bit rate The volume of data measured in bits over time. Equivalent to data rate.

bits per pixel The number of bits used to represent the color or intensity of each pixel in a bitmap. One bit enables only two values (black and white), two bits enable four values, and so on. Bits per pixel is also referred to as color depth or bit depth.

bitstream Digital data, usually encoded, that is designed to be processed sequentially and continuously.

bitstream recorder A device capable of recording a stream of digital data, but not necessarily capable of processing the data.

bits per second A unit used to measure bit rate.

BLER Block error rate. A measure of the average number of raw channel errors when reading or writing a disc.

block In video encoding, an 8×8 matrix of pixels or DCT values representing a small chunk of luma or chroma. In DVD MPEG-2 video, a macroblock is made up of six blocks: four luma and two chroma.

blocking A term referring to the occasional blocky appearance of compressed video (an artifact). Blocking is caused when the compression ratio is high enough that the averaging of pixels in 8×8 blocks becomes visible.

Blue Book The document that specifies the CD Extra interactive music CD format. The original CDV specification was also in a blue book. See Enhanced CD.

Book A The document specifying the DVD physical format (DVD-ROM). Finalized in August 1996.

Book B The document specifying the DVD-Video format. Mostly finalized in August 1996.

Book C The document specifying the DVD-Audio format.

Book D The document specifying the DVD record-once format (DVD-R). Finalized in August 1997.

Book E The document specifying the rewritable DVD format (DVD-RAM). Finalized in August 1997.

B picture or **B frame** One of three picture types used in MPEG video. B pictures are bidirectionally predicted, based on both previous and following pictures. B pictures usually use the least number of bits and they do not propagate coding errors because they are not used as a reference by other pictures.

bps Bits per second. A data rate unit.

brightness Defined by the CIE as the attribute of a visual sensation according to which area appears to emit more or less light. Loosely, it is the intensity of an image or pixel, independent of color; that is, its value along the axis from black to white.

buffer A temporary storage space in the memory of a device that helps smooth data flow.

burst A short segment of the color subcarrier in a composite signal that is inserted to help the composite video decoder regenerate the color subcarrier.

B-Y, R-Y The general term for color-difference video signals carrying blue and red color information where the brightness (Y') has been subtracted from the blue and red RGB signals to create B'-Y' and R'-Y' color-difference signals. Refer to Chapter 3, Technology Primer.

Button A rectangular area appearing on the screen of a Title, used to initiate navigation commands. It is denoted by a highlight state.

byte A unit of data or data storage space consisting of eight bits, commonly representing a single character. Digital data storage is usually measured in bytes, kilobytes, megabytes, etc.

C

caption A textual representation of the audio information in a video program. Captions are usually intended for the hearing impaired and therefore include additional text to identify the person speaking, offscreen sounds, and so on.

CAV Constant angular velocity. Refers to rotating disc systems in which the rotation speed is kept constant, where the pickup head travels over a longer surface as it moves away from the center of the disc. The advantage of CAV is that the same amount of information is provided in one rotation of the disc. Contrast with CLV and ZCLV.

C_b, C_r The components of digital color-difference video signals carrying blue and red color information, where the brightness (Y′) has been subtracted from the blue and red RGB signals to create B′-Y′ and R′-Y′ color-difference signals (refer to Chapter 3, Technology Primer).

CBEMA Computer and Business Equipment Manufacturers Association (Refer to Appendix C, References and Information Sources.)

CBR Constant bit rate. Data compressed into a stream with a fixed data rate. The amount of compression (such as quantization) is varied to match the allocated data rate, but as a result, quality may suffer during high-compression periods. In other words, the data rate is held constant, while quality is allowed to vary. Compare this to VBR.

CCI Copy control information. Information specifying if the content is allowed to be copied.

CCIR Rec. 601 A standard for digital video. The CCIR changed its name to ITU-R, and the standard is now properly called ITU-R BT.601.

CD Short for compact disc, an optical disc storage format developed by Philips and Sony.

CD-DA Compact disc digital audio. The original music CD format, storing audio information as digital PCM data. Defined by the Red Book standard.

CD+G Compact disc plus graphics. A CD variation that embeds graphical data in with the audio data, allowing video pictures to be displayed periodically as music is played. Primarily used for karaoke.

CD-i Compact disc interactive. An extension of the CD format designed around a set-top computer that connects to a TV to provide interactive home entertainment, including digital audio and video, video games, and software applications. Defined by the Green Book standard.

CD-Plus A type of Enhanced CD format using stamped multisession technology.

CD-R An extension of the CD format that enables data to be recorded once on a disc by using dye-sublimation technology. It is defined by the Orange Book standard.

CD-ROM Compact disc read-only memory. An extension of the compact disc digital audio (CD-DA) format that enables computer data to be stored in digital format. Defined by the Yellow Book standard.

CD-ROM XA CD-ROM extended architecture. A hybrid CD that enables interleaved audio and video.

CDV A combination of laserdisc and CD that places a section of CD-format audio on the beginning of the disc and a section of laserdisc-format video on the remainder of the disc.

cell In DVD-Video, a unit of video with a duration that is anywhere from a fraction of a second to several hours long. Cells enable the video to be grouped for sharing content among titles, interleaving for multiple angles, etc.

cell block A group of cells.

CEA The Consumer Electronics Association. A subsidiary of the Electronics Industry Association (EIA). (Refer to Appendix C, References and Information Sources.)

CGMS The Copy Guard Management System. A method of preventing copies or controlling the number of sequential copies allowed. CGMS/A is added to an analog signal (such as line 21 of NTSC). CGMS/D is added to a digital signal, such as IEEE 1394.

challenge key Data used in the authentication key exchange process between a DVD-ROM drive and a host computer, where one side determines if the other side contains the necessary authorized keys and algorithms for passing encrypted (scrambled) data.

channel A part of an audio track. Typically, one channel is allocated for each loudspeaker.

channel bit The bits stored on the disc after being modulated.

channel data The bits physically recorded on an optical disc after error-correction encoding and modulation. Because of the extra information and processing, channel data is larger than the user data contained within it.

chapter In DVD-Video, a division of a title. Technically, it is called a part of title (PTT).

Chorus For Karaoke, it is each song number in the text.

chroma (C') The nonlinear color component of a video signal, independent of the luma. It is identified by the symbol C' (where ' indicates nonlinearity), but it is usually written as C because it's never linear in practice.

chroma subsampling Reducing the color resolution by taking fewer color samples than luminance samples. (See 4:1:1 and 4:2:0.)

chrominance (**C**) The color component (hue and saturation) of light, independent of luminance. Technically, chrominance refers to the linear component of video, as opposed to the transformed nonlinear chroma component.

CIE Commission Internationale de l'Éclairage/International Commission on Illumination. (Refer to Appendix C, References and Information Sources.)

CIF The common intermediate format, which is a video resolution of 352×288.

CIRC Cross-interleaved Reed Solomon code. An error-correction coding method that overlaps small frames of data.

clamping area The area near the inner hole of a disc where the drive grips the disc in order to spin it.

closed captions Textual video overlays that are not normally visible, as opposed to open captions, which are a permanent part of the picture. Captions are usually a textual representation of the spoken audio. In the U.S., the official NTSC Closed Caption standard requires that all TVs larger than 13 inches include circuitry to decode and display caption information stored on line 21 of the video signal. DVD-Video can provide closed caption data, but the subpicture format is preferred for its versatility.

CLUT Color lookup table. An index that maps a limited range of color values to a full range of values such as RGB or YUV.

CLV Constant linear velocity. This refers to a rotating disc system in which the head moves over the disc surface at a constant velocity, requiring that the motor vary the rotation speed as the head travels in and out. The further the head is from the center of the disc, the slower the rotation. The advantage of CLV is that data density remains constant, optimizing the use of the surface area. Contrast this with CAV and ZCLV.

CMI Content management information. This is general information about copy protection and the allowed use of protected content. CMI includes CCI.

codec Coder/decoder. The circuitry or computer software that encodes and decodes a signal.

colorburst See burst.

color depth The number of levels of color (usually including luma and chroma) that can be represented by a pixel. It is generally expressed as a number of bits or a number of colors. The color depth of MPEG video in DVD is 24 bits, although the chroma component is shared across four pixels (averaging 12 actual bits per pixel).

color difference A pair of video signals that contain the color components minus the brightness component, usually B'-Y' and R'-Y' (G'-Y' is not used, since it generally carries less information). The color-difference signals for a black-and-white picture are zero. The advantage of color-difference signals is that the color component can be reduced more than the brightness (luma) component without being visually perceptible.

colorist Someone who operates a telecine machine to transfer film to video. Part of the process involves correcting the video color to match the film.

combo drive A DVD-ROM drive capable of reading and writing CD-R and CD-RW media. It may also refer to a DVD-R, DVD-RW, or DVD+RW drive with the same capability. See RAMbo.

component video A video system containing three separate color component signals, either red/green/blue (RGB) or chroma/color difference ($Y'C_bC_r$, $Y'P_bP_r$, YUV), in analog or digital form. The MPEG-2 encoding system used by DVD is based on color-difference component digital video. Very few televisions have component video inputs.

composite video An analog video signal in which the luma and chroma components are combined (by frequency multiplexing), along with sync and burst. This is also called CVBS. Most televisions and VCRs have composite video connectors, which are usually colored yellow.

compression The process of removing redundancies in digital data to reduce the amount that must be stored or transmitted. Lossless compression removes only enough redundancy so that the original data can be recreated exactly as it was. Lossy compression sacrifices additional data to achieve greater compression.

constant data rate or **constant bit rate** See CBR.

contrast The range of brightness between the darkest and lightest elements of an image.

control area A part of the lead-in area on a DVD containing one ECC block (16 sectors) repeated 192 times. The repeated ECC block holds information about the disc.

CPPM Content Protection for Prerecorded Media. Copy protection for DVD-Audio.

CPRM Content Protection for Recordable Media. Copy protection for writable DVD formats.

CPSA Content Protection System Architecture. An overall copy protection design for DVD.

CPTWG Copy Protection Technical Working Group. The industry body responsible for developing or approving DVD copy protection systems.

CPU Central processing unit. The integrated circuit chip that forms the brain of a computer or other electronic device. DVD-Video players contain rudimentary CPUs to provide general control and interactive features.

crop To trim and remove a section of the video picture in order to make it conform to a different shape. Cropping is used in the pan and scan process, but not in the letterbox process.

Cross field A frame displaying two different overlapped images (fields). For video edited directly from film sources converted on a telecine via the 2-3 pulldown method, one frame (two fields) may have two fields containing entirely different picture data.

cursor A mouse pointer or a text insertion point.

CVBS Composite video baseband signal. This is a standard single-wire video, mixing luma and chroma signals together.

D

DAC Digital-to-analog converter. Circuitry that converts digital data (such as audio or video) to analog data.

DAE Digital audio extraction. Reading digital audio data directly from a CD audio disc.

DAT Digital audio tape. A magnetic audio tape format that uses PCM to store digitized audio or digital data.

data area The physical area of a DVD disc between the lead in and the lead out that contains the stored data content of the disc.

data rate The volume of data measured over time. The rate at which digital information can be conveyed. This is usually expressed as bits per second with notations of kbps (thousand/sec), Mbps (million/sec), and Gbps (billion/sec). Digital audio date rate is generally computed as the number of samples per second times the bit size of the sample. For example, the data rate of uncompressed 16-bit, 48-kHz, two-channel audio is 1536 kbps. The digital video bit rate is generally computed as the number of bits per pixel times the number of pixels per line times the number of lines per frame times the number of frames per second. For example, the data rate of a DVD movie before compression is usually $12 \times 720 \times 480 \times 24 = 99.5$ Mbps. Compression reduces the data rate. Digital data rate is sometimes inaccurately equated with bandwidth.

dB See decibel.

DBS Digital broadcast satellite. The general term for 18-inch digital satellite systems.

DC Direct current. The electrical current flowing in one direction only. Adopted in the video world to refer to a signal with zero frequency. Compare this to AC.

DCC Digital compact cassette. A digital audio tape format based on the popular compact cassette that was abandoned by Philips in 1996.

DCT Discrete cosine transform. An invertible, discrete, orthogonal transformation. A mathematical process used in MPEG video encoding to transform blocks of pixel values into blocks of spatial frequency values with lower-frequency components organized into the upper-left corner, allowing the high-frequency components in the lower-right corner to be discounted or discarded. DCT also stands for digital component technology, a videotape format.

DDWG Digital Display Working Group. See DVI.

decibel (dB) A unit of measurement expressing ratios using logarithmic scales related to human aural or visual perception. Many different measurements are based on a reference point of 0 dB, such as a standard level of sound or power.

decimation A form of subsampling that discards existing samples (pixels, in the case of spatial decimation, or pictures, in the case of temporal decimation). The resulting information is reduced in size but may suffer from aliasing.

decode To reverse the transformation process of an encoding method. The process of converting a digitally encoded signal into its original form. Decoding processes are usually deterministic.

decoder 1) A circuit that decodes compressed audio or video, taking an encoded input stream and producing output such as audio or video. DVD players use the decoders to recreate information that was compressed by systems such as MPEG-2 and Dolby Digital; 2) A circuit that converts composite video to component video or matrixed audio to multiple channels.

delta picture or **delta frame** A video picture based on the changes from the picture before (or after) it. MPEG P pictures and B pictures are examples. Contrast this with key picture.

deterministic A process or model in which the outcome does not depend upon chance, and a given input always produces the same output. Audio and video decoding processes are mostly deterministic.

digital Expressed in digits. A set of discrete numeric values, as used by a computer. Analog information can be digitized by sampling.

digital signal processor (DSP) A digital circuit that can be programmed to perform digital data manipulation tasks such as decoding or audio effects.

digital video noise reduction (DVNR) Digitally removing noise from video by comparing frames in sequence to spot temporal aberrations.

Digital zero The case where all digitally encoded audio amplitude data is equal to zero.

digitize To convert analog information to digital information by sampling.

DIN Deutsches Institut für Normung/German Institute for Standardization (Refer to Appendix C, References and Information Sources.)

directory The part of a disc that indicates which files are stored on the disc and where they are located.

DirectShow A software standard developed by Microsoft for the playback of digital video and audio in the Windows operating system. This has replaced the older MCI and Video for Windows software.

DIN The German Institute for Standardization.

Directory The section of a disk that contains information about which files are stored and where.

disc key A value used to encrypt and decrypt (scramble) a title key on DVD-Video discs.

disc menu The main menu of a DVD-Video disc from which titles are selected. This is also called the system menu or title selection menu.

discrete cosine transform See DCT.

discrete surround sound Audio in which each channel is stored and transmitted separate from and independent of other channels. Multiple independent channels, directed to loudspeakers in front of and behind the listener, enable precise control of the soundfield in order to generate localized sounds and simulate moving sound sources.

display rate The number of times per second the image in a video system is refreshed. Progressive scan systems such as film or HDTV change the image once per frame. Interlace scan systems such as standard television change the image twice per frame, with two fields in each frame. Film has a frame rate of 24 fps, but each frame is shown twice by the projector for a display rate of 48 fps. 525/60 (NTSC) television has a rate of 29.97 frames per second (59.94 fields per second). 625/50 (PAL/SECAM) television has a rate of 25 frames per second (50 fields per second).

Divx Digital Video Express. A short-lived pay-per-viewing-period variation of DVD. The term has been resurrected for DivX, which is a proprietary encode/decode scheme for audio/video content and presentation.

DLT Digital linear tape. A digital archive standard using half-inch tapes, commonly used for submitting a premastered DVD disc image to a replication service.

Dolby Digital A perceptual coding system for audio, developed by Dolby Laboratories and accepted as an international standard. Dolby Digital is the most common means of encoding audio for DVD-Video and is the mandatory audio compression system for 525/60 (NTSC) discs.

Dolby Pro Logic The technique (or the circuit that applies the technique) of extracting surround audio channels from a matrix-encoded audio signal. Dolby Pro Logic is a decoding technique only, but it is often mistakenly used to refer to Dolby Surround audio encoding.

Dolby Surround The standard for matrix encoding surround-sound channels in a stereo signal by applying a set of defined mathematical functions when combining center and surround channels with left and right channels. The center and surround channels can then be extracted by a decoder such as a Dolby Pro Logic circuit that applies the inverse of the mathematical functions. A Dolby Surround decoder extracts surround channels, while a Dolby Pro Logic decoder uses additional processing to create a center channel. The process is essentially independent of the recording or transmission format. Both Dolby Digital and MPEG audio compression systems are compatible with Dolby Surround audio.

downmix To convert a multichannel audio track into a two-channel stereo track by combining the channels with the Dolby Surround process. All DVD players are required to provide downmixed audio output from Dolby Digital audio tracks.

downsampling See subsampling.

Double-sided disc A type of DVD disc on which data is recorded on both sides.

drop frame timecode The method of timecode computation that accounts for the reality of there being only 29.97 frames of video per second. The 0.03 frame is visually insignificant, but mathematically very significant. A one-hour video program will have 107,892 frames of video (29.97 frames per second x 60 seconds x 60 minutes). The drop frame timecode method of accommodating reality was developed where 2 frames are dropped from the numerical count for every minute in an hour except for every tenth minute, when no frames are dropped. See also non-drop frame timecode and timecode.

DRC See dynamic range compression.

driver A software component that enables an application to communicate with a hardware device.

DSD Direct Stream Digital. An uncompressed audio bitstream coding method developed by Sony. It is used as an alternative to PCM.

DSI Data search information. Navigation and search information contained in the DVD-Video data stream. DSI and PCI together make up an overhead of about one Mbps.

DSP Digital signal processor (or processing).

DSVCD Double Super Video Compact Disc. A long-playing variation of SVCD.

DTS Digital Theater Sound. A perceptual audio-coding system developed for theaters. A competitor to Dolby Digital and an optional audio track format for DVD-Video and DVD-Audio.

DTS-ES A version of DTS decoding that is compatible with 6.1-channel Dolby Surround EX. DTS-ES Discrete is a variation of DTS encoding and decoding that carries a discrete rear center channel instead of a matrixed channel.

DTV Digital television. In general, any system that encodes video and audio in digital form. Specifically, the Digital Television System proposed by the ATSC or the digital TV standard proposed by the Digital TV Team founded by Microsoft, Intel, and Compaq.

duplication The reproduction of media. This generally refers to producing discs in small quantities, as opposed to large-scale replication.

DV Digital Video. This usually refers to the digital videocassette standard developed by Sony and JVC.

DVB Digital video broadcast. A European standard for broadcast, cable, and digital satellite video transmission.

DVC Digital video cassette. The early name for DV.

DVCAM Sony's proprietary version of DV.

DVCD Double Video Compact Disc. A long-playing (100-minute) variation of VCD.

DVCPro Matsushita's proprietary version of DV.

DVD An acronym that officially stands for nothing but is often expanded as Digital Video Disc or Digital Versatile Disc. The audio/video/data storage system based on 12- and 8-cm optical discs.

DVD-Audio (DVD-A) The audio-only format of DVD that primarily uses PCM audio with MLP encoding, along with an optional subset of DVD-Video features.

DVD-R A version of DVD on which data can be recorded once. It uses dye sublimation recording technology.

DVD-RAM A version of DVD on which data can be recorded more than once. It uses phase-change recording technology.

DVD-ROM The base format of DVD-ROM stands for read-only memory, referring to the fact that standard DVD-ROM and DVD-Video discs can't be recorded on. A DVD-ROM can store essentially any form of digital data.

DVD-Video (DVD-V) A standard for storing and reproducing audio and video on DVD-ROM discs, based on MPEG video, Dolby Digital and MPEG audio, and other proprietary data formats.

DVI Digital Visual Interface. The digital video interface standard developed by the Digital Display Working Group (DDWG). A replacement for analog VGA monitor interface.

DVNR See digital video noise reduction.

DVS Descriptive video services that provide a narration for blind or sight-impaired viewers.

dye polymer The chemical used in DVD-R and CD-R media that darkens when heated by a high-power laser.

dye-sublimation An optical disc recording technology that uses a high-powered laser to burn readable marks into a layer of organic dye. Other recording formats include magneto-optical and phase-change.

dynamic range The difference between the loudest and softest sound in an audio signal. The dynamic range of digital audio is determined by the sample size. Increasing the sample size does not allow louder sounds; it increases the resolution of the signal, thus allowing softer sounds to be separated from the noise floor (and allowing more amplification with less distortion). Therefore, the dynamic range refers to the difference between the maximum level of distortion-free signal and the minimum limit reproducible by the equipment.

dynamic range compression A technique of reducing the range between loud and soft sounds in order to make dialog more audible, especially when listening at low volume levels. It is used in the downmix process of multichannel Dolby Digital sound tracks.

E

EBU European Broadcasting Union. (See Appendix C, References and Information Sources.)

ECC See error-correction code.

ECD Error-detection and correction code. See error-correction code.

ECMA European Computer Manufacturers Association. (See Appendix C, References and Information Sources.)

EDC A short error-detection code applied at the end of a DVD sector.

edge enhancement When films are transferred to video in preparation for DVD encoding, they are commonly run through digital processes that attempt to clean up the picture. These processes include noise reduction (DVNR) and image enhancement. Enhancement increases the contrast (similar to the effect of the sharpen or unsharp mask filters in Photoshop), but it can tend to overdo areas of transition between light and dark or different colors. This causes a chiseled look or a ringing effect like the haloes you see around streetlights when driving in the rain. Video noise reduction is a good thing when done well, because it can remove scratches, spots, and other defects from the original film. Enhancement, which is rarely done well, is a bad thing. The video may look sharper and clearer to the casual observer, but fine tonal details of the original picture are altered and lost.

EDS Enhanced data services. Additional information in the NTSC vertical interval, such as a time signal.

EDTV Enhanced-definition television. A system that uses existing transmission equipment to send an enhanced signal that looks the same on existing receivers, but carries additional information to improve the picture quality on enhanced receivers. PALPlus is an example of EDTV. Contrast this with HDTV and IDTV.

EFM Eight-to-14 modulation. A modulation method used by CD. The 8/16 modulation used by DVD is sometimes called EFM plus.

EIA Electronics Industry Association. (See Appendix C, References and Information Sources.)

E-IDE Enhanced Integrated Drive Electronics. These are extensions to the IDE standard that provide faster data transfers and enable access to larger drives, including CD-ROM and tape drives, using ATAPI. E-IDE was adopted as a standard by ANSI in 1994. ANSI calls it Advanced Technology Attachment-2 (ATA-2) or Fast ATA.

elementary stream A general term for a coded bitstream such as audio or video. Elementary streams are made up of packs of packets.

emulate To test the function of a DVD disc on a computer after formatting a complete disc image.

emphasis Amplifying a given range of recorded frequency signals in order to reproduce the original signal during playback.

encode To transform data for storage or transmission, usually in such a way that redundancies are eliminated or complexity is reduced. Most compression is based on one or more encoding methods. Data such as audio or video is encoded for efficient storage or transmission and is decoded for access or display.

encoder 1) A circuit or a program that encodes audio or video; 2) A circuit that converts component digital video to composite analog video. DVD players include TV encoders to generate standard television signals from decoded video and audio; 3) A circuit that converts multichannel audio to two-channel matrixed audio.

Enhanced CD A music CD that has additional computer software and can be played in a music player or read by a computer. Also called CD Extra, CD Plus, hybrid CD, interactive music CD, mixed-mode CD, pre-gap CD, or track-zero CD.

entropy coding Variable-length, lossless coding of a digital signal to reduce redundancy. MPEG-2, DTS, and Dolby Digital apply entropy coding after the quantization step. MLP also uses entropy coding.

EQ Equalization of audio.

error-correction code Additional information added to data to enable errors to be detected and possibly corrected. (See Chapter 3, Technology Primer.)

ETSI European Telecommunications Standards Institute. (See Appendix C, References and Information Sources.)

export To convert computer data to another form of data or media.

F

father The metal master disc formed by electroplating the glass master. The father disc is used to make mother discs from which multiple stampers (sons) can be made.

field A set of alternating scan lines in an interlaced video picture. A frame is made of a top (odd) field and a bottom (even) field.

file A collection of data stored on a disc, usually in groups of sectors.

file system A defined way of storing files, directories, and information about such files and directories on a data storage device.

filter 1) To reduce the amount of information in a signal. 2) A circuit or process that reduces the amount of information in a signal. Analog filtering usually removes certain frequencies. Digital filtering (when not emulating analog filtering) usually averages together multiple adjacent pixels, lines, or frames to create a single new pixel, line, or frame. This generally causes a loss of detail, especially with complex images or rapid motion. See letterbox filter. Compare this to interpolate.

FireWire A standard for the transmission of digital data between external peripherals, including consumer audio and video devices. The official name is IEEE 1394, based on the original FireWire design by Apple Computer.

fixed rate Information flow at a constant volume over time. See CBR.

forced display A feature of DVD-Video that enables subpictures to be displayed even if the player's subpicture display mode is turned off. It is also designed to show subtitles in a scene where the language is different from the native language of the film.

First play The function that is automatically executed when a disc is initially inserted into a DVD Video player.

Fixed bit rate A method of compression that records data at an unchanging rate, regardless of image complexity. It is also known as fixed transfer rate.

Forced display A DVD Video function that forces subpictures to display on a screen regardless of the user's subpicture settings.

formatting 1) Creating a disc image. 2) Preparing storage media for recording.

fps Frames per second. A measure of the rate at which pictures are shown to create a motion video image. In NTSC and PAL video, each frame is made up of two interlaced fields.

fragile watermark A watermark designed to be destroyed by any form of copying or encoding other than a bit-for-bit digital copy. The absence of the watermark indicates that a copy has been made.

frame The piece of a video signal containing the spatial detail of one complete image, or the entire set of scan lines. In an interlaced system, a frame contains two fields.

frame doubler A video processor that increases the frame rate (display rate) in order to create a smoother-looking video display. Compare this to line doubler.

frame rate The frequency of discrete images. This is usually measured in frames per second (fps). Film has a rate of 24 frames per second, but it usually must be adjusted to match the display rate of a video system.

frequency The number of repetitions of a phenomenon in a given amount of time. The number of complete cycles of a periodic process occurring per unit of time.

G

G or **Giga** An SI prefix for denominations of one billion (10^9).

G byte One billion (10^9) bytes. Not to be confused with GB or gigabyte (2^{30} bytes).

Galaxy Group The group of companies proposing the Galaxy watermarking format (IBM/NEC, Hitachi/Pioneer/Sony).

GB Gigabyte.

Gbps Gigabits/second. Billions (10^9) of bits per second.

General parameters A DVD Video player function that memorizes a user's operational history to modify the player's behavior.

gigabyte 1,073,741,824 (2^{30}) bytes. (See Introduction, for more information.)

GOF Group of Audio Frames, a data area containing 20 audio frames of Linear PCM audio.

GOP Group of pictures. In MPEG video, one or more I pictures followed by P and B pictures. A GOP is the atomic unit of MPEG video access. GOPs are limited in DVD-Video to maximums of 18 frames for 525/60 and 15 frames for 625/50.

gray market Dealers and distributors who sell equipment without proper authorization from the manufacturer.

Green Book The document developed in 1987 by Philips and Sony as an extension to CD-ROM XA for the CD-i system.

Guide melody The melody for the vocal part that is recorded to assist singers in karaoke.

Guide vocal A leading guide function for singing a song, and the vocal part of karaoke songs in karaoke-equipped DVD Video players. It is also called the Assistant vocal.

H

HAVi A consumer electronics industry standard for interoperability between digital audio and video devices connected via a network in the consumer's home.

HDCD High-definition Compatible Digital. A proprietary method of enhancing audio on CDs.

HDTV High-definition television. A video format with a resolution approximately twice that of conventional television in both the horizontal and vertical dimensions, and a picture aspect ratio of 16:9. Used loosely to refer to the U.S. DTV System. Contrast this with EDTV and IDTV.

H/DTV High-definition/digital television. A combination of acronyms that refers to both HDTV and DTV systems.

hertz See Hz.

hexadecimal Representation of numbers using base 16.

HFS Hierarchical file system. A file system used by Apple Computer's Mac OS operating system.

Highlight A method of display that emphasizes a selected item on a menu screen by increasing the brightness level to show which function is executed.

High Sierra The original file system standard developed for CD-ROM, later modified and adopted as ISO 9660.

horizontal resolution See lines of horizontal resolution.

HQ-VCD High-Quality Video Compact Disc. Developed by the Video CD Consortium (Philips, Sony, Matsushita, and JVC) as a successor to VCD. It has evolved into SVCD.

HRRA Home Recording Rights Association.

HSF See High Sierra.

HTML Hypertext markup language. This is a tagging specification based on the standard generalized markup language (SGML) for formatting text to be transmitted over the Internet and displayed by client software.

hue The color of light or a pixel. The property of color determined by the dominant wavelength of light.

Huffman coding A lossless compression technique of assigning variable-length codes to a known set of values. The values occurring the most frequently are assigned the shortest codes. MPEG uses a variation of Huffman coding with fixed code tables, often called variable-length coding (VLC).

Hz Hertz. A unit of frequency measurement that determines the number of cycles (repetitions) per second.

I

I picture or **I frame** In MPEG video, this is an intra picture that is encoded independent from other pictures (see intraframe). Transform coding (DCT, quantization, and VLC) is used with no motion compensation, resulting in only moderate compression. I pictures provide a reference point for dependent P pictures and B pictures and enable random access into the compressed video stream.

i.Link Trademarked Sony name for IEEE 1394.

IDE Integrated Drive Electronics. An internal bus or standard electronic interface between a computer and internal block storage devices. IDE was adopted as a standard by ANSI in November 1990. ANSI calls it Advanced Technology Attachment (ATA). See E-IDE and ATAPI.

IDTV Improved-definition television. A television receiver that improves the apparent quality of the picture from a standard video signal by using techniques such as frame doubling, line doubling, and digital signal processing.

IEC International Electrotechnical Commission. Refer to Appendix C, References and Information Sources.

IED ID error correction. An error-detection code applied to each sector ID on a DVD disc.

IEEE Institute of Electrical and Electronics Engineers, an electronics standards body.

IEEE 1394 A standard for the transmission of digital data between external peripherals, including consumer audio and video devices. Also known as FireWire.

IFE In-flight entertainment.

I-MPEG Intraframe MPEG. An unofficial variation of MPEG video encoding that uses only intraframe compression. I-MPEG is used by DV equipment.

Import To convert media or data from one form to another to be recognized by a computer system.

Interactive The capability of DVD Video to respond to commands issued by a user, and prompt a user to issue commands.

interframe Something that occurs between multiple frames of video. Interframe compression takes temporal redundancy into account. Contrast this with intraframe.

interlace A video scanning system in which alternating lines are transmitted so that half a picture is displayed each time the scanning beam moves down the screen. An interlaced frame comprises two fields. (See Chapter 3, Technology Primer.)

interleave To arrange data in alternating chunks so that selected parts can be extracted while other parts are skipped, or so that each chunk carries a piece of a different data stream.

interpolate To increase the pixels, scan lines, or pictures when scaling an image or a video stream by averaging adjacent pixels, lines, or frames to create additional inserted pixels or frames. This generally causes a softening of still images and a blurring of motion images because no new information is created. Compare this to filter.

intraframe Something that occurs within a single frame of video. Intraframe compression does not reduce temporal redundancy but enables each frame to be independently manipulated or accessed. See I picture. Compare this to interframe.

inverse telecine The reverse of 2-3 pulldown, where the frames that were duplicated to create 60-fields/second video from 24-frames/second film source are removed. MPEG-2 video encoders usually apply an inverse telecine process to convert 60-fields/second video into 24-frames/second encoded video. The encoder adds information that enables the decoder to recreate the 60-fields/second display rate.

iso isolated camera or recorder.

ISO International Organization for Standardization. (See Appendix C, References and Information Sources.)

ISO 9660 The international standard for the file system used by CD-ROM. ISO 9660 allows filenames of only eight characters plus a three-character extension.

ISRC International Standard Recording Code.

ITU International Telecommunication Union. Refer to Appendix C, References and Information Sources.

ITU-R BT.601 The international standard specifying the format of digital component video. Currently at version 5 (identified as 601-5).

J

Java A programming language with specific features designed for use with the Internet and HTML.

JCIC Joint Committee on Intersociety Coordination.

JEC Joint Engineering Committee of EIA and NCTA.

jewel box The plastic clamshell case that holds a CD or DVD.

jitter A temporal variation in a signal from an ideal reference clock. Many kinds of jitter can occur, including sample jitter, channel jitter, and interface jitter. Refer to Chapter 3, Technology Primer.

JPEG Joint Photographic Experts Group. The international committee that created its namesake standard for compressing still images.

K

k or **Kilo** An SI prefix for denominations of one thousand (10^3). Also used, in capital form, for 1024 bytes of computer data (see kilobyte).

k byte One thousand (10^3) bytes. Not to be confused with KB or kilobyte (2^{10} bytes). Note the small "k."

karaoke Literally, "empty orchestra". The social sensation from Japan where sufficiently inebriated people embarrass themselves in public by singing along to a music track. Karaoke was largely responsible for the success of laserdisc in Japan, thus supporting it elsewhere.

KB Kilobyte.

kbps Kilobits/second. Thousands (10^3) of bits per second.

key picture or **key frame** A video picture containing the entire content of the image (intraframe encoding), rather than the difference between it and another image (interframe encoding). MPEG I pictures are key pictures. Contrast this with delta picture.

kHz Kilohertz. A unit of frequency measurement. It is one thousand cycles (repetitions) per second or 1,000 hertz.

kilobyte 1024 (2^{10}) bytes. (See Introduction, for more information.)

land The raised area of an optical disc.

L

laserdisc A 12-inch (or 8-inch) optical disc that holds analog video (using an FM signal) and both analog and digital (PCM) audio. Laserdisc was a precursor to DVD.

layer The plane of a DVD disc where information is recorded in a pattern of microscopic pits. Each substrate of a disc can contain one or two layers. The first layer, closest to the read-out surface, is layer 0; the second is layer 1.

lead in The physical area that is 1.2 mm or wider preceding the data area on a disc. The lead in contains sync sectors and control data including disc keys and other information.

lead out On a single-layer disc or PTP dual-layer disc, this is the physical area 1.0 mm or wider toward the outside of the disc following the data area. On an OTP dual-layer disc, this is the physical area 1.2 mm or wider at the inside of the disc following the recorded data area (which is read from the outside toward the inside on the second layer).

legacy A term used to describe a hybrid disc that can be played in both a DVD player and a CD player.

letterbox The process or form of video where black horizontal mattes are added to the top and bottom of the display area to create a frame in which to display video using an aspect ratio different than that of the display. The letterbox method preserves the entire video picture, as opposed to pan and scan. DVD-Video players can automatically letterbox an anamorphic widescreen picture for display on a standard 4:3 TV.

letterbox filter The circuitry in a DVD player that reduces the vertical size of anamorphic widescreen video (combining every four lines into three) and adds black mattes at the top and bottom. See filter.

level In MPEG-2, levels specify parameters such as resolution, bit rate, and frame rate. Compare this to profile.

Line 21 The specific part of the NTSC video signal that carries the teletext information for closed captioning.

linear PCM A coded representation of digital data that is not compressed. Linear PCM spreads values evenly across the range from highest to lowest, as opposed to nonlinear (that which is first compressed then expanded, aka companded) PCM that allocates more values to more important frequency ranges.

line doubler A video processor that doubles the number of lines in the scanning system in order to create a display with scan lines that are less visible. Some line doublers convert from an interlaced to a progressive scan.

lines of horizontal resolution Sometimes abbreviated as TVL (TV lines) or LoHR, this is a common but subjective measurement of the visually resolvable horizontal detail of an analog video system, measured in half-cycles per picture height. Each cycle is a pair of vertical lines, one black and one white. The measurement is usually made by viewing a test pattern to determine where the black and white lines blur into gray. The resolution of VHS video is commonly gauged at 240 lines of horizontal resolution, broadcast video at 330, laserdisc at 425, and DVD at 500 to 540. Because the measurement is relative to picture height, the aspect ratio must be taken into account when determining the number of vertical units (roughly equivalent to pixels) that can be displayed across the width of the display. For example, an aspect ratio of 1.33 multiplied by 540 gives 720 pixels.

locale See regional code.

logical An artificial structure or organization of information created for convenience of access or reference, usually different from the physical structure or organization. For example, the application specifications of DVD (the way information is organized and stored) are logical formats.

logical unit A physical or virtual peripheral device, such as a DVD-ROM drive.

L_o/R_o Left only/right only. A stereo signal with no matrixed surround information in which optional downmixing is output in Dolby Digital decoders. It does not change the phase but simply folds surround channels forward into Lf and Rf.

lossless compression Compression techniques that enable the original data to be recreated without loss. Contrast with lossy compression.

lossy compression Compression techniques that achieve very high compression ratios by permanently removing data while preserving as much significant information as possible. Lossy compression includes perceptual coding techniques that attempt to limit the data loss so that it is least likely to be noticed by human perception.

LP Long-playing record. An audio recording on a plastic platter turning at 33 1/3 rpm and read by a stylus.

LPCM See linear PCM.

L_t/R_t Left total/right total. Four surround channels matrixed into two channels. The mandatory downmixing method in Dolby Digital decoders.

luma (Y') The brightness component of a color video image (also called the grayscale, monochrome, or black-and-white component) with nonlinear luminance. The standard luma signal is computed from nonlinear RGB as $Y' = 0.299\,R' + 0.587\,G' + 0.114\,B'$.

luminance (Y) Loosely, the sum of RGB tristimulus values corresponding to brightness. This may refer to a linear signal or (incorrectly) a nonlinear signal.

M

M or **Mega** An SI prefix for denominations of one million (10^6).

Mac OS The operating system used by Apple Macintosh computers.

macroblock In MPEG MP@ML, the four 8×8 blocks of luma information and two 8×8 blocks of chroma information that form a 16×16 area of a video frame.

macroblocking An MPEG artifact. See blocking.

Macrovision An antitaping process that modifies a signal so that it appears unchanged on most televisions but is distorted and unwatchable when played back from a videotape recording. Macrovision takes advantage of the characteristics of AGC circuits and burst decoder circuits in VCRs to interfere with the recording process.

magneto-optical A recordable disc technology using a laser to heat spots that are altered by a magnetic field. Other formats include dye-sublimation and phase-change.

main level (ML) A range of proscribed picture parameters defined by the MPEG-2 video standard, with a maximum resolution equivalent to ITU-R BT.601 ($720 \times 576 \times 30$). See level.

main profile (MP) A subset of the syntax of the MPEG-2 video standard designed to be supported over a large range of mainstream applications such as digital cable TV, DVD, and digital satellite transmission. See profile.

mark The non-reflective area of a writable optical disc. Equivalent to a pit.

master The metal disc used to stamp replicas of optical discs, or the tape used to make additional recordings.

mastering The process of replicating optical discs by injecting liquid plastic into a mold containing a master. This is often used inaccurately to refer to premastering.

matrix encoding The technique of combining additional surround-sound channels into a conventional stereo signal. See Dolby Surround.

matte An area of a video display or motion picture that is covered (usually in black) or omitted in order to create a differently shaped area within the picture frame.

MB Megabyte.

Mbps Megabits/second. Millions (10^6) of bits per second.

M byte One million (10^6) bytes. Not to be confused with MB or megabyte (2^{20} bytes).

megabyte 1,048,576 (2^{20}) bytes. (See Introduction, for more information.)

megapixel An image or display format with a resolution of approximately one million pixels.

memory Data storage used by computers or other digital electronics systems. Read-only memory (ROM) permanently stores data or software program instructions. New data cannot be written to ROM. Random-access memory (RAM) temporarily stores data, including digital audio and video, while it is being manipulated and holds software application programs while they are being executed. Data can be read from and written to RAM. Other long-term memory includes hard disks, floppy disks, digital CD formats (CD-ROM, CD-R, and CD-RW), and DVD formats (DVD-ROM, DVD-R, DVD+R, DVD$\pm$RW and DVD-RAM).

Menu A graphic image provided in a Title to assist in picture, audio, sub-picture and multi-angle selections recorded on a DVD Video disc.

Menu state The condition of a player when a menu is presented.

MHz One million (10^6) Hz.

Microsoft Windows The leading operating system for Intel CPU-based computers developed by Microsoft.

middle area On a dual-layer OTP disc, the physical area 1.0 mm or wider on both layers, adjacent to the outside of the data area.

Millennium Group The group of companies that proposed the Millennium watermarking format that includes Macrovision, Philips, and Digimarc.

mixed mode A type of CD containing both Red Book audio and Yellow Book computer data tracks.

MKB (Media Key Block) A set of keys used in CPPM and CPRM for authenticating players.

MLP (Meridian Lossless Packing) A lossless compression technique (used by DVD-Audio) that removes redundancy from PCM audio signals to achieve a compression ratio of about 2:1 while allowing the signal to be perfectly recreated by the MLP decoder.

MO Magneto-optical rewritable discs.

modulation Replacing patterns of bits with different (usually larger) patterns designed to control the characteristics of the data signal. DVD uses 8/16 modulation, where each set of eight bits is replaced by 16 bits before being written onto the disc.

mosquitoes A term referring to the fuzzy dots that can appear around sharp edges (high spatial frequencies) after video compression. Also known as the Gibbs Effect.

mother The metal discs produced from mirror images of the father disc in the replication process. Mothers are used to make stampers, often called sons.

motion compensation In video decoding, the application of motion vectors to already-decoded blocks in order to construct a new picture.

motion estimation In video encoding, the process of analyzing previous or future frames to identify blocks that have not changed or have changed only their location. Motion vectors are then stored in place of the blocks. This is very computation-intensive and can cause visual artifacts when subject to errors.

motion vector A two-dimensional spatial displacement vector used for MPEG motion compensation to provide an offset from the encoded position of a block in a reference (I or P) picture to the predicted position (in a P or B picture).

MP@ML Main profile at main level. The common MPEG-2 format used by DVD (along with SP@SL).

MP3 MPEG-1 Layer III audio. A perceptual audio coding algorithm. Not supported in DVD-Video or DVD-Audio formats.

MPEG Moving Picture Experts Group. An international committee that developed the MPEG family of audio and video compression systems.

MPEG audio Audio compressed according to the MPEG perceptual encoding system. MPEG-1 audio provides two channels, which can be in Dolby Surround format. MPEG-2 audio adds data to provide discrete multichannel audio. Stereo MPEG audio is one of two mandatory audio compression systems for 625/50 (PAL/SECAM) DVD-Video.

MPEG video Video compressed according to the MPEG encoding system. MPEG-1 is typically used for low data rate video, as on a Video CD. MPEG-2 is used for higher-quality video, especially interlaced video, such as on DVD or HDTV.

MTBF Mean time between failure. A measure of reliability for electronic equipment, usually determined in benchmark testing. The higher the MTBF, the more reliable the hardware.

Mt. Fuji See SFF 8090.

multiangle A DVD-Video program containing multiple angles, allowing different views of a scene to be selected during playback.

multichannel Multiple channels of audio, usually containing different signals for different speakers to create a surround-sound effect.

multilanguage A DVD-Video program containing sound tracks or subtitle tracks for more than one language.

multimedia Information in more than one form, such as text, still images, sound, animation, and video. Usually implies that the information is presented by a computer.

multiplexing Combining multiple signals or data streams into a single signal or stream. This is usually achieved by interleaving at a low level.

MultiRead A standard developed by the Yokohama group, a consortium of companies attempting to ensure that new CD and DVD hardware can read all CD formats (refer to "Innovations of CD" in Chapter 1, The World Before DVD, for a discussion of CD variations).

multisession A technique in write-once recording technology that enables additional data to be appended after data is written in an earlier session.

Multistory A configuration in which one Title contains more than one story.

mux Short for multiplex.

mux_rate In MPEG, the combined rate of all packetized elementary streams (PES) of one program. The mux_rate of DVD is 10.08 Mbps.

N

NAB National Association of Broadcasters.

Navigation The process of manipulating a DVD Video Title toward a particular destination, allowing the user access to the contents and operation functions of a Title. This process is defined and programmed during the authoring process.

Navigation command Instruction used to enable the interactive capabilities of a DVD Video Title, programmed during the authoring process.

NCTA National Cable Television Association.

nighttime mode A Dolby Digital dynamic range compression feature that enables low-volume nighttime listening without losing dialog legibility.

noise Irrelevant, meaningless, or erroneous information added to a signal by the recording or transmission medium or by an encoding/decoding process. An advantage of digital formats over analog formats is that noise can be completely eliminated (although new noise can be introduced by compression).

noise floor The level of background noise in a signal or the level of noise introduced by equipment or storage media, below which the signal can't be isolated from the noise.

Non-drop frame timecode The method of timecode computation where there are 30 numerical frames per second of video. "There are 30 frames of video per second," you say. Wrong. There are only 29.97 frames of video per second. In a mathematical hour, there would be 108,000 frames (30 frames per second × 60 seconds × 60 minutes). So, a mathematical hour is 108 frames longer than an hour of reality. See also drop frame timecode and timecode.

NRZI Non-return to zero, inverted. A method of coding binary data as waveform pulses. Each transition represents a one, while a lack of a transition represents a run of zeros.

NTSC National Television Systems Committee. A committee organized by the Electronic Industries Association (EIA) that developed commercial television broadcast standards for the U.S. The group first established black-and-white TV standards in 1941, using a scanning system of 525 lines at 60 fields per second. A second committee standardized color enhancements using 525 lines at 59.94 fields per second. NTSC refers to the composite color-encoding system. The 525/59.94 scanning system (with a 3.58-MHz color subcarrier) is identified by the letter M and is often incorrectly referred to as NTSC. The NTSC standard is also used in Canada, Japan, and other parts of the world. NTSC is facetiously referred to as meaning "never the same color" because of the system's difficulty in maintaining color consistency.

NTSC-4.43 A variation of NTSC in which a 525/59.94 signal is encoded using the PAL subcarrier frequency and chroma modulation. Also called 60-Hz PAL.

numerical aperture (NA) A unitless measure of the capability of a lens to gather and focus light. NA = n sin 1, where 1 is the angle of the light as it narrows to the focal point. A numerical aperture of 1 implies no change in parallel light beams. The higher the number, the greater the focusing power and the smaller the spot.

OEM Original equipment manufacturer. A computer maker.

operating system The primary software in a computer, containing general instructions for managing applications, communications, input/output, memory, and other low-level tasks. DOS, Windows, Mac OS, Linux, and Unix are examples of operating systems.

opposite path See OTP.

Orange Book The document begun in 1990 that specifies the format of recordable CD. Its three parts define magneto-optical erasable (MO) and write-once (WO) discs, dye-sublimation write-once (CD-R) discs, and phase-change rewritable (CD-RW) discs. Orange Book also added multisession capabilities to the CD-ROM XA format.

OS Operating system.

OSTA Optical Storage Technology Association. Refer to Appendix C, References and Information Sources.

OTP Opposite track path. A variation of DVD dual-layer disc layout where readout begins at the center of the disc on the first layer, travels to the outer edge of the disc, then switches to the second layer and travels back toward the center. Designed for long, continuous-play programs. Also called RSDL. Contrast this with PTP.

out of band In a place not normally accessible.

overscan The area at the edges of a television tube that is covered to hide possible video distortion. Overscan typically covers about four or five percent of the picture.

P

pack A group of MPEG packets in a DVD-Video program stream. Each DVD sector (2,048 bytes) contains one pack.

packet A low-level unit of DVD-Video (MPEG) data storage containing contiguous bytes of data belonging to a single elementary stream such as video, audio, control, etc. Packets are grouped into packs.

packetized elementary stream (PES) The low-level stream of MPEG packets containing an elementary stream, such as audio or video.

PAL Phase alternate line. A video standard used in Europe and other parts of the world for composite color encoding. Various versions of PAL use different scanning systems and color subcarrier frequencies (identified with letters B, D, G, H, I, M, and N), the most common being 625 lines at 50 fields per second, with a color subcarrier of 4.43 MHz. PAL is also said to mean "picture always lousy" or "perfect at last," depending on which side of the ocean the speaker comes from.

palette A table of colors that identifies a subset from a larger range of colors. The small number of colors in the palette enables fewer bits to be used for each pixel. Also called a color look-up table (CLUT).

pan and scan The technique of reframing a picture to conform to a different aspect ratio by cropping parts of the picture. DVD-Video players can automatically create a 4:3 pan and scan version from widescreen anamorphic video by using a horizontal offset encoded with the video.

parallel path See PTP.

parental control function This function automatically compares the upper limit level of permissible Parental Level playback preset by the users in the player with the Parental Level contained in the disc.

Parental ID Instructions contained in a DVD-Video player that identify the Parental Level of a Title.

Parental Level A permissible level to be observed on screen depending on the age of viewers and the nature of the content of a Title.

parental management An optional feature of DVD-Video that prohibits programs from being viewed or substitutes different scenes within a program depending on the parental level set in the player. Parental control requires that Parental Levels and additional material (if necessary) be encoded on the disc.

part of title In DVD-Video, a division of a title representing a scene. Also called a chapter. Parts of titles are numbered 1 to 99.

PCI Presentation control information. A DVD-Video data stream containing details of the timing and presentation of a program (aspect ratio, angle change, menu highlight and selection information, etc). PCI and DSI comprise an overhead of about one Mbps.

PCM An uncompressed, digitally coded representation of an analog signal. The waveform is sampled at regular intervals and a series of pulses in coded form (usually quantized) are generated to represent the amplitude.

PC-TV The merger of television and computers. A personal computer capable of displaying video as a television.

pel See pixel.

perceived resolution The apparent resolution of a display from the observer's point of view, based on viewing distance, viewing conditions, and physical resolution of the display.

perceptual coding Lossy compression techniques based on the study of human perception. Perceptual coding systems identify and remove information that is least likely to be missed by the average human observer.

PGC Program Chain, a group of Cells linked together creating a program sequence.

PGCI Program chain information. Data describing a chain of cells (grouped into programs) and their sector locations that compose a sequential program. PGCI data is contained in the PCI stream.

phase-change A technology for rewritable optical discs using a physical effect in which a laser beam heats a recording material to reversibly change an area from an amorphous state to a crystalline state, or vice versa. Continuous heat just above the melting point creates the crystalline state (an erasure), while high heat followed by rapid cooling creates the amorphous state (a mark). Other recording technologies include dye-sublimation and magneto-optical.

physical format The low-level characteristics of the DVD-ROM and DVD-Video standards, including pits on the disc, the location of data, and the organization of data according to physical position.

PICT picture. A standard graphic format for Macintosh computers.

picture In video terms, a single still image or a sequence of moving images. Picture generally refers to a frame, but for interlaced frames it may refer instead to a field of the frame. In a more general sense, picture refers to the entire image shown on a video display.

picture stop A function of DVD-Video where a code indicates that video playback should stop and a still picture be displayed.

PIP Picture in picture. A feature of some televisions that shows another channel or video source in a small window superimposed in an area of the screen.

pit A microscopic depression in the recording layer of an optical disc. Pits are usually 1/4 of the laser wavelength to cause cancellation of the beam by diffraction.

pit art A pattern of pits stamped onto a disc to provide visual art rather than data. A cheaper alternative to a printed label.

pixel The smallest picture element of an image (one sample of each color component). A single dot in the array of dots that comprise a picture, sometimes abbreviated to pel. The resolution of a digital display is typically specified in terms of pixels (width by height) and color depth (the number of bits required to represent each pixel).

pixel aspect ratio The ratio of width to height of a single pixel. This often means the sample pitch aspect ratio (when referring to sampled digital video). Pixel aspect ratio for a given raster can be calculated as $y/x \times w/h$ (where x and y are the raster horizontal pixel count and vertical pixel count, and w and h are the display aspect ratio width and height). Pixel aspect ratios are also confusingly calculated as $x/y \times w/h$, giving a height-to-width ratio. Refer to Table 9.13.

pixel depth See color depth.

PMMA Polymethylmethacrylate. A clear acrylic compound used in laserdiscs and as an intermediary in the surface transfer process (STP) for dual-layer DVDs. PMMA is sometimes used for DVD substrates.

POP Picture outside picture. A feature of some widescreen displays that uses the unused area around a 4:3 picture to show additional pictures.

Postcommand A navigation command executed after a PGC has been read by a player.

P picture or **P frame** In MPEG video, a "predicted" picture based on the difference from previous pictures. P pictures (along with I pictures) provide a reference for following P pictures or B pictures.

Precommand A navigation command executed before a PGC is read by a player.

premastering The process of preparing data in the final format to create a DVD disc image for mastering. This includes creating DVD control and navigation data, multiplexing data streams together, generating error-correction codes, and performing channel modulation. It also often includes the process of encoding video, audio, and subpictures.

presentation data DVD-Video information such as video, menus, and audio that is presented to the viewer. See PCI.

profile In MPEG-2, profiles specify syntax and processes such as picture types, scalability, and extensions. Compare this to level.

program In a general sense, a sequence of audio or video. In a technical sense for DVD-Video, a group of cells within a program chain (PGC) containing up to 999 Cells.

program chain In DVD-Video, a collection of programs, or groups of cells, linked together to create a sequential presentation.

progressive scan A video scanning system that displays all lines of a frame in one pass. Contrast this with interlaced scan. (See Chapter 3, Technology Primer, for more information.)

psychoacoustic See perceptual encoding.

PTP Parallel track path. A variation of DVD dual-layer disc layout where readout begins at the center of the disc for both layers. This is designed for separate programs (such as a widescreen and a pan and scan version on the same disc side) or programs with a variation on the second layer. PTP is most efficient for DVD-ROM random-access application. Contrast this with OTP.

PUH Pickup head. The assembly of optics and electronics that reads data from a disc.

Q

QCIF Quarter common intermediate format. Video resolution of 176×144.

quantization levels The predetermined levels at which an analog signal can be sampled, as determined by the resolution of the analog-to-digital converter (in bits per sample) or the number of bits stored for the sampled signal.

quantize To convert a value or range of values into a smaller value or smaller range by integer division. Quantized values are converted back (by multiplying) to a value that is close to the original but may not be exactly the same. Quantization is a primary technique of lossless encoding.

QuickTime A digital video software standard developed by Apple Computer for Macintosh (Mac OS) and Windows operating systems. QuickTime is used to support audio and video from a DVD.

QXGA A video graphics resolution of 2048×1536.

R

RAM Random-access memory. This generally refers to solid-state chips. In the case of DVD-RAM, the term was borrowed to indicate the capability to read and write at any point on the disc.

RAMbo drive A DVD-RAM drive capable of reading and writing CD-R and CD-RW media (a play on the word "combo", not Sylvester Stallone).

random access The capability to jump to a point on a storage medium.

raster The pattern of parallel horizontal scan lines that comprise a video picture.

read-modify-write An operation used in writing to DVD-RAM discs. Because data can be written by the host computer in blocks as small as two KB, but the DVD format uses ECC blocks of 32 KB, an entire ECC block is read from the data buffer or disc, modified to include the new data and new ECC data, and then written back to the data buffer and disc.

Raw Uncompressed video data.

Red Book The document first published in 1982 that specifies the original compact disc digital audio format developed by Philips and Sony.

Reed-Solomon An error-correction encoding system that cycles data multiple times through a mathematical transformation to increase the effectiveness of the error correction, especially for burst errors (errors concentrated closely together, as from a scratch or physical defect). DVD uses rows and columns of Reed-Solomon encoding in a two-dimensional lattice called Reed-Solomon product code (RS-PC).

reference picture or **reference frame** An encoded frame that is used as a reference point from which to build dependent frames. In MPEG-2, I pictures and P pictures are used as references.

reference player A DVD player that defines its ideal behavior as specified by the DVD-Video standard.

regional code A code identifying one of the world regions for restricting DVD-Video playback. Refer to Table A.25.

regional management A mandatory feature of DVD-Video to restrict the playback of a disc to a specific geographical region. Each player and DVD-ROM drive include a single regional code, and each disc side can specify in which regions it is permitted to be played. Regional coding is optional; a disc without regional codes will play in all players in all regions.

regional playback control function A system to control the playable geographic region of any DVD Video disc.

replication 1) The reproduction of media such as optical discs by stamping (contrast this with duplication); 2) A process used to increase the size of an image by repeating pixels (to increase the horizontal size) and/or lines (to increase the vertical size) or to increase the display rate of a video stream by repeating frames. For example, a 360×240 pixel image can be displayed at 720×480 size by duplicating each pixel on each line and then duplicating each line. In this case, the resulting image contains blocks of four identical pixels. Obviously, image replication can cause blockiness. A 24-fps video signal can be displayed at 72 fps by repeating each frame three times. Frame replication can cause jerkiness of motion. Contrast this with decimation. See interpolate.

resampling The process of converting between different spatial resolutions or different temporal resolutions. This can be based on a sample of the source information at a higher or lower resolution or it can include interpolation to correct for the differences in pixel aspect ratios or to adjust for differences in display rates.

resolution 1) A measurement of the relative detail of a digital display, typically given in pixels of width and height; 2) The capability of an imaging system to make the details of an image clearly distinguishable or resolvable. This includes spatial resolution (the clarity of a single image), temporal resolution (the clarity of a moving image or moving object), and perceived resolution (the apparent resolution of a display from the observer's point of view). Analog video is often measured as a number of lines of horizontal resolution over the number of scan lines. Digital video is typically measured as a number of horizontal pixels by vertical pixels. Film is typically measured as a number of line pairs per millimeter; 3) The relative detail of any signal, such as an audio or video signal. See lines of horizontal resolution.

Resume A function used when a player returns from a Menu state.

RGB Video information in the form of red, green, and blue tristimulus values. The combination of three values representing the intensity of each of the three colors can represent the entire range of visible light.

ROM Read-only memory.

rpm Revolutions per minute. A measure of rotational speed.

Root menu The lowest level menu contained in a VTS, from which other menus may be accessed.

RS Reed-Solomon. An error-correction encoding system that cycles data multiple times through a mathematical transformation in order to increase the effectiveness of the error correction. DVD uses rows and columns of Reed-Solomon encoding in a two-dimensional lattice, called Reed-Solomon product code (RS-PC).

RS-CIRC See CIRC.

RSDL Reverse-spiral dual-layer. See OTP.

RS-PC Reed-Solomon product code. An error-correction encoding system used by DVD employing rows and columns of Reed-Solomon encoding to increase error-correction effectiveness.

RTFM When all else fails, read the "fine" manual. Yeah, right.

Run length coding Method of lossless compression that codes by analyzing adjacent samples with the same values.

R'-Y', B'-Y' The general term for color-difference video signals carrying red and blue color information, where the brightness (Y') has been subtracted from the red and blue RGB signals to create R'-Y' and B'-Y' color-difference signals. Refer to Chapter 3, Technology Primer.

S

sample A single digital measurement of analog information or a snapshot in time of a continuous analog waveform. See sampling.

sample rate The number of times a digital sample is taken, measured in samples per second, or Hertz. The more often samples are taken, the better a digital signal can represent the original analog signal. The sampling theory states that the sampling frequency must be more than twice the signal frequency in order to reproduce the signal without aliasing. DVD PCM audio enables sampling rates of 48 and 96 kHz.

sample size The number of bits used to store a sample. Also called resolution. In general, the more bits allocated per sample, the better the reproduction of the original analog information. The audio sample size determines the dynamic range. DVD PCM audio uses sample sizes of 16, 20, or 24 bits.

sampling Converting analog information into a digital representation by measuring the value of the analog signal at regular intervals, called samples, and encoding these numerical values in digital form. Sampling is often based on specified quantization levels. Sampling can also be used to adjust for differences between different digital systems. See resampling and subsampling.

saturation The intensity or vividness of a color.

sampling frequency The frequency used to convert an analog signal into digital data.

scaling Altering the spatial resolution of a single image to increase or reduce the size, or altering the temporal resolution of an image sequence to increase or decrease the rate of display. Techniques include decimation, interpolation, motion compensation, replication, resampling, and subsampling. Most scaling methods introduce artifacts.

scan line A single horizontal line traced out by the scanning system of a video display unit. 525/60 (NTSC) video has 525 scan lines, about 480 of which contain the actual picture. 625/50 (PAL/SECAM) video has 625 scan lines, about 576 of which contain the actual picture.

scanning velocity The speed at which the laser pickup head travels along the spiral track of a disc.

SCMS The serial copy management system used by DAT, MiniDisc, and other digital recording systems to control copying and limit the number of copies that can be made from copies.

SCSI Small Computer Systems Interface. An electronic interface and command set for attaching and controlling internal or external peripherals, such as a DVD-ROM drive, to a computer. The command set of SCSI was extended for DVD-ROM devices by the SFF 8090 specification.

SDI See Serial Digital Interface. Also Strategic Defense Initiative, aka Star Wars, which was finally released on DVD so fans can replace their bootleg copies.

SDDI Serial Digital Data Interface. A digital video interconnect designed for serial digital information to be carried over a standard SDI connection.

SDDS Sony Dynamic Digital Sound. A perceptual audio-coding system developed by Sony for multichannel audio in theaters. A competitor to Dolby Digital and an optional audio track format for DVD.

SDMI Secure Digital Music Initiative. Efforts and specifications for protecting digital music.

SDTV Standard-definition television. A term applied to traditional 4:3 television (in digital or analog form) with a resolution of about 700×480 (about 1/3 megapixel). Contrast this with HDTV.

Seamless angle change The ability of DVD-Video to change between multiple points of view within a scene without interrupting normal playback.

seamless playback A feature of DVD-Video where a program can jump from place to place on the disc without any interruption of the video. This enables different versions of a program to be put on a single disc by sharing common parts.

SECAM Sequential couleur avec mémoire/sequential color with memory. A composite color standard similar to PAL but currently used only as a transmission standard in France and a few other countries. Video is produced using the 625/50 PAL standard and is then transcoded to SECAM by the player or transmitter.

sector A logical or physical group of bytes recorded on the disc, the smallest addressable unit. A DVD sector contains 38,688 bits of channel data and 2,048 bytes of user data.

seek time The time it takes for the head in a drive to move to a data track.

Serial Digital Interface (SDI) The professional digital video connection format using a 270-Mbps transfer rate. A 10-bit, scrambled, polarity-independent interface, with common scrambling for both component ITU-R 601 and composite digital video and four groups each of four channels of embedded digital audio. SDI uses standard 75-ohm BNC connectors and coax cable.

Sequential presentation Playback of all Programs in a Title in a specified order determined in the authoring process.

Setup The black level of a video signal.

SFF 8090 The specification number 8090 of the Small Form Factor Committee, an ad hoc group formed to promptly address disk industry needs and to develop recommendations to be passed on to standards organizations. SFF 8090 (also known as the Mt. Fuji specification) defines a command set for CD-ROM- and DVD-ROM-type devices, including implementation notes for ATAPI and SCSI.

Shuffle presentation A function of a DVD-Video player enabling Programs in a Title to play in an order determined at random.

SI Système International (d'Unités)/International System (of Units). A complete system of standardized units and prefixes for fundamental quantities of length, time, volume, mass, etc.

signal-to-noise ratio The ratio of pure signal to extraneous noise, such as tape hiss or video interference. Signal-to-noise ratio is measured in decibels (dB). Analog recordings almost always have noise. Digital recordings, when properly prefiltered and not compressed, have no noise.

simple profile (SP) A subset of the syntax of the MPEG-2 video standard designed for simple and inexpensive applications such as software. SP does not enable B pictures. See profile.

simulate To test the function of a DVD disc in the authoring system without actually formatting an image.

single-sided disc A type of DVD disc on which data is recorded on one side only.

SMPTE The Society of Motion Picture and Television Engineers. An international research and standards organization. This group developed the SMPTE time code, used for marking the position of audio or video in time. Refer to Appendix C, References and Information Sources.

S/N Signal-to-noise ratio. Also called SNR.

son The metal discs produced from mother discs in the replication process. Fathers or sons are used in molds to stamp discs.

space The reflective area of a writable optical disc. Equivalent to a land.

spatial resolution The clarity of a single image or the measure of detail in an image. See resolution.

spatial Relating to space, usually two-dimensional. Video can be defined by its spatial characteristics (information from the horizontal plane and vertical plane) and its temporal characteristics (information at different instances in time).

S/P DIF Sony/Philips digital interface. A consumer version of the AES/EBU digital audio transmission standard. Most DVD players include S/P DIF coaxial digital audio connectors providing PCM and encoded digital audio output.

SP@ML Simple profile at main level. The simplest MPEG-2 format used by DVD. Most discs use MP@ML. SP does not allow B pictures.

Squeezed picture Refers to reducing 16:9 picture data horizontally to conform to a 4:3 image size.

squeezed video See anamorphic.

stamping The process of replicating optical discs by injecting liquid plastic into a mold containing a stamper (father or son). Also (inaccurately) called mastering.

Still In the authoring process, it is a TIFF image not defined as video. In the DVD-Video player, it is a state in which video and subpicture are frozen and audio is muted.

Stop state A condition in which the DVD-Video player is not executing a PGC.

Storage media Materials used to store data.

STP Surface transfer process. A method of producing dual-layer DVDs that sputters the reflective (aluminum) layer onto a temporary substrate of PMMA and then transfers the metalized layer to the already-molded layer 0.

stream A continuous flow of data, usually digitally encoded, designed to be processed sequentially. Also called a bitstream.

subpicture Graphic bitmap overlays used in DVD-Video to create subtitles, captions, karaoke lyrics, menu highlighting effects, etc.

subsampling The process of reducing spatial resolution by taking samples that cover areas larger than the original samples, or the process of reducing temporal resolutions by taking samples that cover more time than the original samples. This is also called downsampling. See chroma subsampling.

substrate The clear polycarbonate disc onto which data layers are stamped or deposited.

subtitle A textual representation of the spoken audio in a video program. Subtitles are often used with foreign languages and do not serve the same purpose as captions for the hearing impaired. See subpicture.

surround sound A multichannel audio system with speakers in front of and behind the listener to create a surrounding envelope of sound and to simulate directional audio sources.

Surround An audio system that produces a vivid stereophonic sound environment by arranging two or more speakers around the listener.

SVCD Super Video Compact Disc. MPEG-2 video on CD. Used primarily in Asia.

SVGA A video graphics resolution of 800×600 pixels.

S-VHS Super VHS (Video Home System). An enhancement of the VHS videotape standard using better recording techniques and Y/C signals. The term S-VHS is often used incorrectly to refer to s-video signals and connectors.

s-video A video interface standard that carries separate luma and chroma signals, usually on a four-pin mini-DIN connector. Also called Y/C. The quality of s-video is significantly better than composite video because it does not require a comb filter to separate the signals, but it's not quite as good as component video. Most high-end televisions have s-video inputs. S-video is often erroneously called S-VHS.

SXGA A video graphics resolution of 1280×1024 pixels.

sync A video signal (or component of a video signal) containing information necessary to synchronize the picture horizontally and vertically. Also, sync is specially formatted data on a disc that helps the readout system identify location and specific data structures.

syntax The rules governing the construction or formation of an orderly system of information. For example, the syntax of the MPEG video encoding specification defines how data and associated instructions are used by a decoder to create video pictures.

system menu The main menu of a DVD-Video disc, from which titles are selected. Also called the title selection menu or disc menu.

System parameters A set of conditions defined in the authoring process used to control basic DVD-Video player functions.

T

T or Tera An SI prefix for denominations of one trillion (10^{12}).

telecine The process (and the equipment) used to transfer film to video. The telecine machine performs 2-3 pulldown by projecting film frames in the proper sequence to be captured by a video camera.

telecine artist The operator of a telecine machine. Also called a colorist.

temporal Relating to time. The temporal component of motion video is broken into individual still pictures. Because motion video can contain images (such as backgrounds) that do not change much over time, typical video has large amounts of temporal redundancy.

temporal resolution The clarity of a moving image or moving object, or the measurement of the rate of information change in motion video. See resolution.

tilt A mechanical measurement of the warp of a disc. This is usually expressed in radial and tangential components, with radial indicating dishing and tangential indicating ripples in the perpendicular direction.

text file A file containing text information.

timecode Information recorded with audio or video to indicate a position in time. This usually consists of values for hours, minutes, seconds, and frames. It is also called SMPTE timecode. Some DVD-Video material includes information to enable the player to search to a specific timecode position. There are two types of timecode — non-drop frame and drop frame. Non-drop frame timecode is based on 30 frames of video per second. Drop frame timecode is based on 29.97 frames of video per second. Truth be told, there are only 29.97 frames of video per second. For short amounts of time, this discrepancy is inconsequential. For longer periods of time, however, it is important. One hour of non-drop frame timecode will be 108 frames longer than one hour of real time. See non-drop frame timecode and drop frame timecode.

title The largest unit of a DVD-Video disc (other than the entire volume or side). A title is usually a movie, TV program, music album, or so on. A disc can hold up to 99 titles, which can be selected from the disc menu. Entire DVD volumes are also commonly called Titles.

title key A value used to encrypt and decrypt (scramble) user data on DVD-Video discs.

Title menu A graphic image presented by the VMG for a user to select a Title.

track 1) A distinct element of audiovisual information, such as the picture, a sound track for a specific language, or the like. DVD-Video enables one track of video (with multiple angles), up to eight tracks of audio, and up to 32 tracks of subpicture; 2) One revolution of the continuous spiral channel of information recorded on a disc.

track buffer The circuitry (including memory) in a DVD player that provides a variable stream of data (up to 10.08 Mbps) to the system decoders of data coming from the disc at a constant rate of 11.08 Mbps. except for breaks, when a different part of the disc is accessed.

track pitch The distance (in the radial direction) between the centers of two adjacent tracks on a disc. The DVD-ROM standard track pitch is 0.74 mm.

transfer rate The speed at which a certain volume of data is transferred from a device such as a DVD-ROM drive to a host such as a personal computer. This is usually measured in bits per second or bytes per second. It is sometimes confusingly used to refer to the data rate, which is independent of the actual transfer system.

transform The process or result of replacing a set of values with another set of values. It can also be a mapping of one information space onto another.

trim See crop.

Trimming The process used for conforming a 16:9 image into a 4:3 image size.

tristimulus A three-valued signal that can match nearly all the colors of visible light in human vision. This is possible because of the three types of photoreceptors in the eye. RGB, $Y'C_bC_r$, and similar signals are tristimulus and can be interchanged by using mathematical transformations (subject to a possible loss of information).

TVL Television line. See lines of horizontal resolution.

TWG Technical Working Group. A general term for an industry working group. Specifically, the predecessor to the CPTWG. It is usually an ad hoc group of representatives working together for a period of time to make recommendations or define standards.

U

UDF Universal Disc Format. A standard developed by the Optical Storage Technology Association designed to create a practical and usable subset of the ISO/IEC 13346 recordable, random-access file system and volume structure format.

UDF Bridge A combination of UDF and ISO 9660 file system formats that provides backward-compatibility with ISO 9660 readers while allowing the full use of the UDF standard.

universal DVD A DVD designed to play in DVD-Audio and DVD-Video players (by carrying a Dolby Digital audio track in the DVD-Video zone).

universal DVD player A DVD player that can play both DVD-Video and DVD-Audio discs.

user The person operating the DVD-Video player. Sometimes referred to as wetware, as in, the operation experienced a wetware breakdown.

user data The data recorded on a disc independent of formatting and error-correction overhead. Each DVD sector contains 2,048 bytes of user data.

UXGA A video graphics resolution of 1600 x 1200.

V

VBI Vertical blanking interval. The scan lines in a television signal that do not contain picture information. These lines are present to enable the electron scanning beam to return to the top, and they are used to contain auxiliary information such as closed captions.

VBR Variable bit rate. Data that can be read and processed at a volume that varies over time. A data compression technique that produces a data stream between a fixed minimum and maximum rate. A compression range is generally maintained, with the required bandwidth increasing or decreasing depending on the complexity (the amount of spatial and temporal energy) of the data being encoded. In other words, the data rate fluctuates while quality is maintained. Compare this to CBR.

VBV Video buffering verifier. A hypothetical decoder that is conceptually connected to the output of an MPEG video encoder. It provides a constraint on the variability of the data rate that an encoder can produce.

VCAP Video capable audio player. An audio player that can read the limited subset of video features defined for the DVD-Audio format. Contrast this with a universal DVD player.

VCD Video Compact Disc. Near-VHS-quality MPEG-1 video on CD. Used primarily in Asia.

VfW See Video for Windows. Not to be confused with Veterans of Foreign Wars, a fine organization composed of persons who occasionally experience wetware problems when operating a VCR, DVD player, or computer.

VGA (Video Graphics Array) A standard analog monitor interface for computers. It is also a video graphics resolution of 640 × 480 pixels.

VHS Video Home System. The most popular system of videotape for home use. Developed by JCV.

Video CD A CD extension based on MPEG-1 video and audio that enables the playback of near-VHS-quality video on a Video CD player, CD-i player, or computer with MPEG decoding capability.

Video for Windows The system software additions used for motion video playback in Microsoft Windows. Replaced in newer versions of Windows by DirectShow (formerly called ActiveMovie).

Video manager (VMG) The disc menu. Also called the title selection menu.

Video title set (VTS) A set of one to ten files holding the contents of a title.

videophile Someone with an avid interest in watching videos or in making video recordings. Videophiles are often very particular about audio quality, picture quality, and aspect ratio to the point of snobbishness. Videophiles never admit to a wetware failure.

VLC Variable length coding. See Huffman coding.

VOB Video object. A small physical unit of DVD-Video data storage, usually a GOP.

volume A logical unit representing all the data on one side of a disc.

VSDA Video Software Dealers Association. Refer to Appendix C, References and Information Sources.

VTS Video Title Set, containing up to ten files with all of the contents of a Title.

W

WAEA World Airline Entertainment Association. Discs produced for use in airplanes contain extra information in a WAEA directory. The in-flight entertainment working group of the WAEA petitioned the DVD Forum to assign region 8 to discs intended for in-flight use.

watermark Information hidden as invisible noise or inaudible noise in a video or audio signal.

wetware see user.

White Book The document from Sony, Philips, and JVC begun in 1993 that extended the Red Book CD format to include digital video in MPEG-1 format. It is commonly called Video CD.

widescreen A video image wider than the standard 1.33 (4:3) aspect ratio. When referring to DVD or HDTV, widescreen usually indicates a 1.78 (16:9) aspect ratio.

window A usually rectangular section within an entire screen or picture.

Windows See Microsoft Windows.

X

XA See CD-ROM XA.

XDS Line 21.

XGA A video graphics resolution of 1024×768 pixels.

XVCD A non-standard variation of VCD.

Y

Y The luma or luminance component of video, which is the brightness independent of color.

Y/C A video signal in which the brightness (luma, Y) and color (chroma, C) signals are separated. This is also called s-video.

$Y'C_bC_r$ A component digital video signal containing one luma and two chroma components. The chroma components are usually adjusted for digital transmission according to ITU-R BT.601. DVD-Video's MPEG-2 encoding is based on 4:2:0 $Y'C_bC_r$ signals. $Y'C_bC_r$ applies only to digital video, but it is often incorrectly used in reference to the $Y'P_bP_r$ analog component outputs of DVD players.

Yellow Book The document produced in 1985 by Sony and Philips that extended the Red Book CD format to include digital data for use by a computer. It is commonly called CD-ROM.

$Y'P_bP_r$ A component analog video signal containing one luma and two chroma components. It is often referred to loosely as YUV or Y', B'-Y', R'-Y'.

YUV In the general sense, any form of color-difference video signal containing one luma and two chroma components. Technically, YUV is applicable only to the process of encoding component video into composite video. See $Y'C_bC_r$ and $Y'P_bP_r$.

Z

ZCLV Zoned constant linear velocity. This consists of concentric rings on a disc within which all sectors are the same size. It is a combination of CLV and CAV.

Index

A

AAC. See Advanced audio coding

AACS. See Advanced Access Content System

AC-3 RF digital audio, 10-13

Address in pregroove, 7-17, 8-9

ADIP. See Address in pregroove

Advanced Access Content System, 5-23

Advanced applications, 9-77 – 9-78

Advanced audio coding, 3-22

Advanced encryption standard, 5-23

Advanced Optical Disc, 2-34 – 2-35

Advanced subtitles, 9-77

Advanced Television Systems Committee, 1-13 – 1-14

Advanced Video Codec, 3-17

Advanced Video Coding, 9-73

Advanced VTS, 9-77

AES. See Advanced encryption standard

AGC. See Automatic gain control

Album text, 9-63

Aliasing, 11-5

Alpha blending, 13-5

AMG. See Audio manager

Ampex, 1-10 – 1-11

Analog audio, 10-12 – 10-13

Analog Hole, 5-25

Analog protection system, 5-7, 5-14 – 5-15, 9-12

Analog recordings, 3-1 – 3-2

Annex J operations, 9-29

AOBs. See Audio objects

AOD. See Advanced Optical Disc

AOPs. See Audio-only players

Apple, 1-20

Apple Macintosh
architecture, 15-11

WebDVD for, 15-25

Application compatibility, 6-19 – 6-20

Application content items, 9-53

Application programming interfaces, 9-86

Application specifications
audio, 9-41 – 9-47
audio formats, 9-59 – 9-61
BD-AV, 9-82 – 9-83
BD-Java, 9-88 – 9-89
BD-MV, 9-84 – 9-87
Blu-ray disc, 9-81 – 9-82
bonus tracks, 9-59
buttons, 9-27 – 9-28
camera angles, 9-49 – 9-50
closed captions, 9-49
data structures, 9-57 – 9-58
development of, 9-1
domains, 9-13 – 9-14
downmixing, 9-62
DVD-AR, 9-65
DVD-Audio, 9-56 – 9-57
DVD file format, 9-2 – 9-3
DVD recording, 9-64
DVD-SR, 9-66
DVD-Video, 9-3 – 9-5
DVD-VR, 9-64 – 9-65
HD DVD, 9-71 – 9-79
HD DVD-Video, 9-71 – 9-74, 9-79
HD DVD-VR, 9-80 – 9-81
HDMV, 9-87 – 9-88
high-frequency audio concerns, 9-61 – 9-62
interactivity on UMD, 9-69 – 9-70
menus, 9-26 – 9-27
navigation data, 9-18 – 9-26
navigation method, 9-59
navigation overview, 9-5 – 9-7
parental management, 9-50 – 9-51
physical data structure, 9-7 – 9-13

G

S

Jim Taylor

Jim is Senior Vice President and General Manager of the Advanced Technology Group at Sonic Solutions, the leading developer of DVD and CD creation software. In addition to writing the first two editions of DVD Demystified, Jim is the author of Everything You Ever Wanted to Know About DVD, a book version of his acclaimed Internet DVD FAQ. Called a "minor tech legend" by E! Online, Jim is recognized worldwide as an expert on DVD and associated technology.

Jim has actively participated in DVD Forum working groups since 1998 and serves as Chairman of the DVD Association. Jim was named one of the 21 most influential DVD executives by DVD Report, received the 2000 DVD Pro Discus Award for Outstanding Contribution to the Industry, was an inaugural inductee into the Digital Media Hall of Fame, and was named one of the pioneers of DVD by One to One magazine. Jim has worked with interactive media for over 25 years, developing educational software, laserdiscs, CD-ROMs, Web sites, and DVDs, and has taught workshops and courses on multimedia, computer-based education, computer applications, and DVD.

Formerly VP of Information Technology at Videodiscovery, an educational multimedia publishing company, Jim then championed the format as Microsoft's DVD Evangelist before joining Daikin US as Chief Technology Officer. Jim Taylor lives on an island near Seattle, Washington.

Mark R. Johnson

Mark is the Vice President of Technology at Technicolor Creative Services (TCS), a Thomson company, and is a member of Technicolor's HD Optical Launch Team. Mark actively participates in both the DVD Forum and Blu-ray Disc Association as a contributing member. As an award-winning DVD author, Mark helped set Technicolor's standards and practices with blockbuster titles such as Disney's "Snow White" and "Beauty and the Beast." Prior to that, he was the "go to" guy at Daikin US where he was the Product Manager for the Scenarist DVD authoring system. Formerly the Chief Technology Officer of DVant Digital, Mark was then known for his breakthroughs in advanced applications of the DVD specification. Mark lives in Pasadena, CA.

Charles G. Crawford

Charles is the Co-Founder of Television Production Services, Inc. (TPS), and Heritage Series, LLC, production companies specializing in traditional and interactive production and title development. He has been involved with DVD technology since 1998 when he wrote the Operations and Reference Manuals for Matsushita/Panasonic's award winning DVD Authoring System. He thought the technology was so cutting edge that he convinced his partner to incorporate DVD title development into their line of products and services. Chuck comes from the world of television broadcasting and production. He has been awarded three national EMMY awards for his technical and directorial expertise and numerous other industry awards for his production skills. Chuck is based in Washington, DC.